Fuzzy Sets in Business Management, Finance, and Economics

Fuzzy Sets in Business Management, Finance, and Economics

Editors

Jorge de Andrés-Sánchez
Laura González-Vila Puchades

MDPI • Basel • Beijing • Wuhan • Barcelona • Belgrade • Manchester • Tokyo • Cluj • Tianjin

Editors
Jorge de Andrés-Sánchez
Rovira i Virgili University
Campus Bellissens
Spain

Laura González-Vila Puchades
University of Barcelona
Spain

Editorial Office
MDPI
St. Alban-Anlage 66 4052
Basel, Switzerland

This is a reprint of articles from the Special Issue published online in the open access journal *Mathematics* (ISSN 2227-7390) (available at: https://www.mdpi.com/journal/mathematics/special_issues/Fuzzy_Sets_Bus_Manag_Financ_Econ).

For citation purposes, cite each article independently as indicated on the article page online and as indicated below:

LastName, A.A.; LastName, B.B.; LastName, C.C. Article Title. *Journal Name* **Year**, *Volume Number*, Page Range.

ISBN 978-3-0365-3004-8 (Hbk)
ISBN 978-3-0365-3005-5 (PDF)

© 2022 by the authors. Articles in this book are Open Access and distributed under the Creative Commons Attribution (CC BY) license, which allows users to download, copy and build upon published articles, as long as the author and publisher are properly credited, which ensures maximum dissemination and a wider impact of our publications.

The book as a whole is distributed by MDPI under the terms and conditions of the Creative Commons license CC BY-NC-ND.

Contents

About the Editors vii

Preface to "Fuzzy Sets in Business Management, Finance, and Economics" ix

Luis A. Perez-Arellano, Fabio Blanco-Mesa, Ernesto Leon-Castro and Victor Alfaro-Garcia
Bonferroni Prioritized Aggregation Operators Applied to Government Transparency
Reprinted from: *Mathematics* **2021**, *9*, 24, doi:10.3390/math9010024 1

Jorge de Andrés-Sánchez, Angel Belzunegui-Eraso and Francesc Valls-Fonayet
Assessing Efficiency of Public Poverty Policies in UE-28 with Linguistic Variables and Fuzzy Correlation Measures
Reprinted from: *Mathematics* **2021**, *9*, 128, doi:10.3390/math9020128 21

Mario Arias-Oliva, Jorge de Andrés-Sánchez and Jorge Pelegrín-Borondo
Fuzzy Set Qualitative Comparative Analysis of Factors Influencing the Use of Cryptocurrencies in Spanish Households
Reprinted from: *Mathematics* **2021**, *9*, 324, doi:10.3390/math9040324 47

Pablo J. Villacorta, Laura González-Vila Puchades and Jorge de Andrés-Sánchez
Fuzzy Markovian Bonus-Malus Systems in Non-Life Insurance
Reprinted from: *Mathematics* **2021**, *9*, 347, doi:10.3390/math9040347 67

Luciano Barcellos-Paula, Iván De la Vega and Anna María Gil-Lafuente
The Quintuple Helix of Innovation Model and the SDGs: Latin-American Countries' Case and Its Forgotten Effects
Reprinted from: *Mathematics* **2021**, *9*, 416, doi:10.3390/math9040416 91

Martha B. Flores-Romero, Miriam E. Pérez-Romero, José Álvarez-García and María de la Cruz del Río-Rama
Fuzzy Techniques Applied to the Analysis of the Causes and Effects of Tourism Competitiveness
Reprinted from: *Mathematics* **2021**, *9*, 777, doi:10.3390/math9070777 115

Miroslav Kelemen, Volodymyr Polishchuk, Beáta Gavurová, Róbert Rozenberg, Juraj Bartok, Ladislav Gaál, Martin Gera and Martin Kelemen, Jr.
Model of Evaluation and Selection of Expert Group Members for Smart Cities, Green Transportation and Mobility: From Safe Times to Pandemic Times
Reprinted from: *Mathematics* **2021**, *9*, 1287, doi:10.3390/math9111287 133

Ranka Gojković, Goran Đurić, Danijela Tadić, Snežana Nestić and Aleksandar Aleksić
Evaluation and Selection of the Quality Methods for Manufacturing Process Reliability Improvement—Intuitionistic Fuzzy Sets and Genetic Algorithm Approach
Reprinted from: *Mathematics* **2021**, *9*, 1531, doi:10.3390/math9131531 155

Lucía Muñoz-Pascual, Carla Curado and Jesús Galende
Fuzzy Set Qualitative Comparative Analysis on the Adoption of Environmental Practices: Exploring Technological- and Human-Resource-Based Contributions
Reprinted from: *Mathematics* **2021**, *9*, 1553, doi:10.3390/math9131553 173

Daniel Edahi Urueta, Pedro Lara, Miguel Ángel Gutiérrez, Sergio Gerardo de-los-Cobos, Eric Alfredo Rincón and Román Anselmo Mora
A Comparative Ranking Model among Mexican Universities Using Pattern Recognition
Reprinted from: *Mathematics* **2021**, *9*, 1615, doi:10.3390/math9141615 . . . . **195**

Fernando Castelló-Sirvent and Pablo Pinazo-Dallenbach
Corruption Shock in Mexico: fsQCA Analysis of Entrepreneurial Intention in University Students
Reprinted from: *Mathematics* **2021**, *9*, 1702, doi:10.3390/math9141702 . . . . **221**

Fabio Blanco-Mesa, Ernesto León-Castro and Jorge Romero-Muñoz
Pythagorean Membership Grade Aggregation Operators: Application in Financial knowledge
Reprinted from: *Mathematics* **2021**, *9*, 2136, doi:10.3390/math9172136 . . . . **253**

Joan Carles Ferrer-Comalat, Salvador Linares-Mustarós and Ricard Rigall-Torrent
Incorporating Fuzzy Logic in Harrod's Economic Growth Model
Reprinted from: *Mathematics* **2021**, *9*, 2194, doi:10.3390/math9182194 . . . . **269**

Tsuen-Ho Hsu and Ling-Zhong Lin
A Multidimensional Fuzzy Quality Function Deployment Design for Brand Experience Assessment of Convenience Stores
Reprinted from: *Mathematics* **2021**, *9*, 2565, doi:10.3390/math9202565 . . . . **289**

Luis Porcuna-Enguix, Elisabeth Bustos-Contell, José Serrano-Madrid and Gregorio Labatut-Serer
Constructing the Audit Risk Assessment by the Audit Team Leader When Planning: Using Fuzzy Theory
Reprinted from: *Mathematics* **2021**, *9*, 3065, doi:10.3390/math9233065 . . . . **313**

About the Editors

Jorge de Andrés-Sánchez has a PhD. in Business Management, is an actuary, and is a Senior Lecturer of Finance at the Rovira i Virgili University (Spain) and a member of its research group, the Social and Business Research Laboratory. He primiarily works in the research field of soft computing in financial and actuarial topics. He is the author of several papers on these topics in academic journals such as Mathematics, Journal of Risk and Insurance, Insurance: Mathematics and Economics, Fuzzy Sets and Systems, etc.

Laura González-Vila Puchades has a PhD. in Economics and Business Sciences and is an actuary. She is currently a Senior Lecturer in the area of Financial Economics and Accounting at the University of Barcelona (Spain) and is a member of its research group studying Actuarial and Financial Modelling. She has auhtored several papers in top-rank journals and has presented at many conferences focussing on insurance and uncertainty.

Preface to "Fuzzy Sets in Business Management, Finance, and Economics"

Since the publication of Lotfi A. Zadeh's seminal paper "Fuzzy Sets" in 1965 in the journal Information and Control, there has been constant growth in theoretical developments and in the practical application of Fuzzy Set Theory and related mathematical tools. These tools have been applied widely, both by industry and academic research, to decision making and economics due to their versatility. On the one hand, they can efficiently represent and handle uncertain and vague information such as subjective judgements, non-precise observations on variables, or ill-defined relations between variables. On the other hand, they make implementing computations or identifying patterns in data much easier. To do so, Fuzzy Set Theory provides a variety of mathematical techniques in fields such as Expert Systems, Soft Computing, Data Analysis, Mathematical Programming, and Multiple Criteria Decision Making. Business management, decision making, or actuarial modelling are some examples of practical applications in these fields. This Special Issue provided a platform for researchers from academia and industry to present novel work in the domain of applied developments for Fuzzy Sets and in methodologies related to business, financial, and economic analysis. We hope that these results will help to foster future research in the fields of economics and in the social sciences.

Jorge de Andrés-Sánchez, Laura González-Vila Puchades
Editors

Article

Bonferroni Prioritized Aggregation Operators Applied to Government Transparency

Luis A. Perez-Arellano [1], Fabio Blanco-Mesa [2], Ernesto Leon-Castro [3,*] and Victor Alfaro-Garcia [4]

1. Unidad Regional Culiacán, Universidad Autonoma de Occidente, Blvd. Lola Beltrán s/n esq. Circuito Vial, Culiacán 80200, Mexico; luis.pereza@uadeo.mx
2. Facultad de Ciencias Económicas y Administrativas, Escuela de Administración de Empresas, Universidad Pedagógica y Tecnológica de Colombia, Tunja 150001, Colombia; fabio.blanco01@uptc.edu.co
3. Faculty of Economics and Administrative Sciences, Universidad Católica de la Santísima Concepción, Concepción 4030000, Chile
4. Facultad de Contaduría y Ciencias Administrativas, Universidad Michoacana de San Nicolas de Hidalgo, Gral. Francisco J. Múgica S/N, C.U., 58030 Morelia, Mexico; victor.alfaro@umich.mx
* Correspondence: eleon@ucsc.cl

Abstract: This article applies the Bonferroni prioritized induced heavy ordered weighted average (OWA) to analyze a series of data and focuses on the Bonferroni average and heavy induced prioritized aggregation operators. The objective of the present work is to present a new aggregation operator that combines the heavy induced prioritized Bonferroni and its formulations and represents the Bonferroni mean with variables that induce an order with vectors that are greater than one. This work develops some extensions using prioritization. The main advantage is that different types of information provided by a group of decision makers to compare real situations are included in this formulation. Finally, an example using the operators to calculate the transparency of the websites of the 32 states of Mexico was performed. The main idea was to visualize how the ranking can change depending on the importance of the five components of the methodology. The main results show that it is possible to detect some important changes depending on the operator and the experts considered.

Keywords: Bonferroni means; prioritized aggregation operators; induced aggregation operators; OWA operator; transparency

1. Introduction

The Organization for Economic Co-operation and Development (OECD) has recently recognized open government initiatives as critical drivers of citizens' trust and key aspects of the modernization, anticorruption, civic freedom, innovation, financial management and human resource management of the public sector of a country [1]. Moreover, a culture of transparency, participation and accountability that conforms to open government yields opportunities for economic growth, as it promotes the creation of businesses, jobs and cost-effective public policies [2]. Nonetheless, the design, creation and implementation of effective open government strategies pose a series of challenges for countries, including their alignment with national plans, strategic visions, public governance and technological resources [3–5].

Transparency and access to information are key issues for the establishment of open governments. Governmental transparency is the ability to determine what is happening inside the government [6]. Moreover, transparency fosters the accountability of actions and offers information to citizens regarding governmental decisions [7], thereby dissuading corruption and promoting efficiency, democracy and legitimacy [8]. In this sense, information is an asset, and while some administrations may use it as a trigger for best practices, others may have a radically different opinion based on their own political, administrative, institutional and demographic contexts [9,10]. These variations based on country contexts constitute the difference between freedom of information laws, their design and operations and the challenges they have for their nations, e.g., Canada and the United Kingdom or the open government of the People's Republic of China [11].

In Mexico, access to information is a citizen's right composed of three elements: normative design, institutional design, and procedures for access to public information and transparency obligations [12]. The National Institute of Transparency (INAI) is a specialized public institution that regulates transparency at the national level, including access to information, personal data protection and the development of methodologies to assess transparency [13]. Additionally, the ranking of transparency websites is measured through five components: institutional arrangements, open data, vertical collaboration, horizontal collaboration, and interface [14]. The main difficulties with this formula are that it takes an average of the results that depend on the state; some of the components are more important than others. Because the calculation is made with the same weights for each subindex for all the states, there is no real evaluation of transparency depending on the specific characteristics and problems of each state.

Recent developments in information technologies have opened the path for assessing decision-making in systemic environments. Expert and intelligent systems have proven effective in subjective, uncertain and highly complex scenarios [15,16]. In this context, to address some of the abovementioned challenges, a combination of several intelligent systems such as the Bonferroni means [17] and the ordered weighted averaging (OWA) operator [18] will be used. A special focus will be placed on the following extensions: (a) the Bonferroni ordered weighted averaging (BON-OWA) operator [19] allows adding information and making multiple comparisons between input arguments and capturing their interrelation to present information, (b) the induced ordered weighted averaging (IOWA) operator [20,21] uses induced variables in the reordering step instead of the traditional reordering based on the value of the arguments of the OWA operator, (c) the prioritized ordered weighted averaging (PrOWA) operator [22] introduces a mechanism for assigning specific weights to the participants in a group decision-making problem, and, finally, (d) the heavy ordered weighted averaging (HOWA) operator [23] features a nonbounded weighting vector that allows the over- or underestimation of results according to the expectation and knowledge of the decision maker.

According to Blanco-Mesa, León-Castro and Merigó [24], aggregation operators allow joining different pieces of information provided by several sources [25], ensuring the inclusion of all the fusion information [26,27] and combining several values into a single value [15,28]. Since the proposal of the BON-OWA operator, several new methodological contributions have been made, among which those developed by Blanco-Mesa, such as (1) the Bonferroni means with distance measures applied to entrepreneurship and human resource management [29,30], (2) the Bonferroni induced operator and heavy operator applied to enterprise risk management and sale forecasting [31,32], (3) the Bonferroni OWA variance used in strategic analysis in enterprise risk management [33], and (4) the Bonferroni covariance OWA used in research and development investment problems [34], stand out as addressing decision-making problems in business management. Recently, a paper has been published that proposed measuring transparency with another aggregation method called the prioritized induced ordered weighted average weighted average (PIOWAWA) operator. This operator considers the degree of importance, reordering and weight factors given to the information in the same formulation by the decision maker and is assessed using a Colombian transparency case [35]. Additionally, formulations have become widespread, and extensions have been proposed with other operators, such as the induced OWA operator (IOWA) [20,21], the heavy OWA operator (HOWA) [23], the OWAWA operator [36] and immediate weights (IWs) [37].

Following the above ideas, it is interesting to explore other operators that can be combined with the Bonferroni means. In that sense, one of the operators that can be extended is the prioritized OWA operator [38]. This operator is characterized by balancing the impact that a decision maker has on decision problems where he or she does not have the same position in the final decision, i.e., this operator assigns an additional impact to some decision makers and less to others. In the case of this research, it is very useful in problems calculating and evaluating the importance of each component because of their interrelationship, their interdependence and the importance that various agents have in this evaluation process.

The objective of this paper is to present a new extension of the BON-OWA operator using the extensions described above in a single formulation. The introduced operator is the Bonferroni prioritized induced heavy OWA (BON-PrIHOWA) operator. The main advantage of this operator is the consideration of a group decision-making problem in a single formulation including a nonlimited to zero weighting vector and an induced weighting vector capable of assigning weights according to the highly complex conditions of the analyzed phenomena. These features allow the analysis of a changing classification according to the additional information provided and the consideration of new scenarios for accurate results. The newly introduced BON-PrIHOWA is used as a method for ranking the transparency portals for the 32 states in Mexico based on experts.

The remainder of this document is organized as follows. In Section 2, we present some of the basic aggregation operators. Section 3 presents the new proposed operator, the BON-PrIHOWA operator. In Section 4, the evaluation of the characteristics of the transparency websites in Mexico based on different experts and aggregation operators are included. Finally, in Section 5, the conclusions of the document are presented.

2. Preliminaries

In this section, we review some of the required basic concepts related to the OWA operator in this article. This operator is supported by criteria that are the bases of a decision that integrate the expectations of the decision makers in the evaluation that he or she makes of the set of actions to be taken [39]. Likewise, the OWA operator has the versatility to add data without losing its mathematical properties. Furthermore, according to the arguments, the qualifications can obtain evaluated alternatives. Thus, operators such as the HOWA, IOWA, PrOWA, PIOWA and IHOWA have been proposed and studied. Additionally, the OWA operator has allowed the development of several extensions that combine new parameters and interactions with other methods and some other extensions [39]. Among these, the BON-OWA, BON-HOWA, BON-IOWA and BON-PrOWA will also be studied to fulfil the purpose of the research. Hence, each of the definitions of the operators mentioned above is presented below.

2.1. OWA Operator and Its Main Extensions

The OWA operator was introduced by Yager [18], and its main feature is that it is possible to obtain the maximum and minimum values according to the operator's rearrangement weight. The purpose of this operator is to obtain a single representative value from the aggregation of a series of data that reflect the predetermined optimism/pessimism parameters. It is defined as follows:

Definition 1. *An OWA operator of dimension n is a mapping of* $OWA : R^n \rightarrow R$ *with a weight vector W of dimension n with* $\sum_{i=1}^{n} w_i = 1$ *and* $w_i \in [0,1]$ *such that:*

$$OWA(a_1, a_2, .., a_n) = \sum_{j=1}^{n} w_j b_j \qquad (1)$$

where b_j *is the jth element and the largest of the collection* $a_1, a_2, \ldots, a_n$.

The fundamental characteristic of the OWA operator is that the rearrangement of the elements or arguments allows argument a_j not to be associated with weight w_j weight if all w_js are associated with the position in the order for aggregation.

Definition 2. *As introduced by Merigo and Gil-Lafuente [21], an IOWA operator of dimension n is an application $IOWA: R^n \to R$ that has an associated weight vector W of dimension n where the sum of the weights is 1, $w_j \in [0,1]$, and an induced set of variables of order are included (u_i). The formula is*

$$IOWA(\langle u_1, a_1 \rangle, \langle u_2, a_2 \rangle, \ldots, \langle u_n, a_n \rangle) = \sum_{j=1}^{n} w_j b_j, \quad (2)$$

where $(b_1, b_2, \ldots, b_n)$ is simply $(a_1, a_2, \ldots, a_n)$ reordered descending or ascending according to the values of u_i. b_j is the a_i value of the OWA pair $<u_i, a_i>$ having the jth largest u_i. u_i is the order inducing variable, and a_i is the argument variable. These operators take argument pairs, called OWA pairs, in which a component is used to induce an order on the second components that are then added.

Among the extensions of the OWA operator that focus on the weight vector is the heavy OWA (HOWA) operator [23]. In this extension, the weight vector is not $\sum_{j=1}^{n} w_j = 1$ but is unbounded; therefore, the weighting vector can be $1 \leq \sum_{j=1}^{n} w_j \leq n$. The definition is as follows:

Definition 3. *An HOWA operator is a mapping $HOWA: R^n \to R$ that is associated with a weight vector w, where $w_j \in [0,1]$ and $1 \leq \sum_{j=1}^{n} w_j \leq n$, such that*

$$HOWA(a_1, a_2, \ldots, a_n) = \sum_{j=1}^{n} w_j b_j, \quad (3)$$

where b_j is the jth largest element of collection a_i. It is also important to note that in some cases, it is possible that the weight vector is $-\infty \leq \sum_{j=1}^{n} w_j \leq \infty$, making it possible to under- or overestimate the results according to the expectations of the decision maker. It is important to note that Yager (2002) also developed a characteristic of the HOWA operator, which is called the beta value. This beta value can be defined as $\beta(W) = (|W| - 1)/(n - 1)$. Note that if $\beta = 1$, we obtain the total operator, and if $\beta = 0$, we obtain the usual OWA operator.

Definition 4. *The prioritized OWA (PrOWA) operator developed by Yager [40] is an aggregation operator that is useful when problem-solving decision makers do not have the same standing in the final decision. Thus, this operator allocates an additional impact to some decision makers and less to others. This operator can be defined as follows (Yager 2008, 2009a). A prioritized OWA (PrOWA) of dimension n is a mapping $PrOWA: R^n \to R$ that has an associated v_k that is the corresponding weight of the jth criterion in the ith category.where $C_i(x) = a_i \in [0,1]$ is the degree of satisfaction with criterion C_i by alternative x.*

$$V_k \in [0,1] \text{ and } \sum_{k=1}^{n} V_k = 1, \quad (4)$$

where $a_{ind(k)}$ is the kth largest element of collection $C_i(x)$.

$$C_{(x)} = \sum_{i=1}^{q} \sum_{h=1}^{n_i} w_{ij} C_{ij}(x), \quad (5)$$

which allows us to obtain ind(j). We calculate this number using the subscript of the associated C_i.

$$\widetilde{R}_K = \sum_{i=1}^{k} r_{ind(i)}, \quad (6)$$

$$r_i = \frac{T_i}{\sum_{j=1}^{n} T_j}, \quad (7)$$

$$v_k = f\left(\widetilde{R}_K\right) - f\left(\widetilde{R}_{K-1}\right), \quad (8)$$

$$C_{(x)} = \sum_{i=1}^{n} v_k \cdot a_{ind(k)}, \quad (9)$$

$$T_1 = 1, \ T_i = C_{i-1} T_{i-1} \quad \text{for } i = 2 \text{ to } n, \quad (10)$$

where b_j is the jth element that has the largest value of u_i; u_i is the induced order of variables; $\hat{v}_{ij}$ is the corresponding weight of the jth criterion in the ith category for each $i = 1, \ldots, q$ and $j = 1, \ldots, i_i$; and $C_{ij}(x)$ measures the satisfaction of the jth criterion in the ith group by alternative $x \in X$ for each $i = 1, \ldots, q$ and $j = 1, \ldots, i_i$.

Definition 5. *A prioritized induced OWA (PIOWA) of dimension n is a mapping PIOWA : $R^n \times R^n \to R$ that has an associated weight vector w of dimension n, where $w_j \in [0,1]$ and $\sum_{j=1}^{n} w_j = 1$, such that*

$$PIOWA(\langle u_1, a_1 \rangle, \langle u_2, a_2 \rangle, \ldots, \langle u_n, a_n \rangle) = \sum_{i=1}^{q} \sum_{h=1}^{n_i} b_j \hat{v}_{ij} C_{ij}(x), \qquad (11)$$

where b_j is the jth element that has the largest value of u_i; u_i is the induced order of variables; $\hat{v}_{ij}$ is the corresponding weight of the jth criterion in the ith category for each $i = 1, \ldots, q$ and $j = 1, \ldots, i_i$; and $C_{ij}(x)$ measure the satisfaction of the jth criterion in the ith group by alternative $x \in X$ for each $i = 1, \ldots, q$ and $j = 1, \ldots, i_i$.

Another extension takes the reordering process of the IOWA operator and the unbounded weighting vector of the HOWA operator. This operator is called the induced heavy OWA (IHOWA) operator. The definition is as follows (Merigó and Casanovas 2011).

Definition 6. *An IHOWA operator of dimension n is a mapping IHOWA : $R^n \times R^n \to R$ that has an associated weighting vector W of dimension n with $w_j \in [0,1]$ and $1 \leq \sum_{j=1}^{n} w_j \leq n$ such that*

$$IHOWA(\langle u_1, a_1 \rangle, \langle u_2, a_2 \rangle, \ldots, \langle u_n, a_n \rangle) = \sum_{j=1}^{n} w_j b_j, \qquad (12)$$

where b_j is the a_i of the IHOWA pair $<u_i, a_i>$ having the jth largest u_i. u_i is the order inducing variable, and a_i is the argument variable.

2.2. Bonferroni-OWA

In relation to soft mathematics and with respect to models that relate to the theory of aggregation [32,38], there is the extension of Bonferroni that allows us to add, organize, and relate information objectively and subjectively simultaneously. These models are the same ones that are applicable in artificial intelligence. This operator is called the BON-OWA. Compared to other models such as traditional statistics, the BON-OWA allows us to obtain important results by treating information simultaneously [29].

Decision-making seeking to reduce uncertainty can improve the results by applying the Bonferroni average since it builds confidence intervals and maintains the global confidence coefficient [17]. The operator is defined as follows:

$$B(a_1, a_2, \ldots, a_n) = \left(\frac{1}{n} \frac{1}{1-n} \sum_{\substack{j=1 \\ j \neq k}}^{n} a_j^{p,q} \right)^{\frac{1}{r+q}} \qquad (13)$$

Definition 7. *The Bonferroni OWA is mean-type aggregation operator. The main characteristics of the Bonferroni average (Bonferroni 1950) are that the arguments a must be greater than or equal to 0, and the parameters p and q must be greater than or equal to 0. The algorithm that combines the OWA operator and the Bonferroni average can be defined as:*

$$BON-OWA(a_1, \ldots, a_n) = \left(\frac{1}{n} \sum_i a_i^p OWA_W(V^i) \right)^{\frac{1}{r+q}}, \qquad (14)$$

where $OWA_W(V^i)$ represents the expression $\left(\frac{1}{n-1}\sum_{\substack{j=1\\j\neq i}}^{n} a_j^q\right)$ with (V_i) being the vector of all a_js except a_i and w being an $n-1$ vector W_i associated with α_i whose components w_{ij} are the OWA weights. Let W be an OWA weighting vector of dimension $n-1$ with components $w_i \in [0,1]$ when $\sum_i w_i = 1$. Then, we can define this aggregation as $OWA_W(V^i) = \left(\sum_{j=1}^{n-1} w_i a_{\pi_k(j)}\right)$, where $a_{\pi_k(j)}$ is the largest element in the tuple V^i and $w_i = \frac{1}{n-1}$ for all i.

Definition 8. *The Bonferroni IOWA (BON-IOWA) (Blanco-Mesa et al. 2019b) is a mean-type aggregation operator that is defined as follows.*

$$BON-IOWA(\langle u_1, a_1\rangle, \ldots, \langle u_n, a_n\rangle) = \left(\frac{1}{n}\sum_i a_i^r IOWA_W(V^i)\right)^{\frac{1}{r+q}}, \qquad (15)$$

where (V^i) is the vector of all a_j except a_j. Let W be an OWA weighing vector of dimension $n-1$ with components $w_i \in [0,1]$ when $\sum_i w_i = 1$, where the weights are associated according to the largest value of u_i, and u_i is the order-inducing variable. Then, we can define this aggregation as $IOWA_W(V^i) = \left(\sum_{j=1}^{n-1} w_i a_{\pi_k(j)}\right)$, where $a_{\pi_k(j)}$ is the largest element in the $n-1$ tuple $V^i = V^i = (\langle u_1, a_1\rangle, \ldots, \langle u_{i-1}, a_{i-1}\rangle, \langle u_{i+1}, a_{i+1}\rangle, \ldots, \langle u_n, a_n\rangle)$.

Definition 9. *The Bonferroni HOWA (BON-HOWA) [31] is a mean-type aggregation operator that has an associated weighting vector W with $w_i \in [0,1]$ and $1 \leq \sum_{j=1}^{n} w_j \leq n$ such that:*

$$BON-HOWA(a_1, \ldots, a_n) = \left(\frac{1}{n}\sum_i a_i^r HOWA_W(V^i)\right)^{\frac{1}{r+q}}, \qquad (16)$$

where (V^i) is the vector of all a_js except a_j. Let W be an OWA weighing vector of dimension $n-1$ with components $w_i \in [0,1]$ when $1 \leq \sum_{j=1}^{n} w_j \leq n$. Thus, the sum of the weights w_j is bounded to n or can be unbounded if the weighting vector $W = -\infty \leq \sum_{j=1}^{n} w_j \leq \infty$. Then, we can define this aggregation as $HOWA_W(V^i) = \left(\sum_{j=1}^{n-1} w_i a_{\pi_k(j)}\right)$, where $a_{\pi_k(j)}$ is the largest element in the $n-1$ tuple $V^i = V^i = (a_1, \ldots, a_{i-1}, a_{i+1}, \ldots, a_n)$.

Definition 10. *The Bonferroni PrOWA (BON-PrOWA) [41] is a mean-type aggregation operator that has an associated weighting vector W:*

$$BON-PrOWA(\langle u_1, a_1\rangle, \langle u_2, a_2\rangle, \ldots, \langle u_n, a_n\rangle) = \frac{1}{n}\left(\sum_{i=1}^{n-1} a_i^r PrOWA_W(V^i)\right)^{\frac{1}{r+q}}, \qquad (17)$$

where (V^i) is the vector of all a_js except a_j. Let W_i be an OWA weighing vector of dimension $n-1$ with components $w_i \in [0,1]$ when $\sum_{j=1}^{n} w_j = 1$. W_i is the vector of weights (associated with the vector V^i) of all w_js except w_i. r and q are parameters such that $r,q \geq 0$. The a_is are the some prioritized a_is, where column "i" is omitted to perform the sorting. A vector of $n-1$ elements remains. r is the exponent of a_i.

Definition 11. *The Bonferroni IHOWA (BON-IHOWA) [31] is a mean-type aggregation operator that has an associated weighting vector W, where $w_i \in [0,1]$ and $1 \leq \sum_{j=1}^{n} w_j \leq n$, such that:*

$$BON-IHOWA(\langle u_1, a_1\rangle, \ldots, \langle u_n, a_n\rangle) = \left(\frac{1}{n}\sum_i a_i^r IHOWA_W(V^i)\right)^{\frac{1}{r+q}}, \qquad (18)$$

where (V^i) is the vector of all a_js except a_js. Let W be an OWA weighting vector of dimension $n-1$ with components $w_i \in [0,1]$ when $1 \leq \sum_{j=1}^{n} w_j \leq n$. The weights are associated according to the largest value of u_i, and u_i is the order-inducing variable. Likewise, the sum of the weights w_j is bounded to n or can be unbounded if the weighting vector $W = -\infty \leq \sum_{j=1}^{n} w_j \leq \infty$. Then, we can define this aggregation as $IHOWAw(V^i) = \left(\sum_{j=1}^{n-1} w_i a_{\pi_k(j)}\right)$, where $a_{\pi_k(j)}$ is the largest element in the $n-1$ tuple $V^i = V^i = (\langle u_1, a_1 \rangle, \ldots, \langle u_{i-1}, a_{i-1} \rangle, \langle u_{i+1}, a_{i+1} \rangle, \ldots, \langle u_n, a_n \rangle)$.

3. New Propositions—Bonferroni Prioritized Induced Heavy OWA Operator

In this section, a new proposition considering the theoretical aspects and the revision of the definitions of each of the methods necessary for its proposal is presented. Here, it is important to mention that the authors of a previous work [35] established an approach that improves the evaluation of the transparency index that considers the degree of importance, reordering and weight factors. This approach seeks to improve the integration of information by considering their interrelationship, their interdependence and the importance of the information and including a nonlimited to zero weighting vector and an induced weighting vector capable of assigning weights according to the highly complex conditions of the analyzed phenomena [35,42]. Thus, this approach offers a better way to understand the information than just the measurement [42]. In this sense, the proposition presented is called the Bonferroni prioritized induced heavy OWA operator (BON-PrIHOWA). From this main proposal, the BON-PrOWA, PrIOWA and PrHOWA are also presented. Each of the propositions is presented below.

Proposition 1. *The Bonferroni PrOWA (BON-PrOWA) is a mean-type aggregation operator that has an associated weighting vector W:*

$$BON - PrOWA(\langle u_1, a_1 \rangle, \langle u_2, a_2 \rangle, \ldots, \langle u_n, a_n \rangle) = \frac{1}{n}\left(\sum_{i=1}^{n-1} a_i^r PrOWA_W(V^i)\right)^{\frac{1}{r+q}}, \quad (19)$$

where (V^i) is the vector of all a_js except a_i. Let W_i be an OWA weighing vector of dimension $n-1$ with components $w_i \in [0,1]$ when $\sum_{j=1}^{n} w_j = 1$. W_i is the vector of weights (associated with the vector V^i) of all w_js except w_i. r and q are parameters such that $r, q \geq 0$, The a_is are some prioritized a_is, where column "i" is omitted to perform the sorting. A vector of $n-1$ elements remains. r is the exponent of a_i. $PrOWA_W(V^i) = \left(\sum_{j=1}^{n-1} w_i a_{\pi_k(j)}\right)$, where $a_{\pi_k(j)}$ is the largest element in the $n-1$ tuple $V^i = V^i = (a_1, \ldots, a_{i-1}, a_{i+1}, \ldots, a_n)$.

Proposition 2. *The Bonferroni PrIOWA is a mean-type aggregation operator that has an associated weighting vector W:*

$$BON - PrIOWA(\langle u_1, a_1 \rangle, \langle u_2, a_2 \rangle, \ldots, \langle u_n, a_n \rangle) = \frac{1}{n}\left(\sum_{i=1}^{n-1} a_i^r PrOWA_W(V^i)\right)^{\frac{1}{r+q}}, \quad (20)$$

where (V^i) is the vector of all a_js except a_i. Let W_i be an OWA weighing vector of dimension $n-1$ with components $w_i \in [0,1]$ when $\sum_{j=1}^{n} w_j = 1$. W_i is the vector of weights (associated with the vector V^i) of all w_js except w_i, the weights are associated according to the largest value of u_i and u_i is the order-inducing variable. r and q are parameters such that $r, q \geq 0$, The a_is are some prioritized a_is, where column "i" is omitted to perform the sorting. A vector of $n-1$ elements remains. r is the exponent of a_i. Then, $PrIOWA_W(V^i) = \left(\sum_{j=1}^{n-1} w_i a_{\pi_k(j)}\right)$, where $a_{\pi_k(j)}$ is the largest element in the $n-1$ tuple $V^i = V^i = (\langle u_1, a_1 \rangle, \ldots, \langle u_{i-1}, a_{i-1} \rangle, \langle u_{i+1}, a_{i+1} \rangle, \ldots, \langle u_n, a_n \rangle)$.

Proposition 3. *The Bonferroni PrHOWA is a mean-type aggregation operator that has an associated weighting vector W:*

$$BON - PrHOWA(\langle u_1, a_1 \rangle, \langle u_2, a_2 \rangle, \ldots, \langle u_n, a_n \rangle) = \frac{1}{n}\left(\sum_{i=1}^{n-1} a_i^r PrOWA_W(V^i)\right)^{\frac{1}{r+q}}, \quad (21)$$

where (V^i) is the vector of all $a_j s$ except a_i. Let W_i be an OWA weighing vector of dimension $n-1$ with components $w_i \in [0,1]$ when $1 \leq \sum_{j=1}^{n} w_j \leq n$. W_i is the vector of weights (associated with the vector V^i) of all $w_j s$ except w_i. Thus, the sum of the weights w_j is bounded to n or can be unbounded if the weighting vector $W = -\infty \leq \sum_{j=1}^{n} w_j \leq \infty$. r and q are parameters such that $r, q \geq 0$. The $a_i s$ are some prioritized $a_i s$, where column "i" is omitted to perform the sorting. A vector of $n-1$ elements remains. r is the exponent of a_i. $PrHOWA_W(V^i) = \left(\sum_{j=1}^{n-1} w_i a_{\pi_k(j)} \right)$, where $a_{\pi_k(j)}$ is the largest element in the $n-1$ tuple $V^i = V^i = (a_1, \ldots, a_{i-1}, a_{i+1}, \ldots, a_n)$.

Proposition 4. *The BON-PrIHOWA on (V^i) is the vector of all $a_j s$ except a_i. Let W_i be an OWA weighing vector of dimension $n-1$ with components $w_i \in [0,1]$, where $1 \leq \sum_{j=1}^{n} w_j \leq n$.*

$$BON - PrIHOWA(\langle u_1, a_1 \rangle, \langle u_2, a_2 \rangle, \ldots, \langle u_n, a_n \rangle) = \frac{1}{n} \left(\sum_{i=1}^{n-1} a_i^r PrIHOWA_W(V^i) \right)^{\frac{1}{r+q}}, \quad (22)$$

where W_i is the vector of weights (associated with the vector V^i) of all $w_j s$ except w_i. Let W be an OWA weighting vector of dimension $n-1$ with components $w_i \in [0,1]$ when $1 \leq \sum_{j=1}^{n} w_j \leq n$, where the weights are associated according to the largest value of u_i and u_i is the order-inducing variable. The induced u_i given to the elements a_i is given in an ascending or a descending manner according to the criteria of each decision maker. Therefore, each element a_i has an associated induced u_i. Likewise, the sum of weights w_j is bounded to n or can be unbounded if the weighting vector $W = -\infty \leq \sum_{j=1}^{n} w_j \leq \infty$. Likewise, r and q are parameters such that $r, q \geq 0$. The $a_i s$ are the same prioritized $a_i s$, where column "i" is omitted to perform the sorting. A vector of $n-1$ elements remains. r is the exponent of a_i. Then, we can define this aggregation as $PrIHOWAw(V^i) = \left(\sum_{j=1}^{n-1} w_i a_{\pi_k(j)} \right)$, where $a_{\pi_k(j)}$ is the largest element in the $n-1$ tuple $V^i = V^i = (\langle u_1, a_1 \rangle, \ldots, \langle u_{i-1}, a_{i-1} \rangle, \langle u_{i+1}, a_{i+1} \rangle, \ldots, \langle u_n, a_n \rangle)$.

4. Evaluation of the Transparency Websites in Mexico

4.1. Aggregation Operators Calculation

The objective of this paper is to use and apply the operators proposed in Section 3 to rank the transparency websites of the states in Mexico. As mentioned previously, government transparency is vital for the development of countries, and therefore, the possibility of using web pages to report and be able to make complaints and reports is of the utmost importance to facilitate interaction with users. In Mexico, the transparency websites are measured and ranked using five components, which are as follows [14]:

(a) Institutional arrangements. Refers to compliance with regulations;
(b) Open data. Refers to the amount of information published;
(c) Vertical collaboration. Measures the use and performance of the portal and the complaints made;
(d) Horizontal collaboration. Measures the use of social networks, blogs and chats;
(e) Interface. Eases the use of the website.

The questionnaire used to measure these websites has 63 items, and within the present investigation, the data from the last evaluation are used, which is that of 2017. The main problem of the actual ranking is that all five components have the same importance to the ranking. Because of that, not all states seek ways to improve their transparency because one good component can improve the final score, even when some components have a score of 0. The qualification of each component for each of the 32 states of Mexico is given in Table A1. Finally, the steps to use the BON-PrOWA operator and other extensions are as follows.

Step 1. Locate different experts that give information regarding each of the components of the ranking of transparency websites. The information that will be requested is (a) weights, (b) heavy weights and (c) induced values. The profile of the experts for this article was as follows: (a) they had minimum of five years of experience within the government sector, specifically in areas related to transparency; and (b) they work or worked directly with government transparency websites.

Step 2. With the information provided by each expert, generate different classifications using the BON-OWA, BON-IOWA, BON-HOWA and BON-IHOWA operators.

Step 3. With the results obtained in Step 3, unify the information of the different experts based on the BON-PrOWA, BON-PrIOWA, BON-PrHOWA and BON-PrIHOWA operators, where the results of each expert are given a specific weight according to their experience in the field.

Step 4. Finally, the results are compared and analyzed.

To more clearly visualize the process to obtain the results, a simplified graph is presented (see Figure 1).

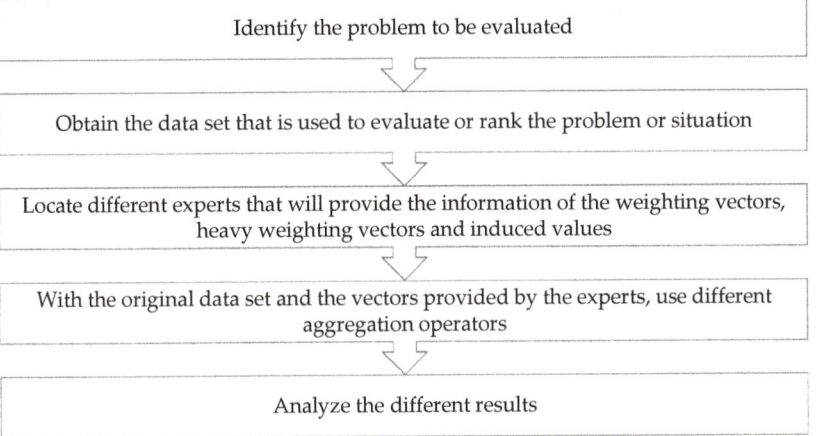

Figure 1. Flowchart of the steps to use the Bonferroni prioritized induced heavy ordered weighted average (BON-PIHOWA) operator.

4.2. Evaluation of the Determinants of Transparency

Step 1. The information was provided by five experts. The conditions for being selected were as follows: (a) must be an active worker in an institution related to transparency and (b) must have more than 10 years in a similar position. The information provided by the experts is given in Tables 1–3.

Table 1. Weights provided by the experts.

Expert	Institutional Arrangements (c_1)	Open Data (c_2)	Vertical Collaboration (c_3)	Horizontal Collaboration (c_4)	Interface (c_5)
e_1	0.15	0.30	0.20	0.20	0.15
e_2	0.10	0.20	0.30	0.30	0.10
e_3	0.10	0.30	0.25	0.25	0.10
e_4	0.15	0.15	0.30	0.20	0.20
e_5	0.10	0.15	0.30	0.30	0.15

Table 2. Heavy weights (heavy weights are the weights that will be used to calculate the heavy ordered weighted average (HOWA) operator. Their difference with the weights in Table 1 is that heavy weights are not bounded to $\sum_{j=1}^{n} w_j = 1$; in this sense, the weighting vector can be from $1 \leq \sum_{j=1}^{n} w_j \leq n$) provided by the experts.

Expert	Institutional Arrangement (c_1)	Open Data (c_2)	Vertical Collaboration (c_3)	Horizontal Collaboration (c_4)	Interface (c_5)
e_1	0.20	0.30	0.20	0.20	0.20
e_2	0.15	0.20	0.30	0.30	0.15
e_3	0.10	0.30	0.30	0.30	0.10
e_4	0.20	0.20	0.30	0.20	0.20
e_5	0.10	0.20	0.30	0.30	0.20

Table 3. Induced values provided by the experts. Induced values are the values that will be used in the induced ordered weighted average (IOWA) operator, instead of a reordering step based on the value of the arguments, in this case, will be based on the induced value determined by the experts, generating a different reordering between the arguments and the weights. Also, the weights used in the Bonferroni induced ordered weighted average (Bon-IOWA) and Bonferroni prioritized induced ordered weighted average (Bon-PIOWA) operators are from Table 1 and for the Bonferroni induced heavy ordered weighted average (Bon-IHOWA) and Bonferroni prioritized induced heavy ordered weighted average (Bon-PIHOWA) operators are from Table 2.

Expert	Institutional Arrangements (c_1)	Open Data (c_2)	Vertical Collaboration (c_3)	Horizontal Collaboration (c_4)	Interface (c_5)
e_1	5	2	1	3	4
e_2	4	2	1	3	5
e_3	3	2	1	4	5
e_4	5	4	2	3	1
e_5	5	2	3	1	4

Step 2. With the information provided in Step 1, generate the results using the BON-OWA, BON-IOWA, BON-HOWA and BON-IHOWA operators to understand the process that has been performed. An example using the information of expert 1 for the state of Zacatecas will be explained in detail, assuming that the process will be the same for all other states and experts. The values of q and p are equal to 1.

The first thing is determine the vectors V^i, and the results are

$$V_1 = (90, 56, 85.71, 60)$$
$$V_2 = (56, 85.71, 60, 100)$$
$$V_3 = (85.71, 60, 100, 90)$$
$$V_4 = (60, 100, 90, 56)$$
$$V_5 = (100, 90, 56, 85.71)$$

Next, the BON-OWA operator is applied. Then, a weight is assigned to each attribute according to a maximum criterion, and the results are

$$V_1 = [(90 \times 0.30) + (56 \times 0.15) + (85.71 \times 0.20) + (60 \times 0.20)] = 64.54$$
$$V_2 = [(56 \times 0.15 + 85.71 \times 0.20 + 60 \times 0.20 + 100 \times 0.30)] = 67.54$$
$$V_3 = (85.71 \times 0.15 + 60 \times 0.15 + 100 \times 0.20 + 90 \times 0.20) = 59.86$$
$$V_4 = (60 \times 0.15 + 100 \times 0.30 + 90 \times 0.20 + 56 \times 0.15) = 65.40$$
$$V_5 = (100 \times 0.30 + 90 \times 0.20 + 56 \times 0.15 + 85.71 \times 0.15) = 69.26$$

$$BON-OWA = \left(\frac{[(64.54 \times 100) + (67.54 \times 90) + (59.86 \times 56) + (65.40 \times 85.71) + (69.26 \times 60)]}{5} \right)^{\frac{1}{1+1}}$$

$$BON-OWA = 71.62$$

All the results for each state and expert are presented in Table A2.

In the case of the calculation for the BON-IOWA operator, the vectors V^i are the same as those used in the BON-OWA operator. The next step is the association of the weights with the attributes that in this case will be performed by using the induced variables instead of the values of the attributes. Here, the results for Zacatecas are the following.

$$V_1 = [(90 \times 0.20) + (56 \times 0.30) + (85.71 \times 0.20) + (60 \times 0.15)] = 64.11$$
$$V_2 = [(56 \times 0.20) + (85.71 \times 0.30) + (60 \times 0.20) + (100 \times 0.15)] = 66.11$$
$$V_3 = [(85.71 \times 0.20) + (60 \times 0.20) + (100 \times 0.15) + (90 \times 0.15)] = 59.86$$
$$V_4 = [(60 \times 0.30) + (100 \times 0.20) + (90 \times 0.15) + (56 \times 0.15)] = 65.40$$
$$V_5 = [(100 \times 0.20) + (90 \times 0.30) + (56 \times 0.15) + (85.71 \times 0.15)] = 68.26$$
$$BON - IOWA = \left(\frac{[(64.11 \times 100) + (66.11 \times 90) + (59.86 \times 56) + (65.40 \times 85.71) + (68.26 \times 60)]}{5} \right)^{\frac{1}{1+1}}$$
$$BON - OWA = 71.29$$

All the results for each state and expert are presented in Table A3.

In the case of the BON-HOWA operator, the vectors V^i are also the same, but the weights will the ones presented in Table 2 and will be ordered with the arguments with a maximum criterion. Therefore, the results for Zacatecas are the following.

$$V_1 = [(90 \times 0.30) + (56 \times 0.20) + (85.71 \times 0.20) + (60 \times 0.20)] = 67.34$$
$$V_2 = [(56 \times 0.20) + (85.71 \times 0.20) + (60 \times 0.20) + (100 \times 0.30)] = 70.34$$
$$V_3 = [(85.71 \times 0.20) + (60 \times 0.20) + (100 \times 0.20) + (90 \times 0.20)] = 67.14$$
$$V_4 = [(60 \times 0.20) + (100 \times 0.30) + (90 \times 0.20) + (56 \times 0.20)] = 71.20$$
$$V_5 = [(100 \times 0.30) + (90 \times 0.20) + (56 \times 0.20) + (85.71 \times 0.20)] = 76.34$$
$$Bon - HOWA = \left(\frac{[(67.34 \times 100) + (70.34 \times 90) + (67.14 \times 56) + (71.20 \times 85.71) + (76.34 \times 60)]}{5} \right)^{\frac{1}{1+1}}$$
$$BON - HOWA = 74.17$$

All the results for each state and expert are presented in Table A4.

Finally, the BON-IHOWA operator is constructed. The vectors V^i are the same as the other operators, but the weights will be the ones in Table 2 and will be ordered based on the induced values of Table 3 The results for Zacatecas are the following.

$$V_1 = [(90 \times 0.20) + (56 \times 0.30) + (85.71 \times 0.20) + (60 \times 0.20)] = 66.91$$
$$V_2 = [(56 \times 0.20) + (85.71 \times 0.30) + (60 \times 0.20) + (100 \times 0.20)] = 68.91$$
$$V_3 = [(85.71 \times 0.20) + (60 \times 0.20) + (100 \times 0.20) + (90 \times 0.20)] = 67.14$$
$$V_4 = [(60 \times 0.30) + (100 \times 0.20) + (90 \times 0.20) + (56 \times 0.20)] = 71.20$$
$$V_5 = [(100 \times 0.20) + (90 \times 0.30) + (56 \times 0.20) + (85.71 \times 0.20)] = 75.34$$
$$BON - IHOWA = \left(\frac{[(66.91 \times 100) + (68.91 \times 90) + (67.14 \times 56) + (71.20 \times 85.71) + (75.34 \times 60)]}{5} \right)^{\frac{1}{1+1}}$$
$$BON - IHOWA = 73.86$$

All the results for each state and expert are presented in Table A5.

Step 3. With all the results obtained in Step 2, the results for the BON-POWA, BON-PIOWA, BON-PHOWA and BON-PIHOWA operators can be obtained. The weights associated with each expert are the following: $e_1 = 0.30$, $e_1 = 0.10$, $e_1 = 0.20$, $e_1 = 0.15$ and $e_1 = 0.25$. The result for each operator for Zacatecas is as follows.

$$BON - POWA = [(71.62 \times 0.30) + (72.77 \times 0.10) + (72.60 \times 0.20) + (71.12 \times 0.15) + (72.47 \times 0.25)] = 72.07$$
$$BON - PIOWA = [(71.29 \times 0.30) + (71.97 \times 0.10) + (71.77 \times 0.20) + (70.77 \times 0.15) + (71.32 \times 0.25)] = 71.39$$
$$BON - PHOWA = [(74.17 \times 0.30) + (75.28 \times 0.10) + (76.06 \times 0.20) + (73.77 \times 0.15) + (75.48 \times 0.25)] = 74.93$$
$$BON - PIHOWA = [(73.86 \times 0.30) + (74.63 \times 0.10) + (75.21 \times 0.20) + (73.44 \times .15) + (74.75 \times 0.25)] = 74.37$$

The results for all the states are presented in Table A6.

4.3. Discussion of the Results

After an analysis of the different results obtained and presented in Tables A2–A6, the main changes that are found are as follows.

Based on the top 10 results of the different aggregation operators and experts, the first four positions do not change at all with the different aggregation operators and experts. In this sense, even when the importance of each component varies, the four best states remain the same: Zacatecas, Oaxaca, Nuevo Leon and Puebla. Then, according to the aggregation operator and expert that we analyze, the ranking can change. For example, in ranks five and six, we usually find the states of San Luis Potosi and Nayarit, respectively, but with the use of the BON-IHOWA operator, the positions change to Nayarit and San Luis Potosi, respectively. The other remaining positions vary, but the states remain the same and are Tlaxcala, Sonora, Yucatan and Queretaro.

Based on the bottom 10 results, the first four positions (as in the case of the top 10) remain the same considering the different aggregation operators and experts. In this sense, the worst states are Chihuahua, Ciudad de Mexico, Aguascalientes and Campeche. Then, the fifth and sixth positions are Tabasco and Guerrero depending on the aggregation operator and expert. Finally, positions seven to ten can change drastically. For example, for expert 1, from the BON-OWA operator, Chiapas is considered among the bottom 10 states and Jalisco is not; however, according to the information provided by expert 2, Jalisco is among the bottom 10 and Chiapas is not. This is important because in this process, it is possible to see that depending on the importance that is given to the information, the states can be or cannot be in the bottom 10 list.

The same analysis can be performed for the states in the middle of the ranking, and they change positions based on the different experts and aggregation operators. First, the top 10 of the lists does not change at all, but it is possible to see some notable changes as the ones explained in the bottom 10 analysis. This information is important for policy-makers and governments to analyze to change and implement public policies according to the deficiency of each state, which can vary depending on the importance given to the components. Additionally, as seen, the ranking changes, and the benefits and government support for the states can be rearranged because of their positions in the ranking.

5. Conclusions

The main objective of this document is to present the new BON-PrIHOWA operator. The main features of this new proposition are that one can combine a nonrestricted to one weighting vector, an induced vector that assigns weights to the attributes and a prioritized vector that unifies the opinions of the decision makers in a group decision-making process, where not all stakeholders have the same importance in the computation.

Additionally, in this document, the main definitions of the BON-PrIHOWA operator are included, and it is important to mention that the BON-PrIHOWA can be reduced to the PrIOWA, PrHOWA, IHOWA, and OWA. This is suggested when the complexity of the problem is minimal and not very extensive. However, the design of this operator, its functionality and its operability are intended for complex phenomena with highly dynamic information. This is the case, e.g., when a combination of expert information is required to assess open government initiatives and public policies.

The complete design of the BON-PrIHOWA operator uses a ranking of transparency websites for Mexico. Among the main results, it was possible to identify that the top and bottom four states remained the same even when the weights, operators and experts changed. This is important because their positions cannot change easily. However, other positions can also change drastically depending on the operator or expert and, because of that, the perception of transparency of the citizens and governors. The main component that changes the ranking is the importance that is given to each component of transparency websites. When the weights assigned to each result are not $\frac{1}{n}$, but rather they depend on the focus and goals of each government, the score can change drastically. This change in the weights is important because not all information can be treated in the same way since the characteristics, objectives and goals of the states are not always the same. They are derived from their demographic, economic, and geographic characteristics, among others, in such a way that treating similar information is not appropriate. The idea of identifying changes in the ranking can improve the public policies that are established because the ranking can be established not only by using the average of the components but also by using a specific operator depending on the individual characteristics of the state.

For future research, more extensions of the OWA operator can be conceived with the use of distance operators [43], Bonferroni means [17,29,44], moving averages [45–48], forgotten effects [47,49], the least square deviation [50,51] or logarithmic operators [52,53]. This is important when the subjectivity and the uncertainty of the decision–making process are presented. With the use of aggregation operators and other fuzzy techniques, it is possible to generate new scenarios based on the expertise and expectations of the decision makers. Additionally, the use of different coefficients to test the similarity between the rankings will be useful to compare rankings in decision-making fields [54]. Finally, these new techniques can be applied in different areas such as economics, finance, engineering, social science and other areas [55] where the idea and characteristics of fuzzy logic and fuzzy sets can be used [56,57].

Author Contributions: Conceptualization, V.A.-G.; Data curation, L.A.P.-A.; Formal analysis, F.B.-M.; Methodology, E.L.-C. All authors have read and agreed to the published version of the manuscript.

Funding: This research received no external funding.

Acknowledgments: Author Leon-Castro acknowledges support from the Chilean government through FONDECYT initiation grant No. 11190056.

Conflicts of Interest: The authors declare no conflict of interest.

Appendix A

Table A1. Ranking of the transparency portal of the states of Mexico.

Ranking	State	Institutional Arrangements	Open Data	Vertical Collaboration	Horizontal Collaboration	Interface	Total
1	Zacatecas	90	56	85.71	60	100	78.34
2	Oaxaca	50	92	100	40	90.91	74.58
3	Nuevo Leon	80	88	85.71	30	81.82	73.11
4	Puebla	80	80	71.43	40	81.82	70.65
5	Nayarit	80	76	71.43	30	72.73	66.03
6	San Luis Potosí	90	80	57.14	30	72.73	65.97
7	Tlaxcala	90	44	71.43	30	81.82	63.45
8	Sonora	80	48	71.43	40	72.73	62.43
9	Yucatán	90	60	57.14	20	81.82	61.79
10	Querétaro	70	52	71.43	40	72.73	61.23
11	Quintara Roo	80	56	42.86	50	63.64	58.50
12	Estado de México	90	60	42.86	40	45.45	55.66
13	Guanajuato	70	60	71.43	20	45.45	53.38
14	Michoacán	90	48	28.57	20	72.73	51.86
15	Sinaloa	80	28	42.86	30	72.73	50.72
16	Coahuila	100	64	28.57	0	54.55	49.42
17	Veracruz	90	44	28.57	0	81.82	48.88
18	Baja California	70	28	57.14	20	63.64	47.76
19	Morelos	50	40	57.14	40	45.45	46.52
20	Hidalgo	90	36	28.57	20	54.55	45.82
21	Colima	90	36	28.57	0	63.64	43.64

Table A1. Cont.

Ranking	State	Institutional Arrangements	Open Data	Vertical Collaboration	Horizontal Collaboration	Interface	Total
22	Jalisco	70	40	28.57	40	27.27	41.17
23	Baja California Sur	50	36	42.86	40	36.36	41.04
24	Tamaulipas	50	32	28.57	30	63.64	40.84
25	Chiapas	80	44	14.29	0	63.64	40.39
26	Durango	80	36	28.57	10	45.45	40.00
27	Guerrero	50	24	57.14	40	27.27	39.68
28	Tabasco	60	40	28.57	0	63.64	38.44
29	Campeche	70	36	14.29	0	54.55	34.97
30	Aguascalientes	40	24	28.57	0	27.27	23.97
31	Ciudad de México	40	20	14.29	0	36.36	22.13
32	Chihuahua	30	12	0	0	45.45	17.49

Table A2. Bon-OWA operator results.

States	e_1	e_2	e_3	e_4	e_5
Zacatecas	71.62	72.77	72.60	71.12	72.47
Oaxaca	67.75	70.10	69.48	67.75	69.63
Nuevo Leon	66.19	68.30	67.87	66.23	68.02
Puebla	63.93	65.61	65.20	64.03	65.41
Nayarit	59.85	61.66	61.31	59.93	61.38
San Luis Potosí	60.08	62.46	61.73	60.46	62.25
Tlaxcala	58.40	60.19	59.86	57.93	59.80
Sonora	57.06	58.27	58.03	56.72	58.05
Yucatán	56.74	59.01	58.63	56.68	58.48
Querétaro	55.72	56.81	56.61	55.45	56.61
Quintara Roo	53.38	54.43	54.15	53.54	54.28
Estado de México	50.73	51.93	51.52	51.09	51.85
Guanajuato	48.68	50.56	50.14	48.53	50.26
Michoacán	47.66	49.94	49.18	47.95	49.67
Sinaloa	46.95	48.07	47.98	46.59	47.67
Coahuila	44.58	48.27	46.95	45.40	47.98
Veracruz	44.81	47.80	46.97	45.00	47.31
Baja California	44.02	45.53	45.23	43.54	45.21
Morelos	42.22	42.51	42.46	42.08	42.45
Hidalgo	42.01	43.56	43.14	42.11	43.26
Colima	39.93	42.36	41.79	39.99	41.87
Jalisco	37.52	38.51	38.19	37.74	38.44
Baja California Sur	37.23	37.49	37.46	37.13	37.43
Tamaulipas	36.91	37.43	37.31	36.84	37.39
Chiapas	36.62	39.77	38.53	37.23	39.55
Durango	36.53	38.33	37.84	36.65	38.05
Guerrero	36.52	36.99	37.10	36.10	36.73
Tabasco	35.23	37.63	37.01	35.34	37.24
Campeche	31.78	34.34	33.39	32.22	34.12
Aguascalientes	21.89	23.15	22.91	21.81	22.92
Ciudad de México	20.31	21.63	21.30	20.37	21.39
Chihuahua	15.36	16.50	15.94	15.71	16.50

Table A3. Bon-IOWA operator results.

States	e_1	e_2	e_3	e_4	e_5
Zacatecas	71.29	71.97	71.77	70.77	71.32
Oaxaca	67.53	69.53	67.46	67.52	68.72
Nuevo Leon	66.13	68.11	66.85	66.14	67.83
Puebla	63.90	65.38	64.43	63.98	65.08
Nayarit	59.74	61.54	60.43	59.78	61.24
San Luis Potosí	59.80	61.88	60.70	60.07	61.43
Tlaxcala	57.93	59.31	58.54	57.67	58.54
Sonora	56.88	57.56	57.31	56.59	57.04
Yucatán	55.93	58.85	56.73	56.21	58.31
Querétaro	55.67	56.30	55.99	55.41	55.88
Quintara Roo	52.82	54.14	53.45	53.10	53.81
Estado de México	49.96	51.48	50.88	50.07	51.20
Guanajuato	48.49	50.13	48.74	48.34	49.81
Michoacán	46.47	49.20	47.53	47.24	48.60
Sinaloa	45.80	47.60	46.52	45.98	46.98
Coahuila	43.93	47.11	45.50	44.54	46.54
Veracruz	43.18	47.20	44.29	44.16	46.46
Baja California	43.70	44.58	44.16	43.36	43.85
Morelos	41.93	42.27	42.14	41.81	42.09
Hidalgo	40.72	43.08	41.67	41.23	42.63
Colima	38.41	41.92	39.63	39.13	41.46
Jalisco	36.98	38.08	37.62	37.10	37.65
Baja California Sur	37.04	37.30	37.21	36.95	37.20
Tamaulipas	36.45	37.27	36.58	36.58	37.15
Chiapas	35.55	38.62	36.68	36.61	37.90
Durango	35.67	37.82	36.73	36.01	37.54
Guerrero	36.16	36.59	36.34	35.73	36.24
Tabasco	34.46	37.23	35.12	34.90	36.67
Campeche	30.78	33.48	31.76	31.64	32.91
Aguascalientes	21.72	22.88	22.31	21.66	22.76
Ciudad de México	19.62	21.41	20.10	20.01	21.08
Chihuahua	14.13	15.94	15.00	14.94	15.66

Table A4. Bon-HOWA operator results.

States	e_1	e_2	e_3	e_4	e_5
Zacatecas	74.17	75.28	76.06	73.77	75.48
Oaxaca	69.82	72.11	73.16	69.82	72.25
Nuevo Leon	68.33	70.38	71.55	68.33	71.00
Puebla	66.24	67.86	68.66	66.35	68.28
Nayarit	61.82	63.57	64.61	61.88	64.08
San Luis Potosí	61.86	64.17	65.01	62.19	64.60
Tlaxcala	59.95	61.70	62.77	59.53	62.01
Sonora	59.00	60.17	60.86	58.68	60.45
Yucatán	58.05	60.26	61.68	58.09	60.66
Querétaro	57.74	58.79	59.44	57.49	59.05
Quintara Roo	55.30	56.31	56.78	55.48	56.49
Estado de México	52.43	53.59	53.96	52.73	53.81
Guanajuato	50.00	51.84	52.81	49.85	52.23
Michoacán	48.60	50.83	51.55	48.96	51.03
Sinaloa	48.10	49.20	50.09	47.85	49.31
Coahuila	44.96	48.62	49.55	45.74	48.97
Veracruz	45.09	48.07	49.30	45.40	48.33
Baja California	45.07	46.54	47.39	44.61	46.77

Table A4. *Cont.*

States	e_1	e_2	e_3	e_4	e_5
Morelos	43.98	44.26	44.44	43.85	44.33
Hidalgo	42.90	44.42	45.09	43.05	44.50
Colima	40.18	42.61	43.81	40.35	42.85
Jalisco	38.71	39.67	39.99	38.92	39.87
Baja California Sur	38.79	39.04	39.24	38.70	39.12
Tamaulipas	38.30	38.80	39.07	38.35	39.00
Chiapas	36.80	39.93	40.51	37.44	40.08
Durango	37.15	38.92	39.69	37.31	39.16
Guerrero	37.66	38.11	38.77	37.27	38.27
Tabasco	35.56	37.94	39.00	35.76	38.24
Campeche	31.94	34.49	35.08	32.42	34.64
Aguascalientes	22.19	23.44	24.16	22.11	23.72
Ciudad de México	20.45	21.76	22.36	20.56	21.89
Chihuahua	15.36	16.50	16.50	15.71	16.50

Table A5. Bon-IHOWA operator results.

States	e_1	e_2	e_3	e_4	e_5
Zacatecas	73.86	74.63	75.21	73.44	74.75
Oaxaca	69.61	71.57	70.72	69.60	71.66
Nuevo Leon	68.27	70.27	70.30	68.25	70.88
Puebla	66.21	67.66	67.71	66.29	68.06
Nayarit	61.71	63.50	63.56	61.73	63.99
San Luis Potosí	61.58	63.69	63.89	61.81	64.08
Tlaxcala	59.50	60.93	61.38	59.28	61.21
Sonora	58.83	59.57	60.06	58.55	59.80
Yucatán	57.25	60.18	59.78	57.64	60.55
Querétaro	57.69	58.33	58.68	57.45	58.58
Quintara Roo	54.76	56.16	56.24	55.06	56.19
Estado de México	51.69	53.40	53.62	51.73	53.39
Guanajuato	49.82	51.55	51.14	49.67	51.94
Michoacán	47.43	50.21	50.19	48.26	50.33
Sinaloa	46.98	48.80	48.95	47.26	48.86
Coahuila	44.31	47.68	48.12	44.89	48.04
Veracruz	43.47	47.51	46.90	44.56	47.78
Baja California	44.75	45.69	46.23	44.44	45.91
Morelos	43.71	44.10	44.22	43.59	44.10
Hidalgo	41.64	44.18	44.02	42.19	44.10
Colima	38.68	42.34	42.00	39.49	42.58
Jalisco	38.19	39.47	39.61	38.30	39.36
Baja California Sur	38.61	38.96	39.03	38.52	38.98
Tamaulipas	37.86	38.75	38.42	38.11	38.85
Chiapas	35.73	38.84	38.81	36.82	39.00
Durango	36.31	38.68	38.82	36.68	38.83
Guerrero	37.31	37.87	38.04	36.91	37.95
Tabasco	34.79	37.56	37.09	35.32	37.87
Campeche	30.94	33.70	33.63	31.84	33.85
Aguascalientes	22.03	23.33	23.51	21.96	23.62
Ciudad de México	19.77	21.56	21.27	20.21	21.69
Chihuahua	14.13	15.94	16.06	14.94	15.94

Table A6. Results of prioritized Bonferroni operators.

States	Bon-POWA	Bon-PIOWA	Bon-PHOWA	Bon-PIHOWA
Zacatecas	72.07	71.39	74.93	74.37
Oaxaca	68.80	68.01	71.32	70.54
Nuevo Leon	67.20	66.90	69.85	69.53
Puebla	64.74	64.46	67.41	67.13
Nayarit	60.72	60.44	63.13	62.83
San Luis Potosí	61.25	60.64	63.46	62.91
Tlaxcala	59.15	58.30	61.14	60.41
Sonora	57.57	57.03	59.80	59.35
Yucatán	57.77	57.02	59.65	58.93
Querétaro	56.19	55.81	58.47	58.14
Quintara Roo	53.89	53.37	56.02	55.60
Estado de México	51.34	50.62	53.24	52.68
Guanajuato	49.53	49.01	51.28	50.77
Michoacán	48.74	47.60	50.07	49.11
Sinaloa	47.39	46.45	48.87	48.07
Coahuila	46.40	45.31	47.37	46.43
Veracruz	46.20	44.77	47.08	45.80
Baja California	44.64	43.87	46.04	45.38
Morelos	42.33	42.03	44.17	43.93
Hidalgo	42.72	41.70	43.91	43.07
Colima	41.04	39.88	41.84	40.81
Jalisco	38.02	37.40	39.38	38.91
Baja California Sur	37.34	37.13	38.97	38.81
Tamaulipas	37.15	36.75	38.69	38.35
Chiapas	38.14	36.83	38.77	37.64
Durango	37.37	36.62	38.36	37.73
Guerrero	36.67	36.19	38.02	37.61
Tabasco	36.35	35.49	37.18	36.38
Campeche	33.01	31.90	33.57	32.62
Aguascalientes	22.47	22.21	23.08	22.84
Ciudad de México	20.92	20.32	21.34	20.79
Chihuahua	15.93	14.99	16.04	15.27

References

1. OECD. *Open Government: The Global Context and the Way Forward*; Organisation for Economic Co-operation and Development: Paris, France, 2016.
2. OECD. *Government at a Glance 2019*; OECD Publishing: Paris, France, 2019.
3. Jaeger, P.T.; Bertot, J.C. Transparency and technological change: Ensuring equal and sustained public access to government information. *Gov. Inf. Q.* **2010**, *27*, 371–376. [CrossRef]
4. Abu-Shanab, E.A. Reengineering the open government concept: An empirical support for a proposed model. *Gov. Inf. Q.* **2015**, *32*, 453–463. [CrossRef]
5. Nam, T. Challenges and concerns of open government: A case of government 3.0 in Korea. *Soc. Sci. Comput. Rev.* **2015**, *33*, 556–570. [CrossRef]
6. Piotrowski, S.J.; Van Ryzin, G.G. Citizen attitudes toward transparency in local government. *Am. Rev. Public Adm.* **2007**, *37*, 306–323. [CrossRef]
7. McDermott, P. Building open government. *Gov. Inf. Q.* **2010**, *27*, 401–413. [CrossRef]
8. Meijer, A.; 't Hart, P.; Worthy, B. Assessing government transparency: An interpretive framework. *Adm. Soc.* **2018**, *50*, 501–526. [CrossRef]
9. Héritier, A. Composite democracy in Europe: The role of transparency and access to information. *J. Eur. Public Policy* **2003**, *10*, 814–833. [CrossRef]
10. Gupta, A. *Transparency in Global Environmental Governance: A Coming of Age?* MIT Press: Cambridge, MA, USA, 2010.
11. Piotrowski, S.J.; Zhang, Y.; Lin, W.; Yu, W. Key issues for implementation of Chinese open government information regulations. *Public Adm. Rev.* **2009**, *69*, S129–S135. [CrossRef]
12. Terrazas-Tapia, R. *IDAIM 2014. Índice del Derecho de Acceso a la Información de México*; Fundar, Centro de Análisis e Investigación: Mexico City, Mexico, 2014.
13. INAI. *Informe de Labores. Utilidad del Acceso a la Información*; Instituto Nacional de Transparencia, Acceso a la Información y Protección de Datos Personales: Mexico City, Mexico, 2019.

14. Sandoval-Almazán, R. Midiendo al gobierno abierto en México: Los portales estatales de transparencia durante el periodo 2015–2016. *Transparencia Combate a la Corrupción y Gobierno Abierto La Experiencia en México* **2017**, *29*, 47–66.
15. Beliakov, G.; Pradera, A.; Calvo, T. *Aggregation Functions: A Guide for Practitioners*; Springer: Heidelberg, Germany, 2007; Volume 221.
16. Yager, R.R.; Kacprzyk, J.; Beliakov, G. *Recent Developments in the Ordered Weighted Averaging Operators: Theory and Practice*; Springer: Berlin/Heidelberg, Germany, 2011; Volume 265.
17. Bonferroni, C. Sulle medie multiple di potenze. *Bollettino dell'Unione Matematica Italiana* **1950**, *5*, 267–270.
18. Yager, R.R. On ordered weighted averaging aggregation operators in multicriteria decisionmaking. *IEEE Trans. Syst. ManCybern.* **1988**, *18*, 183–190. [CrossRef]
19. Yager, R.R. Prioritized OWA aggregation. *Fuzzy Optim. Decis. Mak.* **2009**, *8*, 245–262. [CrossRef]
20. Yager, R.R.; Filev, D.P. Induced ordered weighted averaging operators. *IEEE Trans. Syst. ManCybern. Part B (Cybern.)* **1999**, *29*, 141–150. [CrossRef] [PubMed]
21. Merigó, J.M.; Gil-Lafuente, A.M. The induced generalized OWA operator. *Inf. Sci.* **2009**, *179*, 729–741. [CrossRef]
22. Yager, R.R. Generalized OWA aggregation operators. *Fuzzy Optim. Decis. Mak.* **2004**, *3*, 93–107. [CrossRef]
23. Yager, R.R. Heavy OWA operators. *Fuzzy Optim. Decis. Mak.* **2002**, *1*, 379–397. [CrossRef]
24. Blanco-Mesa, F.; León-Castro, E.; Merigó, J.M. A bibliometric analysis of aggregation operators. *Appl. Soft Comput.* **2019**, *81*, 105488. [CrossRef]
25. Detyniecki, M.; Bouchon-meunier, D.B.; Yager, D.R.; Prade, R.H. Mathematical Aggregation Operators and Their Application to Video Querying. Ph.D. Thesis, Pierre and Marie Curie University, Paris, France, 2000.
26. Herrera, F.; Martínez, L. A 2-tuple fuzzy linguistic representation model for computing with words. *IEEE Trans. Fuzzy Syst.* **2000**, *8*, 746–752.
27. Yu, D. A scientometrics review on aggregation operator research. *Scientometrics* **2015**, *105*, 115–133. [CrossRef]
28. Rickard, J.T.; Aisbett, J. New classes of threshold aggregation functions based upon the Tsallis q-exponential with applications to perceptual computing. *IEEE Trans. Fuzzy Syst.* **2013**, *22*, 672–684. [CrossRef]
29. Blanco-Mesa, F.; Merigó, J.M.; Kacprzyk, J. Bonferroni means with distance measures and the adequacy coefficient in entrepreneurial group theory. *Knowl. Based Syst.* **2016**, *111*, 217–227. [CrossRef]
30. Blanco-Mesa, F.; Merigó, J.M. Bonferroni distances and their application in group decision making. *Cybern. Syst.* **2020**, *51*, 27–58. [CrossRef]
31. Blanco-Mesa, F.; León-Castro, E.; Merigó, J.M. Bonferroni induced heavy operators in ERM decision-making: A case on large companies in Colombia. *Appl. Soft Comput.* **2018**, *72*, 371–391. [CrossRef]
32. Blanco-Mesa, F.; León-Castro, E.; Merigó, J.M.; Xu, Z. Bonferroni means with induced ordered weighted average operators. *Int. J. Intell. Syst.* **2019**, *34*, 3–23. [CrossRef]
33. Blanco-Mesa, F.; León-Castro, E.; Merigó, J.M.; Herrera-Viedma, E. Variances with Bonferroni means and ordered weighted averages. *Int. J. Intell. Syst.* **2019**, *34*, 3020–3045. [CrossRef]
34. Blanco-Mesa, F.; León-Castro, E.; Merigó, J.M. Covariances with OWA operators and Bonferroni means. *Soft Comput.* **2020**, 1–16. [CrossRef]
35. Perez-Arellano, L.A.; Leon-Castro, E.; Blanco-Mesa, F.; Fonseca-Cifuentes, G. The ordered weighted government transparency average: Colombia case. *J. Intell. Fuzzy Syst.* **2020**, 1–13. [CrossRef]
36. Merigo, J.M.; Casanovas, M. Decision-making with distance measures and induced aggregation operators. *Comput. Ind. Eng.* **2011**, *60*, 66–76. [CrossRef]
37. Merigó, J.M.; Gil-Lafuente, A.M.; Gil-Aluja, J. A new aggregation method for strategic decision making and its application in assignment theory. *Afr. J. Bus. Manag.* **2011**, *5*, 4033–4043.
38. Yager, R.R. On generalized Bonferroni mean operators for multi-criteria aggregation. *Int. J. Approx. Reason.* **2009**, *50*, 1279–1286. [CrossRef]
39. Baez-Palencia, D.; Olazabal-Lugo, M.; Romero-Muñoz, J. Toma de decisiones empresariales a través de la media ordenada ponderada. *Inquietud Empresarial* **2019**, *19*, 11–23.
40. Yager, R.R. Prioritized aggregation operators. *Int. J. Approx. Reason.* **2008**, *48*, 263–274. [CrossRef]
41. Merigó, J.M.; Palacios-Marqués, D.; Soto-Acosta, P. Distance measures, weighted averages, OWA operators and Bonferroni means. *Appl. Soft Comput.* **2017**, *50*, 356–366. [CrossRef]
42. Blanco-Mesa, F.; Merigó, J.M.; Gil-Lafuente, A.M. Fuzzy decision making: A bibliometric-based review. *J. Intell. Fuzzy Syst.* **2017**, *32*, 2033–2050. [CrossRef]
43. Hamming, R.W. Error detecting and error correcting codes. *Bell Syst. Tech. J.* **1950**, *29*, 147–160. [CrossRef]
44. Espinoza-Audelo, L.F.; Olazabal-Lugo, M.; Blanco-Mesa, F.; León-Castro, E.; Alfaro-Garcia, V. Bonferroni Probabilistic Ordered Weighted Averaging Operators Applied to Agricultural Commodities' Price Analysis. *Mathematics* **2020**, *8*, 1350. [CrossRef]
45. León-Castro, E.; Avilés-Ochoa, E.; Merigó, J.M.; Gil-Lafuente, A.M. Heavy Moving Averages and Their Application in Econometric Forecasting. *Cybern. Syst.* **2018**, *49*, 26–43. [CrossRef]
46. León-Castro, E.; Avilés-Ochoa, E.; Merigó, J.M. Induced heavy moving averages. *Int. J. Intell. Syst.* **2018**, *33*, 1823–1839. [CrossRef]
47. Olazabal-Lugo, M.; Leon-Castro, E.; Espinoza-Audelo, L.F.; Maria Merigo, J.; Gil Lafuente, A.M. Forgotten effects and heavy moving averages in exchange rate forecasting. *Econ. Comput. Econ. Cybern. Stud. Res.* **2019**, *53*. [CrossRef]

48. Kenny, J.F.; Keeping, E.S. Relative merits of mean, median, and mode. *Math. Stat. Van Nostrans Nj (Ed)* **1962**, 211–212.
49. Alfaro Calderón, G.G.; Godinez Reyes, N.L.; Gómez-Monge, R.; Alfaro-García, V.G.; Gil Lafuente, A.M. Forgotten effects in the valuation of the social well-being index in Mexico's sustainable development. *Fuzzy Econ. Rev.* **2019**, *24*, 67–81. [CrossRef]
50. Hong, D.H.; Han, S. The general least square deviation OWA operator problem. *Mathematics* **2019**, *7*, 326. [CrossRef]
51. Wang, Y.-M.; Luo, Y.; Liu, X. Two new models for determining OWA operator weights. *Comput. Ind. Eng.* **2007**, *52*, 203–209. [CrossRef]
52. Alfaro-García, V.G.; Merigó, J.M.; Gil-Lafuente, A.M.; Kacprzyk, J. Logarithmic aggregation operators and distance measures. *Int. J. Intell. Syst.* **2018**, *33*, 1488–1506. [CrossRef]
53. Zhou, L.G.; Chen, H.y. Generalized ordered weighted logarithm aggregation operators and their applications to group decision making. *Int. J. Intell. Syst.* **2010**, *25*, 683–707. [CrossRef]
54. Sałabun, W.; Urbaniak, K. A new coefficient of rankings similarity in decision-making problems. In *Computational Science—ICCS 2020*; ICCS 2020. Lecture Notes in Computer Science; Springer: Cham, Switzerland, 2020; pp. 632–645.
55. Blanco-Mesa, F. La ciencia de la decisión. *Revista UIS Ingenierías* **2020**, *19*, I–V. [CrossRef]
56. Zadeh, L.A. Fuzzy sets. *Inf. Control.* **1965**, *8*, 338–353. [CrossRef]
57. Zadeh, L.A. Fuzzy logic. *Computer* **1988**, *21*, 83–93. [CrossRef]

Article

Assessing Efficiency of Public Poverty Policies in UE-28 with Linguistic Variables and Fuzzy Correlation Measures

Jorge de Andrés-Sánchez *, Angel Belzunegui-Eraso and Francesc Valls-Fonayet

Social and Business Research Laboratory, Business Administration Department, Rovira i Virgili University, 43002 Tarragona, Spain; angel.belzunegui@urv.cat (A.B.-E.); francesc.valls@urv.cat (F.V.-F.)
* Correspondence: jorge.deandres@urv.cat

Abstract: The present study analyzes the efficiency of social expenditure by EU-28 countries within the period 2014–2018 to reduce poverty. The data are provided by programs European Union Statistics on Income and Living Conditions (EU-SILC) and European System of Integrated Social Protection Statistics (ESSPROS) of Eurostat. We first calculate the Debreu–Farrell (DF) productivity measure similarly to our previous work, published in 2020, for each EU-28 country and rank these poverty policies (PPPs) on the basis of that efficiency index. We also quantify the intensity of the relationship between efficiency and the proportion that each item of social expending suppose within the overall. When evaluating public policies within a given number of years, we have available a longitudinal set of crisp observations (usually annual) for each embedded variable and country. The observed value of variables for any country for the whole period 2014–2018 is quantified as fuzzy numbers (FNs) that are built up by aggregating crisp annual observations on those variables within that period. To rank the efficiency of PPPs, we use the concept of the expected value of an FN. To assess the relation between DF index and the relative effort done in each type of social expense, we interpret Pearson's correlation as a linguistic variable and also use Pearson's correlation index between FNs proposed by D.H. Hong in 2006.

Keywords: fuzzy sets; fuzzy numbers; linguistic variables; fuzzy data analysis; correlation between fuzzy variables; poverty policy; efficiency; Debreu–Farrell productivity index

1. Introduction

This paper assesses public poverty policies (PPPs) in European Union by considering not only attained a diminution of poverty, which obviously is directly linked with social expenditure (SE), but also the productivity of this expenditure. Likewise, we quantify the relationship of every kind of social expenditure (health and sickness benefits, pensions benefits, family and children benefits, ...) with the efficiency of overall SE.

There are many papers on the productivity of public policies from the point of view of fiscal systems [1–3], but also from the perspective of nonmonetary social benefits, such as health and education [4–6]. The productivity of SE to reduce inequality and poverty indexes have also been investigated in for EU-27 countries (EU-28 less Croatia) [7]; within OECD countries [8]; in EU-15 countries [9] and within EU-28 [10–12]. Our analysis shows a new perspective on this topic since we analyze a different period, and moreover, we use fuzzy set theory tools to analyze data.

The key question we address here is the relationship between SE, on one hand, and poverty rates on the other in EU-28. It is well known that there is an inverse relation of SE with poverty and inequality indexes [13,14]. However, ref. [15] showed that despite there is a great negative correlation of SE and poverty levels in EU-28 states, it cannot be concluded that increases in SE lead directly to reductions in poverty. Therefore, ref. [15] suggests that more efforts in social benefits suppose unequal results of poverty policies. Hence, convergence in SE does not imply converging in poverty levels. This fact comes clear in the case of Mediterranean countries as, e.g., Spain where a growth of social expenses could end up absorbed by middle-income households instead lower-income ones due to Mathew effect [16]. Indeed, ref. [17] point out that whereas tax systems and in-cash benefits could generate a diminution of income inequality indexes, they may also produce undesired consequences. It is well known that pensions for the elderly people have a small redistributive effect.

Although family and housing benefits are more progressive than pensions, they have a limited impact since they do not suppose a great proportion of SE. On the other hand, ref. [18] indicates that the results of unemployment policies objected to the Lisbon strategy. That paper outlines that unemployment expenses have not had expected results, and also redistribution programs have not been effective enough in poverty elimination. Likewise, several papers found that the benefits of social assistance policies have not a great effect in many countries [19–23]. Hence, we feel justified assessing productivity of SE on poverty diminutions within EU-28 countries. It could lead us to understand why several countries, after making a similar budgetary effort in social policies, obtain different results in poverty reduction due to the unequal productivity of their social programs.

Our analysis on EU-28 PPPs is done within 2014–2018 by using annual data from the Eurostat programs: European Union Statistics on Income and Living Conditions (EU-SILC) and European System of Integrated Social Protection Statistics (ESSPROS). To assess the results of social policies within a period of several years, a usual practice consists of taking the average value of annual observations as variable observations [8,10,12] or, alternatively, limiting the analysis to a concrete year [7,12]. Clearly, those procedures suppose using limited information. This drawback leads us to propose modeling observations on variables in a period of multiple years by means of fuzzy numbers (FNs). Fuzzy data analysis will allow using all the information in the sample and, in addition, structuring the value of observations in such a way that we may obtain results with an intuitive interpretation. With fuzzy data on evaluated PPPs, we perform two analyses. First, we rank EU-28 countries taking into account the productivity of their PPP. To do it we calculate the fuzzy Debreu–Farrell index for each country and face a problem of FN ordering. Second, we calculate and evaluate the correlation of PPP efficiency with the way SE has been split into items as health expenditure, pensions payments, ... by using fuzzy tools.

The motivations of our research and its novelty can be summarized as follows:

- Periodic assessment of public policies is a must for their improvement. That explains the existence of a wide literature on social policy evaluation, which has been summarized in above paragraphs. Our analysis complement these papers by showing a different perspective on this topic since we cover a different period and in some cases we use a different sample of countries and database. Likewise, we also use a novel (in this kind of analysis) mathematical instruments. Likewise, our results are compared with those from precedent literature.
- The methodology proposed to deal with the variability in longitudinal data supposes a novelty in the field of public policy evaluation. To the best of our knowledge in this field, there is a scarcity of papers that model uncertainty in data by using soft computing tools as fuzzy sets. In most of the studies on this topic, when an observation is given by a set of crisp results, these are reduced to a real number (e.g., the arithmetical mean) to model the observation. Hence, subsequent analysis is done ignoring actual data uncertainty. The use of fuzzy numbers lets modeling and structuring observed uncertainty and also provides a developed mathematical core that allows handling these data linked with their uncertainty in similar way to we do with real numbers.
- The literature on productivity measurement under fuzziness is built up by adapting conventional data envelopment analysis to fuzzy mathematics, and then so-called fuzzy data envelopment analysis (FDEA) methods reach. Our paper also uses a fuzzy efficient frontier to evaluate productivity, but it is built over the basis of the Debreu–Farrell index that is fitted with a regression method. This way to fit efficient frontier is

very common in economics (see [24]) but supposes a novel approach from fuzzy sets literature perspective.

The following section describes the mathematical instruments from fuzzy set theory used in this paper. Methodological aspects of our article: variables, database and methodology that lead to assess the efficiency of SE in poverty reduction are exposed in the third section. In the fourth section, we establish a hierarchy of PPPs in EU-28 by using the concept of the expected value of an FN. Likewise, we evaluate the influence of the composition of SE over the efficiency of PPPs. This last issue is done with the concept of fuzzy correlation and modeling Pearson's correlation coefficient (PCC) as a fuzzy linguistic variable. The last section presents the principal conclusions from our paper.

2. Concepts of Fuzzy Set Mathematics

2.1. Fuzzy Numbers

A fuzzy number (FN) is a fuzzy set $\widetilde{A}$ defined over the set of real numbers, and it is a fundamental concept of FST for representing uncertain quantities. Let us symbolize as $\mu_{\widetilde{A}}(x)$ the membership function of a fuzzy set $\widetilde{A}$. Hence, $\widetilde{A}$ is also a FN if it is normal, i.e., $\max_{x \in X} \mu_{\widetilde{A}}(x) = 1$, and convex, that is, its α-cuts are closed and bounded intervals. Hence, it can be represented as confidence intervals $A = [\underline{A}(\alpha), \overline{A}(\alpha)]$, where $\underline{A}(\alpha)$ ($\overline{A}(\alpha)$) increases (decreases) monotonously respect the membership degree $\alpha \in [0,1]$. A FN $\widetilde{A}$ is a fuzzy quantity that matches "more or less" the real number A, such that $\mu_{\widetilde{A}}(A) = 1$. This paper uses triangular fuzzy numbers (TFNs), that are symbolized as $\widetilde{A} = (A, l_A, r_A)$. Hence, A is the core, and it is the most reliable value: $\mu_{\widetilde{A}}(A) = 1$. Likewise, $l_A, r_A \geq 0$ are the left and right radius and measure the variability of $\widetilde{A}$ respect A. Membership function and α-cuts of a FN are:

$$\mu_{\widetilde{A}}(x) = \begin{cases} \frac{x - A + l_A}{l_A} & A - l_A < x \leq A \\ \frac{A + r_A - x}{r_A} & A < x \leq A + r_A \\ 0 & \text{otherwise} \end{cases} \tag{1a}$$

$$A = [\underline{A}(\alpha), \overline{A}(\alpha)] = [A - l_A(1-\alpha), A + r_A(1-\alpha)] \tag{1b}$$

The hypothesis of a triangular shape for uncertain variables is commonplace in papers on practical applications of FNs. We are aware that this hypothesis may suppose simplifying the complexity of available information. However, we feel that this drawback is balanced by several benefits:

- TFNs are well-adapted to how humans think about imprecise quantities. For example, a prediction as "I expect that for the next two years the GPD growth rate will be 1.5% and deviations no greater than 0.05%" may be quantified in a very natural way as (0.015, 0.005, 0.005). Notice that it is not needed to be a fuzzy set practitioner to interpret and understand the information provided by that FN;
- When the information about a variable is vague and imprecise, as that in this paper, representing the information as simple as possible is desirable, and the linear shape of TFNs meets that requirement;
- TFNs are easier to handle arithmetically than other more complex shapes. From a soft-computing perspective, they provide a good balance between precision on one hand and computational effort and interpretability of results on the other. This fact explains the great deal of literature about approximating triangular shapes to non-TFNs.

In some cases, it will be useful to transform a fuzzy number $\widetilde{A}$ into a crisp equivalent. For example, when we are ranking alternatives from their scores in a variable that are done by FNs. Fuzzy literature provides a great deal of ordering methods (see [25]). In this paper, we will use the concept of the expected value of an FN in [26]. Let be an FN $\widetilde{A}$ and a parameter $\lambda \in [0,1]$ that quantifies the evaluator's optimism grade. The expected value of $\widetilde{A}$ for a given λ is:

$$EV(\widetilde{A}; \lambda) = (1-\lambda) \int_0^1 \underline{A}(\alpha) d\alpha + \lambda \int_0^1 \overline{A}(\alpha) d\alpha \tag{1c}$$

Hence, for a TFN:

$$EV(\widetilde{A}; \lambda) = A - \frac{l_A}{2}(1-\lambda) + \frac{r_A}{2} \tag{1d}$$

2.2. Modeling the Value of the Pearson Correlation Coefficient as a Linguistic Variable

Linguistic variables are variables whose values are sentences from natural or artificial languages named linguistic labels [27]. They are built up by segmenting a universal set in a set of FNs where each one represents a linguistic label. For example, the variable coefficient of correlation has a reference set $[-1, 1]$ and may be partitioned in several linguistic labels as, e.g., "no correlation", "low(−)", "medium(−)", "high(−)" ... }. Hence, "no correlation" may be quantified with the TFN (0, 0.005, 0.005).

Let be a linguistic variable V with a reference set $[V_{min}, V_{max}]$. It is built up by granulating the reference set into J levels (i.e., J fuzzy numbers), $j = 1, 2, \ldots, J$, which in this paper are assumed to be TFNs. Then, by considering $V_{min} = V_1 < V_2 < V_3 < \ldots < V_{J-1} < V_J = V_{max}$ we obtain:

$$\{\widetilde{V}_{k_1} = (V_1, 0, V_2 - V_1); \widetilde{V}_j = (V_j, V_j - V_{j-1}, V_{j+1} - V_j), j = 2, 3, \ldots, J-1; \\ \widetilde{V}_J = (V_J, V_J - V_{J-1}, 0)\}. \tag{2}$$

Notice that it is accomplished that $\sum_j \mu_{\widetilde{V}_j}(x) = 1$ for any crisp value $x \in [V_{min}, V_{max}]$.

The association of a given kind of social expense with the efficiency of PPPs is done by means of a correlation index. As far as decision-making is concerned, it is usual to interpret the value of the correlation coefficient qualitatively by means of linguistic labels as "high (+) correlation" or "weak (−) correlation" that may depend on the context. Table 1 shows three scales exposed in [27] that are used in psychology, political science and medicine.

Table 1. Interpretation of the Pearson's and Spearman correlation coefficients.

Correlation Coefficient		Dancey and Reidy (Psychology)	Quinnipiac University (Politics)	Chan (Medicine)
−1	1	Perfect	Perfect	Perfect
−0.9	0.9	Strong	Very Strong	Very Strong
−0.8	0.8	Strong	Very Strong	Very Strong
−0.7	0.7	Strong	Very Strong	Moderate
−0.6	0.6	Moderate	Strong	Moderate
−0.5	0.5	Moderate	Strong	Fair
−0.4	0.4	Moderate	Strong	Fair
−0.3	0.3	Weak	Moderate	Fair
−0.2	0.2	Weak	Weak	Poor
−0,1	0.1	Weak	Negligible	Poor
0	0	Zero	None	None

Source: [28], Akoglu, H. (2018). User's guide to correlation coefficients. *Turkish Journal of Emergency Medicine*, 18(3), 91–93.

By applying (2) on the scale by the Department of Politics at Quinnipiac University in Figure 1, we built up the fuzzy linguistic variable "correlation coefficient" used in this paper to qualitatively interpret the correlation. It is shown in Table 2.

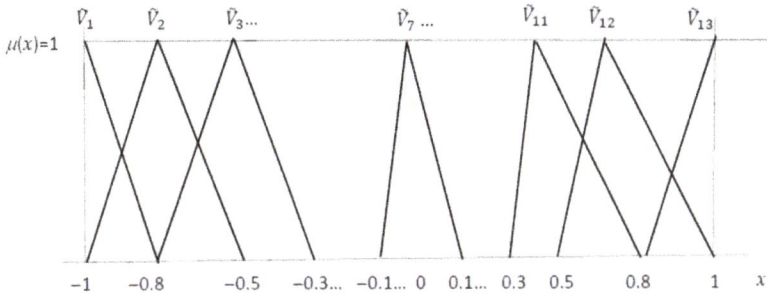

Figure 1. Linguistic variable "coefficient of correlation" built up from the scale by the Department of politics of Quinnipiac University. Source: own elaboration by using [28], Akoglu, H. (2018). User's guide to correlation coefficients. *Turkish Journal of Emergency Medicine, 18*(3), 91–93.

Table 2. Fuzzy linguistic variable "coefficient of correlation" built up from de Quinnipiac University scale.

Negative Correlation		Positive Correlation	
Fuzzy Number	Linguistic Label	Fuzzy Number	Linguistic Label
$\widetilde{V}_1 = (-1, 0, 0.2)$	Perfect (−)	$\widetilde{V}_{13} = (1, 0.2, 0)$	Perfect (+)
$\widetilde{V}_2 = (-0.8, 0.2, 0.3)$	Very strong (−)	$\widetilde{V}_{12} = (0.8, 0.3, 0.2)$	Very strong (+)
$\widetilde{V}_3 = (-0.5, 0.3, 0.2)$	Strong (−)	$\widetilde{V}_{11} = (0.5, 0.2, 0.3)$	Strong (+)
$\widetilde{V}_4 = (-0.3, 0.2, 0.1)$	Moderate (−)	$\widetilde{V}_{10} = (0.3, 0.1, 0.2)$	Moderate (+)
$\widetilde{V}_5 = (-0.2, 0.1, 0.1)$	Weak (−)	$\widetilde{V}_9 = (0.2, 0.1, 0.1)$	Weak (+)
$\widetilde{V}_6 = (-0.1, 0.1, 0.1)$	Negligible (−)	$\widetilde{V}_8 = (0.1, 0.1, 0.1)$	Negligible (+)
$\widetilde{V}_7 = (0, 0.1, 0.1)$	No correlation	$\widetilde{V}_7 = (0, 0.1, 0.1)$	No correlation

Source: own elaboration by using Table 1 in [28], Akoglu, H. (2018). User's guide to correlation coefficients. *Turkish Journal of Emergency Medicine, 18*(3), 91–93.

2.3. Aggregating Crisp Observations by Means of a Triangular Fuzzy Number

In this paper, we capture the uncertainty in data by using FNs. We are aware that Soft Computing Science provides several tools apart from fuzzy sets to represent uncertain data: rough sets, gray sets, intuitionistic and neutrosophic sets ... Using FNs instead other alternatives presents pros and cons. In any case we feel that using TFN in our analysis is suitable for the following reasons:

- Tools as intuitionistic fuzzy sets (IFSs) or neutrosophic fuzzy sets (NFSs) provide an analytical framework to quantify uncertainty more precisely than FNs. Therefore, they are able to capture more nuances from data and its imprecision. For example, NFS state for any element not only a truth membership degree but also an indeterminacy and a falsity degree. However, their estimation implies a greater cost since the number of parameters to fit for each uncertain observation is three times that number than in the case of FNs. On the other hand, gray numbers (GNs) provide a simpler representation of uncertain quantities than FNs. To define a GN is enough to fit its kernel and a grayness measure. Hence, in several circumstances GNs may oversimplify information. For example, GNs suppose a symmetrical structure for a uncertain quantity when perhaps available information does not suggest so. TFNs balances capturing much of the uncertainty in available information (less than, e.g.,

NFS, but more than GNs), but with a smooth shape (less than GNs, but more than NFSs).
- In many circumstances, rough sets also could provide accurate quantification of uncertainty. However, as we will explain below, our problem is more linked to vagueness in observations than with their indiscernibility.
- Our analysis needs implementing arithmetical operations with uncertain variables, ranking them and evaluating Pearson's correlation with uncertain observations. Fuzzy sets literature has developed these questions widely, and so the use of TFNs allows making these analyses similar to real numbers. Likewise, fuzzy arithmetic allows results conserving the triangular shape as well as the uncertainty within data throughout calculations.

Cheng in [29] proposes a method that allows transforming a set of crisp observations on a given variable in an FN. Let us symbolizing as $\{a_1, a_2, \ldots, a_n\}$ the set of crisp observations and $\widetilde{A} = (A, l_A, r_A)$ the TFN than will embed these observations. To fit $\widetilde{A}$ the following steps must be followed:

Step 1. Calculate the distance between ith and jth value as $d_{ij} = |a_i - a_j|$. Of course, $d_{ii} = 0$, $d_{ij} = d_{ji}$. Hence, we can build up a distance matrix $D = [d_{ij}]_{n \times n}$.

Step 2. Calculate the mean distance of ith opinion the other $n - 1$ as:

$$\overline{d}_i = \frac{\sum_{i=1}^{n} d_{ij}}{n-1} \tag{3a}$$

Hence, $\overline{d}_i$ measures the distance of ith opinion to the center of gravity of the opinion pool. Of course, the weight of the value a_i to determine A is decreasing respect to $\overline{d}_i$.

Step 3. Find the matrix $P = [p_{ij}]_{n \times n}$ that indicates the importance of ith opinion over the jth to fix A by doing:

$$p_{ij} = \frac{\overline{d}_j}{\overline{d}_i} \tag{3b}$$

Moreover, so, $p_{ii} = 1$ and $p_{ij} = \frac{1}{p_{ji}}$. Notice that P is obtained from a comparison of distances and so, it ensures its consistency, i.e., that there is a coherent judgment in specifying the pairwise comparison of score importance.

Step 4. Fit coefficients w_i, $i = 1, 2, \ldots, n$, which measure the degree of importance of ith observation to fit $\widetilde{A}$, in such a way $0 \leq w_i \leq 1$, $i = 1, 2, \ldots, n$. These weights are adjusted by taking into account the relative degree of importance of ith observation respect jth, $j = 1, 2, \ldots, n$ (3b). Following [29], if we symbolize as w the vector of weights $n \times 1$, $Pw = nw$, where n is an eigenvalue of P and w an eigenvector. Likewise, given that it must accomplished that $\sum_{i=1}^{n} w_i = 1$, the weights are solved from (3b) by doing:

$$w_i = \frac{1}{\sum_{j=1}^{n} p_{ji}} \tag{3c}$$

Hence, ref. [29] indicates that the consistency of P lead to:

$$p_{ij} = \frac{w_i}{w_j} \tag{3d}$$

Step 5. Calculate the center of $\widetilde{A}$ as:

$$A = \sum_{i=1}^{n} w_i a_i \tag{3e}$$

Step 6. Estimate so-called mean deviation (σ) of the FN $\widetilde{A}$ as a first step to adjusting their spreads. Hence, ref. [29] defines the mean deviation of a FN as $\sigma = \dfrac{\int_{A-l_A}^{A+r_A} |x-A| \mu_{\widetilde{A}}(x) dx}{\int_{A-l_A}^{A+r_A} \mu_{\widetilde{A}}(x) dx}$

and then for a TFN $\sigma = \frac{l_A^2 + r_A^2}{3(l_A + r_A)}$ By using the rate of the left spread respect to the right $\eta = \frac{l_A}{r_A}$:

$$l_A = \frac{3(1+\eta)\eta\sigma}{1+\eta^2} \text{ and } r_A = \frac{3(1+\eta)\sigma}{1+\eta^2} \tag{3f}$$

Notice that σ and η are, in fact, unknown parameters because they are built up from l_A and r_A. Hence, ref. [29] proposes the following approximation for σ, $\hat{\sigma}$:

$$\hat{\sigma} = \sum_{i=1}^{n} w_i |A - a_i| \tag{3g}$$

Step 7. Find the estimate of η, $\hat{\eta}$. By defining as $a^l = \frac{\sum_{\substack{i=1 \\ a_i < A}}^{n} w_i a_i}{\sum_{\substack{i=1 \\ a_i < A}}^{n} w_i}$ and $a^r = \frac{\sum_{\substack{i=1 \\ a_i > A}}^{n} w_i a_i}{\sum_{\substack{i=1 \\ a_i > A}}^{n} w_i}$, $\hat{\eta}$ is:

$$\hat{\eta} = \frac{A - a^l}{a^r - A} \tag{3h}$$

Step 8. Find l_A and r_A by doing

$$l_A = \frac{3(1+\hat{\eta})\hat{\eta}\hat{\sigma}}{1+\hat{\eta}^2} \text{ and } r_A = \frac{3(1+\hat{\eta})\hat{\sigma}}{1+\hat{\eta}^2} \tag{3i}$$

Numerical Application 1

Table 3 shows the values provided by Eurostat within 2014–2018 of the social expenses over GPD, SER = SE/GPD for Belgium. Let us fitting that variable for the quinquennium as a TFN $\widetilde{SER} = (SER, l_{SER}, r_{SER})$ by using (3a)–(3i).

Table 3. Annual social expenses (over GPD) by Belgium in the period 2014–2018.

Year	2014	2015	2016	2017	2018
SER	30	29.8	29.2	28.8	28.8

Source: own elaboration from data provided by EU-SILC (2008–2018) and ESSPROS (2008–2017).

The matrix of distances between observations is exposed in Table 4, and the relative importance of each annual value of SER in the final TFN is provided in Table 5.

Table 4. Matrix of distances to build up the fuzzy number "Belgian SER within 2014–2018".

	2014	2015	2016	2017	2018
2014	0	0.2	0.8	1.2	1.2
2015	0.2	0	0.6	1	1
2016	0.8	0.6	0	0.4	0.4
2017	1.2	1	0.4	0	0
2018	1.2	1	0.4	0	0

Source: own elaboration from data provided by EU-SILC (2008–2018) and ESSPROS (2008–2017).

Table 5. Relative importance of social expenses over GPD (SER) in each year in the triangular fuzzy numbers (TFN) "Belgian SER in 2014–2018".

	2014	2015	2016	2017	2018
2014	1	0.824	1.294	0.765	0.765
2015	1.214	1	1.571	0.929	0.929
2016	0.773	0.636	1	0.591	0.591
2017	1.308	1.077	1.692	1	1
2018	1.308	1.077	1.692	1	1

Source: own elaboration from data provided by EU-SILC (2008–2018) and ESSPROS (2008–2017).

Then, the vector of weights is: $w = (0.178, 0.217, 0.138, 0.233, 0.233)$ and, therefore SER = 29.286. To fit the spreads, l_{SER} and r_{SER} we find that $\hat{\sigma} = 0.478$, $SER^l = 28.89$ and $SER^r = 29.89$. Hence, $\hat{\eta} = \frac{29.286-28.89}{29.89-29.286} = 0.654$, $l_{SER} = \frac{3(1+0.654) \cdot 0.654 \cdot 0.478}{1+0.654^2} = 1.085$ and $r_{SER} = \frac{3(1+0.654) \cdot 0.478}{1+0.654^2} = 1.660$. Hence, annual SER by Belgium for 2014–2018 is fitted as $\widetilde{SER} = (29.286\%, 1.085\%, 1.660\%)$.

Notice that quantifying Belgian SER as the TFN $\widetilde{SER} = (29.286\%, 1.085\%, 1.660\%)$ is very suited to the intuition that comes after a visual inspection of Table 3 "Belgium SER has been around 29% within 2014–2018". Of course, more sophisticated representations of uncertainty as IFSs or NFSs can capture a greater amount of information. However, the cost of fitting these kind of sets is much greater and not very reliable with the information available for our analysis (identical to that in Table 3 for Belgium SER). On the other hand, if information came from an extended and structured questionnaire submitted to experts, surely NFSs will provide a better representation of that information than FNs.

Belgium SER in Table 3 admits a gray number representation. Following the exposition in [30] and taking into account that SER in Table 3 is within the interval [28.8, 30] since is the discrete set {28.8, 28.8, 29.2, 29.8, 30}, the kernel of SER is $S\hat{E}R = (28.8 + 28.8 + \ldots + 30)/5 = 29.32$. Grayness degree can be estimated by taking into account that SER for any country must be within [0, 100] and so its value is $(30 - 28.8)/100 = 0.012$. By using notation in [30], SER is $29.32_{(0.012)}$. Notice that this parameterization is simpler than $\widetilde{SER} = (29.286\%, 1.085\%, 1.660\%)$, but on the other hand, the TFN captures the asymmetric distribution of values around the gravity center of the data that GN does not.

Due to the kind of data that we will use in our analysis, we feel that using TFN parameterization from [29] provides an adequate compromise between applying the principle of parsimony in vagueness modeling and avoiding unnecessary loss of information.

2.4. Correlation Coefficients for Fuzzy Data

Pearson's correlation coefficient (PCC) is a real-valued function in $\Re^{2n}$ of the pairwise observations over the variables X and Y: $\{(x_1, y_1); (x_2, y_2); \ldots; (x_n, y_n)\}$. Hence, PCC between X and Y is estimated as:

$$corr_{X,Y} = f(x_1, \ldots, x_n; y_1, \ldots, y_n) = \frac{\sum_{i=1}^{n}\left(x_i - \frac{\sum_{i=1}^{n} x_i}{n}\right)\left(y_i - \frac{\sum_{i=1}^{n} y_i}{n}\right)}{\sqrt{\sum_{i=1}^{n}\left(x_i - \frac{\sum_{i=1}^{n} x_i}{n}\right)^2 \sum_{i=1}^{n}\left(y_i - \frac{\sum_{i=1}^{n} y_i}{n}\right)^2}} \quad (4a)$$

i.e., $corr_{X,Y}$ is a function $f(x_1, \ldots, x_n; y_1, \ldots, y_n)$. Hence, if pairwise observations are given by FNs $\{(\tilde{X}_1, \tilde{Y}_1); (\tilde{X}_2, \tilde{Y}_2); \ldots; (\tilde{X}_n, \tilde{Y}_n)\}$, $corr_{X,Y}$ induce a FN:

$$\widetilde{corr}_{X,Y} = f(\tilde{X}_1, \ldots, \tilde{X}_n; \tilde{Y}_1, \ldots, \tilde{Y}_n) = \frac{\sum_{i=1}^{n}\left(\tilde{X}_i - \frac{\sum_{i=1}^{n} \tilde{X}_i}{n}\right)\left(\tilde{Y}_i - \frac{\sum_{i=1}^{n} \tilde{Y}_i}{n}\right)}{\sqrt{\sum_{i=1}^{n}\left(\tilde{X}_i - \frac{\sum_{i=1}^{n} \tilde{X}_i}{n}\right)^2}\sqrt{\sum_{i=1}^{n}\left(\tilde{Y}_i - \frac{\sum_{i=1}^{n} \tilde{Y}_i}{n}\right)^2}} \quad (4b)$$

Fuzzy literature has proposed two ways to estimate PCC when the observations are done by FNs (FPCC). The first approach to FPCC, ref. [31], applies Zadeh's extension principle to (4b). So:

$$\mu_{\widetilde{corr}_{X,Y}}(z) = \max_{z = f(x_1,\ldots,x_n;y_1,\ldots,y_n)} \min[\mu_{\widetilde{X}_1}x_1),\ldots,\mu_{\widetilde{X}_n}x_n);\mu_{\widetilde{Y}_1}y_1),\ldots,\mu_{\widetilde{Y}_n}y_n)] \quad (5a)$$

Notice that it is often difficult computing the membership function of $\widetilde{corr}_{X,Y}$. Following [32], it may be easier computing $corr_{X,Y\alpha}$ such as:

$$corr_{X,Y\alpha} = \left[\underline{corr_{X,Y}}(\alpha), \overline{corr_{X,Y}}(\alpha)\right]$$
$$= \left\{z = f(x_1,\ldots,x_n;y_1,\ldots,y_n) \middle| x_j \in \left[\underline{X}_j(\alpha), \overline{X}_j(\alpha)\right], y_j \right. \quad (5b)$$
$$\left. \in \left[\underline{Y}_j(\alpha), \overline{Y}_j(\alpha)\right], j = 1,2,\ldots,n\right\}$$

Hence, in (5b) $\underline{corr_{X,Y}}(\alpha)$ ($\overline{corr_{X,Y}}(\alpha)$) are the global minimum (maximum) of $f(\cdot)$ within the rectangular domain in (5b).

$$\underline{corr_{X,Y}}(\alpha) = \min_{j,k}\{f(V_j), f(E_k)\} \text{ and } \overline{corr_{X,Y}}(\alpha) = \max_{j,k}\{f(V_j), f(E_k)\} \quad (5c)$$

Being the vector in $\Re^{2n}$ $V_j, j = 1, 2, \ldots, 2^{2n}$ a vertex of (5b), $f(E_k)$ $k = 1, 2, \ldots, K$ an extreme point of the function and E_k an interior point of (5b). Hence, to find the lower (upper) extreme of α-cuts, a nonlinear minimizing (maximizing) mathematical program must be solved.

The second approach to fit fuzzy correlation uses the weakest T-norm (Tw-norm) in [33] instead of the min operator. So:

$$T_W(a,b) = \begin{cases} a \text{ if } b = 1 \\ b \text{ if } a = 1 \\ 0 \text{ otherwise} \end{cases} \quad (6a)$$

where $T_W(a,b) \leq min(a,b)$.

Since the max-operator is still the T-conorm to apply the use of the norm (6a), suppose we reformulate the membership function of the correlation between X and Y as:

$$\mu_{\widetilde{corr}_{X,Y}}(z) = \max_{z = f(x_1,\ldots,x_n;y_1,\ldots,y_n)} T_W[\mu_{\widetilde{X}_1}x_1),\ldots,\mu_{\widetilde{X}_n}x_n);\mu_{\widetilde{Y}_1}y_1),\ldots,\mu_{\widetilde{Y}_n}y_n)] \quad (6b)$$

Tw-norm lets obtaining less uncertain results than min-norm. Likewise, Tw-norm allows an easier computation of (4b) when the observations are LR fuzzy numbers [34] since $\widetilde{corr}_{X,Y}$ will conserve L-R shape. In the particular case of TFNs, the calculation of $\widetilde{corr}_{X,Y}$. is developed in [35]. In [33] following arithmetical rules to handle arithmetically two TFNs $\widetilde{A} = (A, l_A, r_A)$ and $\widetilde{B} = (B, l_B, r_B)$ are stated:

$$\widetilde{A} + \widetilde{B} = (A, l_A, r_A) + (B, l_B, r_B) = (A + B, max(l_A, l_B), max(r_A, r_B)) \quad (6c)$$

$$\widetilde{A} - \widetilde{B} = (A, l_A, r_A) - (B, l_B, r_B) = (A - B, max(l_A, r_B), (r_A, l_B)) \quad (6d)$$

$$\lambda \widetilde{A} = \lambda(A, l_A, r_A) = \begin{cases} (\lambda A, \lambda l_A, \lambda r_A), \lambda > 0 \\ (\lambda A, -\lambda r_A, \lambda l_A), \lambda \leq 0 \end{cases} \quad (6e)$$

$$\sqrt{\widetilde{A}} \approx \left(\sqrt{A}, \frac{l_A}{\sqrt{A}}, \frac{r_A}{\sqrt{A}}\right), A - l_A > 0 \quad (6f)$$

$$\frac{1}{\widetilde{A}} \approx \left(\frac{1}{A}, \frac{r_A}{A^2}, \frac{l_A}{A^2}\right), A - l_A > 0 \quad (6g)$$

$$\tilde{A} \cdot \tilde{B} = (A, l_A, r_A) \cdot (B, l_B, r_B) =$$
$$= \begin{cases} (A \cdot B, max(A \cdot l_B, B \cdot l_A), max(A \cdot r_B, B \cdot r_A)) \; if \; A, B \geq 0 \\ (A \cdot B, -max(A \cdot l_B, B \cdot l_A), -max(A \cdot r_B, B \cdot r_A)) \; if \; A, B \leq 0 \\ (A \cdot B, max(-A \cdot r_B, B \cdot l_A), max(-A \cdot l_B, B \cdot r_A)) \; if \; A \leq 0, B \geq 0 \\ (A \cdot B, -max(A \cdot l_B, -B \cdot r_A), -max(A \cdot r_B, -B \cdot l_A)) \; if \; A, B \leq 0 \end{cases} \quad (6h)$$

$$\frac{\tilde{A}}{\tilde{B}} = (A, l_A, r_A) \left(\frac{1}{B}, \frac{r_B}{B^2}, \frac{l_B}{B^2} \right) =$$
$$= \begin{cases} \left(A \cdot B, max\left(\frac{A \cdot r_B}{B^2}, B \cdot l_A \right), max\left(\frac{A \cdot l_B}{B^2}, B \cdot r_A \right) \right) \; if \; A, B \geq 0 \\ \left(A \cdot B, max\left(-A \frac{A \cdot l_B}{B^2}, B \cdot l_A \right), max\left(-\frac{A \cdot r_B}{B^2}, B \cdot r_A \right) \right) \; if \; A \leq 0, B \geq 0 \end{cases} \quad (6i)$$

being $B - l_B > 0$

Hence, to fit $\widetilde{corr}_{X,Y} = (corr_{X,Y}, l_{corr_{X,Y}}, r_{corr_{X,Y}})$. We must evaluate (4b) with (6c)–(6i). Of course, FPCC may be interpreted qualitatively by using the linguistic variable defined in Table 2. If *min* T-norm is used, the compatibility grade of $\widetilde{corr}_{X,Y}$ with the *j*th linguistic label $\tilde{V}_j$ C($\widetilde{corr}_{X,Y}, \tilde{V}_j$) can be found by using the *max-min* rule as:

$$C(\widetilde{corr}_{X,Y}, \tilde{V}_j) = \max_x \min \left[\mu_{\widetilde{corr}_{X,Y}} x), \mu_{\tilde{V}_j} x) \right] \quad (7a)$$

On the other hand, if Tw-norm is used and so the correlation is calculated by following (6c)–(6h), the compatibility between $\widetilde{corr}_{X,Y}$ and $\tilde{V}_j$ is measured by using a *max*-Tw rule:

$$C(\widetilde{corr}_{X,Y}, \tilde{V}_k) = max \left[\mu_{\widetilde{corr}_{X,Y}} V_j), \mu_{\tilde{V}_j} corr_{X,Y}) \right] \quad (7b)$$

In both cases, we can find the closest linguistic label in Table 1 to $\widetilde{corr}_{X,Y}, \tilde{V}_k$, by doing:

$$\tilde{V}_k = argmax\{C(\widetilde{corr}_{X,Y}, \tilde{V}_j)\}_{k=1 \leq j \leq n}$$

Numerical Application 2

We fit for 2014–2018 FPCC between social expenses over GPD (SER) and the percentual diminution of poverty risk index (RRP) in EU-28 countries. Fuzzy observations of these variables are shown in Table 6. Likewise, we also fit crisp PCC by considering as observations the core triangular shapes SER and RRP (3e).

Table 6. Fuzzy observations on social expenses over GPD, relative reduction of poverty-at risk index and Debreu–Farrell measure in the period 2014–2018.

Country	$\widetilde{SER}_i = (SER_i, l_{SER_i}, r_{SER_i})$			$\widetilde{RRP}_i = (RRP_i, l_{RRP_i}, r_{RRP_i})$			$\widetilde{DF}_i = (DF_i, l_{DF_i}, r_{DF_i})$		
	SER_i	l_{SER_i}	r_{SER_i}	RRP_i	l_{RRP_i}	r_{RRP_i}	DF_i	l_{DF_i}	r_{DF_i}
Belgium	29.29	1.09	1.66	41.43	8.19	5.15	0.590	0.030	0.003
Denmark	32.82	2.10	3.22	51.86	5.74	3.77	0.728	0.046	0.033
Germany	29.52	0.82	0.51	33.36	0.23	0.59	0.538	0.056	0.040
Ireland	15.82	2.38	3.95	53.42	3.61	5.92	1.000	0.000	0.000
Greece	25.67	1.72	0.88	16.13	0.23	2.36	0.295	0.017	0.063
Spain	24.09	1.34	2.36	25.09	3.63	6.03	0.399	0.031	0.028
France	34.22	0.60	0.36	44.15	2.49	0.95	0.607	0.074	0.118
Italy	29.28	0.85	1.29	21.37	1.02	0.09	0.302	0.038	0.031
Luxembourg	22.19	0.74	1.03	40.35	3.39	1.50	0.656	0.039	0.064
Netherlands	29.59	1.00	1.54	42.73	5.62	6.58	0.599	0.037	0.054
Austria	29.66	0.94	0.49	44.28	4.07	2.69	0.623	0.019	0.054
Portugal	25.19	1.65	2.56	24.65	3.30	5.08	0.454	0.020	0.049
Finland	31.24	2.21	1.18	54.51	1.52	4.48	0.774	0.110	0.089

Table 6. Cont.

Country	$\widetilde{SER}_i=(SER_i,l_{SER_i},r_{SER_i})$			$\widetilde{RRP}_i=(RRP_i,l_{RRP_i},r_{RRP_i})$			$\widetilde{DF}_i=(DF_i,l_{DF_i},r_{DF_i})$		
	SER_i	l_{SER_i}	r_{SER_i}	RRP_i	l_{RRP_i}	r_{RRP_i}	DF_i	l_{DF_i}	r_{DF_i}
Sweden	28.98	1.04	0.64	45.71	3.08	2.18	0.767	0.052	0.014
UK	26.62	1.01	1.68	42.18	4.67	2.36	0.680	0.060	0.037
Bulgaria	17.39	1.45	1.29	21.12	3.60	6.38	0.400	0.061	0.106
Czechia	18.90	0.51	0.82	41.84	4.38	1.45	0.661	0.049	0.071
Estonia	16.00	0.46	0.91	24.94	4.62	6.02	0.456	0.065	0.019
Croatia	21.76	0.21	0.10	29.59	6.13	7.69	0.485	0.090	0.095
Cyprus	19.39	2.47	1.71	36.53	0.22	2.55	0.637	0.055	0.070
Latvia	14.93	0.22	0.51	20.57	5.13	2.69	0.386	0.099	0.057
Lithuania	15.47	0.58	0.86	23.33	0.36	3.96	0.464	0.007	0.080
Hungary	18.76	1.97	1.36	44.72	4.12	6.88	0.817	0.065	0.079
Malta	16.66	2.22	1.69	30.64	0.49	1.71	0.517	0.072	0.047
Poland	19.79	1.29	1.02	29.95	6.79	8.98	0.497	0.096	0.110
Romania	14.73	0.23	0.41	14.56	3.32	4.90	0.303	0.109	0.079
Slovenia	23.20	2.21	1.43	42.85	1.19	1.99	0.636	0.077	0.108
Slovakia	18.17	0.59	0.39	32.42	5.35	8.34	0.492	0.057	0.088

Source: own elaboration from data provided by EU-SILC (2008–2018) and ESSPROS (2008–2017). Variables SER and RRP are expressed over 100 and DF over 1.

The α-cut representation of two FPCCs is given in Table 7. Of course, crisp PCC is simply the 1-cut of both FPCCs, i.e., 0.4585. Hence, FPCC generalizes the results of crisp PCC since this last is the core of FPCCs. Likewise, α-cuts of FPCCs can be understood as an structured set of simulations that range from maximum fuzziness scenario (generated by the 0-cut of fuzzy estimates of SER and RRP) to maximum reliability situation (that comes from the cores of the observations on SER and RRP).

Table 7. α-cut representation of [31,34] FPPC between SER and relative reduction of poverty in EU-28 countries within the period 2014–2018.

	Max–Min Correlation		Max-T_w Correlation	
α	$corr_{X,Y}(\alpha)$	$\overline{corr}_{X,Y}(\alpha)$	$corr_{X,Y}(\alpha)$	$\overline{corr}_{X,Y}(\alpha)$
0	−0.0643	0.7635	0.4196	0.4819
0.25	0.0762	0.7040	0.4294	0.4761
0.5	0.2140	0.6332	0.4391	0.4702
0.75	0.3427	0.5511	0.4488	0.4644
1	0.4585	0.4585	0.4585	0.4585

Source: own elaboration from data provided by EU-SILC (2008–2018) and ESSPROS (2008–2017).

FPCC in [31] does not preserve the triangular shape of input data. On the other hand, by using FPCC [34], we obtain $\widetilde{corr}_{X,Y}$ = (0.4585, 0.0389, 0.0413). Table 8 shows that the closest linguistic label for both correlations is "strong (+) relation". However, max-min correlation is extremely imprecise since embed values from −0.0643 (no correlation) to 0.7635 (very strong (+)), and so it is compatible with 4 linguistic levels in a truth level above 0.5. Those levels vary from "negligible (+) correlation" to "very strong (+) correlation". On the other hand, the correlation [34] is clearly less uncertain and allows a better balance between maintaining all the information in the sample, which is not made by conventional PCC and providing a useful value to obtain conclusions.

Table 8. Qualitative interpretation of max-min, max-Tw-conorm and crisp correlation between SER and relative reduction of poverty in EU 28 countries within the period 2014–2018 by using $C(\widetilde{corr}_{X,Y}, \widetilde{V}_k)$.

Linguistic Label	Max–Min	Max-Tw	Crisp	Linguistic Label	Max–Min	Max-Tw	
1 Perfect (−)	0	0	0	8 Negligible (+)	0.42	0	0
2 Very strong (−)	0	0	0	9 Weak (+)	0.58	0	0
3 Strong (−)	0	0	0	10 Moderate (+)	0.78	0.21	0.21
4 Moderate (−)	0	0	0	11 Strong (+)	0.92	0.79	0.79
5 Weak (−)	0.0	0	0	12 Very strong (+)	0.44	0	0
6 Negligible (−)	0.1	0	0	13 Perfect (+)	0	0	0
7 No correlation	0.26	0	0	$argmax[C(\widetilde{corr}_{X,Y}, \widetilde{V}_j)]$	Strong (+) correlation	Strong (+) correlation	Strong (+) correlation

Source: own elaboration from data provided by EU-SILC (2008–2018) and ESSPROS (2008–2017).

2.5. Literature Revision

Evaluating the productivity of a set of entities (in this case, countries) with a Debreu–Farrell efficient frontier, as we do in this paper, is very common in standard economic literature [24], but not at all fuzzy literature. On the other hand, productivity evaluation has been developed extensively within fuzzy literature by means of fuzzy data envelopment analysis (DEA) methods. Zhou and Xu [36] show that only in year 2018 more than 700 papers on fuzzy DEA were published. Without aim to be extensive, let us point out some applications in Economics and Finance. Wu et al. [37] deal with the evaluation of bank efficiency; [38,39] are devoted with the assessment of sustainability in energy and transportation policies and [40] use fuzzy DEA to examine profits by foreign investment in transition Economies.

Similar problems as we address in this paper have been analyzed fruitfully by using fuzzy multiple criteria decision-making (FMCDM). However, usually these methods need using expert opinions as a input. Our study does not use this information. Fuzzy literature has provided a great deal of methods on this issue by mixing existing tools to represent uncertain quantities (fuzzy sets, hesitant fuzzy sets, IFSs ...) with well-known Multiple Criteria Decision-Making schedules (AHP, PROMETHEE, ELECTREE, TOPSIS, ...). A panoramic review on this matter can be consulted in [41]. Some applications in areas linked with public decision-making are energy policies [41,42], environmental decisions [43,44], healthcare evaluation [45,46], urbanism [47], public infrastructure management [48,49], assessment transparency by public organisms [50] or general economic policy analysis [51].

Fuzzy Pearson's correlation coefficient has been used in several areas of Economics and Finance. Hence, ref. [52,53] use FPCC to model interactions between asset return in portfolio management and [31,54,55] apply FPCC in several business administration issues as, e.g., capital budgeting problems. Likewise, ref. [56,57] analyze the relationship between variables embedded in education policy. In [35,58] FPCC is used to evaluate attributes of hotel services. Finally, ref. [59] uses FCCC to state the linkage between price index and exchange rates in China.

3. Data and Methodology

3.1. Data Description

The data we have used is provided by Eurostat programs EU-SILC and ESSPROS and embeds EU-28 countries in 2018 (i.e., it is included Great Britain). The data have annual periodicity and comprise the period 2014–2018. From the database, we directly obtain observations on the following variables for every country and year. Concretely:

1. $ARPR(0)_{i,t}$ = At-risk-of-poverty rate before social transfers including pensions for the ith country at year t;
2. $ARPR(1)_{i,t}$ = At-risk-of-poverty rate after social transfers for the ith country at year t;
3. $GI_{i,t}$ = Income inequality (measured as the Gini index) before social transfers for the ith country at year t;
4. $SER_{i,t}$ = Ratio social expenses/GDP before social transfers for the ith country at year t.

Likewise, our analysis also needs the observations on the proportion that each kind of social benefit supposes over whole social expending according to EU-SILC classification. These items are defined over the basis of eight protection functions linked with a set of needs [60]:

- Sickness/healthcare benefits ($Sick_{i,t}$)—Include, for example, medical assistance or the provision of pharmaceutical products;
- Disability benefits ($Dis_{i,t}$)—Pensions, goods and services for disabled persons;
- Old age benefits ($Old_{i,t}$)—Basically retirement pensions;
- Survivors' benefits ($Surv_{i,t}$)—An example are survivors' pensions;
- Family/children benefits ($Fam_{i,t}$)—Support programs linked to pregnancy assistance, childbirth, etc.;
- Unemployment benefits ($Une_{i,t}$)—Include unemployment in-cash benefits, but also vocational training services provided by public agencies;
- Housing benefits ($Hou_{i,t}$)—Interventions and programs from public agencies to help households reaching housing expenses;
- Social exclusion benefits not elsewhere classified (n.e.c.) ($SocE_{i,t}$)—A miscellanea of public interventions that may include, e.g., rehabilitation of drug abusers, etc.

From the variables indicated above, we derivate for each country and year the diminution of poverty and the productivity that SE has reached in such diminution. Following [10], we measure poverty reduction in relative terms. Hence, for the ith country at year t we obtain:

$$RRP_{i,t} = \frac{ARPR(0)_{i,t} - ARPR(1)_{i,t}}{ARPR(0)_{i,t}}, \ i = 1, 2, \ldots, 28 \qquad (8)$$

Hence, $RRP_{i,t}$ ranks from 0 (and so $ARPR(0)_{i,t} = ARPR(1)_{i,t}$), to 1 (if poverty is completely eliminated).

When analyzing the productivity of SE in reducing poverty risk, we seek to determine to what extent the diminution of poverty (the assessed output) is adequate to the initial situation of poverty and SER (inputs). To measure the efficiency of public spending in achieving poverty reduction for the ith country in a year t we follow [10] that quantifies efficiency by means of a Debreu–Farrell coefficient, $DF_{i,t}$. Hence, we consider SE and, more concretely, its quantification by means of its ratio with GPD (SER) as the main input. We also use the GI before transfers as second input variable to reflect the social status of the population before executing the SE. Hence, GI is not strictly an input, but a contextual variable. Likewise, GI is clearly linked to economic context, social and demographic structure and public policy priorities of every state. A greater retired people supposes a larger population that depends on pensions. Likewise, higher unemployment rates imply a greater number of citizens with small (or null) personal income. The method used to evaluate the productivity of PPP in [10] is based on fitting the efficient productive frontier by mixing corrected least-squares method (CLS) and logit regression. Hence, for a year t it is estimated:

$$logit(RRP_{i,t}) = \beta_{0,t} + \beta_{1,t} SER_{i,t} + \beta_{2,t} GI_{i,t} + \varepsilon_{i,t} \qquad (9)$$

where the error term accomplishes $\varepsilon_{i,t} \geq 0$, $i = 1, 2, \ldots, 28$. After adjusting the value of $\beta_{0,t}, \beta_{1,t}$ and $\beta_{2,t}$ with corrected least squares, $\beta_{0,t}^F, \beta_{1,t}^F, \beta_{2,t}^F$, the estimate of the productive frontier value of $RRP_{i,t}$, $RRP_{i,t}^F$ is:

$$RRP_{i,t}^F = \frac{1}{1 + exp(-\beta_{0,t}^F - \beta_{0,t}^F SER_{i,t} - \beta_{0,t}^F GI_{i,t})} \qquad (10a)$$

Hence, Debreu–Farrell efficiency measure for ith country in the year t, $DF_{i,t}$, is the ratio between its attained RRP ($RRP_{i,t}$) in (8) and frontier value of RRP in (10a), $RRP_{i,t}^F$:

$$DF_{i,t} = \frac{RRP_{i,t}}{RRP_{i,t}^F} \qquad (10b)$$

where $0 \leq DF_{i,t} \leq 1$. Hence, $DF_{i,t} = 1$ imply full efficiency and $DF_{i,t} = 0$, complete nonefficiency.

Eurostat database provides the values of the variables related to the social protection benefits and poverty of EU-28 countries with an annual periodicity. Therefore, the variables RRP and DF are calculated with this periodicity. To evaluate the results of social policies within a period of more than one year (e.g., a quinquennium), a usual practice consists of taking for the variables the average of their annual observations [8,10,12]. Other papers reduce the analysis to a concrete year [7,11]. A complete analysis consists of repeating it in every year of the period of interest as it is done in Lefevre et al. (2010). Alternatively, we propose quantifying the variables in a period of multiple years by means of TFNs that are built up from observed longitudinal point values of those variables.

Our analysis is done by using SER, the proportion that each kind of social expense suppose over whole SE, the relative reduction of poverty RRP (8) and Debreu–Farrell efficiency index (10) of all countries throughout 2014–2018. We fit for the ith country the value of any variable "A" (e.g., SER) for the whole period 2014–2018 as an FN $\tilde{A}_i = (A_i, l_{A_i}, r_{A_i})$ $i = 1, 2, \ldots, 28$. They are fitted from the point observations in each year of the period that we are analyzing, $\{a_{2014}, a_{2015}, a_{2016}, a_{2017}, a_{2018}\}$ by following the method in Section 2.3. Figure 2a summarizes all the process followed to fit TFNs to the observations on variables embedded in the study. Table 6 shows the fuzzy estimates of SER ($\widetilde{SER}_i$), RRP ($\widetilde{RRP}_i$) and efficiency index DF, ($\widetilde{DF}_i$) in EU-28 countries. Fuzzy values of the proportion that each item of social spending (Sick, Dis, Old, ...) suppose in overall SE are in Table A1 of annex.

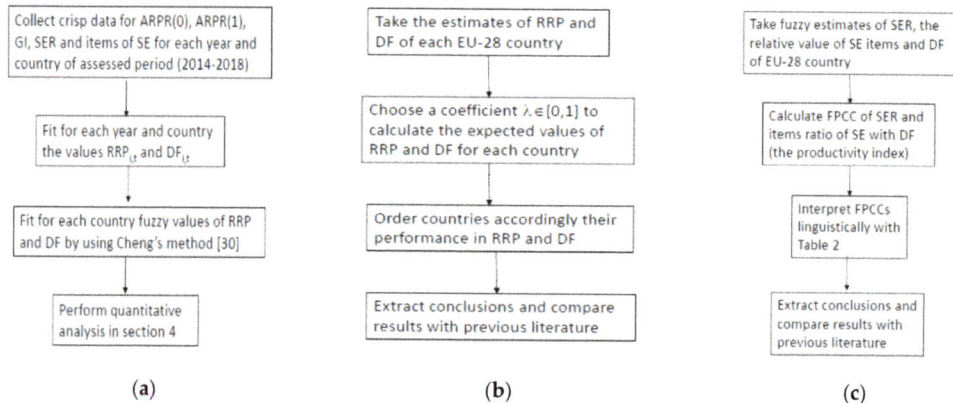

Figure 2. Flowcharts for the analysis of poverty policies in UE-28. (**a**) Fitting fuzzy estimates of variables for the while period 2014–2018. (**b**) Methodology used to rank PPPs. (**c**) Methodology followed to measure the influence of SER and its composition in the productivity of public poverty policies (PPPs).

3.2. Methodology

We first rank UE28 states by considering the efficiency of their PPP. To make this assessment, we defuzzify the values of DF with (1d) and state their hierarchy. Likewise, we relate our results with those in [7,10]. The flowchart of this analysis is depicted in Figure 2b.

The second analysis tries to determine the sign of the relation between the eight items that [60] differentiates in social expending and the efficiency of PPP. Figure 2c shows how we have implemented this assessment on European PPPs. As it is stated in [16] and also checked in [7,8,10], despite the clear negative linkage between SER and poverty indexes in EU-28, it cannot be concluded that a poverty reduction is reached automatically by increasing SER. Hence, we first measure the intensity of the relation between the effort in social policies (i.e., SER) with the results in reducing poverty (RRP) and with PPP efficiency measured by DF. Subsequently we investigate why two different countries with a similar SER will obtain different reductions of poverty indexes. Following [7,10] we perform this analysis with the FPCC of the proportion that each kind of social benefit supposes in overall social expenditure. Concretely we use FPCC in [34] (Equations (4b) and (6a)–(6i)) instead max-min fuzzy correlation due to the reasons exposed above. We have used the fuzzy version of PPC instead other correlation measure as, e.g., Spearman correlation by two reasons. First, we pretend comparing our results with those in [7,10] that evaluate the PPPs of the same set of countries, and they use Pearson's correlation coefficient. To allow our results to be fully comparable, the same correlation measure must be used. Likewise, calculating Spearman correlation requires an early defuzzification of triangular observations in order to rank them, e.g., by calculating their expected value. Subsequently Spearman correlation index is a real valued number since that comes from applying a conventional PPC on the crisp rank of variables. Therefore, the fuzzy uncertainty of data is waived in that correlation measure.

Within EU-28, we can differentiate two types of countries whose history and political evolution from II World War to the end of the 20th century XX has been notably different. One on hand we have EU-15 countries, basically Western Europe countries, which are part of European Union from 20th-century. On the other hand, we find former communist republics plus Cyprus and Malta that belonged progressively in European Union during the 21st century. Table 9 shows the mean value of $\widetilde{SE}_i$, $\widetilde{RRP}_i$ and $\widetilde{DF}_i$, $i = 1, 2, \ldots, 28$ in EU-28, but also, separately, the average value of EU-15 countries and non-EU-15 states. Those mean values have been obtained by using Max-Tw norm convolution in such a way that for a variable A, the mean value $\widetilde{A}^M$ is:

$$\widetilde{A}^M = \left(A^M, l_{A^M}, r_{A^M}\right) = \left(\frac{\sum_{i=1}^n A_i}{n}, \frac{\max_i l_{A_i}}{n}, \frac{\max_i r_{A_i}}{n}\right) \quad (11)$$

Table 9. Mean values of SER, poverty risk index (RRP) and Debreu–Farrell index (DF) in EU-28, EU-15 and non-EU-15 in the period 2014–2018.

	$\widetilde{SER}^M$			$\widetilde{RRP}^M$			$\widetilde{DF}^M$		
	SER^M	l_{SER^M}	r_{SER^M}	RRP^M	l_{RRP^M}	r_{RRP^M}	DF^M	l_{DF^M}	r_{DF^M}
EU-28	23.190	0.088	0.141	34.796	0.292	0.321	0.563	0.004	0.004
Non EU-15	18.089	0.190	0.131	30.236	0.522	0.691	0.519	0.008	0.008
EU-15	27.612	0.165	0.263	38.748	0.546	0.598	0.601	0.007	0.008

Source: own elaboration from data provided by EU-SILC (2008–2018) and ESSPROS (2008–2017). Variables SER and RRP are expressed over 100 and E over 1.

In Table 9 it can be checked that the mean value of EU-15 and non-EU-15 in SER, RRP and efficiency of PPP is completely different. EU-15 countries present a mean value of SER 10 points above non-EU-15 countries. Likewise, Table 9 shows that whereas EU-15 countries rarely have a value for SER below 25%, non-EU-15 states with SER greater than 20% are an exception. Consequently, the mean reduction of poverty in EU-15 countries is clearly above non-EU-15 states. It is also remarkable that the mean DF is notably greater in EU-15 countries than in non-EU-15 countries.

4. Results

4.1. Ranking Public Poverty Policies by the EU-28 States

The results of PPPs efficiency are given in 10a,10b and 2 of the annex A. It can be checked in Table 10b that all hierarchies in RRP and DF index present a correlation close to 1. This fact does not depend on the coefficient λ used to defuzzify $\widetilde{RRP}_i$ and

$\widetilde{DF}_i$, $i = 1, 2, \ldots, 28$. Hence, by applying fuzzy linguistic interpretation of correlation proposed in Section 2.2, we can conclude that all the expected values of RRP and DF present a perfect(+) correlation.

Table 10a. Ranking social policy in UE-28 countries on the basis of the expected value of RRP and DF in Table 2 for several values of λ.

Criteria	Relative Reduction of Poverty (RRP)				Debreu–Farrell Ratio (DF)			
Country	λ = 1	λ = 0.5	λ = 0	V–F	λ = 1	λ = 0.5	λ = 0	V–F
Belgium	12	13	13	13	14	14	13	19
Denmark	3	3	3	5	5	5	5	7
Germany	16	15	15	14	15	15	15	10
Ireland	2	2	2	10	1	1	1	3
Greece	27	27	27	21	26	26	26	21
Spain	20	20	20	25	24	24	23	28
France	7	6	5	4	12	12	14	14
Italy	25	25	24	19	27	27	27	27
Luxembourg	13	12	12	11	7	8	7	5
Netherlands	9	9	9	3	13	13	12	8
Austria	6	7	8	8	11	9	9	16
Portugal	22	21	22	17	21	21	20	18
Finland	1	1	1	6	3	4	4	11
Sweden	5	4	4	12	4	3	3	6
United Kingdom	10	10	10	15	8	6	6	17
Bulgaria	24	24	25	27	23	23	24	26
Czechia	11	11	11	1	6	7	7	1
Estonia	21	23	23	26	22	22	22	23
Croatia	19	19	19	22	20	20	21	22
Cyprus	14	14	14	20	10	9	10	25
Latvia	26	26	26	28	25	25	25	24
Lithuania	23	22	21	23	19	18	18	15
Hungary	4	5	6	2	2	2	2	2
Malta	17	17	16	18	17	16	16	20
Poland	18	18	18	16	18	19	19	13
Romania	28	28	28	24	28	28	28	12
Slovenia	8	8	7	9	9	11	11	9
Slovakia	15	16	17	7	16	17	17	4

Source: own elaboration from data provided by EU-SILC (2008–2018) and ESSPROS (2008–2017). V–F stands for the rank in [10] within the period 2007–2015.

Table 10b. Spearman correlations between the expected values of RRP and/or DF for several values of the index of optimism λ.

	RRP (λ = 1)	RRP (λ = 0.5)	RRP (λ = 0)	RRP (V–F)	DF (λ = 1)	DF (λ = 0.5)	DF (λ = 0)	DF (V–F)
RRP (λ = 1)	1							
RRP (λ = 0.5)	0.996	1						
RRP (λ = 0)	0.991	0.997	1					
RRP (V–F)	0.830	0.831	0.824	1				
DF (λ = 1)	0.943	0.947	0.941	0.784	1			
DF (λ = 0.5)	0.939	0.944	0.937	0.757	0.994	1		
DF (λ = 0)	0.935	0.939	0.932	0.761	0.992	0.997	1	
DF (V–F)	0.642	0.652	0.642	0.795	0.718	0.680	0.692	1

Source: own elaboration from data provided by EU-SILC (2008–2018) and ESSPROS (2008–2017). Note: As V–F, we symbolize the value of RRP and DF obtained by Valls-Fonayet et al. (2020) within the period 2007–2015.

The results that we have obtained are similar to those in [7,9,12]. Better performances are attained by Anglo-Saxon welfare states (Ireland and Great Britain), Scandinavian welfare states (Finland, Sweden and Denmark) and some Visegrad pact countries like Hungary and Czechia. The less efficient PPPs are those from the Mediterranean welfare states (Italy, Greece, Spain) and some Mediterranean and Baltic former communist republics as Romania, Bulgaria, Latvia or Estonia. In intermediate positions, we find continental welfare states (as, e.g., France, Belgium, etc.) and a heterogeneous set of non-UE-15 as, e.g., Cyprus, Malta or Slovakia. Table 10b shows that the hierarchies in Valls-Fonayet et al. in [10] and those in our paper present a Spearman correlation coefficient that Table 2 labels as very strong (+).

4.2. Fuzzy Assessment of the Relation between Social Expense Effort and Efficiency in Poverty Reduction

Tables 11 and 12 show the relation of the volume of social expenses with the reduction of poverty and its efficiency. As we expected, in the whole EU-28, the relation between SER and RRP is positive (strong (+)). However, the behavior of that relation is completely different in EU-15 and non-EU 15. In non-EU 15 countries it is very strong (+), i.e., there is a direct relation between the volume of social expenses and poverty reduction. On the other hand, in EU-15 countries that relation is negligible. Hence, this result is in accordance with the statement in [16] that showed that despite there is a strong negative correlation between SER and poverty rates in several European countries, it cannot be concluded that increases in SER lead directly to reductions in poverty.

Table 11. Correlations of SER with RRP and DF efficiency index.

	EU-28			Non-EU-15 Countries			EU-15 Countries		
	$corr_{X,Y}$	$l_{corr_{X,Y}}$	$r_{corr_{X,Y}}$	$corr_{X,Y}$	$l_{corr_{X,Y}}$	$r_{corr_{X,Y}}$	$corr_{X,Y}$	$l_{corr_{X,Y}}$	$r_{corr_{X,Y}}$
RRP	0.459	0.039	0.023	0.715	0.100	0.078	0.125	0.075	0.091
DF	0.194	0.062	0.038	0.569	0.086	0.078	−0.204	0.131	0.079

Source: own elaboration from data provided by EU-SILC (2008–2018) and ESSPROS (2008–2017).

Table 12. Value of the compatibility indexes (7b) of the correlations between SER with RRP and DF and the labels of linguistic variable "correlation coefficient".

	EU-28		Non-EU-15		EU-15	
Linguistic Label	SER vs. RRP	SER vs. DF	SER vs. RRP	SER vs. DF	SER vs. RRP	SER vs. DF
Perfect (−)	0	0	0	0	0	0
Very Strong (−)	0	0	0	0	0	0
Strong (−)	0	0	0	0	0	0
Moderate (−)	0	0	0	0	0	0.267
Weak (−)	0	0	0	0	0	0.970
Negligible (−)	0	0	0	0	0	0.207
No corr	0	0	0	0	0	0
Negligible (+)	0	0.065	0	0	0.750	0
Weak (+)	0	0.935	0	0	0.250	0
Moderate (+)	0.207	0	0	0	0	0
Strong (+)	0.793	0	0.283	0.771	0	0
Very Strong (+)	0	0	0.717	0.229	0	0
Perfect (+)	0	0	0	0	0	0
argmax	Strong (+)	Weak (+)	Very strong (+)	Strong (+)	Negligible (+)	Weak (−)

Source: own elaboration from data provided by EU-SILC (2008–2018) and ESSPROS (2008–2017).

Likewise, in whole EU-28 countries, the relation between the volume of social expenditure and its efficiency is weak (+). Again, the relation between these variables in non-EU-15 countries and EU-15 states is completely different. In EU-15 countries, that relation is weak (−), whereas in non-EU-15 countries is strong (+). Notice that the value of SER in non-EU-15 countries is substantially lower than in EU-15 countries. Therefore, its marginal productivity of SER is clearly positive when social expenses are relatively low, and so a greater diminution of poverty comes fair from increasing SE. However, when a critical level of SER is reached, the relationship between increases in SER and diminutions in RRP is not so direct. This fact motivates a detailed analysis of the influence of the social expenditure composition over the value of DF reached by every PPP.

From Tables 13 and 14a, we can state that in overall EU-28, there is not any SE item with a high positive relation with DF. Hence, the greater positive relation is reached by the expenses in family/children and social exclusion with a moderate (+) relation. Those results are in accordance with [19–23] where it is pointed out the limited impact in poverty reduction of social assistance policies as those for family and children or housing in a great deal of countries due to its reduced value. To consult those values, see Table A1.

On the other hand, pension expenses (old age and survivors, not disability) are strong (−) correlated with DF measure. That finding is in accordance with [17], where it is indicated that benefits for the elderly people generally have a low redistributive impact. However, Tables 13 and 14b,14c shows that those patterns are not uniform within EU-28.

Table 13. Value of the correlations between DF and the proportion that each kind of social expense suppose in overall SER.

	EU-28			Non-EU-15			EU-15		
Item	$corr_{X,Y}$	$l_{corr_{X,Y}}$	$r_{corr_{X,Y}}$	$corr_{X,Y}$	$l_{corr_{X,Y}}$	$r_{corr_{X,Y}}$	$corr_{X,Y}$	$l_{corr_{X,Y}}$	$r_{corr_{X,Y}}$
Sick	0.234	0.038	0.034	−0.008	0.078	0.115	0.443	0.053	0.041
Dis	0.117	0.037	0.026	−0.434	0.133	0.093	0.361	0.022	0.041
Old	−0.529	0.039	0.032	−0.128	0.133	0.140	−0.645	0.050	0.051
Surv	−0.429	0.032	0.028	0.229	0.097	0.085	−0.811	0.052	0.049
Fam	0.333	0.052	0.063	−0.090	0.166	0.108	0.590	0.058	0.105
Une	−0.202	0.133	0.094	−0.202	0.205	0.165	−0.209	0.246	0.176
Hou	−0.083	0.041	0.061	−0.068	0.103	0.098	−0.140	0.077	0.113
SocE	0.273	0.053	0.024	0.277	0.052	0.045	0.209	0.116	0.052

Source: own elaboration from data provided by EU-SILC (2008–2018) and ESSPROS (2008–2017).

Table 14a. Value of the compatibility indexes (7b) of the correlations between DF and the proportion that each kind of social expense suppose in overall SE and the labels of the linguistic variable "correlation coefficient" (EU-28).

	Sick	Dis	Old	Surv	Fam	Une	Hou	SocH
Perfect (−)	0.000	0.000	0.000	0.000	0.000	0.000	0.000	0.000
Very strong (−)	0.000	0.000	0.096	0.000	0.000	0.000	0.000	0.000
Strong (−)	0.000	0.000	0.904	0.647	0.000	0.000	0.000	0.000
Moderate (−)	0.000	0.000	0.000	0.353	0.000	0.261	0.000	0.000
Weak (−)	0.000	0.000	0.000	0.000	0.000	0.984	0.000	0.000
Negligible (−)	0.000	0.000	0.000	0.000	0.000	0.230	0.828	0.000
No corr	0.000	0.000	0.000	0.000	0.000	0.000	0.172	0.000
Negligible (+)	0.000	0.830	0.000	0.000	0.000	0.000	0.000	0.000
Weak (+)	0.663	0.170	0.000	0.000	0.000	0.000	0.000	0.265
Moderate (+)	0.337	0.000	0.000	0.000	0.836	0.000	0.000	0.735
Strong (+)	0.000	0.000	0.000	0.000	0.164	0.000	0.000	0.000
Very strong (+)	0.000	0.000	0.000	0.000	0.000	0.000	0.000	0.000
Perfect (+)	0.000	0.000	0.000	0.000	0.000	0.000	0.000	0.000
Argmax	Weak (+)	Negligible (+)	Strong (−)	Strong (−)	Moderate (+)	Weak (−)	Negligible (−)	Moderate (+)

Source: own elaboration from data provided by EU-SILC (2008–2018) and ESSPROS (2008–2017).

Table 14b. Value of the compatibility indexes (7b) of the correlations between DF and the proportion that each kind of social expense suppose in overall SE and the labels of the linguistic variable "correlation coefficient" (Non-EU-15 countries).

	Sick	Dis	Old	Surv	Fam	Une	Hou	SocH
Perfect (−)	0.000	0.000	0.000	0.000	0.000	0.000	0.000	0.000
Very strong (−)	0.000	0.000	0.000	0.000	0.000	0.000	0.000	0.000
Strong (−)	0.000	0.668	0.000	0.000	0.000	0.000	0.000	0.000
Moderate (−)	0.000	0.332	0.000	0.000	0.000	0.520	0.000	0.000
Weak (−)	0.000	0.000	0.461	0.000	0.337	0.992	0.000	0.000
Negligible (−)	0.079	0.000	0.788	0.000	0.940	0.505	0.691	0.000
No corr	0.921	0.000	0.037	0.000	0.456	0.018	0.336	0.000
Negligible (+)	0.063	0.000	0.000	0.000	0.000	0.000	0.000	0.000
Weak (+)	0.000	0.000	0.000	0.705	0.000	0.000	0.000	0.233
Moderate (+)	0.000	0.000	0.000	0.295	0.000	0.000	0.000	0.767
Strong (+)	0.000	0.000	0.000	0.000	0.000	0.000	0.000	0.000
Very strong (+)	0.000	0.000	0.000	0.000	0.000	0.000	0.000	0.000
Perfect (+)	0.000	0.000	0.000	0.000	0.000	0.000	0.000	0.000
Argmax	No corr	Strong (−)	Negligible (−)	Weak (+)	Negligible (−)	Weak (−)	Negligible (−)	Moderate (+)

Source: own elaboration from data provided by EU-SILC (2008–2018) and ESSPROS (2008–2017).

Table 14c. Value of the compatibility indexes (7b) of the correlations between DF and the proportion that each kind of social expense suppose in overall SE and the labels of the linguistic variable "correlation coefficient" (EU-15 countries).

	Sick	Dis	Old	Surv	Fam	Une	Hou	SocH
Perfect (−)	0.000	0.000	0.000	0.055	0.000	0.000	0.000	0.000
Very strong (−)	0.000	0.000	0.483	0.945	0.000	0.000	0.000	0.000
Strong (−)	0.000	0.000	0.517	0.000	0.000	0.000	0.000	0.000
Moderate (−)	0.000	0.000	0.000	0.000	0.000	0.633	0.000	0.000
Weak (−)	0.000	0.000	0.000	0.000	0.000	0.961	0.395	0.000
Negligible (−)	0.000	0.000	0.000	0.000	0.000	0.556	0.605	0.000
No corr	0.000	0.000	0.000	0.000	0.000	0.150	0.000	0.000
Negligible (+)	0.000	0.000	0.000	0.000	0.000	0.000	0.000	0.062
Weak (+)	0.000	0.000	0.000	0.000	0.000	0.000	0.000	0.912
Moderate (+)	0.284	0.694	0.000	0.000	0.000	0.000	0.000	0.088
Strong (+)	0.716	0.306	0.000	0.000	0.700	0.000	0.000	0.000
Very Strong (+)	0.000	0.000	0.000	0.000	0.300	0.000	0.000	0.000
Perfect (+)	0.000	0.000	0.000	0.000	0.000	0.000	0.000	0.000
Argmax	Strong (+)	Moderate (+)	Strong (−)	Very Strong (−)	Strong (+)	Weak (−)	Negligible (−)	Weak (+)

Source: own elaboration from data provided by EU-SILC (2008–2018) and ESSPROS (2008–2017).

There are clear differences in the relation between SE items and DF of PPP within EU-15 with respect to non-EU-15 countries. In non-EU-15 countries, six of eight types of expenses are low correlated with the productivity measure. Only social exclusion miscellanea (moderate(+)) and disability benefits (strong (−)) show a significant correlation. Notice that, in general, non-EU-15 countries present lower levels of SER than EU-15 states. Hence, it seems that the efficiency of social expenses within non-EU-15 is improved by simply increasing them. Hence, except in the case of disability benefits, increasing a given type of social expense may not lead to a greater result in poverty reduction than increasing any other.

On the other hand, in EU-15, the correlations of each type of social expense and DF is often more intense. Likewise, we can check that not necessarily a given kind of expense has the same sign in its correlation with DF within EU-15 and non-EU-15 countries. That is the case of Sick, Dis, Old and Surv. As [10] we also obtain that Sick, Fam and Dis expenses have a significant positive relationship with the efficiency pf PPPs (strong in two first cases and moderate in the third). Likewise, as [10] we find that benefits due to old age or survival have a significant negative relation with DF (strong and very strong, respectively). Likewise, we must point out that in the case of Une, Hou and Soc, the sign and the intensity of the correlation are essentially the same in EU-15 and non-EU-15 countries. Hence, the relation of DF with housing benefits is negligible, with unemployment is weak (−) and lastly, with Soc is weak/moderate(+) in both EU-15 non-EU-15 countries. Notice that results obtained for the correlation between unemployment benefits and DF are according with Cantillon [18] who outlines that financial aid for the unemployed has not the desired effects in reducing poverty.

5. Conclusions

This article evaluates the efficiency of public poverty policies (PPPs) in EU-28 countries on the basis of their effort measured as social expenditure over GPD (SE). We perform this analysis for the quinquennium 2014–2018 from annual observations on variables provided by programs EU-SILC and ESSPROS of Eurostat. To obtain a single observation for the whole period, 2014–2018 for a given variable and country longitudinal observations are aggregated by means of triangular fuzzy numbers with the method in [3].

As far as the ranking of PPPs is concerned, we have ordered a set of fuzzy efficiency indexes by using their expected value. The results that we have obtained are similar to those in [7,9,10]. Better performances are attained by Anglo-Saxon welfare states (Ireland and Great Britain), Scandinavian welfare states (Finland, Sweden and Denmark) and some Visegrad pact countries like Hungary and Czechia. The less efficient countries are the Mediterranean welfare states (Italy, Greece, Spain) and some Mediterranean and Baltic former communist republics as Romania, Bulgaria, Latvia or Estonia. In intermediate positions, we find continental welfare states (as, e.g., France, Germany . . . .) and a heterogeneous set of non-UE-15 states as, e.g., Cyprus, Malta or Slovakia.

To measure the relation between the efficiency of PPP with SER or with the effort done in a concrete type of social benefit, we have used the fuzzy correlation index in [34] instead that in [31] since this last may provide too uncertain outputs. Likewise, we interpret the correlation index qualitatively as a linguistic variable. We have observed that the relation between the volume of social expending and the poverty diminution despite positive, is different between EU-15 countries (that have greater SE) and non-EU-15 countries. Hence, in EU-15 countries the results are in accordance with [16] that showed that in several European countries increases in social expenses do not lead directly to reductions in poverty.

The relation between the rate of each item of social benefits with Debreu–Farrell measure also shows different behavior in EU-15 and non-EU-15 countries. In non-EU-15 countries, six of eight types of expenses are low correlated with the productivity measure. Only social exclusion miscellanea (moderate(+)) and disability benefits (strong (−)) show a significant correlation. On the other hand, in EU-15 the correlations of each type of social expense and DF is often more intense. As [10] we have found that sickness/healthcare, family/children and disability benefits expenses have a significant positive relation with the efficiency index. We have also checked that benefits due to old age and survivors have a negative strong significant relation with the efficiency of PPP.

In the case of unemployment, housing and social exclusion, the sign and the intensity of the correlation are essentially the same in EU-15 and non-EU-15 countries. Hence, the relation of DF with housing benefits is negligible, with unemployment weak(−) and lastly with social exclusion weak/moderate(+).

We have also discussed the application of other tools connected with fuzzy sets as NFSs, rough sets or GNs to quantify the observations in our problem. Instruments as IFSs or NFSs provide a more complete capture of uncertainty than FNs. However, their adjustment has a greater cost than in the case of FNs. On the other hand, GNs provide more parsimonious representations of uncertain quantities than FNs. To define a GN, it is enough to estimate its kernel and a grayness measure. Therefore, in some circumstances it can be considered that information is too simplified by GNs. Our paper evaluates poverty policies of EU-28 countries within the quinquennium 2014–2018 in such a way that for each variable/country we actually have available five annual real valued observations. By using [29] we aggregate annual observations into one TFN observation that is addressed to the whole quinquennium. We feel justified the use of TFNs because they let modeling vague observations as smooth as possible without any loss of information.

We are aware that our study has limitations. First, it is done in a concrete period with a limited sample of countries. Hence, the conclusions in our paper must be carefully interpreted since they do not necessarily apply automatically to countries/periods out of the sample. Likewise, evaluating poverty policies by using exclusively Eurostat database and performing its analysis by means of fuzzy arithmetic has limitations. It may be of interesting complementing information in Eurostat database with experts' opinion that may be extracted from structured questionnaires and/or interviews. The use of tools to deal with this kind of information that are beyond fuzzy numbers as, e.g., NFSs or hesitant fuzzy sets is fully justified. Hence, further research on the evaluation of PPPs by the use of experts' opinions and the application of Fuzzy Multicriteria Decision Methods can be a suitable complement to the methodology presented in this paper.

Author Contributions: A.B.-E. and F.V.-F. have developed the revision of the literature on poverty public policies. J.d.A.-S. has developed the exposition and application of fuzzy data analysis tools on the sample. All the authors have contributed to extracting conclusions from empirical results and relating them with revised literature. All authors have read and agreed to the published version of the manuscript.

Funding: This research has partially founded by the Social Inclusion Chair of University Rovira i Virgili.

Institutional Review Board Statement: Not applicable.

Informed Consent Statement: Not applicable.

Data Availability Statement: The data used in this study are available in https://ec.europa.eu/eurostat/data/database.

Acknowledgments: Authors acknowledge the suggestions of two anonymous referees that have improved the paper.

Conflicts of Interest: The authors declare no conflict of interest.

Appendix A

Table A1. Proportion that 8 items of social benefits suppose in overall social expenditures (percentage).

Country	Sickness/Healthcare			Disability			Old Age			Survivors			Family/Children			Unemployment			Housing			Social Exclusion		
	Center	Left	Right	Center	Left	Right	Center	Left	Right	Center	Left	Right	Center	Left	Right	Center	Left	Right	Center	Left	Right	Center	Left	Right
Belgium	27.019	0.097	1.07	8.458	1.18	0.755	38.439	4.508	5.208	6.525	0.306	0.206	7.498	0.177	0.22	5.492	1.221	0.939	1.51	0.017	0.037	2.425	0.543	0.371
Bulgaria	21.282	1.172	0.708	16.181	0.733	0.197	38.731	0.81	1.333	0.783	0.13	0.257	11.139	0.252	0.156	2.625	0.24	0.408	1.345	0.522	0.289	5.165	0.626	0.386
Denmark	35.144	0.089	0.166	8.207	0.56	0.986	32.301	0.113	0.174	6.312	0.314	0.505	11.403	0.182	0.246	4.681	1.027	0.766	2.173	0.041	0.021	0.979	0.401	0.262
Germany	36.908	3.55	2.537	5.443	0.248	0.41	31.465	0.824	1.301	2.723	0.077	0.022	9.053	0.873	0.61	2.792	0.254	0.164	0.335	0.363	0.204	0.769	0.056	0.085
Ireland	19.421	0.883	1.824	4.371	0.086	1.028	55.4	0.986	2.994	10.107	0.418	1.121	4.874	1.626	2.953	10.12	5.252	3.744	3.534	0.673	0.99	1.285	1.358	0.538
Greece	26.683	0.711	1.001	7.174	0.071	0.208	41.002	3.669	2.356	9.84	0.049	0.12	5.368	0.162	0.261	3.736	0.197	0.332	0.054	0.148	0.096	0.999	0.028	0.013
Spain	28.564	0.39	0.258	6.449	0.048	0.021	40.17	0.02	0.194	5.384	0.192	0.262	7.68	0.229	0.368	8.345	2.059	3.276	0.443	0.045	0.027	3.068	0.239	0.347
France	23.132	0.086	0.276	5.713	0.2	0.175	48.987	0.093	0.039	9.492	0.123	0.31	4.1	0.192	0.057	2.895	0.45	1.114	0.054	0.032	0.019	2.728	0.967	0.627
Italy	24.977	0.575	0.855	10.857	0.568	0.856	31.362	2.845	1.614	7.743	0.672	0.457	15.439	0.174	0.25	3.581	1.412	0.963	0.106	0.129	0.081	2.257	0.106	0.064
Luxembourg	33.766	1.342	1.998	9.16	0.846	0.081	38.346	0.294	0.196	3.933	0.548	0.34	3.977	0.853	0.486	2.61	1.117	1.114	0.858	0.422	0.357	4.941	0.818	0.293
Netherlands	25.528	0.546	1.059	6.516	0.426	0.644	44.421	0.186	0.263	5.871	0.365	0.507	9.523	0.15	0.094	4.641	1.412	0.766	0.396	0.196	0.1	1.989	0.292	0.698
Austria	24.994	2.525	1.759	7.223	0.417	0.293	50.339	0.655	0.994	7.6	0.063	0.154	4.834	0.519	0.347	1.235	0.309	0.472	1.641	0.102	0.146	0.905	0.125	0.073
Portugal	23.023	1.057	1.958	9.943	0.814	1.393	40.985	3.512	4.921	2.702	0.083	0.135	10.02	0.15	0.52	2.887	0.122	0.073	0.224	0.066	0.109	2.926	0.15	0.404
Finland	26.145	0.396	0.685	10.241	0.661	1.302	43.582	0.801	1.188	1.091	0.207	0.315	10.386	0.281	0.394	7.816	2.316	1.193	0.253	0.778	1.093	3.278	1.379	1.034
Sweden																								
United Kingdom	32.463	0.775	0.427	6.324	0.762	0.518	42.446	0.469	0.724	0.311	0.023	0.033	9.968	0.399	0.54	3.623	0.238	0.104	1.498	0.126	0.177	2.318	0.217	0.162
Czechia	28.053	1.645	2.532	7.451	0.271	0.44	43.932	1.386	0.951	5.422	0.128	0.167	10.631	0.573	0.264	8.919	6.385	4.871	0.847	0.046	0.068	1.47	0.517	0.282
Croatia	32.281	2.131	1.574	6.464	0.323	0.399	43.783	0.265	0.373	3.312	0.277	0.377	8.811	0.102	0.343	3.005	0.309	0.215	0	0	0	1.332	0.705	0.503
Estonia	29.601	1.311	0.65	11.503	0.281	0.525	42.104	2.793	4.276	0.352	0.065	0.098	12.929	2.022	1.362	3.545	0.413	0.579	1.969	0.111	0.046	0.554	0.083	0.202
Cyprus	32.987	1.502	0.516	11.046	1.498	2.377	33.636	0.464	0.179	8.981	0.809	1.221	8.79	0.304	0.57	6.132	0.219	0.11	2.55	0.176	0.097	1.498	0.704	0.453
Latvia	17.912	2.278	1.598	4.116	1.522	1.059	48.577	1.015	1.585	7.227	0.11	0.081	6.816	0.28	0.496	5.723	0.423	0.679	0.13	0.049	0.026	6.696	1.184	0.696
Lithuania	25.047	1.514	2.511	9.071	0.255	0.322	48.129	2.182	3.453	1.258	0.049	0.084	10.755	0.88	0.078	6.681	3.123	2.625	1.83	0.351	0.212	0.741	0.099	0.174
Hungary	30.315	2.63	1.584	9.3	0.336	0.117	43.377	3.021	4.716	2.718	0.54	0.348	8.026	1.122	2.132	4.318	0.693	0.451	0.534	0.19	0.26	2.005	0.616	1.132
Malta	27.18	2.944	1.617	6.364	1.235	1.77	44.766	0.64	1.243	5.518	0.859	0.604	11.949	0.114	0.174	5.949	1.157	1.669	1.585	0.501	0.342	0.565	0.109	0.169
Poland	33.927	2.388	1.649	3.647	0.235	0.34	43.323	1.827	1.161	8.274	0.107	0.19	5.963	1.002	1.743	1.785	0.198	0.341	1.894	1.02	1.514	1.302	0.318	0.558
Romania	22.952	0.65	1.455	7.553	1.217	0.845	47.42	4.599	1.978	9.328	1.08	1.8	11.102	8.806	5.041	5.668	0.101	0.245	0.379	0.149	0.082	0.623	0.158	0.219
Slovenia	27.116	0.806	1.307	6.907	1.606	1.212	50.284	0.694	0.065	4.467	0.319	0.526	9.513	2.525	1.649	3.954	1.788	2.842	0	0	0	1.056	0.466	0.248
Slovakia	33.015	3.168	1.867	5.309	1.698	1.314	41.847	0.631	0.486	6.169	0.9	0.614	7.963	0.667	1.008	0.591	0.396	0.615	0.103	0	0.05	3.103	0.157	0.253
	31.739	0.838	2.036	8.809	0.282	0.189	40.67	0.362	0.477	4.996	0.238	0.356	9.088	0.374	0.14	2.598	0.432	0.664	0.107	0.016	0.005	1.56	0.54	0.808

Source: own elaboration from data provided by EU-SILC (2008–2018) and ESSPROS (2008–2017).

Table 2. Expected value of fuzzy RRP (over 100) and E (over 1) for several values of the index of λ.

	Relative Reduction of Poverty (RRP)			Debreu–Farrell Ratio (DF)		
Country	λ = 1	λ = 0.5	λ = 0	λ = 1	λ = 0.5	λ = 0
Belgium	39.91	38.62	37.34	0.577	0.576	0.575
Denmark	50.88	49.93	48.99	0.722	0.713	0.705
Germany	33.54	33.39	33.25	0.53	0.52	0.51
Ireland	54.58	53.10	51.62	1	1	1
Greece	17.20	16.61	16.02	0.318	0.302	0.287
Spain	26.29	24.78	23.28	0.398	0.391	0.384
France	43.38	43.14	42.91	0.629	0.600	0.570
Italy	20.91	20.88	20.86	0.299	0.291	0.283
Luxembourg	39.41	39.03	38.66	0.669	0.653	0.637
Netherlands	43.21	41.57	39.92	0.608	0.594	0.581
Austria	43.59	42.92	42.25	0.641	0.627	0.614
Portugal	25.54	24.27	23.00	0.469	0.456	0.444
Finland	55.99	54.87	53.75	0.764	0.741	0.719
Sweden	45.26	44.72	44.17	0.748	0.745	0.741
United Kingdom	41.03	40.44	39.85	0.669	0.659	0.650
Bulgaria	22.51	20.92	19.32	0.423	0.396	0.370
Czechia	40.38	40.01	39.65	0.672	0.654	0.637
Estonia	25.64	24.14	22.63	0.433	0.428	0.424
Croatia	30.37	28.45	26.53	0.488	0.464	0.440
Cyprus	37.70	37.06	36.42	0.645	0.627	0.610
Latvia	19.35	18.68	18.01	0.365	0.351	0.337
Lithuania	25.13	24.14	23.15	0.501	0.481	0.461
Hungary	46.10	44.38	42.66	0.824	0.804	0.785
Malta	31.25	30.82	30.40	0.505	0.493	0.481
Poland	31.05	28.80	26.56	0.504	0.477	0.449
Romania	15.35	14.13	12.90	0.288	0.268	0.249
Slovenia	43.25	42.75	42.26	0.652	0.625	0.598
Slovakia	33.92	31.83	29.75	0.508	0.486	0.464

Source: own elaboration from data provided by EU-SILC (2008–2018) and ESSPROS (2008–2017).

References

1. Bárcena-Martín, E.; Blanco-Arana, M.; Pérez-Moreno, S. Social Transfers and Child Poverty in European Countries. Pro-poor Targeting or Pro-child Targeting? *J. Soc. Policy* **2018**, 1–20. [CrossRef]
2. Marchal, S.; Marx, I.; Van Mechelen, N. The Great Wake-Up Call? Social Citizenship and MinimumIncome Provisions in Europe in Times of Crisis. *J. Soc. Policy* **2014**, *43*, 247–267. [CrossRef]
3. Smith, K.D.; Shone, B. Progressive State Taxes and Welfare. *Poverty Public Policy* **2016**, *8*, 430–437. [CrossRef]
4. Afonso, A.; St Aubyn, M. Non-parametric Approaches to Education and Health Efficiency in OECD Countries. *J. Appl. Econ.* **2005**, *8*, 227–246. [CrossRef]
5. Clements, B. How efficient is education spending in Europe? *Eur. Rev. Econ. Financ.* **2002**, *1*, 3–26.
6. Kapsoli, J.; Teodoru, I.R. *Benchmarking Social Spending Using Efficiency Frontiers*, International Monetary Fund Working Paper 17/197; International Monetary Fund: Washington, DC, USA, 2017.
7. Herrmann, P.; Tausch, A.; Heshmati, A.; Bajalan, C. *Efficiency and Effectiveness of Social Spending*; IZA Discussion Papers, No. 3482; Institute for the Study of Labor (IZA): Bonn, Germany, 2008; Available online: http://nbn-resolving.de/urn:nbn:de:101:1-2008052722 (accessed on 10 September 2020).
8. Afonso, A.; Schuknecht, L.; Tanzi, V. Income distribution determinants and public spending efficiency. *J. Econ. Inequal.* **2010**, *8*, 367–389. [CrossRef]
9. Lefebvre, M.; Coelli, T.; Pestieau, P. On the Convergence of Social Protection Performance in the European Union. *CESifo Econ. Stud.* **2010**, *56*, 300–322. [CrossRef]
10. Valls Fonayet, F.; Belzunegui Eraso, Á.; De Andrés Sánchez, J. Efficiency of Social Expenditure Levels in Reducing Poverty Risk in the EU-28. *Poverty Public Policy* **2020**, *12*, 43–62. [CrossRef]
11. Antonelli, M.A.; De Bonis, V. Social Spending, Welfare and Redistribution: A Comparative Analysis of 22 European Countries. *Mod. Econ.* **2017**, *8*, 1291–1313. [CrossRef]
12. Andrés-Sánchez, J.; Belzunegui-Eraso, Á.; Valls-Fonayet, F. Pattern recognition in social expenditure and social expenditure performance in EU 28 countries. *Acta Oeconomica* **2020**, *70*, 37–61. [CrossRef]

13. Ferrarini, T.; Nelson, K.; Palme, J. Social transfers and poverty in middle- and high-income countries—A global perspective. *Glob. Soc. Policy* **2016**, *16*, 22–46. [CrossRef]
14. Oxley, H.; Dang, T.-T.; Förster, M.; Pellizari, M. Income inequalities and poverty among children and households with children in selected OECD countries. In *Child Well-Being, Child Poverty and Child Policy in Modern Nations: What Do We Know*; Vleminckx, K., Smeeding, T., Eds.; Policy Press: Bristol, UK, 2001; pp. 371–405.
15. Cantillon, B.; Marx, I. Van den Bosch, K. *The Puzzle of Egalitarianism: About the Relationships between Employment, Wage Inequality, Social Expenditures and Poverty*; Working Papers No 337; Luxembourg Income Study: Luxembourg, 2002.
16. Wade, R.H. On the causes of increasing world poverty and inequality, or why the Matthew effect prevails. *New Political Econ.* **2004**, *9*, 163–188. [CrossRef]
17. European Economy. *Report on Public Finances in EMU*; Publications Office of the European Union: Luxemburg, 2018.
18. Cantillon, B. *The Paradox of the Social Investment State. Growth, Employment and Poverty in the Lisbon Era*; Working Paper No. 11/03; University of Antwerp: Antwerp, Belgium, 2011.
19. Cincinnato, S.; Nicaise, I. *Minimum Income Schemes: Panorama and Assessment. A Study of National Policies (Belgium)*; European Commission: Brussels, Belgium, 2009.
20. Bogdanov, G.; Zahariev, B. *Analysis of the Situation in Relation to Minimum Income Schemes in Bulgaria. A Study of National Policies*; European Commission: Brussels, Belgium, 2009.
21. Ruoppila, S.; Lamminmäki, S. *Minimum Income Schemes. A Study of National Policies*; European Commission: Brussels, Belgium, 2009.
22. Legros, M. *Minimum Income Schemes. From Crisis to Another, The French Experience of Means-Tested Benefits*; European Commission: Brussels, Belgium, 2009.
23. Radu, M. *Analysis of the Situation in Relation to Minimum Income Schemes in Romania. A Study of National Policies*; European Commission: Brussels, Belgium, 2009.
24. Greene, W.H. The econometric approach to efficiency analysis. In *The Measurement of Productive Efficiency and Productivity Growth*; Fried, H.O., Knox Lovell, C.A., Schmidt, S.S., Eds.; Oxford Scholarship: Oxford, UK, 2008; pp. 92–250.
25. Brunelli, M.; Mezei, J. How different are ranking methods for fuzzy numbers? A numerical study. *Int. J. Approx. Reason.* **2013**, *54*, 627–639. [CrossRef]
26. Campos, L.M.; González, A. A subjective approach for ranking fuzzy numbers. *Fuzzy Sets Syst.* **1989**, *29*, 145–153.
27. Zadeh, L.A. The concept of a linguistic variable and its application to approximate reasoning—I. *Inf. Sci.* **1975**, *8*, 199–249. [CrossRef]
28. Akoglu, H. User's guide to correlation coefficients. *Turk. J. Emerg. Med.* **2018**, *18*, 91–93. [CrossRef]
29. Cheng, C.B. Group opinion aggregation based on a grading process: A method for constructing triangular fuzzy numbers. *Comput. Math. Appl.* **2004**, *48*, 1619–1632. [CrossRef]
30. Liu, S.; Fang, Z.; Zang, Y.; Forrest, J. General grey numbers and their operations. *Grey Syst. Theory Appl.* **2012**, *2*, 341–349. [CrossRef]
31. Liu, S.T.; Kao, C. Fuzzy measures for correlation coefficient of fuzzy numbers. *Fuzzy Sets Syst.* **2002**, *128*, 267–275. [CrossRef]
32. Dong, W.; Shah, H.C. Vertex method for computing functions of fuzzy variables. *Fuzzy Sets Syst.* **1987**, *24*, 65–78. [CrossRef]
33. Mesiar, R. Shape preserving additions of fuzzy intervals. *Fuzzy Sets Syst.* **1997**, *86*, 73–78. [CrossRef]
34. Hong, D.H. Fuzzy measures for a correlation coefficient of fuzzy numbers under TW (the weakest t-norm-based fuzzy arithmetic operations. *Inf. Sci.* **2006**, *176*, 150–160. [CrossRef]
35. Ban, A.I.; Ban, O.I.; Tuse, D.A. Derived fuzzy importance of attributes based on the weakest triangular norm-based fuzzy arithmetic and applications to the hotel services. *Iran. J. Fuzzy Syst.* **2016**, *13*, 65–85. [CrossRef]
36. Zhou, W.; Xu, Z. An Overview of the Fuzzy Data Envelopment Analysis Research and Its Successful Applications. *Int. J. Fuzzy Syst.* **2020**, *22*, 1037–1055. [CrossRef]
37. Wu, D.S.; Yang, Z.J.; Liang, L. Efficiency analysis of crossregion bank branches using fuzzy data envelopment analysis. *Appl. Math. Comput.* **2006**, *181*, 271–281. [CrossRef]
38. Lee, S.K.; Mogi, G.; Hui, K.S. A fuzzy analytic hierarchy process (AHP)/data envelopment analysis (DEA) hybrid model for efficiently allocating energy R&D resources: In the case of energy technologies against high oil prices. *Renew. Sustain. Energy Rev.* **2013**, *21*, 347–355.
39. Gupta, P.; Mehlawat, M.K.; Aggarwal, U.; Charles, V. An integrated AHP–DEA multi-objective optimization model for sustainable transportation in mining industry. *Resour. Policy* **2018**, 101180. [CrossRef]
40. Aydin, N.; Zortuk, M. Measuring efficiency of foreign direct investment in selected transition economies with fuzzy data envelopment analysis. *Econ. Comput. Econ. Cybern. Stud. Res.* **2014**, *48*, 273–286.
41. Kaya, İ.; Çolak, M.; Terzi, F. A comprehensive review of fuzzy multi criteria decision making methodologies for energy policy making. *Energy Strategy Rev.* **2019**, *24*, 207–228. [CrossRef]
42. Sengar, A.; Sharma, V.; Agrawal, R.; Dwivedi, D.; Dwivedi, P.; Joshi, K.; Barthwal, M. Prioritization of barriers to energy generation using pine needles to mitigate climate change: Evidence from India. *J. Clean. Prod.* **2020**, *275*, 123840. [CrossRef]
43. Kumar, A.; Wasan, P.; Luthra, S.; Dixit, G. Development of a framework for selecting a sustainable location of waste electrical and electronic equipment recycling plant in emerging economies. *J. Clean. Prod.* **2020**, 122645. [CrossRef]

44. Rani, P.; Mishra, A.R.; Mardani, A.; Cavallaro, F.; Alrasheedi, M.; AlRashidi, A. A novel approach to extended fuzzy TOPSIS based on new divergence measures for renewable energy sources selection. *J. Clean. Prod.* **2020**, *257*, 120352. [CrossRef]
45. Büyüközkan, G.; Çifçi, G. A combined fuzzy AHP and fuzzy TOPSIS based strategic analysis of electronic service quality in healthcare industry. *Expert Syst. Appl.* **2012**, *39*, 2341–2354. [CrossRef]
46. Teresa, L.; Liern, V.; Pérez-Gladish, B. A multicriteria assessment model for countries' degree of preparedness for successful impact investing. *Manag. Decis.* **2019**, *58*, 2455–2471. [CrossRef]
47. Chen, Y.; Yoo, S.; Hwang, J. Fuzzy multiple criteria decision-making assessment of urban conservation in historic districts: Case study of Wenming Historic Block in Kunming City, China. *J. Urban Plan. Dev.* **2017**, *143*, 105016008. [CrossRef]
48. Zyoud, S.H.; Fuchs-Hanusch, D. An Integrated Decision-Making Framework to Appraise Water Losses in Municipal Water Systems. *Int. J. Inf. Technol. Decis. Mak. (IJITDM)* **2020**, *19*, 1293–1326. [CrossRef]
49. Sari, I.U.; Behret, H.; Kahraman, C. Risk governance of urban rail systems using fuzzy AHP: The case of Istanbul. *Int. J. Uncertain. Fuzziness Knowl.-Based Syst.* **2012**, *20* (Suppl. 01), 67–79. [CrossRef]
50. Perez-Arellano, L.A.; Blanco-Mesa, F.; León-Castro, E.; Alfaro-García, V. Bonferroni Prioritized Aggregation Operators Applied to Government Transparency. *Mathematics* **2021**, *9*, 24. [CrossRef]
51. González, A.; Ortigoza, E.; Llamosas, C.; Blanco, G.; Amarilla, R. Multi-criteria analysis of economic complexity transition in emerging economies: The case of Paraguay. *Socio-Econ. Plan. Sci.* **2019**, *68*, 100617. [CrossRef]
52. Zmeškal, Z. Value at risk methodology of international index portfolio under soft conditions (fuzzy-stochastic approach). *Int. Rev. Financ. Anal.* **2005**, *14*, 263–275. [CrossRef]
53. Nguyen, T.T.; Gordon-Brown, L.; Khosravi, A.; Creighton, D.; Nahavandi, S. Fuzzy portfolio allocation models through a new risk measure and fuzzy sharpe ratio. *IEEE Trans. Fuzzy Syst.* **2014**, *23*, 656–676. [CrossRef]
54. Tsao, C.T. Fuzzy net present values for capital investments in an uncertain environment. *Comput. Oper. Res.* **2012**, *39*, 1885–1892. [CrossRef]
55. Wu, B.; Sha, W.-S.; Chen, J.-C. Correlation Evaluation with Fuzzy Data and its Application in the Management Science. In *Econometrics of Risk*; Springer: Cham, Switzerland, 2015; pp. 273–285. [CrossRef]
56. Wu, B.; Lai, W.; Wu, C.L.; Tienliu, T.K. Correlation with fuzzy data and its applications in the 12-year compulsory education in Taiwan. *Commun. Stat.-Simul. Comput.* **2016**, *45*, 1337–1354. [CrossRef]
57. Chang, D.-F. Student Mobility in Higher Education Explained by Cultural and Technological Awareness in Taiwan. Multicultural Awareness and Technology in Higher Education: Global Perspectives. *IGI Glob.* **2014**, 66–85. [CrossRef]
58. Kumar, M. Evaluation of the intuitionistic fuzzy importance of attributes based on the correlation coefficient under weakest triangular norm and application to the hotel services. *J. Intelligent Fuzzy Syst.* **2019**, *36*, 3211–3223. [CrossRef]
59. Yuan, J.; Li, X.; Shi, Y.; Chen, F.T.S.; Ruan, J.; Zhu, Y. Linkages Between Chinese Stock Price Index and Exchange Rates—An Evidence From the Belt and Road Initiative. *IEEE Access* **2020**, *8*, 95403–95416. [CrossRef]
60. Eurostat. Social Protection Statistics. Available online: https://ec.europa.eu/eurostat/statistics-explained/index.php?title=Archive:Social_protection_statistics&direction=next&oldid=503877#Expenditure_on_pensions (accessed on 10 September 2020).

Article

Fuzzy Set Qualitative Comparative Analysis of Factors Influencing the Use of Cryptocurrencies in Spanish Households

Mario Arias-Oliva [1,*], Jorge de Andrés-Sánchez [2] and Jorge Pelegrín-Borondo [3]

1. Department of Management and Marketing, Complutense University of Madrid, 28040 Madrid, Spain
2. Social and Business Research Laboratory, University Rovira i Virgili, 43002 Tarragona, Spain; jorge.deandres@urv.cat
3. Economics and Business Department, University of La Rioja, 26006 Logroño, Spain; jorge.pelegrin@unirioja.es
* Correspondence: mario.arias@ucm.es

Citation: Arias-Oliva, M.; de Andrés-Sánchez, J.; Pelegrín-Borondo, J. Fuzzy Set Qualitative Comparative Analysis of Factors Influencing the Use of Cryptocurrencies in Spanish Households. *Mathematics* **2021**, *9*, 324. https://doi.org/10.3390/math9040324

Academic Editor: Basil Papadopoulos
Received: 20 January 2021
Accepted: 3 February 2021
Published: 6 February 2021

Publisher's Note: MDPI stays neutral with regard to jurisdictional claims in published maps and institutional affiliations.

Copyright: © 2021 by the authors. Licensee MDPI, Basel, Switzerland. This article is an open access article distributed under the terms and conditions of the Creative Commons Attribution (CC BY) license (https://creativecommons.org/licenses/by/4.0/).

Abstract: This paper assesses the variables influencing the expansion of cryptocurrency (crypto for short) use in households. To carry on the study we apply a consumer-behavior focus and so-called fuzzy set Qualitative Comparative Analysis (fsQCA). In a previous research, that was grounded on Unified Theory of Acceptance and Use of Technology (UTAUT) and Partial Least Squares (PLS), we found that main factors to explain the intention to use of cryptos by individuals were performance expectancy (in fact, it was the main factor), effort expectancy and facilitating conditions. We did not found evidences about the relevance of social influence, perceived risk and financial literacy. This study revisits these results by applying fsQCA instead PLS. Empirical research on factors influencing cryto use is relatively scarce due to the novelty of blockchain techs, so the present paper expands the literature on this topic by using an original analytical tool in this context. The main contribution of this paper consists in showing empirically that fsQCA provides a complementary and enriching perspective to interpret data about the use of cryptos. We obtain again that the most relevant factor to explain the intention of using cryptocurrencies is perceived expectancy and that also effort expectancy and facilitation conditions are relevant. But also fsQCA has allowed us discovering that despite social influence, perceived risk and financial literacy were not significant in the PLS model, they impact on the intention to use cryptocurrencies when are combined with other factors. Social influence acts as an "enable factor" for the rest of explanatory variables and it is linked positively with intention to use cryptos. Also financial literacy is relevant because its lack is a sufficient condition for the non-acceptance of that blockchain tech. Likewise we have checked that perceived risk influences the intention of using cryptos. However, this influence may be positive or negative depending of the circumstances.

Keywords: cryptocurrencies; bitcoin; blockchain; fintech; unified theory of acceptance and use of technology; intention to use; fuzzy sets; fuzzy set qualitative comparative analysis

1. Introduction

The origin of blockchain tech and cryptocurrencies (cryptos for short) dates back to 2008. That year, Satoshi Nakamoto posted a paper to a cryptography forum entitled "Bitcoin: A Peer-to-Peer Electronic Cash System" [1]. That post described a decentralized peer-to-peer monetary system, whose motivation was that "a purely peer-to-peer version of electronic cash would allow online payments to be sent directly from one party to another without going through a financial institution" [1]. Thus, in 2009 the first cryptocurrency, so-called bitcoin, was created. Cryptocurrencies are digital currencies based on blockchain technology, which employ cryptographic techniques. They are also non-fiat digital currencies i.e., digital currencies that are not linked to any underlying asset, have no intrinsic value, and do not suppose a liability from any economic agent [2].

Glasser et al. [3] differentiate the following uses of cryptos: as a speculative digital asset and as a currency. They conclude that the use of cryptos is biased to speculation.

So, [4] estimates that above 50% of crypto platforms users have utilized them only in speculative trades. On the other hand, about 46% of cryptocurrency platforms users have employed cryptos as a transactional medium at least once in a year.

Following the statistics by CoinmarketCap [5], the capitalization of crypto markets has grown approximately 100 times from last week of April 2013 (US $1.37 billions) to last week of 2020 (US $140 billions). In both dates bitcoin was the most capitalized cryptocurrency but whereas 4th week of April 2013 that value was US $1.3 billion and concentrated above 94% of whole market, capitalization in last week of 2020 grew to US $488 billion, but supposed slightly less than 50% of overall market. The magnitude of growth rate is still more impressive for the 10th currency. On April 2013 that currency was Mincoin with a capitalization value US $118.657. In last week of December 2020 this place was occupied by Polkadlot (US $4.6 billions). That is to say, the value of Polkadlot at the end of 2020 is 3.5 times overall cryptocurrency market on 2013 April. Likewise, the number of actually negotiated cryptos in market has increased within that period from 10 last week of 2003 April to more than 2000 in last week of 2020.

A simple bibliographic search on Web of Knowledge database shows that from bitcoin creation to middle 2010, papers on this topic were scarce or null. Table 1 shows the results of the simple search "cryptos" AND several terms as e.g., "prices" in Social Sciences publications. Until 2017, research on cryptos grew slowly and likewise at year 2017 it still was scarce [6]. At 2018 there is a breakpoint in the number of published papers that reaches a maximum in 2019.

Table 1. Number of papers indexed in Web of Science within 2014–2010 for the search "cryptos AND".

Year	∅	Prices	Regulation	Investment	Markets
2014	10	2	1	4	7
2015	24	2	2	4	16
2016	38	4	4	15	24
2017	101	16	16	19	55
2018	293	67	50	53	176
2019	541	146	68	80	377
2020	400	131	42	40	258

Source: Own elaboration from database Web of Knowledge.

Cryptos generate many opportunities as e.g., fast, efficient, and anonymous transactions and moreover are non-intermediated. However, they also have drawbacks, such as their risk and price volatility, clearly greater than those of a conventional currency; the great technological and financial knowledge needed for their handle and the fuzzy social perception about holding them. Taking into account these considerations, this paper assesses factors that influence the acceptance of cryptos by households from the framework provided by Technology Acceptance Models (TAMs).

This research has been made with the same sample of adults from Spain as Arias-Oliva et al. [7]. Literature on the application of TAMs on this topic is not so widespread and, due to the reasons above, all very recent. Let us point out apart from [7–22]. Despite all reviewed literature is based in the use of TAMs, the final configuration of hypothesis to test and the use/user of cryptos under consideration have different nuances. They may come from how TAM is applied but also due to the use of cryptocurrency tested: a generic intention to use by individuals [7–20]; as a payment method in commercial transactions [14–21] or as a purely way of investment [19]. This paper uses the configuration of hypotheses in [7] and also tests the intention to use by households, i.e., their motivation may be either as payment method or as an investment asset.

With the exception of [19], all reviewed papers on the acceptance of cryptos use Partial Least Squares (PLS) to test the influence of factors. This paper supposes a novelty in this context since uses fuzzy set Qualitative Comparative Analysis (fsQCA) developed by Ragin in [23,24] to extend the results that we reached by using PLS in [7]. This methodology is very used in sociological studies, but also there is a great deal of applications in management and marketing (see [25] for an extended survey). As far as similar questions to ours are concerned, fsQCA has been applied instead PLS in the assessment of new technologies acceptance [26,27] and also as a complementary method to PLS in management issues [28,29]. The use of fsQCA provides a complementary approach to correlational methods to deal with causality. Conventional regression is variable-oriented, i.e., it is focused in fitting the mean effect of every variable on the output. On the other hand fsQCA is case-oriented. It is based on measuring the membership degree of each case in the set of attributes and the outcome set [23,28]. Thus, with fsQCA we cannot quantify with a coefficient the influence of a given variable over the output but we can discover how input variables are combined to produce or not produce an output [26].

The rest of the paper is structured as follows. In the second section we built up our hypotheses over the basis of existing literature. Subsequently we present our materials and methods. Fourth section describes results from analytical tools. We finally outline conclusions and future research lines.

2. A Model to Explain the Acceptance of Cryptocurrencies by Households

The model and hypothesis that we propose to explain the variables influencing crypto acceptance are those we used in [7]. So, our theoretical ground is the Unified Theory of Acceptance and Use of Technology (UTAUT) [30] and its extension UTAUT2 [31]. These models are widely accepted by academic community to explain how an emerging technology is adopted in a society. Both are based on Technology Acceptance Models [32,33], Theory of Reasoned Action [34] and Theory of Planned Behavior [35]. UTAUT models postulate a direct and positive influence of performance expectancy (PE), effort expectation (EE), social norm (SN), and facilitating conditions (FC) on the intention to use (IU) a tech. Likewise, as we do in [7] we include in our model behavioral research findings about how perception of risk and financial literacy affect the IU financial products.

Performance Expectancy (PE) is defined in [30] as the expectation of a person about the influence of using a technology on his/her performance. It is widely accepted that current electronic payment systems are slow, insecure, inefficient, uncollaborative and non-global [36]. So, crypto use has potential to solve these drawbacks [37]. From the emerging of bitcoin, business sphere has integrated progressively cryptos into their activities. The first purchase with bitcoins was done in 2010 to buy two pizzas [38]. Nowadays it is possible to use bitcoins in some 18,500 businesses around the world [39].

Bitcoin is only one of more than 8000 cryptos on the market. That number does not include all cryptos, just those quoted on the market to be traded. The volatility of cryptos opens enormous psychological thresholds in prices [40]. That variety of currencies allows crypto portfolios accessing to wide risk-return configurations. Likewise, as it pointed out by Liu and Tsyvisnsky [41] factors influencing the price of cryptos are different to those linked to the price of conventional financial assets and so, their returns are uncorrelated with those from stock and bonds. For example, cryptos are not influenced by economic cycles. Therefore, they are very suitable to implement so-called alternative portfolio management strategies or as shelter investment in recession periods. Moreover, the anonymity provided by the use of cryptos allows public to keep the confidentiality of their savings and movement of funds. The other side of their usefulness is that they make easier criminal acts as e.g., tax evasion, money laundering or contraband transactions [42].

PE is possibly the variable that literature finds as the most relevant in FinTech acceptance. Some evidences in this way are [43] for the use of a payment authentication system based on biometrics; [44] in the behavioral intention to adopt plastic money; [45] on the use of financial websites; [46] for adopting online banking. In the use of m-banking [26,47–51] obtain similar results. Regarding specific literature on cryptos results in [7–12,14,15,18,20,21] also suggests that PE is the key factor to explain intention to use, so the following hypothesis can be stated:

Hypothesis 1 (H1). *Performance expectancy influences positively intention to use cryptos.*

The second variable tested is effort expectancy (EE) that [30] define as the ease extend linked with using a given tech. In this regard [52] states that major part of interviewed people feel that blockchain is not an easy technology to use. In this way [53] outlines that making transaction with bitcoins is a great challenge for many people. In fact [54] finds that non-users of bitcoin felt incapable of using it and so, it is possibly the greater barrier to the widespread use of cryptos.

There are many evidences of the positive relation EE with the adoption of new financial technologies. Some evidences in this way are [55] for electronic ways of micro crowdfunding and also abovementioned papers [26,43–45,47–50,56]. Within cryptos and blockchain research we find [9–12,15,16,18,20] confirming so. We have to point out that in [20] EE may be assimilated to the construct "web quality" in this paper. In [7] we also found for this relation a positive sign but with a weak statistical significance, so the following hypothesis is proposed:

Hypothesis 2 (H2). *Effort expectancy is positively linked with the intention of using cryptos.*

Social influence (SI) is the degree to which people feel that close persons think that they have to use a specific technology [30]. At this regard persons from cryptocurrency community participate in collaborative works giving help to the rest [57]. Social support generates trust and commitment and so is linked positively to the intention of using a technology [15]. Despite there are less evidences of the link between SI and IU, there are still a great number of findings in this way. Within e-banking we can remark [43,48,49,51,55,56]. Regarding cryptos let us outline that [8,10,12,14,17–20,22]. In our paper [7] we also found a positive relation but it was not significant, so the following hypothesis is tested:

Hypothesis 3 (H3). *Social influence is positively linked with the intention of using cryptos.*

Facilitating conditions (FC) are the degree to which an individual considers that he/she has the necessary infrastructure to run a specific technology [30]. It is clear that operating with cryptos needs being technologically equipped and, likewise, a minimum level of computer comprehension and knowledge is required [58,59]. As far as FindTech is concerned [26,46,47,56] found a positive significant relation. Regarding cryptos and bitcoin [10,16,17,19,21] show that FC influences cryptos IU. In [7] we had also identified FC to be a determinant factor, so the following hypothesis is made:

Hypothesis 4 (H4). *Facilitating conditions are related positively with the intention of using cryptos.*

Despite not being explicitly considered in UTAUT2, perceived risk (PR) is considered in many papers as a key barrier to using new techs. From a behavioral research perspective, Faqih in [60] defines PR as consumers' belief about the degree of uncertainty and non-desired consequences due to putting to work a product. It has been considered a key variable of consumers' purchase intention [61,62], as well as a predictor of technology adoption [63]. Likewise, standard financial economics state that the risk of an asset is a key variable to make a financial decision. Therefore, market risk, that is, the risk of losses due to the diminution of cryptos price is clearly greater than of conventional currencies [58]. It is well-known that determinants of cryptos return are completely different from those for stocks and bonds [41,64]. Likewise, volatility, deflation and speculative bubbles are more probable in crypto market than in conventional currency market due to cryptos have no supervision from any Central Bank and their intrinsic value is null [65]. These reasons explains why an ING study on bitcoin opinion found that 29% of Europeans had intention of never investing in cryptos since they had the perception that stocks are a less risky investment tool [58]. On the other hand, cryptos are not linked with any country and so, they are not subject to country-risk. Likewise, following [4], national currency-focused fund transfer systems and B2B payment platforms are more exposed to the risk of exchange rate than cryptocurrency-focused merchant services. The reason is that the latter often also deal exchanges with cryptos in their payment activities.

Likewise, cryptos (specially the small ones) are clearly subjected to liquidity and counterparty risk. For example it is often very difficult changing cryptos with local currencies in many countries as e.g., Latin American states [4].

Undoubtedly, a big deal of risk in cryptos use is operational risk that in several papers [14,15,20] is identified as the main determinant of trustiness. Following [4], the greatest risk factor for small exchanges and second one within large exchanges context are security breaks whose consequence may be a permanent loss of funds. These problems may come from possible cyberattacks, the irreversibility of agreements or the impossibility of key recovering [15]. Likewise from large exchanges point of view the lack of regulation is also a source of risk. This question seems to be less relevant in small exchanges. In these agreements a great drawback comes from the difficulties with maintaining banking relationships. However, for large exchanges this risk seems to be under control [4]. Small trades are more distressed about fraud than large exchanges. The reason could be that they are addressed more often than large exchanges but also that fraud has a greater patrimonial impact due to their constrained budget [4].

Several studies analyze the influence of PR on the IU financial technologies [25,46,48,66]. However, in [66] it is stated that while the direct influence of PR on the IU m-banking is normally small, it supposes a key factor in pre-adoption process. In cryptocurrency context [14,15,17] consider that variable relevant to explain crypto use. Some papers [14,15,18,20,21] reveal that trustiness and perceived security, that are linked with operational risk, are relevant to decide about the use of cryptos. Thus, the following hypothesis is tested:

Hypothesis 5 (H5). *Perceived risk is related negatively with the intention of using cryptos.*

The last factor tested in our paper is financial literacy (or financial knowledge), FL. Following [67] people's financial literacy consists in the level of their knowledge about financial concepts and in the degree of confidence on their skills to apply that knowledge in real-world situations.

Financial knowledge has been proved to be influent in adopting new financial techs. Whereas [68] conclude that persons with low financial literacy are consistently less likely to trade stocks, the survey by [69] shows that a great financial knowledge implies a higher propensity to participate in financial markets and investing in shares. In [67] it is pointed out that financial knowledge is associated with more saving planning, active participation in stock markets and rational choices of financial products. On the other hand, lower financial knowledge implies poorer financial decisions as, e.g., more expensive loans. In [70] it is outlined that financial knowledge effectively impacts in financial decisions as those related to credit cards use, mortgage loans, etc. In a cryptocurrency context [13] observes that in Japan the IU cryptos as a payment method is positively and significantly linked with FL. However, we did not find in [7] that FL to be relevant in the IU. The following hypothesis is put forth:

Hypothesis 6 (H6). *Financial literacy positively influences the IU cryptos.*

3. Materials and Methods
3.1. Data Collection

The database used in this paper is that in [7]. The sample has 402 answers from people over 20 years old from Spain, with a university degree and at least basic skills on Internet. The survey was answered between August 1 and September 10 2018. The responses are distributed between men and women as 53/47. In [7] it is outlined that this rate is due to the later incorporation in Spain of women to higher education. The composition of sample by ages is depicted in Figure 1. In [7] it is stated that this distribution is in accordance to the age configuration of Spanish population. So, as we expected, the greater proportion of answers come from people between the ages of 41 and 50. Notice that individuals under 21 years old were not included because usually they have already not obtained a university degree. Arias-Oliva et al. in [7] point out that these patterns are similar to the distribution of the Spanish population and so, the sample is adequate to represent target population.

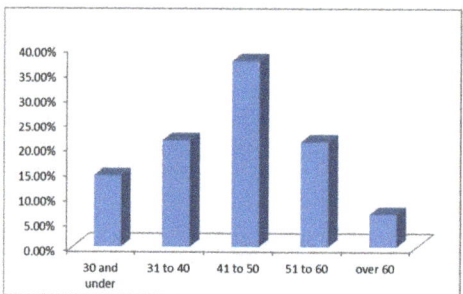

Figure 1. Sample distribution by age.

Figure 2 shows the distribution of net monthly household income. In [7] it is checked that this distribution implies a higher average income than those for whole Spanish population. However, this fact is not surprising since our survey was answered by college-educated people, who are more likely to earn higher salaries.

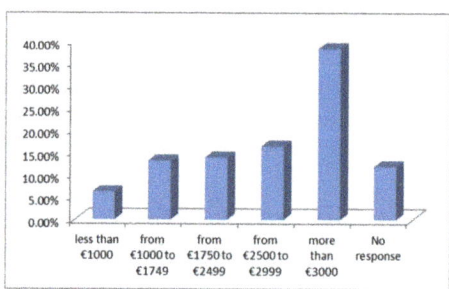

Figure 2. Sample distribution by monthly household income.

3.2. Measurement Scales

Our measurement scales are the same as [7] and are widely used by empirical papers about technology acceptance. Table 2 shows the questionnaire outlining constructs, items, and theoretical foundations of each one. The responses were done over a Likert scale from 0 to 10.

Table 2. Constructs, items and their theoretical foundation.

Construct/Item	Theoretical Foundation
Intention to Use	TAM2 scale (Venkatesh & Davis, 2000)
I intend to use cryptocurrencies	TAM2 scale (Venkatesh & Davis, 2000)
I predict that I will use cryptocurrencies	Adapted from the UTAUT2 scale (Venkatesh et al., 2012)
Performance Expectancy	
Using cryptocurrencies will increase opportunities to achieve important goals for me	
Using cryptocurrencies will help me achieve my goals more quickly	Adapted from the UTAUT2 scale (Venkatesh et al., 2012)
Using cryptocurrencies will increase my standard of living	Adapted from the UTAUT2 scale (Venkatesh et al., 2012)

Table 2. Cont.

Construct/Item	Theoretical Foundation
Effort Expectancy It will be easy for me to learn how to use cryptocurrencies Using cryptocurrencies will be clear and understandable for me It will be easy for me to use cryptocurrencies It will be easy for me to become an expert in the use of cryptocurrencies	Adapted from the UTAUT2 scale (Venkatesh et al., 2012) Adapted from the UTAUT2 scale (Venkatesh et al., 2012)
Social Influence The people who are important to me will think that I should use cryptocurrencies The people who influence me will think that I should use cryptocurrencies People whose opinions I value would like me to use cryptocurrencies	Adapted from the UTAUT2 scale (Venkatesh et al., 2012) Adapted from the UTAUT2 scale (Venkatesh et al., 2012)
Facilitating Conditions I have the necessary resources to use cryptocurrencies I have the necessary knowledge to use cryptocurrencies Cryptocurrencies are compatible with other technologies that I use I can get help if I have difficulty using cryptocurrencies	Adapted from the UTAUT2 scale (Venkatesh et al., 2012) Faqih (2016) based on Shim et al. (2011)
Perceived Risk Using cryptocurrencies is risky There is too much uncertainty associated with the use of cryptocurrencies Compared with other currencies/investments, cryptocurrencies are riskier	Faqih (2016) based on Shim et al. (2011) Based on Hasting et al. (2013)
Financial Literacy I have a good level of financial knowledge I have a high capacity to deal with financial matters	Based on Hasting et al. (2013)

Source: [7].

3.3. Quantitative Methodology

The research used the following sequential process:

Stage 1. Measurement model analysis.

To explore potential existence of dimensions in the scales we have run principal component analysis with Varimax rotation. Subsequently we have performed an assessment on reliability, convergent and discriminant validity of scales.

Stage 2. Test hypothesis H1–H6.

In [7] we used for this analysis a PLS regression. It implies calculating R2, Q2, path coefficients, and linked statistical significance degree. On the other hand, in this paper we use Qualitative Comparative Analysis (QCA) [24] and fuzzy set QCA (fsQCA) in [23] to evaluate such hypothesis.

There is a great deal of applications of fsQCA in management and marketing [25]. As far as similar fields to ours are concerned, fsQCA has been applied instead PLS in the assessment of new technologies acceptance [26,27] and also as a complementary method to PLS [28,29].

Any correlational method in general and PLS in particular assumes symmetrical relations between variables and measures the net effect of each variable on assessed output. On the other hand fsQCA allows discovering the combinatorial effects of variables over the output as well as taking into account that these interactions could be asymmetrical [28]. So, to test hypothesis in Section 2, PLS and fsQCA follow different ways. With PLS we find an average value (a coefficient) for the influence each factor on output variables and then we test its statistical relevance from its t-ratio. When applying fsQCA we find the logical implicates that combining the presence/absence of input variables suit better output results by using Boolean logic. Subsequently consistency and coverage measures inform about the relevance of discovered logical implicates and compare them with initial hypothesis. Figures 3 and 4 depict a graphical comparison about how PLS and fsQCA test hypothesis under the framework exposed in Section 2.

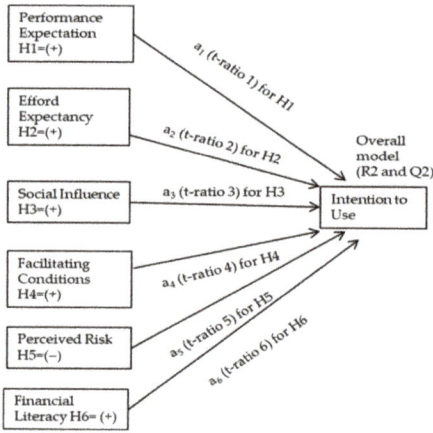

Figure 3. Hypothesis tested and analytical methodology with PLS in [7]. Source: Arias-Oliva, M.; Pelegrín-Borondo, J.; Matías-Clavero, G. Variables influencing cryptocurrency use: a technology acceptance model in Spain. *Frontiers in Psychology* **2019**, 10, 475.

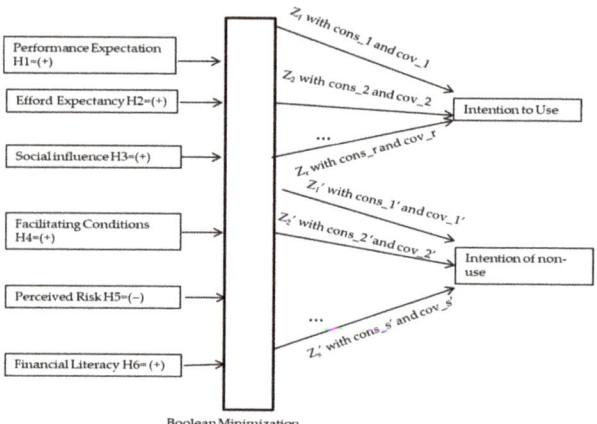

Figure 4. Hypothesis tested and analytical methodology with fsQCA.

Figure 3 represents how PLS works. Coefficients a_1, a_2, ..., a_6 quantify the sign of average influence of factors individually. Their t-ratios allow testing their particular statistical significance. To assess goodness of whole model, R2 and Q2 must be used. Figure 4 shows how fsQCA process empirical data. After a Boolean minimization, logical implicates Z_1, Z_2, .., Z_r, Z_1', Z_2', .., Z_s' are found. They embed the presence/absence of at least one input variable. Consistency and coverage measures summarize the significance and empirical relevance of every logical implicate. To test hypothesis, the configuration of these implicates must be interpreted accordingly.

Therefore, two measures are of interest for a given logical implicate:

- Consistency (cons) of the causal combination that measures the degree in which membership in a combination of causes (recipe) is a subset of outcome set.
- Coverage (cov) that measures the proportion of the outcome explained by each recipe.

As it is usually advised by literature [26] we analyze the influence of input variables not only on the outcome (IU) but also on its negation (~IU). In our case, a facilitating condition may consist in having a powerful PC and a good internet connection. The influence of this fact in crypto use may be great, little or negligible. On the other hand, it comes fair that the negation of such condition ensures not operating with cryptos. So, if we symbolize as "~" the negation of a variable, we evaluate:

$$IU = f(PE, EE, CP, FC, PR, FL), \qquad (1)$$

$$\sim IU = f(PE, EE, CP, FC, PR, FL), \qquad (2)$$

where $f(\cdot)$ symbolizes Boolean function. The implementation of fsQCA is done by using fsqca 3.1 software [71] and follows the following steps:

Step 1. Find the factorial punctuation for of the jth observation ($j = 1, 2, \ldots$, 402 in our sample) in ith variable (IU, PE, EE, SI, FC, PR and FL). We symbolize as f_{ij} those values.

Step 2. Built up the membership function for input variables PR EE, SI, FC, PR and FL and output variable IU by normalizing within [0,1] their standardized factorial punctuation in step 1. So, for a variable i, the membership value for the jth observation X_{ij} is:

$$m_{X_{ij}} = \frac{f_{ij} - \min_j\{f_{ij}\}}{\max_j\{f_{ij}\} - \min_j\{f_{ij}\}} \qquad (3)$$

The membership degree within the negated fuzzy set, ~Xi is defined then as: $m_{\sim X_{ij}} = 1 - m_{X_{ij}}$.

Step 3. Built up a Boolean truth table which is composed by so-called "min terms". We state for the value of the jth observation in ith variable two unique the possible values: true ($m'_{X_{ij}} = 1$) and false ($m'_{X_{ij}} = 0$). So:

$$m'_{X_{ij}} = \begin{cases} 1 & m_{X_{ij}} \geq 0.5 \\ 0 & \text{otherwise} \end{cases} \qquad (4)$$

Step 4. Maintain only those min terms whose consistency to produce the output are at least a threshold ε. Ragin [71] suggests $\varepsilon = 0.8$.

Step 5. By applying Quine-McCluskey algorithm [72] find essential prime implicates in truth table. These implicates conform so-called Qualitative Comparative Analysis complex solution (CQA-CS). That algorithm supposes implementing the following steps:

- 5.1. Order min terms in ascending order and group them from the number of 1 in their binary representation.
- 5.2. Compare observed min terms by pairs successively. If a change in a one-bit position exists, "erase" in the couple of those min terms the differed bit position and maintain the rest of the bits as they are.
- 5.3. Repeat second step until all prime implicates are obtained.
- 5.4. Formulate the prime implicate table. Prime implicates are placed by rows and min terms in column wise. Identify the cells corresponding to the min terms that are covered in each prime implicate.
- 5.5. Identify essential prime implicates from the observation of each column. If a min term is covered only by one prime implicate, then it is an essential prime implicate. Those implicates also belong to simplified Boolean function.

- 5.6. Simplify prime implicate table by deleting the row of each essential prime implicate and the columns linked to min terms that are covered in that essential prime implicate. Repeat step 5 for reduced prime implicate table.
- 5.7. This process is finished when all min terms are over.

Let us remark that essential prime implicates are Boolean products whose factors may be X_j or $\sim X_j$.

Step. 6. CQA-CS is usually hard to interpret and is build up with no more assumption than data. So, fsqca 3.1 also offers a parsimonious solution (QCA-PS). It is fitted from any remainder over non observed configuration of variables in order to make solution as easy as possible regardless whether it constitute an "easy" or "difficult counterfactual" case [71].

Step 7. To continue the minimization process, [23] proposes using simplifying assumptions that should be theoretically well-founded about how a given condition is causally related to the outcome. It must be supposed for non-observed configurations, if an input variable contributes to output exclusively when it is present, absent or in both cases. This step lets us obtaining so-called intermediate solution (QCA-IS).

Step 8. Let be a possible prime implicate (configuration or recipe) Z that without a loss of generality we built as:

$$Z = X_1 * X_2 * \ldots * X_r \tag{5}$$

where $1 \leq r \leq n$, n is the number of output variables and "*" stands for the Boolean product. So, we can obtain for the jth observation:

$$m_{Z_j} = \min\{m_{X_{1,j}}; m_{X_{2,j}}; \ldots; m_{X_{r,j}}\} \tag{6}$$

So, the consistency of recipe Z in producing output Y is:

$$Cons_{Z \to Y} = \frac{\sum_j \min\{m_{Z_j}; m_{Y_j}\}}{\sum_j m_{Z_j}} \tag{7}$$

where Y_j stands for the value of jth observation in output variable (Intention to Use in our case). Consistency may be understood similar as a statistical measure of significance [73]. It is widely accepted that to consider Z as sufficient condition, consistency must be above 0.8.

Subsequently, the coverage of recipe Z to produce Y is:

$$Cov_{Z \to Y} = \frac{\sum_j \min\{m_{Z_j}; m_{Y_j}\}}{\sum_j m_{Y_j}} \tag{8}$$

Coverage provides a measure of empirical relevance. Its analogous statistical measure is R2. A consistency above 0.8 for a recipe (6) implies that the combination is "almost always" necessary or sufficient [23]. It can be considered that a condition is completely sufficient when cons > 0.9 y cov > 0.5 (see [74]).

Step 9. Interpret intermediate and parsimonious solutions. At this point there is no a unified rule about what solution must be interpreted. Ragin in [23] suggests using QCA-IS since it supposes a compromise between simplicity of CQA-PS and complexity of CQA-CS. As it is pointed out by Thiem in [75] empirical studies are usually done over this solution. However, also in [75] it is advised searching causal relations using exclusively QCA-PS instead intermediate solution. That paper argues that QCA-CS and QCA-IS introduces matching counterfactual data with which QCA supplements the empirical observations. These artificial data may induce inferences that violate the actual causal structure that had generated the empirical data in the first place and that QCA is meant to uncover.

4. Results

4.1. Results from Mesurement Model

The analysis of the measurement model was done in [7] and quantitative results are summarized in Tables A1–A3 of annex. Table A1 shows basic descriptive statistics of our sample. Results in Table A2 show that factorial analysis detects only one dimension in all scales. Bartlett sphericity test has always a *p*-value (<0.0001), whereas Kaiser-Meyer-Olkin statistic, was always (≥ 0.5). The percentage of variance explained by the factors was in all dimensions (>70%), which confirms scales suitability. Regarding the evaluation of the measurement model, Table A2 of annex suggests that results on this concern are correct.

Constructs present a composite reliability and Cronbach's alpha always >0.7, confirming so that reliability was correct (see Table A3 of annex). Average variance extracted (AVE) in all scales is greater than 0.5. Therefore the convergent validity criterion was thus met. The HTMT values were correct in all cases (<0.9). Likewise, the square root of the AVE was higher than the correlations between constructs, i.e., discriminant validity criterion has been also accomplished (see Table A4).

4.2. Results from Fuzzy Set Comparative Qqualitative Analysis

A previous step to develop fsQCA implies implementing so-called necessity analysis [73]. It consists in stating cons and cov in (7) and (8) between each individual input (in affirmed and negated form) and the output (also affirmed and negated). Results are in Table 3. This analysis lets stating the degree in which and individual factor that is affirmed/negated is necessary to induce the output (or the negated output).

Table 3. Necessity analysis on IU and ~IU.

Necessity Analysis on IU			Necessity Analysis on ~IU		
Variable	Cons	Coverage	Variable	Cons	Coverage
PE	0.826	0.901	PE	0.310	0.586
~PE	0.620	0.341	~PE	0.948	0.904
EE	0.870	0.661	EE	0.505	0.666
~EE	0.561	0.395	~EE	0.743	0.908
SI	0.649	0.801	SI	0.314	0.672
~SI	0.735	0.382	~SI	0.907	0.817
FC	0.875	0.644	FC	0.522	0.665
~FC	0.545	0.397	~FC	0.721	0.909
PR	0.882	0.446	PR	0.842	0.736
~PR	0.478	0.635	~PR	0.366	0.843
FL	0.792	0.601	FL	0.538	0.707
~FL	0.545	0.397	~FL	0.696	0.853

Table 3 suggests that there is an asymmetry in the explanation of IU and ~IU. Regarding IU are "almost always necessary or sufficient" PE, EE, FC and FL. When analysing ~IU, again negated PE seems to be the main cause (cons and cov > 0.9) but also ~SI (e.g., news about bad experiences with cryptos) presents great cons and cov. Notice that SI did not present high cons to produce IU. PR (and not ~PR) has a cons > 0.8 in IU. Then, the presence of perceived risk can incentive some people to use crypto. These persons act as risk lovers, e.g., they buy cryptos as a method for betting, or as an investment with a high expected return due to its great volatility. This fact might suggest rejecting H5. However, the empirical weight of that finding (cov < 0.5) diminishes its relevance. When analyzing the influence of PR on the negation of IU, it can be checked that PR (and not ~PR) causes also ~IU (cons > 0.8, cov > 0.7). Therefore consistency and coverage of PR on ~IU outline PR as a relevant factor to reject the use of cryptos.

Table 4 shows configurations of QCA-IS and QCA-PA of model (1). To generate intermediate solution we have supposed for non-covered Boolean configurations of outputs (19 over $2^6 = 64$ possible configurations) that PE, EE, SI, FC and FL cause IU only when they are present. It is according to our hypothesis, findings in literature described in Section 2 and with necessity analysis in Table 3. So, due the contradiction between H5 and necessity analysis of PR, we suppose that either presence or absence of PR may cause IU. Analysis of intermediate solution lets appreciating that:

- PE is within the configurations with greater cons (always > 0.9) and cov (for PE*EE and PE*PR cov > 0.7). So, it seems that perceived usefulness is the most influent factor for using cryptos.
- PE, EE, SI, FC and PL are at least in one recipe with cons > 0.8 and cov > 0.5 and in affirmed form. This fact is in accordance with postulated hypotheses. At this regard, SI seems to have great influence on IU since it appears in 4 recipes.
- PR is within the recipe EE*~PR*FL (cons > 0.8 but cov < 0.5) negated as we expected. However this is the recipe of QCA-IS with poorer cov.
- PR appears in two configurations with "+" sign: SI*PR*FL and PE*PR both with cons > 0.8 and cov > 0.5. This fact violates H5. On the other hand these recipes have a suitable interpretation from a risk lover perspective. Configuration SI*PR*FL might reveal a person with financial knowledge (FL) whose environment is devoted to risky investments (SI*PR). The recipe PE*PR might reveal individuals that may use cryptos as a bet method.
- Analysis of QCA-PS reveals similar patterns as QCA-IS.
- PE is the unique input factor that may cause IU without the coincidence of any other variable. Likewise, it presents the greater cons (>0.9) and coverage (>0.8). So, it seems that the presence of PE could cause IU without the help of any other factor.
- EE, FC and PL cause IU in configurations whose cons > 0.8 and cov > 0.5 but always when they are combined with affirmed SI. Again, SI seems to be most influent factor than EE, FC and PL.
- Perceived risk only appears in one recipe with cons > 0.8 but with the lower cov (<0.5). In this case its form (negated) is as expected in H5.

Table 4. QCA-IS and QCA-PS for the model IU = f(PE, EE, SI, FC, PR, FL).

Intermediate Solution	Raw Coverage	Unique Coverage	Consistency
PE*EE	0.761	0.022	0.923
EE*SI	0.621	0.003	0.859
PE*PR	0.758	0.040	0.905
EE*~PR*FL	0.416	0.010	0.838
PE*SI*FL	0.594	0.001	0.936
SI*FC*PR	0.597	0.001	0.834
SI*PR*FL	0.558	0.001	0.840
Cons = 0.859 Cov = 0.790			
Parsimonious Solution	**Raw Coverage**	**Unique Coverage**	**Consistency**
PE	0.826	0.163	0.901
EE*SI	0.621	0.002	0.859
SI*FC	0.624	0.001	0.834
SI*FL	0.577	0.001	0.840
EE*~PR*FL	0.416	0.010	0.838
Cons = 0.869 Cov = 0.788			

The results of fsQCA over IU = f(PE, EE, SI, FC, PR, FL) suggest the acceptance of H1, H2, H3, H4, and H6. On the other hand, the assessment of H5 must be nuanced since its rejection would not imply that PR does not influence IU. Some causal configurations suggest that PR may stimulate IU to some users when it is present and to others if it is absent.

Table 5 shows configurations of QCA-IS and QCA-PA in (2), i.e., for the explanation of ~IU. To generate intermediate solution in ~IU we have supposed for non-covered Boolean configurations that PE, EE, SI, FC and FL cause IU only when they are absent. It is congruent to that we done for intermediate solution in Table 4 and with the results from necessity analysis in Table 3. The same argument applies to justify supposing that PR can be either present or absent to cause ~IU. After checking Tables 4 and 5 we find that the explanation of IU and ~IU by causal recipes of input factors is far to be symmetrical. So, QCA-IS for ~IU shows that:

- The absence of financial literacy (cons = 0.853 and cov = 0.696) and facilitating conditions (cons > 0.9 and cov > 0.7) are sufficient conditions by their selves for the intention of non-using cryptos.
- PE, EE and SI also take part of recipes in their negated form as we expected from H1, H2 and H3 respectively. In all those essential prime implicates cons > 0.85.
- PR is in two recipes with a contradictory sign. It appears as ~PE*PR (cons > 0.9, con > 0.8). It is, so, affirmed as we expected from H5. On the other hand, it is also negated in ~SI*~PR with a consistency close to 0.9 but with a relatively low coverage (0.365).
- Analysis of QCA-PS reveals similar patterns as QCA-IS:
- The simple negation of PE, EE, FC and FL cause the intention of non-using cryptos. This finding fits H1, H2, H4 and H6. In all cases cons > 0.85 and coverage $\geq$ 0.69. Now, the most influent variable seems to be the negation of PE (cons > 0.9 and cov > 0.9).
- In the recipe ~SI*~PR, social influence is negated as we postulated in H3. On the other hand, the presence of ~PR implies a violation of H5. In any case, despite in this recipe cons = 0.882, its empirical relevance is weak (cov = 0.362).

Table 5. QCA-IS and QCA-PS for the model ~IU = f(PE, EE, SI, FC, PR, FL).

Intermediate Solution	Raw Coverage	Unique Coverage	Consistency
~FL	0.696	0.017	0.853
~FC	0.721	0.009	0.909
~SI*~PR	0.362	0.004	0.882
~EE*~SI	0.709	0.007	0.923
~PE*PR	0.809	0.089	0.910
Cons = 0.942			
Cov = 0.835			
Parsimonious Solution	Raw Coverage	Unique Coverage	Consistency
~FL	0.696	0.005	0.853
~FC	0.721	0.002	0.909
~EE	0.743	0.004	0.908
~PE	0.948	0.111	0.904
~SI*~PR	0.362	0.001	0.882
Cons = 0.976			
Cov = 0.829			

So, it seems again that, with the exception of H5, we cannot reject any hypothesis in Section 2.

5. Discussion and Conclusions

There are not so much empirical studies about variables influencing crypto acceptance because of the novelty of blockchain techs. This paper contributes to literature on cryptos and employs an original analytical tool in this context, fsQCA. We have found that complementing UTAUT modelling with fsQCA lets discover relations between variables that influence crypto use that we did not in [7] by using conventional PLS. So, whereas [7] only found relevant three variables to explain the use of cryptos, we have discovered that all factors are relevant. Likewise, with the exception of the hypothesis on PR, all other are confirmed. As in [7], PE is revealed as the most decisive variable to explain IU. EE is also a relevant variable when it is combined with SI and also its absence could be a sufficient condition for ~IU. Similar considerations can be made about FC, that was relevant in [7], but also for FL, that in [7] was not. Let us remark the importance of social influence to explain IU. Its sole presence or absence is never relevant for IU or ~IU. However it is present with the expected "sign" in recipes that allow EE, FC, PR or FL to be relevant. So, it seems that SI acts as a facilitator factor to induce the other input variables to be relevant for IU.

Evidences do not support that the influence of PR on using cryptos is necessary negative. However this fact does not imply that it is not relevant to explain IU. Depending on the context PR may have a positive or negative influence. We find configurations with PR (i.e., its presence affects positively using cryptos) but also with ~PR (i.e., not perceiving risk also influences positively using cryptos), so the configuration in QCA-IS (SI*FC*PR) may explain behaviors of people with FL that from information by close people (SI) use cryptos consciously as a risky asset. Likewise the configuration PE*PR may explain utilizing cryptos as a bet method. On the other hand the configuration EE*~PR*FL explains the behavior of persons with financial knowledge that consider cryptos easy to manage and that, in their context, its use have low risk.

The main objective of this paper was complementing conclusions about IU cryptos in [7] by applying fsQCA instead PLS. As any regression methodology, PLS allows quantifying average influence of each factor over IU by means of a coefficient. Likewise, it supposes a symmetrical impact of the presence/absence of an input variable over the presence/absence of output. By using fsQCA we cannot quantify in a coefficient the average weight of one factor into the intention to use crypto and this fact suppose a drawback. Likewise fsQCA is quite sensitive to how membership functions are built up and also to outliers. On the other hand, fsQCA can discern how factors are combined to produce (or not produce) IU, so the application of conventional correlational techniques in [7] led us to conclude that SI was not a significant factor. However, fsQCA allow checking the important role of SI to induce IU since its presence lets EE, FC and FL to be relevant. Likewise fsQCA lets discover asymmetrical relations between variables. This fact allows us detecting that FC alone is not enough to generate IU but its sole absence influences decisively the intention of not using cryptos. Likewise, the use of fsQCA lets us stating that the non-significance of PR in [7] (i.e., the coefficient that quantify the influence of PR on IU was not significantly different from 0) does not mean that PR is irrelevant to explain IU. It may be caused by the balance of responses from people averse to risk and risk-lovers.

We are aware that this study has some limitations. These constraints are an incentive to conduct further research. This paper is focused on a concrete population segment: college-educated adults with Internet skills. Future studies should focus on other household segments but also on other economic agents (small business, transnational corporations, institutional investors ...). Another constraint is that this research is circumscribed to Spain. Conclusions might be slightly different if the survey had a wider geographical extension or if it were answered in another country, so the use of an international database may allow improving the conclusions that we have extracted in this work. Other issue that could be assessed in future research is the sustainability of blockchain mining. In [76] it is stated that the mining process requires intensive computation resources with large energy consumption. Therefore sustainability factors can impact on the development of blockchain technology. It is an emerging technology that is evolving continuously. Therefore, the findings of this research should be interpreted under above considerations.

Lastly, let us point out that the use of alternative analytical tools to fsQCA based on fuzzy sets as fuzzy correlation indexes [77,78] and fuzzy multiple criteria decision making methods [19] might be also object in future application on this matter.

Author Contributions: M.A.-O. and J.P.-B. have collected data have built up model to validate and have validated scales. J.d.A.-S. have implemented fsqca. All authors have contributed equally to extract conclusions from data and in literature revision. All authors have read and agreed to the published version of the manuscript.

Funding: This research was funded by the COBEMADE research group at University of La Rioja (REGI 20/40) and Programa RIS3 La Rioja. CAR-PID2019-105764RB-I00.

Institutional Review Board Statement: With regard to ethics approval: (1) all participants were given detailed written information about the study and procedure; (2) no data directly or indirectly related to the subjects' health were collected and, thus, the Declaration of Helsinki was not generally mentioned when the subjects were informed; (3) the anonymity of the collected data was ensured at all times; and (4) no permission was obtained from a board or committee ethics approval, it was not required as per applicable institutional and national guidelines and regulations (5) voluntary completion of the questionnaire was taken as consent for the data to be used in research, informed consent of the participants was implied through survey completion.

Informed Consent Statement: Informed consent was obtained from all subjects involved in the study.

Data Availability Statement: Survey supporting the study can be obtained by demanding it to any author.

Acknowledgments: Authors acknowledge helpful comments of anonymous reviewers.

Conflicts of Interest: Authors declare having no conflicts of interest.

Appendix A

Table A1. Descriptive statistics of our questionnaire.

Construct/Item	μ	Q2	σ	Q1	Q3	Q3–Q1
Intention to Use						
I intend to use cryptocurrencies	3.20	2	3.37	0	6	6
I predict that I will use cryptocurrencies	4.12	3	3.42	1	7	6
Performance Expectancy						
Using cryptocurrencies will increase opportunities to achieve important goals for me	3.95	3	3.28	1	7	6
Using cryptocurrencies will help me achieve my goals more quickly	3.34	3	3.03	1	5	4
Using cryptocurrencies will increase my standard of living	3.13	2.5	2.89	0	5	5
Effort Expectancy	3.08	2	2.90	0	5	5
It will be easy for me to learn how to use cryptocurrencies	4.99	5	3.06	2	7	5
Using cryptocurrencies will be clear and understandable for me	5.13	5	3.00	3	7.25	4.25
It will be easy for me to use cryptocurrencies	3.96	4	2.86	2	6	4
It will be easy for me to become an expert in the use of cryptocurrencies	2.82	2	2.62	0	5	5
Social Influence						
The people who are important to me will think that I should use cryptocurrencies	3.03	3	2.68	0	5	5
he people who influence me will think that I should use cryptocurrencies	3.04	3	2.72	0	5	5
People whose opinions I value would like me to use cryptocurrencies	4.91	5	3.19	2	7	5
Facilitating Conditions						
I have the necessary resources to use cryptocurrencies	3.95	3.5	3.25	1	7	6
I have the necessary knowledge to use cryptocurrencies	5.69	6	3.00	4	8	4
Cryptocurrencies are compatible with other technologies that I use	5.29	5	3.09	3	8	5
I can get help if I have difficulty using cryptocurrencies	6.93	7	2.72	5	9	4
Perceived Risk						
Using cryptocurrencies is risky	7.66	8	2.55	7	10	3
There is too much uncertainty associated with the use of cryptocurrencies	7.12	8	2.72	5	10	5
Compared with other currencies/investments, cryptocurrencies are riskier	5.07	5	2.90	3	7	4
Financial Literacy						
I have a good level of financial knowledge	5.45	6	2.76	3	8	5
I have a high capacity to deal with financial matters	3.70	3	3.04	1	6	5

Note: μ stands for mean Q1, Q2 and Q3 for the 1st, 2nd and 3rd quartile and σ for the standard deviation.

Table A2. Standarized loadings and *t*-values.

Construct/Item	Loading (*t*-Value)
Intention to Use	
I intend to use cryptocurrencies	0.90 (52.16)
I predict that I will use cryptocurrencies	0.91 (48.22)
Performance Expectancy	
Using cryptocurrencies will increase opportunities to achieve important goals for me	0.97 (69.60)
Using cryptocurrencies will help me achieve my goals more quickly	0.93 (69.35)
Using cryptocurrencies will increase my standard of living	0.92 (55.40)
Effort Expectancy	
It will be easy for me to learn how to use cryptocurrencies	0.89 (38.66)
Using cryptocurrencies will be clear and understandable for me	0.95 (58.16)
It will be easy for me to use cryptocurrencies	0.94 (62.97)
It will be easy for me to become an expert in the use of cryptocurrencies	0.94 (49.45)
Social Influence	
The people who are important to me will think that I should use cryptocurrencies	0.91 (43.21)
The people who influence me will think that I should use cryptocurrencies	0.93 (48.28)
People whose opinions I value would like me to use cryptocurrencies	0.99 (70.56)
Facilitating Conditions	
I have the necessary resources to use cryptocurrencies	0.79 (23.27)
I have the necessary knowledge to use cryptocurrencies	0.88 (32.08)
Cryptocurrencies are compatible with other technologies that I use	0.78 (21.61)
I can get help if I have difficulty using cryptocurrencies	0.77 (20.66)
Perceived Risk	
Using cryptocurrencies is risky	0.90 (6.30)
There is too much uncertainty associated with the use of cryptocurrencies	0.65 (5.98)
Compared with other currencies/investments, cryptocurrencies are riskier	0.87 (6.65)
Financial Literacy	
I have a good level of financial knowledge	1.00 (62.58)
I have a high capacity to deal with financial matters	0.92 (33.56)

Source: [7].

Table A3. Construct reliability, Cronbach alpha and convergent reliability.

Construct	Composite Reliability	Cronbach's Alpha	AVE
Intention to Use (IU)	0.898	0.897	0.814
Performance Expectancy (PE)	0.960	0.960	0.889
Effort Expectancy (EE)	0.962	0.962	0.864
Social Influence (SI)	0.959	0.959	0.887
Facilitating Conditions (FC)	0.878	0.878	0.645
Perceived Risk (PR)	0.850	0.851	0.658
Financial Literacy (FL)	0.956	0.955	0.916

Source: [7].

Table A4. Divergent validity.

Construct	IU	PE	EE	SI	FC	PR	FL
Intention to Use (IU)	**0.902**	0.896	0.640	0.680	0.674	0.120	0.282
Performance Expectancy (PE)	0.896	**0.943**	0.557	0.739	0.565	0.137	0.237
Effort Expectancy (EE)	0.640	0.557	**0.930**	0.493	0.767	0.088	0.450
Social Influence (SI)	0.680	0.739	0.494	**0.942**	0.566	0.089	0.239
Facilitating Conditions (FC)	0.673	0.565	0.767	0.563	**0.803**	0.094	0.489
Perceived Risk (PR)	−0.123	−0.137	−0.090	−0.084	0.047	**0.817**	0.284
Financial Literacy (FL)	0.282	0.237	0.450	0.239	0.493	0.286	**0.957**

Source: [7].

Table A5. PLS estimates in [7] for the hypothesis in Section 2.

	R2	Q2	Direct Effect	p-Value	Correlation	Variance Explained
Intention to Use (IU)	0.848	0.654				
Performance Expectancy (PE)			0.764	0.000	0.896	68.45%
Effort Expectancy (EE)			0.078	0.070	0.640	4.99%
Social Influence (SI)			−0.041	0.244	0.680	−2.79%
Facilitating Conditions (FC)			0.220	0.000	0.673	14.81%
Perceived Risk (PR)			−0.017	0.278	−0.123	0.21%
Financial Literacy (FL)			−0.028	0.226	0.282	−0.79%

Source: [7].

References

1. Nakamoto, S. *Bitcoin: A Peer-to-Peer Electronic Cash System*; Manubot: Montreal, QC, Canada, 2019; Available online: https://git.dhimmel.com/bitcoin-whitepaper (accessed on 23 December 2020).
2. World Bank Group. *Distributed Ledger Technology (DLT) and Blockchain*; World Bank Group: Washington, DC, USA, 2017. [CrossRef]
3. Glaser, F.; Zimmermann, K.; Haferkorn, M.; Webber, M.C.; Siering, M. Bitcoin-asset or Currency? Revealing Users' Hidden Intentions. ECIS. Available online: https://papers.ssrn.com/sol3/papers.cfm?abstract_id=2425247 (accessed on 9 December 2020).
4. Hileman, G.; Rauchs, M. Global Blockchain Benchmarking Study. *SSRN Electron. J.* **2017**. [CrossRef]
5. CoinmarketCap. Available online: https://coinmarketcap.com/2020 (accessed on 30 December 2020).
6. Holub, M.; Johnson, J. Bitcoin research across disciplines. *Inf. Soc.* **2018**, *34*, 114–126. [CrossRef]
7. Arias-Oliva, M.; Pelegrín-Borondo, J.; Matías-Clavero, G. Variables Influencing Cryptocurrency Use: A Technology Acceptance Model in Spain. *Front. Psychol.* **2019**, *10*, 475. [CrossRef]
8. Kim, J.; Gim, G. A study on factors affecting the intention to accept blockchain technology. *J. Inf. Technol. Serv.* **2017**, *16*, 1–20.
9. Shahzad, F.; Xiu, G.; Wang, J.; Shahbaz, M. An empirical investigation on the adoption of cryptocurrencies among the people of mainland China. *Technol. Soc.* **2018**, *55*, 33–44. [CrossRef]
10. Walton, A.J.; Johnston, K.A. Exploring Perceptions of Bitcoin Adoption: The South African Virtual Community Perspective. *Interdiscip. J. Inf. Knowl. Manag.* **2018**, *13*, 165–182. [CrossRef]
11. Mazumbari, L.; Mutambara, E. Predicting FinTech innovation adoption in South Africa: The case of cryptocurrency. *Afr. J. Econ. Manag. Stud.* **2019**, *11*, 30–50.
12. Fujiki, H. Who adopts crypto assets in Japan? Evidence from the 2019 financial literacy survey. *J. Jpn. Int. Econ.* **2020**, *58*, 101107. [CrossRef]
13. Mendoza-Tello, J.C.; Mora, H.; Pujol-López, F.A.; Lytras, M.D. Social commerce as a driver to enhance trust and intention to use cryptocurrencies for electronic payments. *IEEE Access* **2018**, *6*, 50737–50751. [CrossRef]
14. Mendoza-Tello, J.C.; Mora, H.; Pujol-López, F.A.; Lytras, M.D. Disruptive innovation of cryptocurrencies in consumer acceptance and trust. *Inf. Syst. E Bus. Manag.* **2019**, *17*, 195–222. [CrossRef]
15. Nuryyev, G.; Wang, Y.-P.; Achyldurdyyeva, J.; Jaw, B.-S.; Yeh, Y.-S.; Lin, H.-T.; Wu, L.-F. Blockchain Technology Adoption Behavior and Sustainability of the Business in Tourism and Hospitality SMEs: An Empirical Study. *Sustainability* **2020**, *12*, 1256. [CrossRef]
16. Sheel, A.; Nath, V. Blockchain technology adoption in the supply chain (UTAUT2 with risk)–evidence from Indian supply chains. *Int. J. Appl. Manag. Sci.* **2020**, *12*, 324–346. [CrossRef]
17. Yoo, Y.H.; Park, H.S. A Study on User's Acceptance of Blockchain-based Copyright Distribution Platforms and Its Usage. *J. Ind. Distrib. Bus.* **2019**, *10*, 59–72.
18. Albayati, H.; Kim, S.K.; Rho, J.J. Accepting financial transactions using blockchain technology and cryptocurrency: A customer perspective approach. *Technol. Soc.* **2020**, *62*, 101320. [CrossRef]
19. Gupta, S.; Gupta, S.; Mathew, M.; Sama, H.R. Prioritizing intentions behind investment in cryptocurrency: A fuzzy analytical framework. *J. Econ. Stud.* **2020**. ahead of print. [CrossRef]
20. Gil-Cordero, E.; Cabrera-Sánchez, J.P.; Arrás-Cortés, M.J. Cryptocurrencies as a Financial Tool: Acceptance Factors. *Mathematics* **2020**, *8*, 1974. [CrossRef]
21. Roussou, I.; Stiakakis, E.; and Sifaleras, A. An empirical study on the commercial adoption of digital currencies. *Inf. Syst. E Bus. Manag.* **2019**, *17*, 223–259. [CrossRef]
22. Schaupp, L.C.; Festa, M. Cryptocurrency Adoption and the Road to Regulation. In *Proceedings of the 19th Annual International Conference on Digital Government Research: Governance in the Data Age*; Association for Computing Machinery: New York, NY, USA, 2018; pp. 1–9.
23. Ragin, C.C. Using qualitative comparative analysis to study causal complexity. *Health Serv. Res.* **1999**, *34*, 1225–1239.
24. Ragin, C. *Redesigning Social Inquiry: Fuzzy Sets and Beyond*; Chicago University Press: Chicago, IL, USA, 2008.
25. Kraus, S.; Ribeiro-Soriano, D.; Schüssler, M. Fuzzy-set qualitative comparative analysis (fsQCA) in entrepreneurship and innovation research–the rise of a method. *Int. Entrep. Manag. J.* **2018**, *14*, 15–33. [CrossRef]

26. Veríssimo, J.M.C. Enablers and restrictors of mobile banking app use: A fuzzy set qualitative comparative analysis (fsQCA). *J. Bus. Res.* **2016**, *69*, 5456–5460. [CrossRef]
27. Jenson, I.; Leith, P.; Doyle, R.; West, J.; Miles, M.P. Testing innovation systems theory using Qualitative Comparative Analysis. *J. Bus. Res.* **2016**, *69*, 1283–1287. [CrossRef]
28. Leischnig, A.; Henneberg, S.C.; Thornton, S.C. Net versus combinatory effects of firm and industry antecedents of sales growth. *J. Bus. Res.* **2016**, *69*, 3576–3583. [CrossRef]
29. Kaya, B.; Abubakar, A.M.; Behravesh, E.; Yildiz, H.; Mert, I.S. Antecedents of innovative performance: Findings from PLS-SEM and fuzzy sets (fsQCA). *J. Bus. Res.* **2020**, *114*, 278–289. [CrossRef]
30. Venkatesh, V.; Morris, M.G.; Davis, G.B.; Davis, F.D. User Acceptance of Information Technology: Toward a Unified View. *Mis Q.* **2003**, *27*, 425. [CrossRef]
31. Venkatesh, V.; Thong, J.Y.L.; Xu, X. Consumer Acceptance and Use of Information Technology: Extending the Unified Theory of Acceptance and Use of Technology. *Mis Q.* **2012**, *36*, 157–178. [CrossRef]
32. Davis, F.D. Perceived Usefulness, Perceived Ease of Use, and User Acceptance of Information Technology. *Mis Q.* **1989**, *13*, 319–340. [CrossRef]
33. Venkatesh, V.; Davis, F.D. A Theoretical Extension of the Technology Acceptance Model: Four Longitudinal Field Studies. *Manag. Sci.* **2000**, *46*, 186–204. [CrossRef]
34. Fishbein, M.; Ajzen, I. *Belief, Attitude, Intention, and Behavior: An. Introduction to Theory and Research*; Adison-Wesley: Boston, MA, USA, 1975. [CrossRef]
35. Ajzen, I. The theory of planned behavior. *Organ. Behav. Hum. Decis. Process.* **1991**, *50*, 179–211. [CrossRef]
36. Board of Governors of the Federal Reserve System Home Page. Available online: http://www.federalreserve.gov (accessed on 18 June 2013).
37. Deloitte State-Sponsored Cryptocurrency: Adapting the Best of Bitcoin's Innovation to the Payments Ecosystem. Available online: https://www2.deloitte.com/content/dam/Deloitte/au/Documents/financial-services/deloitte-au-fs-state-sponsored--cryptocurrency-180516.pdf (accessed on 2 February 2018).
38. Bort, J. Is Bitcoin Pizza Day Thanks to These Two Pizzas Worth $5 Million Today. Business Insider 22 May. Available online: https://www.businessinsider.es/2014 (accessed on 29 November 2019).
39. Coinmap. 2020. Available online: https://coinmap.org/welcome/ (accessed on 3 January 2021).
40. Borondo, J.P.; Oliva, M.A.; Menorca, L.G.; Ayensa, E.J. Pricing policies in hotels: A psychological threshold research in online and offline channels. *Int. J. Internet Mark. Advert.* **2015**, *9*, 161. [CrossRef]
41. Liu, Y.; Tsyvinski, A. *Risks and Returns of Cryptocurrency*; No. w24877; National Bureau of Economic Research: Cambridge, MA, USA, 2018.
42. Bloomberg, J. Using Bitcoin or Other Cryptocurrency to Commit Crimes? Law Enforcement is onto You. Forbes. Bussines Wire 2017 $16.3 Billion Global Blockchain Technology Market Analysis & Trends —Industry Forecast to 2025—Research and Markets | Business Wire. Available online: https://www.businesswire.com/news/home/20170130005684/en/16.3-Bil-lion-Global-Blockchain-Technology-Market-Analysis (accessed on 13 December 2020).
43. Kim, S.Y.; Lee, S.H.; Chi, Y.D.; Im, E.T.; Gim, G.Y. A study on the factors affecting the intention to use biometrics in payment services. *Int. J. Bank Mark.* **2018**, *36*, 170–183. [CrossRef]
44. Makanyeza, C.; Mutambayashata, S. Consumers' acceptance and use of plastic money in Harare, Zimbabwe: Application of the unified theory of acceptance and use of technology. *Int. J. Bank Mark.* **2018**, *36*, 379–392. [CrossRef]
45. Torres, J.A.S.; Arroyo-Cañada, F.-J.; Sandoval, A.V.; Sánchez-Alzate, J.A. E-banking in Colombia: Factors favouring its acceptance, online trust and government support. *Int. J. Bank Mark.* **2018**, *36*, 170–183. [CrossRef]
46. Khan, I.U.; Hameed, Z.; Khan, S.U. Understanding Online Banking Adoption in a Developing Country. *J. Glob. Inf. Manag.* **2017**, *25*, 43–65. [CrossRef]
47. Nisha, N. Exploring the Dimensions of Mobile Banking Service Quality. *Int. J. Bus. Anal.* **2016**, *3*, 60–76. [CrossRef]
48. Kishore, S.V.K.; Sequeira, A.H. An Empirical Investigation on Mobile Banking Service Adoption in Rural Karnataka. *Sage Open* **2016**, *6*, 62158244016633731. [CrossRef]
49. Farah, M.F.; Hasni, M.J.S.; Abbas, A.K. Mobile-banking adoption: Empirical evidence from the banking sector in Pakistan. *Int. J. Bank Mark.* **2018**, *36*, 1386–1413. [CrossRef]
50. Warsame, M.H.; Ireri, E.M. Moderation effect on mobile microfinance services in Kenya: An extended UTAUT model. *J. Behav. Exp. Financ.* **2018**, *18*, 67–75. [CrossRef]
51. Hussain, M.; Mollik, A.T.; Johns, R.; Rahman, M.S. M-payment adoption for bottom of pyramid segment: An empirical investigation. *Int. J. Bank Mark.* **2019**, *37*, 362–381. [CrossRef]
52. Baur, A.W.; Bühler, J.; Bick, M.; Bonorden, C.S. Cryptocurrencies as a Disruption? Empirical Findings on User Adoption and Future Potential of Bitcoin and Co. In *Proceedings of the Mining Data for Financial Applications*; Springer: Berlin/Heidelberg, Germany, 2015; pp. 63–80.
53. Krombholz, K.; Judmayer, A.; Gusenbauer, M.; Weippl, E.; Grosslags, J.; Preneel, B. The Other Side of the Coin: User Experiences with Bitcoin Security and Privacy. In *Mining Data for Financial Applications*; Springer: Berlin/Heidelberg, Germany, 2017; Volume 9603, pp. 555–580.

54. Gao, X.; Clark, G.D.; Lindqvist, J. Of Two Minds, Multiple Addresses, and One Ledger. In Proceedings of the 2016 CHI Conference on Human Factors in Computing Systems, San Jose, CA, USA, 7–12 May 2016; pp. 1656–1668.
55. Moon, Y.; Hwang, J. Crowdfunding as an Alternative Means for Funding Sustainable Appropriate Technology: Acceptance Determinants of Backers. *Sustainability* **2018**, *10*, 1456. [CrossRef]
56. Mahfuz, M.A.; Khanam, L.; Mutharasu, S.A. The influence of website quality on m-banking services adoption in Bangladesh: Applying the UTAUT2 model using PLS. In Proceedings of the 2016 International Conference on Electrical, Electronics and Optimization Techniques (ICEEOT), Chennai, India, 3–5 March 2016; pp. 2329–2335.
57. Hajli, N. The role of social support on relationship quality and social commerce. *Technol. Forecast. Soc. Chang.* **2014**, *87*, 17–27. [CrossRef]
58. Yermack, D. Is Bitcoin a real currency? An economic appraisal. In *Handbook of Digital Currency*; Academic Press: New York, NY, USA, 2015; pp. 31–43.
59. Beikverdi, A.; Song, J. Trend of centralization in Bitcoin's distributed network. In Proceedings of the En 2015 IEEE/ACIS 16th International Con-ference on Software Engineering, Artificial Intelligence, Networking and Parallel/Distributed Computing (SNPD), Takamatsu, Japan, 1–3 June 2015; pp. 1–6.
60. Faqih, K.M. An empirical analysis of factors predicting the behavioral intention to adopt Internet shopping technology among non-shoppers in a developing country context: Does gender matter? *J. Retail. Consum. Serv.* **2016**, *30*, 140–164. [CrossRef]
61. Salisbury, W.D.; Pearson, R.A.; Pearson, A.W.; Miller, D.W. Perceived security and World Wide Web purchase intention. *Ind. Manag. Data Syst.* **2001**, *101*, 165–177. [CrossRef]
62. Kannungo, S.; Jain, V. Relationship between risk and intention to purchase in an online context: Role of gender and product category. In Proceedings of the 13th European Conference on Information Systems, The European IS Profession in the Global Networking Environment 2004, ECIS 2004, Turku, Finland, 14–16 June 2004.
63. Featherman, M.S.; Pavlou, P.A. Predicting e-services adoption: A perceived risk facets perspective. *Int. J. Hum. Comput. Stud.* **2003**, *59*, 451–474. [CrossRef]
64. Ahmed, W.M. Is there a risk-return trade-off in cryptocurrency markets? The case of Bitcoin. *J. Econ. Bus.* **2020**, *108*, 105886. [CrossRef]
65. Dwyer, G.P. The economics of Bitcoin and similar private digital currencies. *J. Financ. Stab.* **2015**, *17*, 81–91. [CrossRef]
66. Shaikh, A.A.; Glavee-Geo, R.; Karjaluoto, H. How Relevant Are Risk Perceptions, Effort, and Performance Expectancy in Mobile Banking Adoption? *Int. J. E Bus. Res.* **2018**, *14*, 39–60. [CrossRef]
67. Stolper, O.A.; Walter, A. Financial literacy, financial advice, and financial behavior. *J. Bus. Econ.* **2017**, *87*, 581–643. [CrossRef]
68. Van Rooij, M.; Lusardi, A.; Alessie, R. Financial literacy and stock market participation. *J. Financ. Econ.* **2011**, *101*, 449–472. [CrossRef]
69. Lusardi, A.; Mitchell, O.S. The Economic Importance of Financial Literacy: Theory and Evidence. *J. Econ. Lit.* **2014**, *52*, 5–44. [CrossRef] [PubMed]
70. Hastings, J.S.; Madrian, B.C.; Skimmyhorn, W.L. Financial Literacy, Financial Education, and Economic Outcomes. *Annu. Rev. Econ.* **2013**, *5*, 347–373. [CrossRef] [PubMed]
71. Ragin, C. *User's Guide to Fuzzy-Set/Qualitative Comparative Analysis 3*. Department of Sociology; University of California: Irvine, CA, USA, 2018.
72. McCluskey, E.J. Minimization of Boolean functions. *Bell Syst. Tech. J.* **1956**, *35*, 1417–1444. [CrossRef]
73. Thiem, A. Set-relational fit and the formulation of transformational rules in fsQCA. *Compasss Wp. Ser.* 2010, 2010, p. 61. Available online: http://www.compasss.org/wpseries/Thiem2010.pdf (accessed on 15 December 2020).
74. Legewie, N. An introduction to applied data analysis with qualitative comparative analysis. *Forum Qual. Sozi Alforschung/Forum Qual. Soc. Res.* **2013**, *14*, 3. [CrossRef]
75. Thiem, A. Beyond the Facts: Limited Empirical Diversity and Causal Inference in Qualitative Comparative Analysis. *Sociol. Methods Res.* **2019**. [CrossRef]
76. Krause, M.J.; Tolaymat, T. Quantification of energy and carbon costs for mining cryptocurrencies. *Nat. Sustain.* **2018**, *1*, 711–718. [CrossRef]
77. De Andrés-Sánchez, J.; Belzunegui-Eraso, A.; Valls-Fonayet, F. Assessing Efficiency of Public Poverty Policies in UE-28 with Linguistic Variables and Fuzzy Correlation Measures. *Mathematics* **2021**, *9*, 128. [CrossRef]
78. Ban, A.I.; Ban, O.I.; Tuse, D.A. Derived fuzzy importance of attributes based on the weakest triangular norm-based fuzzy arithmetic and applications to the hotel services. *Iran. J. Fuzzy Syst.* **2016**, *13*, 65–85.

Article

Fuzzy Markovian Bonus-Malus Systems in Non-Life Insurance

Pablo J. Villacorta [1], Laura González-Vila Puchades [2],* and Jorge de Andrés-Sánchez [3]

[1] Department of Computer Science and Artificial Intelligence, Higher Technical School of Computer and Telecommunication Engineering, University of Granada, Cuesta del Hospicio s/n, 18071 Granada, Spain; pjvi@decsai.ugr.es

[2] Department of Mathematics for Economics, Finance and Actuarial Science, Faculty of Economics and Business, University of Barcelona, Avinguda Diagonal 690, 08034 Barcelona, Spain

[3] Social and Business Research Laboratory, Campus Bellisens, Rovira i Virgili University, Avinguda de la Universitat 1, 43204 Reus, Spain; jorge.deandres@urv.cat

* Correspondence: lgonzalezv@ub.edu; Tel.: +34-934020415

Citation: Villacorta, P.J.; González-Vila Puchades, L.; de Andrés-Sánchez, J. Fuzzy Markovian Bonus-Malus Systems in Non-Life Insurance. *Mathematics* **2021**, *9*, 347. https://doi.org/10.3390/math9040347

Academic Editor: Michael Voskoglou

Received: 24 December 2020
Accepted: 3 February 2021
Published: 9 February 2021

Publisher's Note: MDPI stays neutral with regard to jurisdictional claims in published maps and institutional affiliations.

Copyright: © 2021 by the authors. Licensee MDPI, Basel, Switzerland. This article is an open access article distributed under the terms and conditions of the Creative Commons Attribution (CC BY) license (https://creativecommons.org/licenses/by/4.0/).

Abstract: Markov chains (MCs) are widely used to model a great deal of financial and actuarial problems. Likewise, they are also used in many other fields ranging from economics, management, agricultural sciences, engineering or informatics to medicine. This paper focuses on the use of MCs for the design of non-life bonus-malus systems (BMSs). It proposes quantifying the uncertainty of transition probabilities in BMSs by using fuzzy numbers (FNs). To do so, Fuzzy MCs (FMCs) as defined by Buckley and Eslami in 2002 are used, thus giving rise to the concept of Fuzzy BMSs (FBMSs). More concretely, we describe in detail the common BMS where the number of claims follows a Poisson distribution under the hypothesis that its characteristic parameter is not a real but a triangular FN (TFN). Moreover, we reflect on how to fit that parameter by using several fuzzy data analysis tools and discuss the goodness of triangular approximates to fuzzy transition probabilities, the fuzzy stationary state, and the fuzzy mean asymptotic premium. The use of FMCs in a BMS allows obtaining not only point estimates of all these variables, but also a structured set of their possible values whose reliability is given by means of a possibility measure. Although our analysis is circumscribed to non-life insurance, all of its findings can easily be extended to any of the abovementioned fields with slight modifications.

Keywords: bonus-malus system; fuzzy number; fuzzy transition probability; fuzzy Markov chain; fuzzy stationary state

1. Introduction

1.1. Motivation

A bonus-malus system (BMS) is a common method for posteriori ratemaking in non-life insurance. It is based on partitioning the insurer's portfolio into a finite number of classes: bonus and malus classes. A typical case is automobile third-party liability insurance [1]. In a BMS, policyholders do not have a fixed price for their contracts throughout periods (e.g., the mathematical expectation of claims value per period). Their membership into a concrete BMS class is reviewed each period according to the number of claims in the previous one. Claim-free years are rewarded by discounts or bonuses on a base-premium; at-fault accidents are penalized by surcharges called maluses. Some overviews on how BMS are applied in different countries can be found in [2–4].

Following [5], in most commercial BMSs, by knowing the insured's class in the current period and fitting the statistical distribution for the number of claims per period, it is possible to determine the probabilities of the insured's class in the next period. Therefore, these BMSs are Markovian. For that reason, the academic literature on BMSs uses extensively MCs for their modeling ([1,5–11]). Therefore, a key question in a BMS is fitting the value of the one-step transition probability matrix. Following [12,13], if full knowledge of the probabilities of this matrix is not available, they have to be estimated somehow with the uncertainty that any estimation procedure involves. Uncertainty may be due to randomness, hazard, vagueness, incomplete information, etc. In our paper, we consider that the claiming process is probabilistic, but the uncertainty about the parameter that governs this random behavior is captured by means of a fuzzy number (FN) and, as a consequence, fuzzy Markov chains (FMCs) will be used.

1.2. Novelties

Although other hypotheses can be taken, such as, for example, considering that the number of claims in a period, N, follows a negative binomial distribution ([14]), academic literature on BMSs usually assumes that N is a Poisson random variable (RV). So, $N \sim Po(\lambda)$ and the parameter λ, the claim frequency, is perfectly known and can be interpreted as a risk measure of the policy. However, in a more realistic approach, some authors like [15] use intervals to quantify the uncertainty about the parameter of a distribution function that governs a risk variable. This is also the case within a BMS framework of [11], who model the uncertainty about λ by means of a modal interval. One extended way to combine randomness and uncertainty of parameters of distribution functions consists in modeling these parameters as FNs. It has been done both for continuous RVs ([16–18]), and in the discrete case ([19]). Following this approach, [20,21] and also [15] model risk financial parameters with FNs. In the actuarial field, FNs have been used to capture the uncertainty of insurance pricing variables ([22]) but also to model parameters that quantify risks. In this regard, we can point out [23] in a non-life insurance context, [24] to interpret the parameter that quantifies the dependence in a Farlie-Gumbel-Morgestein copula, and [25–28] in life insurance pricing. Since any interval can be seen as the α-cut of a FN, even in the case of improper intervals ([29]), in this work we consider that λ is fitted by means of a FN and, more particularly, by a triangular fuzzy number (TFN). So, this paper builds up a framework to model Markovian BMSs that embed the standard case, where the risk parameter λ is crisp, but also the method developed in [11] that quantifies this parameter as a modal interval. Standard BMSs provide point values for the stationary state and the mean asymptotic premium. Modal BMSs as introduced in [11] allow obtaining these variables as modal intervals whose lower and upper bounds may be understood as pessimistic/optimistic scenarios. Our method generalizes both types of BMSs since it quantifies variables related to BMS as FNs. On the one hand, these FNs can be understood as a set of crisp outcomes with an associated possibility measure. On the other, these FNs can be interpreted as a set of intervals that come from pessimistic/optimistic scenarios and are structured by means of possibility levels. Figure 1 shows a graphical synthesis of the methodological framework developed in this paper.

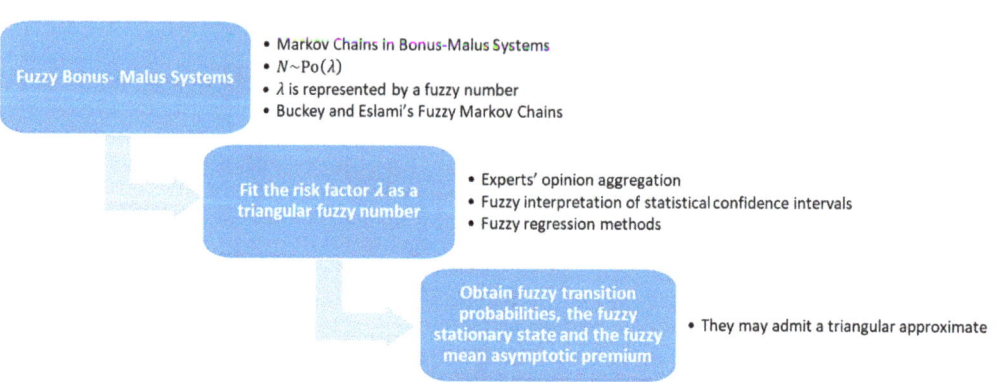

Figure 1. Graphical representation of our fuzzy bonus-malus systems (FBMS) model.

Other more complex forms of FNs, such as generalized FNs (GFNs) or intuitionistic FNs (IFNs), could be considered to quantify uncertain probabilities. Tools like GFNs or IFNs provide a more complete capture of uncertainty than FNs. However, their adjustment has a greater cost than in the case of triangular FNs since they incorporate more parameters and their computational handle may be more expensive as well. Therefore, using TFNs supposes a balance between the simplicity of crisp or modal interval probabilities and more complex representations of uncertain quantities such as GFNs or IFNs.

It should be noted that there are several scientific fields in which MCs are in use. In the field of economics and finance, we can observe applications within Leontief's input-output model, credit risk measurement, asset price volatility modeling, life insurance, etc. In addition, MCs have shown their usefulness in many other areas: industrial engineering (e.g., queuing theory), computer science (e.g., computer performance evaluation and web search engines), healthcare (e.g., pandemics transmission or evolution of ICU patients), etc. Hence, although our developments are carried out within a non-life insurance context, most of the results can be applied to any problem modeled by means of MCs when the transition probabilities (or the parameters that define them) are not precisely known.

The paper is organized as follows. Section 2 describes briefly how BMSs work. Section 3 shows the basic concepts of FNs and FMCs used throughout the paper. In Section 4, a methodologic approach is proposed to fit a fuzzy BMS (FBMS) when the number of claims within a period, N, follows a Poisson distribution with fuzzy parameter λ. This methodology is applied to the Irish BMS. A sensitivity analysis is conducted in Section 5. Finally, in Section 6, the work ends with a summary of its main contributions and potential extensions.

2. Markovian Bonus-Malus Systems in Non-Life Insurance

A BMS is a usual way to deal with risk aversion and moral hazard in some types of insurance, e.g., automobile third-party liability insurance [1]. BMSs classify insureds in r classes in such a way that the percentage of the base-premium to be paid by the ith class, b_i, satisfies $b_i < b_{i+1}, i = 1, 2, \ldots, r-1$. In a BMS, the transition between classes is governed by a set of rules defined over the insured's number of claims in the current period. To summarize, it can be said that every BMS is determined by three elements (see Table 1):

- The initial class, where new insureds are assigned, i_0.
- The premium scale $b = (b_1, \ldots, b_r)$.
- The transition rules, that is to say, the rules that define the conditions for an insured in one class to be transferred to another class in the next period.

Table 1. Elements of a bonus-malus system (BMS).

Class	Premium Level	Class after k Claims			
i	b_i	$k=0$	$k=1$	$\ldots$	$k \geq n$
r	b_r				
$\vdots$	$\vdots$		Transition rules		
i_0	b_{i_0}				
$\vdots$	$\vdots$				
1	b_1				

Source: Own elaboration based on [10].

Let us model the insured's class at time t as a discrete stochastic process $(X_t)_{t \in \mathbb{N}}$, being its state space the classes $S = \{1, 2, \ldots, r\} \subset \mathbb{N}$. Furthermore, as it is usually done in the literature (e.g., [10]), we consider that the BMS is a finite MC, i.e., $p(X_{n+1} = i_{n+1} | X_0 = i_0, X_1 = i_1, \ldots, X_n = i_n) = p(X_{n+1} = i_{n+1} | X_n = i_n)$, $\forall i_0, i_1, \ldots, i_{n+1} \in S$. An insurer uses a finite Markovian BMS when the following conditions hold [1]:

- There exists a finite number of classes such that each insured stays in one class through each period.
- The premium for each insured depends only on the class where they stay.
- The class for a given period is determined by the class in the preceding period and the number of claims reported in that period.

A finite MC is said to be homogeneous if $p(X_{m+h} = j | X_m = i)$ does not depend on m. In this case, transition probabilities $p_{ij} = p(X_{t+1} = j | X_t = i)$, i.e., probabilities of moving from class i to class j in one-step (period), can be collected in a transition matrix $P = (p_{ij})$ with order r. The elements p_{ij} satisfy $p_{ij} \geq 0$ and $\sum_{j=1}^{r} p_{ij} = 1$.

If $\left(p_i^{(0)}\right)_{i \in S}$ denote the probabilities of initially being in state $i \in S$, the probabilities of being in state $i \in S$ after n periods, $\left(p_i^{(n)}\right)_{i \in S}$ are:

$$p_i^{(n)} = p_i^{(n-1)} \cdot P = p_i^{(n-2)} \cdot P \cdot P = \ldots = p_i^{(0)} \cdot P^n \quad (1)$$

where P^n is represented by $P^{(n)} = \left(p_{ij}^{(n)}\right)$ and $p_{ij}^{(n)}$ are the probabilities of moving from state i to state j in n period. From Equation (1), it follows that:

$$p_{ij}^{(n)} = f_{ij}^{(n)}(p_{11}, \ldots, p_{1r}, p_{21}, \ldots, p_{2r}, \ldots, p_{1r}, \ldots, p_{rr}) \quad (2)$$

for some continuous functions $f_{ij}^{(n)}$, i.e., the elements in P^n are some functions of the elements in P.

A homogeneous MC is regular if each state is accessible from any other state, either in one step or more, i.e., there exists $n \in \mathbb{N}$ such that $p_{ij}^{(n)} > 0 \; \forall \, i, j$. One of the features that characterize regular MCs is its stationary distribution, which represents the probability of the chain being at each state after a large number of periods, namely, $\lim_{n \to \infty} P^n = \pi$, where the rows of π are identical. So, any regular MC with transition matrix P has a stationary distribution, π, such that:

$$\pi = \pi \cdot P \quad (3)$$

The vector $\pi = (\pi_j)_{j=1,2,\ldots,r}$ can be interpreted as the probability that an insured belongs to class j, $j = 1, 2, \ldots, r$ after n periods, $n \to \infty$. That vector does not depend on the insured's initial class, i_0. So, two main outputs in a BMS are:

- The stationary distribution of X_t, π, as defined in Equation (3).
- The mean asymptotic premium, b^*, i.e., the average premium paid by the insured in that stationary distribution, defined as:

$$b^* = \sum_{j=1}^{r} b_j \pi_j \quad (4)$$

The mean asymptotic premium, b^* is a concept of the utmost importance because it has been intensively used to assess the efficiency of a BMS (e.g., [1,6,30,31]).

BMSs consider the number of claims, N, as a discrete RV. In our paper, as it is commonplace in actuarial literature, N is supposed to follow a Poisson distribution with parameter λ, $N \sim Po(\lambda)$, ([1,7–12]). Therefore:

$$p(N = k) = \frac{\lambda^k}{k!} e^{-\lambda} \quad (5)$$

where $p(\cdot)$ stands for a probability measure.

Poisson RVs are often used in actuarial modeling due to their interesting arithmetical properties. Furthermore, the risk parameter λ can be fitted specifically to each insured taking into account relevant rating factors (e.g., gender, age, and social status) by using a generalised linear model (GLM) ([5,8,32]).

If N is modeled with (5), p_{ij} is a function of the risk parameter λ, $h_{ij}(\lambda)$, in such a way that the BMS probabilities in the one-step transition matrix are:

$$h_{ij}(\lambda) = p_{ij} = \sum_{k=0}^{\infty} \frac{\lambda^k}{k!} e^{-\lambda} t_{ij}(k) \qquad (6)$$

where $t_{ij}(k) = 1$ if k causes the transition from i to j and 0 otherwise.

Numerical application 1. Let $\lambda = 0.04$ in Equation (5) for the Irish BMS in Table 2. The transition matrix, P, that corresponds to this BMS is:

$$\begin{pmatrix} p(N=0) & 0 & p(N=1) & 0 & 0 & p(N\geq 2) \\ p(N=0) & 0 & 0 & p(N=1) & 0 & p(N\geq 2) \\ 0 & p(N=0) & 0 & 0 & p(N=1) & p(N\geq 2) \\ 0 & 0 & p(N=0) & 0 & 0 & p(N\geq 1) \\ 0 & 0 & 0 & p(N=0) & 0 & p(N\geq 1) \\ 0 & 0 & 0 & 0 & p(N=0) & p(N\geq 1) \end{pmatrix} \qquad (7)$$

Table 2. Irish bonus-malus system.

Class	Premium Level	Class after k Claims		
i	b_i	$k=0$	$k=1$	$k \geq 2$
6	100	5	6	6
5	90	4	6	6
4	80	3	6	6
3	70	2	5	6
2	60	1	4	6
1	50	1	3	6

Source: [10].

From (3), $\pi = (0.916232, 0.037394, 0.038921, 0.003861, 0.002523, 0.001069)$ and, by considering Equation (4), the mean asymptotic premium is $b^* = 51.423$.

In this paper, we will consider that the risk parameter λ cannot be determined precisely. Uncertainty may be the result of different causes: stochastic variability, inaccuracy, incomplete information, etc. Stochastic variability can be described by using RVs or stochastic processes, but inaccuracy and incomplete information can be captured by means of intervals or FNs. Given that any interval can be interpreted as the α-cut of a FN, in this work it is assumed that λ is a FN.

3. Fuzzy Numbers and Fuzzy Markov Chains

3.1. Fuzzy Numbers

A fuzzy number is a fuzzy set $\tilde{A}$ on the referential set $\mathbb{R}$ that satisfies (i) $\tilde{A}$ is normal, (ii) $\tilde{A}$ is convex, and (iii) the α-cuts of $\tilde{A}$, $A(\alpha) = [\underline{A}(\alpha), \overline{A}(\alpha)]$, are closed and bounded (compact) intervals $\forall \alpha \in [0,1]$. The lower and upper bounds of the FN $A(\alpha)$ are:

$$A(\alpha) = [\underline{A}(\alpha), \overline{A}(\alpha)] = \left[\inf_{x \in \mathbb{R}} \{\mu_{\tilde{A}}(x) \geq \alpha\}, \sup_{x \in \mathbb{R}} \{\mu_{\tilde{A}}(x) \geq \alpha\} \right] \qquad (8)$$

FNs can be interpreted as the extension of the concept of a real number.

A triangular fuzzy number represented as $\tilde{A} = (A_L/A_C/A_U)$, is a FN whose α-cuts, $\forall \alpha \in [0,1]$, are, from Equation (8):

$$A(\alpha) = [\underline{A}(\alpha), \overline{A}(\alpha)] = [A_L + (A_C - A_L)\alpha, A_U - (A_U - A_C)\alpha] \tag{9}$$

from where, if needed, the membership of $\tilde{A}$ could be obtained. The core of $\tilde{A}$ is A_C and can be understood as the most reliable value of this TFN. i.e., the possibility of A_C is 1. The support of $\tilde{A}$ is $[A_L, A_U]$. TFNs are used in countless practical applications including actuarial ones [22] because they are easy to handle arithmetically and they are well adapted to the way humans think of uncertain quantities. Moreover, when the information about a variable is vague and imprecise, the parsimony principle leads us to represent that information as simply as possible. The linear shape of TFNs meets that requirement. For instance, the uncertain quantity "approximately 0.04" can be represented in a very natural way as the TFN $(0.038/0.04/0.042)$.

Likewise, let it be a TFN $\tilde{A}$:

$$\tilde{A} > x \text{ (or } \geq x) \text{ if } A_L > x \, (A_L \geq x) \, \forall x \in \mathbb{R} \tag{10}$$

$$\tilde{A} < x \text{ (or } \leq x) \text{ if } A_U < x \, (A_L \leq x) \, \forall x \in \mathbb{R} \tag{11}$$

Let f be a continuous real-valued function of n-real variables $x = (x_1, x_2, \ldots, x_n)$, $f(x) = f(x_1, x_2, \ldots, x_n)$. If x_js are not crisp numbers, but FNs $\tilde{A}_j$ with α-cuts $A_j(\alpha) = [\underline{A}_j(\alpha), \overline{A}_j(\alpha)]$, $j = 1, 2, \ldots, n$, a FN $\tilde{B}$ is induced via f such that $\tilde{B} = f(\tilde{A}_1, \tilde{A}_2, \ldots, \tilde{A}_n)$. It is often difficult to obtain a closed expression for the membership function of $\tilde{B}$. However, following [33], the α-cuts of $\tilde{B}$, $B(\alpha)$, in the usual case where $A_1(\alpha), \ldots, A_n(\alpha)$ are not interactive, i.e., the variables that they quantify have an independent behavior, can be obtained as:

$$B(\alpha) = \{f(x) | x = (x_1, x_2, \ldots, x_n) \in Dom(\alpha)\} \tag{12}$$

where $Dom(\alpha)$ stands for the rectangular domain:

$$Dom(\alpha) = \left\{ x_j \in [\underline{A}_j(\alpha), \overline{A}_j(\alpha)], j = 1, 2, \ldots, n \right\} \tag{13}$$

So, the lower (upper) bounds of $B(\alpha)$, $\underline{B}(\alpha)$ ($\overline{B}(\alpha)$), are the global minimum (maximum) of $f(x)$ within the rectangular domain in Equation (13), that is to say:

$$\underline{B}(\alpha) = \min_{j,k}\{f(V_j), f(E_k)\} \text{ and } \overline{B}(\alpha) = \max_{j,k}\{f(V_j), f(E_k)\} \tag{14}$$

being V_j, $j = 1, 2, \ldots, 2^n$, a vertex of the domain (13) and $f(F_k)$, $k = 1, 2, \ldots, K$, an extreme value of the function f within this domain that takes this value at point $x = E_k$.

Therefore, if $f(x_1, x_2, \ldots, x_n)$ is monotonic, the lower and upper bounds of $B(\alpha)$, $\underline{B}(\alpha)$ and $\overline{B}(\alpha)$, are in one of the 2^n vertexes of (13). Without loss of generality, let us suppose that f increases with respect to x_j, $j = 1, 2, \ldots, m$, $m \leq n$, and decreases in the last $n - m$ variables, [34] demonstrates that:

$$(B(\alpha) = [\underline{B}(\alpha), \overline{B}(\alpha)] = [f(\underline{A}_1(\alpha), \underline{A}_2(\alpha), \ldots, \underline{A}_m(\alpha), \overline{A}_{m+1}(\alpha), \overline{A}_{m+2}(\alpha), \ldots \overline{A}_n(\alpha)), \tag{15}$$
$$f(\overline{A}_1(\alpha), \overline{A}_2(\alpha), \ldots \overline{A}_m(\alpha), \underline{A}_{m+1}(\alpha), \underline{A}_{m+2}(\alpha), \ldots, \underline{A}_n(\alpha))]$$

If $A_1(\alpha), \ldots, A_n(\alpha)$ are interactive, (15) cannot be used to evaluate $B(\alpha)$. However, according to [35], the general formulation to obtain the lower and upper bounds of $B(\alpha)$ from Equations (12)–(14) is still valid but now the number of vertexes, j, is less than 2^n. In [35], the authors study the role of interactive fuzzy variables in decision-making problems and analyze some particular cases. Concretely, when $f(x)$ is the mathematical expectation function, x_j is the probability of the jth outcome, and it is quantified as a FN, the domain in Equation (13) turns into:

$$Dom(\alpha) = \left\{ x_j \in \left[\underline{A}_j(\alpha), \overline{A}_j(\alpha)\right], \sum_{j=1}^n x_j = 1 \right\} \quad (16)$$

and Equation (14) becomes:

$$\underline{B}(\alpha) = \min_j \{f(V_j)\} \text{ and } \overline{B}(\alpha) = \max_j \{f(V_j)\} \quad (17)$$

due to the fact that $f(x)$ is a linear function.

It is worth noting that the result $\widetilde{B}$ of evaluating a non-linear f with the TFNs $\widetilde{A}_j$ is not necessarily a TFN. However, often $\widetilde{B}$ admits a good triangular approximation through the secant approach. It builds up the shape of the triangular approximate FN $\widetilde{B}'$ to $\widetilde{B}$ by means of the secant lines that unite the 0-cut and the 1-cut of $\widetilde{B}$. Such that $\widetilde{B}'$ is a TFN as:

$$\widetilde{B} \approx \widetilde{B}' = (B_L/B_C/B_U) = \left(\underline{B}(0)/\underline{B}(1) = \overline{B}(1)/\overline{B}(0)\right) \quad (18)$$

This approximation, as shown in [36], works pretty well for nonlinear monotonic functions of TFNs such as product, division, power, etc. Likewise, [37,38] show that this approach fits satisfactorily common actuarial and financial calculations with TFN parameters, e.g., the present value of a stream of fuzzy cash-flows. Keeping the triangular shape of the initial data when handling FNs is quite interesting. According to [39], complex shapes of FNs can generate problems with calculations in computer work or interpreting results intuitively. [40] state that a triangular approximate is a kind of defuzzification that is richer than just transforming a FN into a crisp representative value. If defuzzification is carried out too early, a great loss of information occurs, so it is preferable to drag all the fuzzy information in the calculations for as long as possible. The triangular approximation involves a compromise between simplification in computation and interpretation, and not oversimplifying the value of fuzzy parameters. In addition, TFNs have a very intuitive interpretation and, therefore, from the insurance industry point of view, a triangular approximate to actuarial variables and parameters could be very useful in decision-making processes.

3.2. Fuzzy Markov Chains

Fuzzy set literature has provided three approaches to MC under fuzziness, namely FMC. The first one, due to [41], supposes fuzzy probabilities and proposes calculating the matrices P^n, $\forall n > 1$, by applying Zadeh's extension principle [42]. The second approach, in [43], consists of defining the matrix that governs the transition between states by means of a fuzzy relation. In the third one, [2], like [41], suppose that the probabilities of the one-step transition matrix are FNs. However, Buckley and Eslami's framework of FMCs uses restricted matrix multiplication to operate with probabilities in such a way that the constraint of being a well-formed probability distribution always holds. That is to say, they take into account the interdependence between the probabilities of a distribution function, similarly to Equations (16) and (17). This paper follows this last approach.

Let us assume that some probabilities p_{ij} in the one-step transition matrix $P = (p_{ij})$ are uncertain and are quantified by means of the FNs $\widetilde{p}_{ij}$, with α-cuts $p_{ij}(\alpha)$. Now we have a fuzzy transition matrix $\widetilde{P} = (\widetilde{p}_{ij})$, with $0 \leq \widetilde{p}_{ij} \leq 1$. See Equations (10) and (11). Of course, some elements in $\widetilde{P}$ may be crisp since crisp numbers are a particular case of a FN. FMCs defined by [2] have uncertainty in the transition probabilities but not in the set of outcomes, that is discrete. So, the following constraint on $\widetilde{p}_{ij}$ is added: $\sum_{j=1}^r p_{ij}(1) = 1$.

To compute the n-period transition matrix, $\widetilde{P}^{(n)}$, and the fuzzy stationary distribution, $\widetilde{\pi}$, the following process is implemented:

Step 1. For a given $\alpha \in [0,1]$, obtain the matrix of intervals $P(\alpha) = (p_{ij}(\alpha)) = \left[\underline{p_{ij}(\alpha)}, \overline{p_{ij}(\alpha)}\right]$.

Step 2. Define the domain of a row i of this matrix, $Dom_i(\alpha)$, as:

$$Dom_i(\alpha) = \left(\times_{j=1}^{r} p_{ij}(\alpha)\right) \cap D \qquad (19)$$

where $\times$ stands for the cartesian product and $D = \left\{(x_1, x_2, \ldots, x_r) \left| \sum_{j=1}^{r} p_{ij} = 1, p_{ij} \geq 0 \right. \right\}$.

Step 3. Define the domain of the matrix $P(\alpha)$, for the given $\alpha \in [0,1]$, as:

$$Dom(\alpha) = \left(\times_{j=1}^{r} Dom_i(\alpha)\right) \qquad (20)$$

This domain defines a set of matrices that satisfy that each row sums up 1 with a possibility level of at least α. So, each matrix $P = (p_{ij})$, $p_{ij} \in Dom(\alpha)$, is a crisp MC.

Step 4. Since $Dom_i(\alpha)$ in Equation (19) are compact sets, $Dom(\alpha)$ in Equation (20) is also compact. So, any continuous function applied to its elements has a compact image. Then, if Equation (2) is applied, such an image for that α is a compact interval. This interval is set as the α-cut of the FN $\widetilde{p}_{ij}^{(n)}$, i.e., $p_{ij}^{(n)}(\alpha) = f_{ij}^{(n)}(Dom(\alpha))$, which is surely normal [2].

To determine $p_{ij}^{(n)}(\alpha)$ we must find the lower and upper bounds, $\underline{p_{ij}^{(n)}}(\alpha)$ and $\overline{p_{ij}^{(n)}}(\alpha)$ by solving:

$$\underline{p_{ij}^{(n)}}(\alpha) = \min\left\{f_{ij}^{(n)}(p) \middle| p \in Dom(\alpha)\right\} \qquad (21)$$

and

$$\overline{p_{ij}^{(n)}}(\alpha) = \max\left\{f_{ij}^{(n)}(p) \middle| p \in Dom(\alpha)\right\} \qquad (22)$$

Notice that Equations (21) and (22) can be easily solved in low-dimensional problems. However, in more complex problems it is necessary to use an algorithm (see, e.g., [44]) or a heuristic constrained optimization technique [45].

Finally, by performing Steps 1–4 $\forall \alpha \in [0,1]$, the FNs $\widetilde{p}_{ij}^{(n)}$ can be obtained.

In regards to the fuzzy stationary state, $(\widetilde{\pi}_1, \ldots, \widetilde{\pi}_r)$, its α-cuts $\pi_j(\alpha) = \left[\underline{\pi_j(\alpha)}, \overline{\pi_j(\alpha)}\right]$ can be determined from Equation (3) as:

$$\underline{\pi_j}(\alpha) = \min\{w_j | w = wP, (p_{11}, \ldots, p_{rr}) \in Dom(\alpha)\} \qquad (23)$$

$$\overline{\pi_j}(\alpha) = \max\{w_j | w = wP, (p_{11}, \ldots, p_{rr}) \in Dom(\alpha)\} \qquad (24)$$

4. Implementing a Markovian Fuzzy Bonus-Malus System Governed by a Fuzzy Poisson Discrete Random Variable

In this Section, we propose an integral methodology to develop a FBMS under the hypothesis that $N \sim Po(\widetilde{\lambda})$. It embeds the fitting of the risk parameter λ as a TFN, the obtaining of fuzzy transition matrix and the triangular approximate of the stationary distribution calculated by using Equations (23) and (24), and also the determination of the fuzzy mean asymptotic premium in Equation (4).

Step 1. Fit the risk factor λ as a TFN.

We point out three different options to estimate this parameter.

Option 1

Given that λ has an intuitive interpretation since it is the mean number of claims in one period, it may be quantified as a FN based on experts' opinions. For example, an expert may judge that a concrete type of driver generates approximately one claim every 5 years and so the TFN $\widetilde{\lambda} = \widetilde{0.2} = (0.18/0.2/0.22)$ can be considered. Imprecise or subjective quantitative predictions can often come from a pool of experts, leading to a set of fuzzy quantifications. This set of fuzzy opinions can be aggregated simply by their arithmetic mean or other more sophisticated methods (see [46–48] for full details).

Option 2

Papers [49,50] consider a standard $1 - \alpha$ statistical confidence interval as the observed α-cut of the FN, for some increasing values of $\varepsilon \leq \alpha < 1$, where ε is an arbitrary value near 0 (it is often chosen to be 0.001, 0.005 or 0.01). In [50] it is suggested that by placing those confidence intervals one on top of the other, a FN close to triangular-shaped is obtained. So, we point out two alternatives to apply that idea:

(a) Given that λ is the mean value of a Poisson RV, the interval estimates of λ can be used as the α-cuts of $\widetilde{\lambda}$. Let us denote as N^* the mean number of claims in a pool of similar contracts, S_N the standard deviation of N and n the number of policies in the pool. The $1 - \alpha$ statistical confidence interval for the mean number of claims is:

$$\left[N^* - t_{(1-\frac{\alpha}{2}, n-1)} \frac{S_N}{\sqrt{n}}, N^* + t_{(1-\frac{\alpha}{2}, n-1)} \frac{S_N}{\sqrt{n}} \right] \quad (25)$$

where $t_{(\frac{\alpha}{2}, n-1)}$ stands for the $(1 - \frac{\alpha}{2})$-percentile of a Student t with $n - 1$ degrees of freedom and S_N the standard deviation of the sample. So, $\widetilde{\lambda}$ can be fitted through its α-cuts by doing, from Equation (25):

$$\lambda(\alpha) = [\underline{\lambda}(\alpha), \overline{\lambda}(\alpha)] = \left[N^* - t_{(1-\frac{\alpha}{2}, n-1)} \frac{S_N}{\sqrt{n}}, N^* + t_{(1-\frac{\alpha}{2}, n-1)} \frac{S_N}{\sqrt{n}} \right] \quad (26)$$

(b) Papers [49,51] propose making fuzzy predictions from statistical linear regression models. In [49] it is stated that a $1 - \alpha$ statistical confidence interval of coefficients adjusted with a linear regression may be interpreted as the α-cut of a FN for these coefficients. Therefore, let us suppose that a GLM estimate of λ is determined, as usual, by:

$$\ln \lambda = \sum_{i=1}^{m} a_i x_i \quad (27)$$

being a_i, $i = 1, \ldots, m$, the coefficients and x_i the explanatory variables (e.g., age, gender, and driving experience in a car insurance context) that are crisp non-negative observations (in fact, they are usually modeled as dichotomic variables). For the estimate of each coefficient, it is possible to generate a FN $\widetilde{a}_i$ whose α-cuts, $a_i(\alpha)$, $i = 1, \ldots, m$, are:

$$a_i(\alpha) = [\underline{a_i}(\alpha), \overline{a_i}(\alpha)] = \left[a_i^* - t_{(\frac{\alpha}{2}, n-m-1)} S_{a_i}, a_i^* + t_{(\frac{\alpha}{2}, n-m-1)} S_{a_i} \right] \quad (28)$$

where a_i^* is the GLM point estimate of a_i and S_{a_i} the standard deviation of that estimate. So, from the fuzzy function $\widetilde{\lambda} = e^{\sum_{i=1}^{m} \widetilde{a}_i x_i}$, and bearing in mind Equations (15) and (28), the following FN $\widetilde{\lambda}$ is induced:

$$\lambda(\alpha) = [\underline{\lambda}(\alpha), \overline{\lambda}(\alpha)] = \left[e^{\sum_{i=1}^{m} \underline{a_i}(\alpha) x_i}, e^{\sum_{i=1}^{m} \overline{a_i}(\alpha) x_i} \right] \quad (29)$$

A similar approach may be developed from the results in [51]. However, in this case, it must be taken into account that their approach to making fuzzy predictions from a statistical regression is built up from the interval predictions of residuals instead of using interval estimates of coefficients. So $\widetilde{\lambda} = e^{\sum_{i=1}^{m} a_i x_i + \widetilde{\varepsilon}}$ where $\widetilde{\varepsilon}$ is a fuzzy error term induced from the residuals of the conventional regression and, so:

$$\lambda(\alpha) = [\underline{\lambda}(\alpha), \overline{\lambda}(\alpha)] = \left[e^{\sum_{i=1}^{m} a_i x_i + \underline{\varepsilon}(\alpha)}, e^{\sum_{i=1}^{m} a_i x_i + \overline{\varepsilon}(\alpha)} \right] \qquad (30)$$

Equations (26), (29) and (30) do not give a TFN. However, $\widetilde{\lambda}$ can be approximated as a TFN simply by using Equation (18).

Option 3

Fuzzy Regression Methods (FRMs) have been applied in several actuarial issues to fit relevant variables [52] for a comprehensive description of application areas). In this way, [53] fits the term structure of interest rates, [54,55] predicts claim provisions, and [25,28] adjusts the Lee-Carter mortality law.

To fit $\widetilde{\lambda}$, the fuzzy extension of the log-Poisson regression by [55] may be used. It combines the conventional Poisson GLM and the minimum fuzziness principle by [56]. In this case, the coefficients in Equation (27) are supposed to be TFNs $\widetilde{a}_i = (a_{iL}/a_{iC}/a_{iU})$, $i = 1, \ldots, m$. These coefficients are fitted in two stages. At the first stage, the centres a_{iC} are adjusted as in a conventional log-Poisson regression for $i = 1, \ldots, m$. At the second stage, the spreads of $\widetilde{a}_i$, $a_{iC} - a_{iL}$ and $a_{iU} - a_{iC}$ and, consequently, a_{iL} and a_{iU}, $i = 1, \ldots, m$, are fitted by solving a quadratic programming problem that minimizes the fuzziness of the system allowing that estimates on the dependent variable contain its observed values.

Once the parameters $\widetilde{a}_i$ have been estimated, obtaining $\widetilde{\lambda} = e^{\sum_{i=1}^{m} \widetilde{a}_i x_i}$ is straightforward (see Equation (29)).

Let us remark again that although $\widetilde{a}_i$ is a TFN, $\widetilde{\lambda}$ is not. Nevertheless, $\widetilde{\lambda}$ can be approximated as a TFN $\widetilde{\lambda}' = (\lambda_L/\lambda_C/\lambda_U)$ with Equation (18):

$$\lambda_L = e^{\sum_{i=1}^{m} a_{iL} x_i}, \lambda_C = e^{\sum_{i=1}^{m} a_{iC} x_i} \text{ and } \lambda_U = e^{\sum_{i=1}^{m} a_{iU} x_i} \qquad (31)$$

Step 2. Obtain the fuzzy transition matrix.

We now suppose that after performing any of the options in Step 1, and the corresponding triangular approximate, the risk parameter is given as the TFN $\widetilde{\lambda} = (\lambda_L/\lambda_C/\lambda_U)$, with α-cuts, $\forall \alpha \in [0,1]$, $\lambda(\alpha) = [\underline{\lambda}(\alpha), \overline{\lambda}(\alpha)] = [\lambda_L + (\lambda_C - \lambda_L)\alpha, \lambda_U - (\lambda_U - \lambda_C)\alpha]$.

Fuzzy transitions probabilities come from the fuzzified version of Equation (6):

$$h_{ij}(\widetilde{\lambda}) = \widetilde{p}_{ij} = \sum_{k=0}^{\infty} \frac{\widetilde{\lambda}^k}{k!} e^{-\widetilde{\lambda}} t_{ij}(k) \qquad (32)$$

To obtain the α-cuts of $\widetilde{p}_{ij}$ by using Equation (15), it is necessary to determine the sign of the first derivative of $h_{ij}(\lambda)$. Let us show the case of the Irish BMS whose transition matrix is Expression (7), and p_{ij} is either zero, $p(N=0)$, $p(N=1)$, $p(N \geq 1)$ or $p(N \geq 2)$. Then:

$$p(N = 0) = e^{-\lambda} \qquad (33)$$

$$p(N = 1) = \lambda e^{-\lambda} \qquad (34)$$

$$p(N \geq 1) = 1 - e^{-\lambda} \qquad (35)$$

$$p(N \geq 2) = 1 - (1 + \lambda)e^{-\lambda} \qquad (36)$$

So, in Equations (33)–(36), $\frac{\partial p(N=0)}{\partial \lambda} < 0$, $\frac{\partial p(N=1)}{\partial \lambda} > 0$, $\frac{\partial p(N \geq 1)}{\partial \lambda} > 0$, $\frac{\partial p(N \geq 2)}{\partial \lambda} > 0$ and therefore:

$$p(N=0)(\alpha) = \left[e^{-\overline{\lambda}(\alpha)}, e^{-\underline{\lambda}(\alpha)} \right] \qquad (37)$$

$$p(N=1)(\alpha) = \left[\underline{\lambda}(\alpha)e^{-\underline{\lambda}(\alpha)}, \overline{\lambda}(\alpha)e^{-\overline{\lambda}(\alpha)}\right] \quad (38)$$

$$p(N \geq 1)(\alpha) = \left[1 - e^{-\underline{\lambda}(\alpha)}, 1 - e^{-\overline{\lambda}(\alpha)}\right] \quad (39)$$

$$p(N \geq 2)(\alpha) = \left[1 - (1+\underline{\lambda}(\alpha))e^{-\underline{\lambda}(\alpha)}, 1 - (1+\overline{\lambda}(\alpha))e^{-\overline{\lambda}(\alpha)}\right] \quad (40)$$

Similarly, any other probabilities for different FBMSs could be calculated. Notice that the FNs whose α-cuts are Equations (37)–(40) do not have a triangular shape but they admit a triangular approximation by using the secant approach described in Section 3.1. If this is done, we obtain:

$$\widetilde{p}(N=0) \approx \left(e^{-\lambda_U}/e^{-\lambda_C}/e^{-\lambda_L}\right) \quad (41)$$

$$\widetilde{p}(N=1) \approx \left(\lambda_L e^{-\lambda_L}/\lambda_C e^{-\lambda_C}/\lambda_U e^{-\lambda_U}\right) \quad (42)$$

$$\widetilde{p}(N \geq 1) \approx \left(1 - e^{-\lambda_L}/1 - e^{-\lambda_C}/1 - e^{-\lambda_U}\right) \quad (43)$$

$$\widetilde{p}(N \geq 2) \approx \left(1 - (1+\lambda_L)e^{-\lambda_L}/1 - (1+\lambda_C)e^{-\lambda_C}/1 - (1+\lambda_U)e^{-\lambda_U}\right) \quad (44)$$

Numerical Application 2. Example 3 in [11] (p. 846) considers $\lambda = [0.038, 0.042]$ in Equation (6), and obtains the modal interval version of this crisp transition matrix:

$$P = \begin{pmatrix} p(N \geq 1) & p(N=0) & 0 \\ p(N \geq 1) & 0 & p(N=0) \\ p(N \geq 1) & 0 & p(N=0) \end{pmatrix}$$

Let us suppose that this interval is the 0-cut of the fuzzy estimate of a triangular $\widetilde{\lambda}$ in a Poisson FBMS (i.e., $\lambda(0) = [0.038, 0.042]$) and $\lambda(1) = 0.04$, that is to say, $\widetilde{\lambda} = (0.038/0.04/0.042)$. By considering Expression (32) and using Equations (41)–(44), the fuzzy transition matrix, $\widetilde{P}$, which corresponds to a FMC, is:

$$\widetilde{P} = \begin{pmatrix} (0.037287/0.039211/0.041130) & (0.958870/0.960789/0.962713) & 0 \\ (0.037287/0.039211/0.041130) & 0 & (0.958870/0.960789/0.962713) \\ (0.037287/0.039211/0.041130) & 0 & (0.958870/0.960789/0.962713) \end{pmatrix}.$$

From this matrix, elements different from 0 in the associated matrix $P(\alpha) = (p_{ij}(\alpha))$ are, from Equation (9):

$$p(N=0)(\alpha) = p_{12}(\alpha) = p_{23}(\alpha) = p_{33}(\alpha) = [0.958870 + 0.001919\alpha, 0.962713 - 0.001924\alpha]$$

$$p(N \geq 1)(\alpha) = p_{11}(\alpha) = p_{21}(\alpha) = p_{31}(\alpha) = [0.037287 + 0.001924\alpha, 0.041130 - 0.001919\alpha].$$

Numerical Application 3. In example 4 of [11] (p. 848), it is considered the risk factor $\lambda = [0.038, 0.042]$ for an Irish BMS. Like in our numerical application above, again, this interval is the 0-cut of the triangular fuzzy estimate for $\widetilde{\lambda}$ and $\lambda(1) = 0.04$, i.e., $\widetilde{\lambda} = (0.038/0.04/0.042)$. So, the triangular approximates by Equations (41)–(44) to the probabilities of the transition matrix in Expression (7) and induced by Equation (32) are:

$$\widetilde{p}_{11} = \widetilde{p}_{21} = \widetilde{p}_{32} = \widetilde{p}_{43} = \widetilde{p}_{54} = \widetilde{p}_{65} = \widetilde{p}(N=0) \approx (0.958870/0.960789/0.962713)$$

$$\widetilde{p}_{13} = \widetilde{p}_{24} = \widetilde{p}_{35} = \widetilde{p}(N=1) \approx (0.036583/0.038432/0.040273)$$

$$\widetilde{p}_{46} = \widetilde{p}_{56} = \widetilde{p}_{66} = \widetilde{p}(N \geq 1) \approx (0.037287/0.039211/0.041130)$$

$$\widetilde{p}_{16} = \widetilde{p}_{26} = \widetilde{p}_{36} = \widetilde{p}(N \geq 2) \approx (0.000704/0.000779/0.000858)$$

$$p_{ij} = 0 \text{ otherwise}$$

Let us remark that approximates in Equations (41)–(44) produce small errors of the real values by Equations (37)–(40). Table 3 shows that when approximating $\widetilde{p}(N \geq 1)$ with

Equation (43), the errors incurred on the lower and upper bounds of its α-cuts are negligible since they are never over 0.001%. Moreover, notice that we measure the performance of the calculations on a scale of eleven grades of possibility. Following [38], this scale provides sufficient discernment without being excessive since we are using imprecise data and, therefore, more precision is not necessary for a FN representation.

Table 3. α-cuts of $\widetilde{p}(N \geq 1)$, it's triangular approximate, $\widetilde{p}'(N \geq 1)$, and errors.

	$p(N \geq 1)(\alpha)$		$p'(N \geq 1)(\alpha)$		Error *	
α	$\underline{p(N \geq 1)}(\alpha)$	$\overline{p(N \geq 1)}(\alpha)$	$\underline{p'(N \geq 1)}(\alpha)$	$\overline{p'(N \geq 1)}(\alpha)$	$\underline{err}(\alpha)$	$\overline{err}(\alpha)$
1	0.03921	0.03921	0.03921	0.03921	0.000%	0.000%
0.9	0.03902	0.03940	0.03902	0.03940	0.000%	0.000%
0.8	0.03883	0.03959	0.03883	0.03959	0.001%	0.001%
0.7	0.03863	0.03979	0.03863	0.03979	0.001%	0.001%
0.6	0.03844	0.03998	0.03844	0.03998	0.001%	0.001%
0.5	0.03825	0.04017	0.03825	0.04017	0.001%	0.001%
0.4	0.03806	0.04036	0.03806	0.04036	0.001%	0.001%
0.3	0.03786	0.04055	0.03786	0.04055	0.001%	0.001%
0.2	0.03767	0.04075	0.03767	0.04075	0.001%	0.001%
0.1	0.03748	0.04094	0.03748	0.04094	0.000%	0.000%
0	0.03729	0.04113	0.03729	0.04113	0.000%	0.000%

Source: Own elaboration. * $\underline{err}(\alpha) = \frac{|\underline{p(N \geq 1)}(\alpha) - \underline{p'(N \geq 1)}(\alpha)|}{\underline{p(N \geq 1)}(\alpha)}$ and $\overline{err}(\alpha) = \frac{|\overline{p(N \geq 1)}(\alpha) - \overline{p'(N \geq 1)}(\alpha)|}{\overline{p(N \geq 1)}'(\alpha)}$.

Step 3. Determine the fuzzy stationary distribution function.

Once the fuzzy transition matrix associated with the FBMS has been obtained, to determine the fuzzy stationary state, Steps 1 to 4 in Section 3.2 should be applied and, therefore, optimization problems in Equations (23) and (24) must be solved. Notice that although the probabilities $\widetilde{\pi}_j$, $j = 1, \ldots, r$, are obtained by solving complex optimization problems, the results of the numerical applications 4 and 5, that have been obtained with the R package FuzzyStatProb by [13] (see Figures 1 and 2) suggest that its triangular approximate by using Equation (18), $\widetilde{\pi}'_j = (\pi_{jL}/\pi_{jC}/\pi_{jU}) = \left(\underline{\pi_j}(0)/\underline{\pi_j}(1) = \overline{\pi_j}(1)/\overline{\pi_j}(0)\right)$ provides a satisfactory fitting.

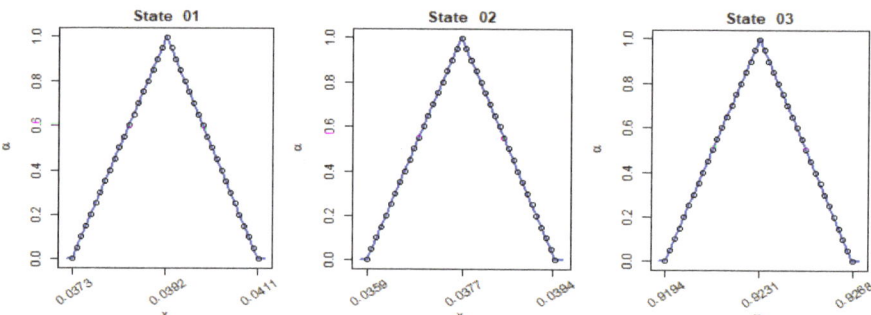

Figure 2. Results of Numerical Application 2. Source: Own elaboration.

Numerical Application 4. Now we compute the fuzzy stationary distribution $(\widetilde{\pi}_1, \widetilde{\pi}_2, \widetilde{\pi}_3)$ for the fuzzy transition matrix in numerical application 2. In order to do so, we use the R package FuzzyStatProb described in [13], which is based on the use of Equations (23) and (24). The pseudo-codes and codes used are included in Appendices A and B, respectively, and the result of $(\widetilde{\pi}_1, \widetilde{\pi}_2, \widetilde{\pi}_3)$ in Figure 2.

It should be remarked that the probabilities obtained by [11] (p. 847) are intervals whose values are the 0-cuts of the probabilities in our FMC.

Numerical Application 5. Let us consider the Irish BMS in numerical application 3. Table 4 shows the supports and cores of the fuzzy stationary state $\tilde{\pi}_j$, $j = 1, \ldots, 6$ when considering fuzzy probabilities in Equations (37)–(40). The pseudo-codes and the codes of the R package FuzzyStatProb that have been used to get these results are included in Appendices A and B, respectively. Figure 3 depicts the graphical representation of $\tilde{\pi}_j$, $j = 1, \ldots, 6$.

Table 4. Results of the Irish BMS when $\tilde{\lambda} = (0.038/0.04/0.042)$—supports and cores.

Stationary Probabilities	$\alpha = 0$	$\alpha = 1$
$\tilde{\pi}_1$	[0.912318, 0.920394]	0.916232
$\tilde{\pi}_2$	[0.035705, 0.039075]	0.037394
$\tilde{\pi}_3$	[0.037080, 0.040717]	0.038921
$\tilde{\pi}_4$	[0.003519, 0.004186]	0.003861
$\tilde{\pi}_5$	[0.002275, 0.002758]	0.002523
$\tilde{\pi}_6$	[0.000954, 0.001190]	0.001069

Source: Own elaboration.

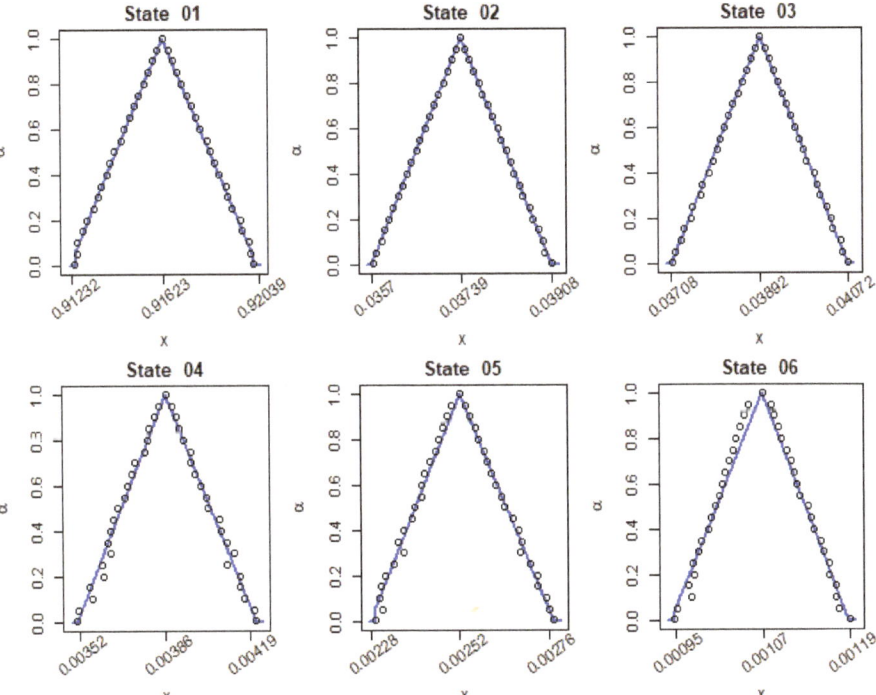

Figure 3. Results of the Irish BMS when $\tilde{\lambda} = (0.038/0.04/0.042)$—graphical representation. Source: Own elaboration.

The shape of the fuzzy stationary distribution suggests that their triangular approximate must work quite well. Table 5 shows that the relative deviations of the lower and upper bounds of the approximate of $\tilde{\pi}_4$, $\pi'_4(\alpha)$, with respect to the respective bounds of $\pi_4(\alpha)$ are always under 1%. Therefore, the above intuition is confirmed in the case of $\tilde{\pi}_4$. We have also observed that this fact also applies to the other stationary probabilities in the numerical application. So, it can be written:

$$\tilde{\pi}_1 \approx (0.912318/0.916232/0.920394).$$

$$\tilde{\pi}_2 \approx (0.035705/0.037394/0.039075).$$

$$\tilde{\pi}_3 \approx (0.037080/0.038921/0.040717).$$

$$\tilde{\pi}_4 \approx (0.003519/0.003861/0.004186).$$

$$\tilde{\pi}_5 \approx (0.002275/0.002523/0.002758).$$

$$\tilde{\pi}_6 \approx (0.000954/0.001069/0.001190).$$

Table 5. α-cuts of $\tilde{\pi}_4$, it's triangular approximate, $\tilde{\pi}'_4$, and errors.

	$\pi_4(\alpha)$		$\pi'_4(\alpha)$		Error *	
α	$\underline{\pi_4}(\alpha)$	$\overline{\pi_4}(\alpha)$	$\underline{\pi'_4}(\alpha)$	$\overline{\pi'_4}(\alpha)$	$\underline{err}(\alpha)$	$\overline{err}(\alpha)$
1	0.00386	0.00386	0.00386	0.00386	0.000%	0.000%
0.9	0.00381	0.00390	0.00383	0.00389	0.345%	0.188%
0.8	0.00379	0.00393	0.00379	0.00393	0.092%	0.082%
0.7	0.00375	0.00398	0.00376	0.00396	0.368%	0.405%
0.6	0.00371	0.00400	0.00372	0.00399	0.442%	0.145%
0.5	0.00368	0.00403	0.00369	0.00402	0.289%	0.194%
0.4	0.00364	0.00408	0.00366	0.00406	0.324%	0.474%
0.3	0.00360	0.00411	0.00362	0.00409	0.537%	0.484%
0.2	0.00356	0.00413	0.00359	0.00412	0.655%	0.272%
0.1	0.00354	0.00418	0.00355	0.00415	0.406%	0.550%
0	0.00352	0.00419	0.00352	0.00419	0.000%	0.000%

Source: Own elaboration. * $\underline{err}(\alpha) = \frac{|\underline{\pi_4}(\alpha) - \underline{\pi'_4}(\alpha)|}{\underline{\pi_4}(\alpha)}$ and $\overline{err}(\alpha) = \frac{|\overline{\pi_4}(\alpha) - \overline{\pi'_4}(\alpha)|}{\overline{\pi_4}(\alpha)}$.

It is worth pointing out that the intervals fitted for the stationary state in [11] (p. 848) are the 0-cuts of $\tilde{\pi}_j$, $j = 1, \ldots, 6$, in the fuzzy version of the Irish BMS.

Two considerations are worth highlighting:

- FBMSs generalize the results of crisp and modal interval BMSs as can be checked by comparing the results of numerical applications 1 and 5. The results of the crisp case are exactly the 1-cut of estimates from FBMSs whereas the estimates obtained by [11] (p. 848) coincide, except in the order of the interval lower and upper bounds in some cases, with the 0-cut of the results by our FBMSs. Likewise, operating by means of α-cuts allows obtaining the simulations of intermediate scenarios between that of maximum fuzziness (generated by the 0-cut of $\tilde{\lambda}$) and that with maximum reliability (that comes from $\lambda(1)$), as well as their grade of possibility. This information can be extremely useful to the decision-maker since it makes easier the sensitivity analysis for each possible value of Poisson parameter λ.

- Although a TFN $\tilde{\lambda}$ does not produce triangular probabilities $\tilde{p}_{ij}^{(n)}$ and $\tilde{\pi}_j$, their triangular approximates work pretty well. We consider this result interesting for two reasons:

  (a) The calculations can be done easily with less computational effort. For example, in the two first rows of Table 5, we have performed the calculations on a scale with eleven grades of possibility. So, 20 optimization programs have been solved for a single probability (10 minimizing programs for the lower bounds of α-cuts $\alpha = 0, 0.1, \ldots, 0.9$, and other 10 maximizing programs for the respective upper bounds). Likewise, obtaining the 1-cut implies nothing but solving a conventional Markov chain. This computational effort is reduced drastically by using the triangular approximate in Equation (18), which leads us to obtain the results in columns 3 and 4 of Table 5. In this case, it is enough to solve 2 optimization programs (1 minimizing program for the lower bound of the 0-cut and 1 maximizing for the upper one) and also, of course, evaluating a conventional BMS in λ_C. The interest in this result is amplified by the fact that the Irish BMS is relatively simple (there are 6 classes) and so it embeds only

36 $\widetilde{p}_{ij}^{(n)}$ and 6 probabilities $\widetilde{\pi}_j$. However, BMSs often have more than 20 classes (e.g., Belgian or German BMSs).

(b) From the perspective of an actuary, a triangular approximate of the fuzzy probabilities can be very useful. A TFN provides an estimate of the most feasible, minimum, and maximum probability that can be interpreted intuitively without any knowledge of fuzzy set theory (FST). Therefore, the triangular approximates presented in this paper could facilitate the use of FBMSs in the insurance industry.

Step 4. Obtain the mean asymptotic premium, b^*.

In order to obtain the asymptotic mean premium, we have to evaluate the fuzzy version of Equation (4), $\widetilde{b}^* = \sum_{j=1}^{r} b_j \widetilde{\pi}_j$. Bearing in mind Equations (12)–(14), (16) and (17), we first consider the domain:

$$Dom(\alpha) = \left\{ \pi_j \in \left[\underline{\pi}_j(\alpha), \overline{\pi}_j(\alpha)\right], \sum_{j=1}^{r} \pi_j = 1 \right\} \quad (45)$$

and then the set $\left\{ b^* = \sum_{j=1}^{r} b_j \pi_j \,\middle|\, \pi_j \in Dom(\alpha) \right\}$. The α-cuts of $\widetilde{b}^*$, $b^*(\alpha) = \left[\underline{b}^*(\alpha), \overline{b}^*(\alpha)\right]$, are obtained by solving:

$$\underline{b}^*(\alpha) = \min \left\{ b^* = \sum_{j=1}^{r} b_j \pi_j \,\middle|\, \pi_j \in \left[\underline{\pi}_j(\alpha), \overline{\pi}_j(\alpha)\right], \sum_{j=1}^{r} \pi_j = 1 \right\} \quad (46)$$

$$\overline{b}^*(\alpha) = \max \left\{ b^* = \sum_{j=1}^{r} b_j \pi_j \,\middle|\, \pi_j \in \left[\underline{\pi}_j(\alpha), \overline{\pi}_j(\alpha)\right], \sum_{j=1}^{r} \pi_j = 1 \right\} \quad (47)$$

which are linear programming problems and so solvable, e.g., with the simplex algorithm. In fact, the problem to solve in this case is the same as that in [35].

A triangular approximate for $\widetilde{b}^*$, $\widetilde{b}^{*'} = (b_L^*/b_C^*/b_U^*)$, can be obtained by using the 0-cut and the 1-cut obtained from Equations (46) and (47) or, alternatively, if these results have not been previously calculated, by considering the TFNs $(\pi_{jL}/\pi_{jC}/\pi_{jU})$, $j = 1, \ldots, r$, and solving the following linear problems:

$$b_L^* = \min \left\{ b^* = \sum_{j=1}^{r} b_j \pi_j \,\middle|\, \pi_j \in [\pi_{jL}, \pi_{jU}], \sum_{j=1}^{r} \pi_j = 1 \right\} \quad (48)$$

$$b_U^* = \max \left\{ b^* = \sum_{j=1}^{r} b_j \pi_j \,\middle|\, \pi_j \in [\pi_{jL}, \pi_{jU}], \sum_{j=1}^{r} \pi_j = 1 \right\} \quad (49)$$

In this latter way, 20 linear problems that come from Expressions (46) and (47) when $\widetilde{b}^*$ is performed with a scale of eleven grades of possibility are reduced to 2 linear programs. Moreover:

$$b_C^* = \sum_{j=1}^{r} b_j \pi_{jC} \quad (50)$$

Numerical Application 6. Let us consider again the Irish BMS in numerical application 3, i.e., $N \sim Po((0.038/0.04/0.042))$. By using the premium level of each class (see Table 2), we obtain the α-cuts of the fuzzy mean asymptotic premium, $\widetilde{b}^*$, from Equations (45)–(47). These results are in Table 6, which also show the α-cuts of its triangular approximate, $\widetilde{b}^{*'}$, by Equations (48)–(50). From that table, it can be seen that $\widetilde{b}^*$ is practically triangular since the errors by $b'(\alpha)$ in fitting $b(\alpha)$ are negligible. Notice that the triangular approximate $\widetilde{b}^{*'}$ provides a straightforward generalization of both the point estimate by a crisp BMS, 51.423, as well as the modal interval estimate in [11] (p. 849), [51.344, 51.498].

Table 6. α-cuts of $\widetilde{b}^*$, it's triangular approximate $\widetilde{b}^{*\prime}$ and its errors.

	$b^*(\alpha)$		$b^{*\prime}(\alpha)$		Error *	
α	$\underline{b}^*(\alpha)$	$\overline{b}^*(\alpha)$	$\underline{b}^{*\prime}(\alpha)$	$\overline{b}^{*\prime}(\alpha)$	$\underline{err}(\alpha)$	$\overline{err}(\alpha)$
1	51.423	51.423	51.423	51.423	0.000%	0.000%
0.9	51.415	51.430	51.415	51.430	0.000%	0.000%
0.8	51.407	51.438	51.407	51.438	0.000%	0.000%
0.7	51.399	51.445	51.399	51.445	0.000%	0.000%
0.6	51.391	51.453	51.391	51.453	0.000%	0.000%
0.5	51.383	51.460	51.383	51.460	0.000%	0.000%
0.4	51.375	51.468	51.375	51.468	0.000%	0.000%
0.3	51.367	51.475	51.367	51.475	0.000%	0.000%
0.2	51.359	51.483	51.359	51.483	0.000%	0.000%
0.1	51.352	51.491	51.352	51.491	0.000%	0.000%
0	51.344	51.498	51.344	51.498	0.000%	0.000%

Source: Own elaboration. * $\underline{err}(\alpha) = \frac{|\underline{b}^*(\alpha) - \underline{b}^{*\prime}(\alpha)|}{\underline{b}^*(\alpha)}$ and $\overline{err}(\alpha) = \frac{|\overline{b}^*(\alpha) - \overline{b}^{*\prime}(\alpha)|}{\overline{b}^*(\alpha)}$.

5. Sensitivity Analysis

In this Section, we evaluate the sensitivity of the errors of the triangular approximates seen in Section 4 with respect to the parameter λ. The following assumptions are considered:

- The core of $\widetilde{\lambda}$ may be low (0.04), medium (0.5), or high (0.96).
- The uncertainty of $\widetilde{\lambda}$, which can be measured by its spreads, is symmetrical, i.e., left and right spreads are equal. This uncertainty can take two possible values: 0.002 or 0.015.
- We use the Irish BMS in Table 2.

Only results for $\widetilde{p}(N \geq 1)$ and $\widetilde{\pi}_4$ are shown. Furthermore, in order to avoid very long calculations, we have performed them on a scale with five grades of possibility. However, it can be verified that for other transition and stationary probabilities, and for a greater scale of grades of possibility, the conclusions to be drawn are practically the same. Table 7 shows that:

- The goodness of triangular approximates is always better than acceptable as can be checked in Table 7. In the worst case, for $\widetilde{\pi}_4$, $\alpha = 0.5$ and risk parameter $\widetilde{\lambda} = 0.025/0.04/0.055$, errors are below 5%.
- Errors (as defined for Tables 3 and 5) increase with respect to the uncertainty of risk parameter and decrease with respect to the value of its core.

Table 8 shows the mean asymptotic premiums for the cores of $\widetilde{\lambda}$ considered in Table 7 and the most uncertain scenario (left and right spread equal to 0.015). It can be checked that triangular approximates $\widetilde{b}^{*\prime}$ always reach a practically perfect match to $\widetilde{b}^*$, i.e., errors (as defined for Table 6) are very close to 0.

Table 7. α-cuts of $\widetilde{p}(N \geq 1)$ and $\widetilde{\pi}_4$, their triangular approximate and errors for different parameters $\widetilde{\lambda}$.

	$\widetilde{\lambda} = (0.038/0.04/0.042)$							
	$\widetilde{p}'(N \geq 1) = (0.03729/0.03921/0.04113)$				$\widetilde{\pi}'_4 = (0.00352/0.00386/0.00419)$			
	$p(N \geq 1)(\alpha)$		Error		$\pi_4(\alpha)$		Error	
α	$\underline{p(N \geq 1)}(\alpha)$	$\overline{p(N \geq 1)}(\alpha)$	$\underline{err}(\alpha)$	$\overline{err}(\alpha)$	$\underline{\pi_4}(\alpha)$	$\overline{\pi_4}(\alpha)$	$\underline{err}(\alpha)$	$\overline{err}(\alpha)$
1	0.03921	0.03921	0.000%	0.000%	0.00386	0.00386	0.000%	0.000%
0.75	0.03873	0.03969	0.001%	0.001%	0.00376	0.00395	0.044%	0.041%
0.5	0.03825	0.04017	0.001%	0.001%	0.00367	0.00405	0.060%	0.054%
0.25	0.03777	0.04065	0.001%	0.001%	0.00358	0.00415	0.046%	0.039%
0	0.03729	0.04113	0.000%	0.000%	0.00352	0.00419	0.000%	0.000%
	$\widetilde{\lambda} = (0.025/0.04/0.055)$							
	$\widetilde{p}'(N \geq 1) = (0.02469/0.03921/0.05351)$				$\widetilde{\pi}'_4 = (0.00153/0.00386/0.00717)$			
	$p(N \geq 1)(\alpha)$		Error		$\pi_4(\alpha)$		Error	
α	$\underline{p(N \geq 1)}(\alpha)$	$\overline{p(N \geq 1)}(\alpha)$	$\underline{err}(\alpha)$	$\overline{err}(\alpha)$	$\underline{\pi_4}(\alpha)$	$\overline{\pi_4}(\alpha)$	$\underline{err}(\alpha)$	$\overline{err}(\alpha)$
1	0.03921	0.03921	0.000%	0.000%	0.00386	0.00386	0.000%	0.000%
0.75	0.03560	0.04281	0.057%	0.047%	0.00318	0.00460	2.985%	1.955%
0.5	0.03198	0.04639	0.085%	0.058%	0.00257	0.00540	4.958%	2.208%
0.25	0.02834	0.04996	0.072%	0.040%	0.00202	0.00626	4.757%	1.420%
0	0.02469	0.05351	0.000%	0.000%	0.00153	0.00717	0.000%	0.000%
	$\widetilde{\lambda} = (0.498/0.05/0.502)$							
	$\widetilde{p}'(N \geq 1) = (0.39226/0.39347/0.39468)$				$\widetilde{\pi}'_4 = (0.15617/0.15618/0.15618)$			
	$p(N \geq 1)(\alpha)$		Error		$\pi_4(\alpha)$		Error	
α	$\underline{p(N \geq 1)}(\alpha)$	$\overline{p(N \geq 1)}(\alpha)$	$\underline{err}(\alpha)$	$\overline{err}(\alpha)$	$\underline{\pi_4}(\alpha)$	$\overline{\pi_4}(\alpha)$	$\underline{err}(\alpha)$	$\overline{err}(\alpha)$
1	0.39347	0.39347	0.000%	0.000%	0.15618	0.15618	0.000%	0.000%
0.75	0.39317	0.39377	0.000%	0.000%	0.15617	0.15618	0.000%	0.000%
0.5	0.39286	0.39408	0.000%	0.000%	0.15617	0.15618	0.001%	0.001%
0.25	0.39256	0.39438	0.000%	0.000%	0.15617	0.15618	0.000%	0.000%
0	0.39226	0.39468	0.000%	0.000%	0.15617	0.15618	0.000%	0.000%
	$\widetilde{\lambda} = (0.485/0.5/0.515)$							
	$\widetilde{p}'(N \geq 1) = (0.38430/0.39347/0.40250)$				$\widetilde{\pi}'_4 = (0.15596/0.15618/0.15625)$			
	$p(N \geq 1)(\alpha)$		Error		$\pi_4(\alpha)$		Error	
α	$\underline{p(N \geq 1)}(\alpha)$	$\overline{p(N \geq 1)}(\alpha)$	$\underline{err}(\alpha)$	$\overline{err}(\alpha)$	$\underline{\pi_4}(\alpha)$	$\overline{\pi_4}(\alpha)$	$\underline{err}(\alpha)$	$\overline{err}(\alpha)$
1	0.39347	0.39347	0.000%	0.000%	0.15618	0.15618	0.000%	0.000%
0.75	0.39119	0.39574	0.003%	0.003%	0.15615	0.15618	0.024%	0.023%
0.5	0.38890	0.39800	0.004%	0.004%	0.15611	0.15620	0.032%	0.030%
0.25	0.38661	0.40025	0.003%	0.003%	0.15604	0.15623	0.024%	0.022%
0	0.38430	0.40250	0.000%	0.000%	0.15596	0.15625	0.000%	0.000%
	$\widetilde{\lambda} = (0.958/0.96/0.962)$							
	$\widetilde{p}'(N \geq 1) = (0.61634/0.61711/0.61787)$				$\widetilde{\pi}'_4 = (0.09820/0.09850/0.09879)$			
	$p(N \geq 1)(\alpha)$		Error		$\pi_4(\alpha)$		Error	
α	$\underline{p(N \geq 1)}(\alpha)$	$\overline{p(N \geq 1)}(\alpha)$	$\underline{err}(\alpha)$	$\overline{err}(\alpha)$	$\underline{\pi_4}(\alpha)$	$\overline{\pi_4}(\alpha)$	$\underline{err}(\alpha)$	$\overline{err}(\alpha)$
1	0.61711	0.61711	0.000%	0.000%	0.09850	0.09850	0.000%	0.000%
0.75	0.61692	0.61730	0.000%	0.000%	0.09842	0.09857	0.000%	0.000%
0.5	0.61672	0.61749	0.000%	0.000%	0.09835	0.09864	0.000%	0.000%
0.25	0.61653	0.61768	0.000%	0.000%	0.09827	0.09872	0.000%	0.000%
0	0.61634	0.61787	0.000%	0.000%	0.09820	0.09879	0.000%	0.000%

Table 7. Cont.

	$\widetilde{\lambda} = (0.0945/0.96/0.0975)$								
	$\widetilde{p}'(N \geq 1) = (0.61132/0.61711/0.62281)$				$\widetilde{\pi}'_4 = (0.09628/0.09850/0.10074)$				
	$p(N \geq 1)(\alpha)$		Error		$\pi_4(\alpha)$		Error		
α	$\underline{p(N \geq 1)}(\alpha)$	$\overline{p(N \geq 1)}(\alpha)$	$\underline{err}(\alpha)$	$\overline{err}(\alpha)$	$\underline{\pi_4}(\alpha)$	$\overline{\pi_4}(\alpha)$	$\underline{err}(\alpha)$	$\overline{err}(\alpha)$
1	0.61711	0.61711	0.000%	0.000%	0.09850	0.09850	0.000%	0.000%
0.75	0.61567	0.61854	0.001%	0.001%	0.09794	0.09905	0.003%	0.003%
0.5	0.61422	0.61997	0.002%	0.002%	0.09738	0.09962	0.004%	0.003%
0.25	0.61278	0.62139	0.001%	0.001%	0.09683	0.10018	0.003%	0.003%
0	0.61132	0.62281	0.000%	0.000%	0.09628	0.10074	0.000%	0.000%

Source: Own Elaboration.

Table 8. α-cuts of $\widetilde{b}^*$, it's triangular approximate and errors for different parameters $\widetilde{\lambda}$.

	$\widetilde{\lambda}=(0.025/0.04/0.055)$				$\widetilde{\lambda}=(0.485/0.5/0.515)$				$\widetilde{\lambda}=(0.0945/0.96/0.0975)$			
	$\widetilde{b}^{*'}=(50.836/51.422/52.073)$				$\widetilde{b}^{*'}=(80.662/81.396/82.098)$				$\widetilde{b}^{*'}=(93.058/93.246/93.427)$			
	$b^*(\alpha)$		Error		$b^*(\alpha)$		Error		$b^*(\alpha)$		Error	
α	$\underline{b^*}(\alpha)$	$\overline{b^*}(\alpha)$	$\underline{err}(\alpha)$	$\overline{err}(\alpha)$	$\underline{b^*}(\alpha)$	$\overline{b^*}(\alpha)$	$\underline{err}(\alpha)$	$\overline{err}(\alpha)$	$\underline{b^*}(\alpha)$	$\overline{b^*}(\alpha)$	$\underline{err}(\alpha)$	$\overline{err}(\alpha)$
1	51.422	51.422	0.000%	0.000%	81.396	81.396	0.000%	0.000%	93.246	93.246	0.000%	0.000%
0.75	51.270	51.579	0.012%	0.012%	81.215	81.574	0.004%	0.004%	93.200	93.292	0.001%	0.001%
0.5	51.121	51.739	0.016%	0.016%	81.033	81.751	0.005%	0.005%	93.153	93.338	0.001%	0.001%
0.25	50.977	51.904	0.012%	0.012%	80.848	81.925	0.004%	0.004%	93.106	93.383	0.001%	0.001%
0	50.836	52.073	0.000%	0.000%	80.662	82.098	0.000%	0.000%	93.058	93.427	0.000%	0.000%

Source: Own Elaboration.

6. Summary and Further Research

BMSs are often modelled by means of MCs with crisp probabilities. In this paper, it is considered that transition probabilities of Markovian BMSs are not crisp but uncertain. This uncertainty is captured by using a FN, thus giving rise to the concept of FBMSs. FBMSs modeling is based on the concept of FMC by Buckley and Eslami in [12]. As a result, conventional BMSs can be understood as a particular case of our model where transition probabilities are singletons. The model in [11] represents the uncertainty by means of modal intervals. Since its results can be interpreted as the 0-cuts of ours, that model can also be seen as a particular case of our FBMS.

We assume, as it is often done in actuarial literature, that the number of claims in a period is a Poisson RV. Nonetheless, due to uncertainty, its parameter λ is not a real number but a TFN. So, to implement the model presented in the paper, it is necessary, firstly, to structure available information of the behavior of that RV. From this information, the Poisson parameter can be fitted by means of a TFN. Three alternatives to do so are proposed. Subsequently, by using α-cut arithmetic, transition probabilities, the stationary distribution function, and the mean asymptotic premium of the FBMS are obtained by means of their α-cuts. The lower and upper bounds of these α-cuts can be understood as the result of a sensitivity analysis of the BMS that evaluates two extreme scenarios with possibility α. That output can be very useful in actuarial decision-making processes since it provides a set of sensitivity analyses that is structured on the basis of their grade of reliability.

Although the mean number of claims, λ, is assumed to be a TFN, the outputs from our FBMS do not maintain that shape. However, in the numerical applications developed within the framework of the Irish BMS, we have verified that all the outputs obtained from a triangular λ are well approximated by a TFN that maintains the support and core of the original FN. This result is quite interesting. On the one hand, other more complex shapes of FNs can produce drawbacks in information modeling, such as problems with calculations in computer implementation. In this regard, we have observed that the number of optimizing problems to be solved in order to obtain transition probabilities, the stationary distribution, and the mean asymptotic premium is reduced drastically. Likewise, TFNs are very attractive from an insurance decision-making perspective since TFNs admit a very intuitive interpretation even without any knowledge of FST. At least, a TFN provides an estimate of the maximum, minimum, and most feasible values of a variable. Therefore, we feel that the triangular approximations introduced in this document would make it easier to use FMCs in the implementation of BMSs by the insurance industry.

Our methodologic approach can be extended, with the necessary adaptations, to other assumptions for the RV number of claims. Likewise, as far as we are concerned, there are several topics that may be the object of further research. Firstly, a wider investigation on how to apply a fuzzy Poisson regression in a BMS context must be carried out. Secondly, a more in-depth evaluation of the goodness of triangular approximations to BMS probabilities and the mean asymptotic premium is needed. In this respect, a wider range for the values of λ, a greater number of classes in the BMS, and other methods to fit triangular approximates must be tested. Thirdly, it is also needed to extend our model to the case in which fuzzy uncertainty in the BMS does not only appear in the number of claims but also their cost. Moreover, to model λ, instead of TFNs, other types of FNs, such as GFNs or IFNs, could be considered. We are aware that these tools allow capturing uncertainty with more nuances than FNs. However, their fitting has a greater cost than TFNs since it implies adjusting more parameters. Additionally, implementing computational operations with them is more expensive. This last issue is crucial in our context, especially in complex BMSs like, e.g., the German one. So, we feel that applying FNs suppose a balance between the simplicity of crisp or interval probabilities and more complex representations of uncertain quantities such as GFNs or IFNs. Finally, to evaluate the efficiency of a BMS, it is usually calculated the elasticity of the mean premium (4) with respect to the risk parameter λ. To do so, numerical simulations for point values of λ within the reference interval [0,1] are implemented (see, e.g., [2]). The use of fuzzy logic may be of interest in this concern. For example, that reference interval can be granulated into linguistic labels such as "low risk", "medium risk" and so on, similarly to that proposed by [27] and [57]. Therefore, elasticity evaluations may be made on the basis of linguistic labels instead of point values on [0,1]. Fuzzy linguistic Markov chains, presented by [58], may be the starting point for this.

Author Contributions: All authors have contributed equally to all sections and stages of this work. All authors have read and agreed to the published version of the manuscript.

Funding: This research received no external funding. The APC has been funded by the University of Barcelona.

Institutional Review Board Statement: Not applicable.

Informed Consent Statement: Not applicable.

Data Availability Statement: The data presented in this study are available on request from the corresponding author.

Acknowledgments: Authors acknowledge helpful suggestions of anonymous reviewers.

Conflicts of Interest: The authors declare no conflict of interest.

Appendix A. Pseudo-Codes of Numerical Applications 2 and 4

```
# ————————————————
# Numerical Application 2
# ————————————————
```
$\widetilde{p}(N \geq 1) \leftarrow (0.037287/0.039211/0.041130)$

$$\widetilde{p}(N=0) \leftarrow (0.958870/0.960789/0.962713)$$

$$\widetilde{P} \leftarrow \begin{pmatrix} \widetilde{p}(N \geq 1) & \widetilde{p}(N=0) & 0 \\ \widetilde{p}(N \geq 1) & 0 & \widetilde{p}(N=0) \\ \widetilde{p}(N \geq 1) & 0 & \widetilde{p}(N=0) \end{pmatrix}$$

$$(\widetilde{\pi}_j) \leftarrow \text{FuzzyStationaryDistribution}(\widetilde{P})$$

Plot$(\widetilde{\pi}_j)$ for $j = 1, 2, 3$

\#——————————

\# Numerical Application 4

\#——————————

$$\widetilde{p}(N=0) \leftarrow (0.958870/0.960789/0.962713)$$
$$\widetilde{p}(N=1) \leftarrow (0.036583/0.038432/0.040273)$$
$$\widetilde{p}(N \geq 1) \leftarrow (0.037287/0.039211/0.041130)$$
$$\widetilde{p}(N \geq 2) \leftarrow (0.000704/0.000779/0.000858)$$

$$\widetilde{p} \leftarrow \begin{pmatrix} \widetilde{p}(N=0) & 0 & \widetilde{p}(N=1) & 0 & 0 & \widetilde{p}(N \geq 2) \\ \widetilde{p}(N=0) & 0 & 0 & \widetilde{p}(N=1) & 0 & \widetilde{p}(N \geq 2) \\ 0 & \widetilde{p}(N=0) & 0 & 0 & \widetilde{p}(N=1) & \widetilde{p}(N \geq 2) \\ 0 & 0 & \widetilde{p}(N=0) & 0 & 0 & \widetilde{p}(N \geq 1) \\ 0 & 0 & 0 & \widetilde{p}(N=0) & 0 & \widetilde{p}(N \geq 1) \\ 0 & 0 & 0 & 0 & \widetilde{p}(N=0) & \widetilde{p}(N \geq 1) \end{pmatrix}$$

$(\widetilde{\pi}_j) \leftarrow \text{FuzzyStationaryDistribution}(\widetilde{P})$

Plot$(\widetilde{\pi}_j)$ for $j = 1, 2, \ldots, 6$

Appendix B. R Codes of Numerical Applications 2 and 4

```
# ——————————
# Numerical Application 2
# ——————————
library(FuzzyNumbers)
library(FuzzyStatProb)
a = TriangularFuzzyNumber(0.037287, 0.039210, 0.041130)
b = TriangularFuzzyNumber(0.958870, 0.960790, 0.962713)
zero = TriangularFuzzyNumber(0, 0, 0)
allnumbers = list(a = a, b = b, zero = zero)
transitions = matrix(data = c("a", "b", NA, "a", NA, "b", "a", NA, "b"), nrow = 3, byrow = T)
states = c("01", "02", "03")
rownames(transitions) = states
colnames(transitions) = states
stationary = fuzzyStationaryProb(data = transitions, options = list(regression = "linear", fuzzynumbers = allnumbers))
m <- matrix(1:3, nrow = 1, ncol = 3, byrow = TRUE)
layout(mat = m, heights = c(0.25, 0.25, 0.25, 0.25))
for (state in states){
  cat("State", state, "\n")
  fz = stationary$fuzzyStatProb[[state]]
  acuts = stationary$acuts[[state]]
  print(acuts[acuts$y == 0.001,])
  print(acuts[acuts$y == 0.999,])
  par(mar = c(4, 4, 2, 1))
  plot(fz, col = "blue", main = paste("State", state),
       cex.lab = 1.1, lwd = 2, xaxt = "n")
  left = supp(fz)[1]
  right = supp(fz)[2]
  center = core(fz)[1]
  at = round(c(left, right, center), digits = 4)
```

```
        axis(1, at = at, labels = FALSE)
        text(x = at, y = par("usr")[3] - 0.1,
            labels = at, srt = 35, xpd = NA)
        points(acuts)
        print("—————————")
    }
#————————————————
# Numerical Application 4
#————————————————
    pN0 = TriangularFuzzyNumber(0.958870, 0.960789, 0.962713)
    pN1 = TriangularFuzzyNumber(0.036583, 0.038432, 0.040273)
    pNgt1 = TriangularFuzzyNumber(0.037287, 0.039211, 0.041130)
    pNgt2 = TriangularFuzzyNumber(0.000704, 0.000779, 0.000858)
    allnumbers2 = list(pN0 = pN0, pN1 = pN1, pNgt1 = pNgt1, pNgt2 = pNgt2)
    transitions2 = matrix(data = c("pN0", NA, "pN1", NA, NA, "pNgt2", "pN0",
    NA, NA, "pN1", NA, "pNgt2", NA, "pN0", NA, NA, "pN1", "pNgt2", NA, NA,
    "pN0", NA, NA, "pNgt1", NA, NA, NA, "pN0", NA, "pNgt1", NA, NA,
    NA, NA, "pN0", "pNgt1"), nrow = 6, byrow = T)
    states2 = c("01", "02", "03", "04", "05", "06")
    rownames(transitions2) = states2
    colnames(transitions2) = states2
    stationary2 = fuzzyStationaryProb(data = transitions2, options = list(regression =
"linear", fuzzynumbers = allnumbers2))
    m <- matrix(1:6, nrow = 2, ncol = 3, byrow = TRUE)
    layout(mat = m, heights = c(0.25, 0.25, 0.25, 0.25))
    for (state in states2){
        cat("State", state, "\n")
        fz = stationary2$fuzzyStatProb[[state]]
        acuts = stationary2$acuts[[state]]
        print(acuts[acuts$y == 0.001,])
        print(acuts[acuts$y == 0.999,])
        par(mar = c(4, 4, 2, 1))
        plot(fz, col = "blue", main = paste("State", state),
            cex.lab = 1.1, lwd = 2, xaxt = "n")
        left = supp(fz)[1]
        right = supp(fz)[2]
        center = core(fz)[1]
        at = round(c(left, right, center), digits = 5)
        axis(1, at = at, labels = FALSE)
        text(x = at, y = par("usr")[3] - 0.1,
            labels = at, srt = 35, xpd = NA)
        points(acuts)
        print("—————————")
    }
```

References

1. Lemaire, J. *Bonus-Malus Systems in Automobile Insurance*; Kluwer Academic Publishers: Norwell, MA, USA, 1995.
2. Lemaire, J.; Zi, H. A comparative analysis of 30 bonus-malus systems. *ASTIN Bull.* **1994**, *24*, 287–309. [CrossRef]
3. Meyer, U. Third Party Motor Insurance in Europe. Comparative Study of the Economical-Statistical Situation. In *Personal Communications*; University of Bamberg: Bamberg, Germany, 2000.
4. Dodu, D. Comparative Analysis of Bonus Malus Systems in Italy and Central and Eastern Europe. 2017. Available online: https://www.milliman.com/en/insight/comparative-analysis-of-bonus-malus-systems-in-italy-and-central-and-eastern-europe (accessed on 7 October 2020).
5. Pitrebois, S.; Denuit, M.; Walhin, J.F. Fitting the Belgian bonus-malus system. *Belg. Actuar. Bull.* **2003**, *3*, 58–62.
6. de Pril, N. The Efficiency of a Bonus-Malus System. *ASTIN Bull.* **1978**, *10*, 59–72. [CrossRef]

7. Bonsdorff, H. On the convergence rate of bonus-malus systems. *ASTIN Bull.* **1992**, *22*, 217–223. [CrossRef]
8. Pitrebois, S.; Denuit, M.; Walhin, J.F. Bonus-malus scales in segmented tariffs: Gilde & Sundt's work revisited. *Aust. Actuar. J.* **2004**, *10*, 107–125.
9. Niemiec, M. Bonus-Malus Systems as Markov Set-Chains. *ASTIN Bull.* **2007**, *37*, 53–65. [CrossRef]
10. Asmussen, S. Modeling and performance of Bonus-Malus systems: Stationary versus age-correction. *Risks* **2014**, *2*, 49–73. [CrossRef]
11. Adillon, R.; Lambert, J.; Mármol, M. Modal interval probability: Application to Bonus-Malus Systems. *Int. J. Uncertain. Fuzziness Knowl. Based Syst.* **2020**, *28*, 837–851. [CrossRef]
12. Buckley, J.J.; Eslami, E. Fuzzy Markov Chains: Uncertain Probabilities. *Mathw. Soft Comput.* **2002**, *9*, 33–41.
13. Villacorta, P.J.; Verdegay, J.L. FuzzyStatProb: An R Package for the Estimation of Fuzzy Stationary Probabilities from a Sequence of Observations of an Unknown Markov Chain. *J. Stat. Softw.* **2016**, *71*, 1–27. [CrossRef]
14. Denuit, M.; Marchal, X.; Pitrebois, S.; Walhin, J.F. *Actuarial Modelling of Claims Counts: Risk Classification, Credibility and Bonus-Malus Systems*; Wiley: Chichester, UK, 2007.
15. Vernic, R. On risk measures and capital allocation for distributions depending on parameters with interval or fuzzy uncertainty. *Appl. Soft Comput.* **2018**, *64*, 199–215. [CrossRef]
16. Wierzchoń, S.T. Randomness and Fuzziness in A Linear Programming Problem. In *Combining Fuzzy Imprecision with Probabilistic Uncertainty in Decision Making. Lecture Notes in Economics and Mathematical Systems*; Kacprzyk, J., Fedrizzi, M., Eds.; Springer: Berlin, Germany, 1988; Volume 310, pp. 227–239. [CrossRef]
17. Buckley, J.J. Uncertain probabilities III: The continuous case. *Soft Comput.* **2004**, *8*, 200–206. [CrossRef]
18. Buckley, J.J.; Eslami, E. Uncertain probabilities II: The continuous case. *Soft Comput.* **2004**, *8*, 193–199. [CrossRef]
19. Buckley, J.J.; Eslami, E. Uncertain probabilities I: The discrete case. *Soft Comput.* **2003**, *7*, 500–505. [CrossRef]
20. Zmeškal, Z. Value at risk methodology under soft conditions approach (fuzzy-stochastic approach). *Eur. J. Oper. Res.* **2005**, *161*, 337–347. [CrossRef]
21. Moussa, A.M.; Kamdem, J.S.; Terraza, M. Fuzzy value-at-risk and expected shortfall for portfolios with heavy-tailed returns. *Econ. Model.* **2014**, *39*, 247–256. [CrossRef]
22. Shapiro, A.F. Fuzzy logic in insurance. *Insur. Math. Econ.* **2004**, *35*, 399–424. [CrossRef]
23. Huang, T.; Zhao, R.; Tang, W. Risk model with fuzzy random individual claim amount. *Eur. J. Oper. Res.* **2009**, *192*, 879–890. [CrossRef]
24. Kemaloglu, S.A.; Shapiro, A.F.; Tank, F.; Apaydin, A. Using fuzzy logic to interpret dependent risks. *Insur. Math. Econ.* **2018**, *79*, 101–106. [CrossRef]
25. Koissi, M.C.; Shapiro, A.F. Fuzzy formulation of the Lee–Carter model for mortality forecasting. *Insur. Math. Econ.* **2006**, *39*, 287–309. [CrossRef]
26. Shapiro, A.F. Modeling future lifetime as a fuzzy random variable. *Insur. Math. Econ.* **2013**, *53*, 864–870. [CrossRef]
27. Andrés-Sánchez, J.D.; Puchades, L.G.V. Some computational results for the fuzzy random value of life actuarial liabilities. *Iran. J. Fuzzy Syst.* **2017**, *14*, 1–25. [CrossRef]
28. Andrés-Sánchez, J.D.; Puchades, L.G.V. A Fuzzy-Random Extension of the Lee–Carter Mortality Prediction Model. *Int. J. Comput. Intell. Syst.* **2019**, *12*, 775–794. [CrossRef]
29. Jorba, L.; Adillon, R. A Generalization of Trapezoidal Fuzzy Numbers Based on Modal Interval Theory. *Symmetry* **2017**, *9*, 198. [CrossRef]
30. Loimaranta, K. Some Asymptotic Properties of Bonus Systems. *ASTIN Bull.* **1972**, *6*, 233–245. [CrossRef]
31. Heras, A.; Vilar, J.L.; Gil, J.A. Asymptotic Fairness of Bonus-Malus Systems and Optimal Scales of Premiums. *Geneva Pap. Risk Insur. Theory* **2002**, *27*, 61–82. [CrossRef]
32. Kafková, S. Bonus-malus systems in vehicle insurance. *Procedia Econ. Financ.* **2015**, *23*, 216–222. [CrossRef]
33. Dong, W.; Shah, H.C. Vertex method for computing functions of fuzzy variables. *Fuzzy Sets Syst.* **1987**, *24*, 65–78. [CrossRef]
34. Buckley, J.J.; Qu, Y. On using α-cuts to evaluate fuzzy equations. *Fuzzy Sets Syst.* **1990**, *38*, 309–312. [CrossRef]
35. Dong, W.; Wong, F.S. Interactive fuzzy variables and fuzzy decisions. *Fuzzy Sets Syst.* **1989**, *29*, 1–19. [CrossRef]
36. Kaufmann, A. Fuzzy Subsets Applications in O.R. and Management. In *Fuzzy Set Theory and Applications*; Jones, A., Kaufmann, A., Zimmermann, H.-J., Eds.; Springer: Dordrecht, The Netherlands, 1986; pp. 257–300. [CrossRef]
37. Heberle, J.; Thomas, A. Combining chain-ladder reserving with fuzzy numbers. *Insur. Math. Econ.* **2014**, *55*, 96–104. [CrossRef]
38. Jiménez, M.; Rivas, J.A. Fuzzy number approximation. *Int. J. Uncertain. Fuzziness Knowl. Based Syst.* **1998**, *6*, 69–78. [CrossRef]
39. Grzegorzewski, P.; Pasternak-Winiarska, K. Natural trapezoidal approximations of fuzzy numbers. *Fuzzy Sets Syst.* **2014**, *250*, 90–109. [CrossRef]
40. Grzegorzewski, P.; Mrówka, E. Trapezoidal approximations of fuzzy numbers. *Fuzzy Sets Syst.* **2005**, *153*, 115–135. [CrossRef]
41. Kleyle, R.M.; de Korvin, A. Constructing one-step and limiting fuzzy transition probabilities for finite Markov chains. *J. Intell. Fuzzy Syst.* **1998**, *6*, 223–235.
42. Zadeh, L.A. Fuzzy sets. *Inf. Control.* **1965**, *8*, 338–353. [CrossRef]
43. Avrachenkov, K.E.; Sánchez, E. Fuzzy Markov Chains: Specificities and Properties. In Proceedings of the 8th IPMU'2000 Conference, Madrid, Spain, 3–7 July 2000; pp. 1851–1856.
44. Li, G.; Xiu, B. Fuzzy Markov Chains Based on the Fuzzy Transition Probability. In Proceedings of the 26th Chinese Control and Decision Conference, Changsha, China, 31 May–2 June 2014; pp. 4351–4356. [CrossRef]

45. Buckley, J.J. *Fuzzy Probabilities: New Approach and Applications, Volume 115 of Studies in Fuzziness and Soft Computing*, 2nd ed.; Springer: Heidelberg, Germany, 2005.
46. Bardossy, A.; Duckstein, L.; Bogardi, I. Combination of fuzzy numbers representing expert opinions. *Fuzzy Sets Syst.* **1993**, *57*, 173–181. [CrossRef]
47. Hsu, H.M.; Chen, C.T. Aggregation of fuzzy opinions under group decision making. *Fuzzy Sets Syst.* **1996**, *79*, 279–285. [CrossRef]
48. Tsabadze, T. A method for aggregation of trapezoidal fuzzy estimates under group decision-making. *Fuzzy Sets Syst.* **2015**, *266*, 114–130. [CrossRef]
49. Buckley, J.J. Fuzzy statistics: Regression and prediction. *Soft Comput.* **2005**, *9*, 769–775. [CrossRef]
50. Sfiris, D.S.; Papadopoulos, B.K. Non-asymptotic fuzzy estimators based on confidence intervals. *Inf. Sci.* **2014**, *279*, 446–459. [CrossRef]
51. Al-Kandari, M.; Adjenughwure, K.; Papadopoulos, K. A Fuzzy-Statistical Tolerance Interval from Residuals of Crisp Linear Regression Models. *Mathematics* **2020**, *8*, 1422. [CrossRef]
52. Andrés-Sánchez, J.D. Fuzzy Regression Analysis: An Actuarial Perspective. In *Fuzzy Statistical Decision-Making. Studies in Fuzziness and Soft Computing*; Kahraman, C., Kabak, Ö., Eds.; Springer: Cham, Switzerland, 2016; Volume 343, pp. 175–201. [CrossRef]
53. Shapiro, A.F. Fuzzy Regression and the Term Structure of Interest Rates Revisited. In Proceedings of the 14th International AFIR Colloquium, Boston, MA, USA, 8–9 November 2004; pp. 29–45.
54. Apaydin, A.; Baser, F. Hybrid fuzzy least-squares regression analysis in claims reserving with geometric separation method. *Insur. Math. Econ.* **2010**, *47*, 113–122. [CrossRef]
55. Woundjiagué, A.; Bidima, M.L.D.M.; Mwangi, R.W. An Estimation of a Hybrid Log-Poisson Regression Using a Quadratic Optimization Program for Optimal Loss Reserving in Insurance. *Adv. Syst.* **2019**, *2019*, 1393946. [CrossRef]
56. Ishibuchi, H.; Nii, M. Fuzzy regression using asymmetric fuzzy coefficients and fuzzified neural networks. *Fuzzy Sets Syst.* **2001**, *119*, 273–290. [CrossRef]
57. Andrés-Sánchez, J.D.; Puchades, L.G.V.; Zhang, A. Incorporating fuzzy information in pricing substandard annuities. *Comput. Ind. Eng.* **2020**, *145*, 106475. [CrossRef]
58. Villacorta, P.J.; Verdegay, J.L.; Pelta, D. Towards Fuzzy Linguistic Markov Chains. In Proceedings of the 8th Conference of the European Society for Fuzzy Logic and Technology (EUSFLAT-13), Milan, Italy, 11–13 September 2013. [CrossRef]

Article

The Quintuple Helix of Innovation Model and the SDGs: Latin-American Countries' Case and Its Forgotten Effects

Luciano Barcellos-Paula [1,2,*], Iván De la Vega [1,2] and Anna María Gil-Lafuente [3]

1 Departamento Académico de Posgrado en Negocios, CENTRUM Católica Graduate Business School, Lima 15023, Peru; idelavega@pucp.edu.pe
2 Departamento Académico de Posgrado en Negocios, Pontificia Universidad Católica del Perú, Lima 15088, Peru
3 Department of Business Administration, University of Barcelona, 08034 Barcelona, Spain; amgil@ub.edu
* Correspondence: lbarcellosdepaula@pucp.edu.pe

Abstract: The sustainable development of countries is associated with a set of actions that must be implemented in the long term. In this process, society must be a valid partner in the decisions that are made. Studies show the interrelationship between the Sustainable Development Goals (SDGs), which increases uncertainty and makes decision-making more difficult. On the other hand, the Quintuple Helix of Innovation Model (QHIM) provides an analytical framework to explain the systems' interactions. The motivation of the study lies in knowing the relationships between the variables that affect SDGs. The manuscript aims to broaden the discussion on sustainable development and propose two models to support decision making. The first one suggests 20 indicators linked to the QHIM with the SDGs in Latin American countries. The second identifies the forgotten effects through the application of a Fuzzy Logic algorithm. The main contribution is to know these effects and to support decision-making. The research carried out can be classified as applied, with the explanatory objective and the combined approach (quantitative-qualitative), modeling and simulation, and case study methods. The QHIM results indicate that Chile leads the ranking, followed by Brazil, Mexico, Peru, and Colombia. Also, it reveals the importance of correctly identifying cause-effects by seeking harmony between systems. A limitation would be the number of variables used. The study indicates promising lines of research.

Keywords: SDGs; The Quintuple Helix of Innovation Model; sustainability; Latin America; knowledge systems; Forgotten Effects Theory; Fuzzy Logic

Citation: Barcellos-Paula, L.; De la Vega, I.; Gil-Lafuente, A.M. The Quintuple Helix of Innovation Model and the SDGs: Latin-American Countries' Case and Its Forgotten Effects. *Mathematics* **2021**, *9*, 416. https://doi.org/10.3390/math9040416

Academic Editor: Jorge de Andres Sanchez

Received: 22 December 2020
Accepted: 12 February 2021
Published: 20 February 2021

Publisher's Note: MDPI stays neutral with regard to jurisdictional claims in published maps and institutional affiliations.

Copyright: © 2021 by the authors. Licensee MDPI, Basel, Switzerland. This article is an open access article distributed under the terms and conditions of the Creative Commons Attribution (CC BY) license (https://creativecommons.org/licenses/by/4.0/).

1. Introduction

Scientific studies point out the need to act in a strategic and socially responsible way towards sustainable development [1,2]. In this sense, innovation plays a crucial role in achieving this goal [3,4]. On the other hand, the search for lasting solutions for the planet requires balancing objectives from several interest groups and strengthening relationships between institutions [5]. Therefore, countries must have a critical mass of researchers in various knowledge areas [6].

In 1987 the Brundtland Report defined sustainable development as one in which "present needs must be met without compromising the future of future generations" [7] and recognized the importance of the commitment of all to achieve this goal.

In 2015 this theme gained greater relevance with the 17 Sustainable Development Goals (SDGs) [8]. Most importantly, it invites us to create a more sustainable, secure, and prosperous planet for humanity. To achieve the SDGs, individuals, businesses, governments, and non-governmental organizations must commit to sustainable development [5]. Therefore, working together with diverse organizations allows us to remember stakeholders' importance in generating long-term value for both business and society [9].

In this sense, the Quintuple Helix of Innovation Model (QHIM) provides an analytical framework to explain the interactions among the actors of a society that seeks, in theory, to progress [10]. The proposed model is composed of political, educational, economic, environmental, and social systems. Each helix represents a knowledge subsystem that functions as a spiral connecting with the other systems, which, in turn, have a national, regional, and global reach.

Therefore, humanity must find solutions to address significant challenges, such as harmony and cooperation among the five systems towards sustainable development. The uncertainty caused by constant and intense change, which increases decision-making, must also be considered. For these reasons, the primary motivation lies in knowing the relationships between the systems and the variables that affect sustainable development.

As a methodological alternative, the algorithms based on "Fuzzy Logic" [11] contribute to solving problems of the real world when they are dedicated to solving complex systems reducing the uncertainty in decision making [12–14].

In this context, the manuscript aims to broaden the discussion on sustainable development and propose two models to support decision making. The first one suggests 20 indicators linked to the QHIM with the SDGs in Latin American countries. The second identifies the forgotten effects through the application of a Fuzzy Logic algorithm. The main contribution of the study is in knowing these effects and supporting decision making. The main contribution of the study is in knowing these effects and supporting decision making. The results reveal the importance of correctly identifying cause-effects by seeking harmony between systems. A limitation of this research is the number of variables used.

This research can be classified as applied, with the explanatory objective and the combined approach (qualitative-quantitative), modeling and simulation, and case study methods [15]. The combination of the two methods generates an added value to the research since, on the one hand, simulation allows to inform and understand a real-world problem and propose solutions adjusted to the identified needs [16]. On the other hand, the case study method is empirical research that finds a contemporary phenomenon within its real-life context [17]. As a result, the combined research method supports the model's validation and generates interesting theoretical and practical implications.

The document is structured as follows. Section 2 presents the materials and methods. Section 3 shows the results of the simulation applying the Forgotten Effects Theory. Section 4 presents the discussions of the results. Finally, Section 5 details the conclusions followed by the references.

2. Materials and Methods

This section is organized into four parts to explain the methodology of the study. First, it explains the QHIM and SDGs' theoretical framework. Following, it discusses Latin American countries' case analysis concerning sustainable development applying the QHIM. Third, it explains the algorithm used in the simulation. Finally, it details the simulation process carried out to identify sustainable development's forgotten effects.

2.1. QHIM and SDGs: Theoretical Framework

The QHIM results from the continuous development of approaches that seek to explain the dynamic interactions of social actors at different scales, the country level being one of them. Scientific studies confirm the importance of the QHIM in integrating the five systems to achieve sustainable development [10,18]. Table A1 presents the evolution of the models related to the study, definitions, and scope.

The previous theoretical basis on which the QHIM was founded comes from several approaches, the most relevant being the Triple Helix for Development, first published by Etzkowitz and Leydesdorff in 1998 [19].

These approaches have contributed to designing the theoretical base model used for this study, such as the Triple Helix and its evolution towards the Fourth and Fifth Helix of

Innovation. This last model was finally selected to explain its relationship with the SDGs. It adds social capital and environmental capital to the Triple Helix model and was therefore considered the most appropriate for this study.

By examining the longitudinal evolution of the models related to the one used in this study, we could indicate that the Triple Helix approach operates under the construction of a socio-institutional fabric that leverages the network of interactive business-university-government agent relationships. This approach describes and analyzes these agents' relationships, and they were examined, considering the dynamic processes generated among the participants and materialized through initiatives that seek innovative solutions [19–23]. This model has been evolving steadily and today is related to the promotion of new modes of an action directed towards the market and also to propose solutions to problems that are fundamental of a social nature [24–26].

The Quadruple Helix model of innovation is an evolution made by a team of specialists integrating central elements of other approaches such as the Triple Helix, Mode 1 and Mode 2, and the National Innovation Systems. This process includes the attributes related to the new actor that the authors incorporate and call Social Capital.

These characteristics are related to the media and the vision of a change process towards a knowledge economy. Including this new actor as a fourth interconnected subsystem, taking into account the media to support disseminating knowledge in a given society, also integrates other aspects such as culture with its values, experience, and traditions [27]. These attributes are relevant since they can favor or condition a given country's potential development, and society has a relevant weight in decisions [28].

The QHIM has as its central purpose to include the natural environment as a new subsystem of knowledge. This approach's logic is based on generating innovation ecosystems that include nature as a central component, giving it the same weight as the other four helices [10]. The natural environment serves to preserve, survive, and vitalize humanity and create new green technologies.

The search for sustainable development of the planet as a central idea is a reality in this proposal. It speaks of social ecology, and the center of gravity of the discussion is global climate change. This aspect allows us to determine that this approach is a proposal before creating the SDGs [29]. The abuse of renewable and non-renewable natural resources is no longer conceived without global society's participation in the substantive decisions on the impact this generates on the planet. Figure 1 presents the model used in this study, and Table A1 provides the definition.

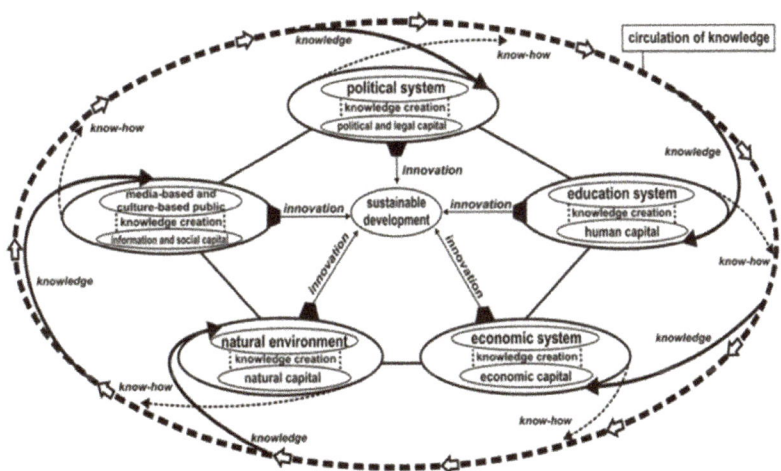

Figure 1. The Quintuple Helix Innovation Model (QHIM). Source: [10].

A literature review was performed on SDGs and highlights that the increasingly constant and intense changes brought about by climate change and social inequality were a warning to humanity's future in recent decades. Sustainable development became the main route to meet these challenges. In this sense, the United Nations intensified the orientations and policies towards sustainable development with various guidelines over time, such as the Brundtland Report [30], Global Compact [31], Millennium Development Goals (MDGs) [32], Paris Agreement [33], and the SDGs [34]. Consequently, sustainable development must be considered a priority and strategic in the countries' policies [2].

Currently, the focus is on Agenda 2030 through the SDGs, which in general terms, is a set of objectives, goals, and actions that aim to guide governments, academics, entrepreneurs, and society as a whole towards a fairer and better world [5,35]. Through scientific studies, the academic sector also contributes to the SDGs, seeking to explore this theme, which is complex and depends on the harmony and integration of systems to achieve effective results [3,36].

The research identified three gaps related to the SDGs that could increase uncertainty [37] and hinder the implementation of measures and problem-solving.

The first gap is in indicator assessment because countries have autonomy in the implementation of the SDGs, which will require the collection of quality, accessible, and timely data. SDG assessment results can be ambiguous and confusing due to the lack of a well-designed conceptual framework of indicators [37]. Other authors warn that applying indicators in an inconsistent or uncoordinated manner can cause serious problems [38,39]. Therefore, consensus on the indicator framework and its use are needed.

The second gap is the lack of understanding between the MDGs and the SDGs [40]. The MDGs focused on countries, whereas the SDGs should be global. For this reason, new methods can help their implementation and systems thinking. The same study states that the danger is prioritizing individual goals without understanding the possible positive interactions between them [40].

The third gap is to understand the correlation between the SDGs in decision making [36,40]. For example, the decision-maker must understand that responding to the threat of climate change (SDG13) influences natural resource management (SDGs 14 and 15) and food production (SDG2). Conversely, climate stability (SDG13) and preventing ocean acidification (SDG14) will support sustainable food production and fisheries (SDG2) [41]. Other examples would be gender equality (SDG10) or improving health (SDG3), which help eradicate poverty (SDG1), and fostering peace and inclusive societies (SDG16), which will reduce inequalities (SDG10) and help economies thrive (SDG8) [8]. However, decision-makers cannot correctly identify interacting variables, which can harm the environment and compromise the SDGs' scope [42]. It is essential to understand sustainable development from a broad and systemic approach, which considers each stakeholder's importance to achieving a more socially just, inclusive, economically viable, and environmentally friendly development.

Along these lines, other studies sought to understand this complexity, reduce uncertainty and facilitate SDG-related decision making through modeling and simulation. For example, in a case study on sustainable tourism in Brazil [14], photovoltaic energy investments in Tanzania [42], and different models, including both scenario analysis and quantitative modeling [43]. However, there is no scientific research on the application of QHIM with the 20 indicators proposed in this study. Also, there are no studies on the Forgotten Effects Theory considering the QHIM and SDGs. In this sense, the study seeks to reduce the identified gaps and contribute to sustainable development with the proposed models. Consequently, the manuscript is novel and useful to various stakeholders, such as governments, society, and academia.

In this context, the present research intends to advance the frontier of knowledge on sustainable development, relating the QHIM with the SDGs through a case study in Latin America and contributing an algorithm in decision making. The next subsection is dedicated to case analysis.

2.2. QHIM: The Latin-American Countries' Case

This subsection is dedicated to five Latin American countries' case analysis concerning sustainable development applying the QHIM. The countries analyzed were Brazil, Chile, Colombia, Peru, and Mexico. The QHIM was the model chosen to carry out the case study because it is scientifically based on the importance of integrating the five systems to achieve sustainable development [10,18].

However, the model has some drawbacks associated with the choice of indicators and data homogenization. Useful tips to overcome the drawbacks are: (i) use official databases, (ii) the indicators must present transversal characteristics. In this way, it will be possible to compare each helix, and according country's real situation, (iii) use the same period, (iv) create a single value scale, and (v) validate the indicators with experts.

The study focused on this region of the world, but the model presented is generic, which means that it can compare any country. Brazil and Mexico were selected for this case study because they are the two countries with the largest Latin American populations. Colombia and Peru have an intermediate population concerning the first two and Chile, the latter being the least densely populated of the five.

Also, this region of the world presents, in general terms, short-term policies, low investment in Research and Development (R&D), and a low number of scientists per million inhabitants. These countries also have low scientific and technological production, economies with high percentages of informality and unemployment, and inefficient use of renewable and non-renewable resources. Finally, society's low participation as an "auditor" of the activities carried out in the political, educational, business, and environmental spheres is evident [6,44]. This selection shows that the indicators applied are useful, regardless of the size of the country analyzed.

The data from official sources correspond to the period between 2000 and 2017. It should be noted that only some indicators had data until 2019, so a period was chosen in which all the information was available. The study uses 20 indicators that represent the QHIM as criteria for analysis. Each helix was assigned four indicators that are associated with the SDGs. Ten experts in the field of sustainability validated the indicators. The 20 indicators present transversal characteristics that allow a generic comparison of each helix and close the relationship with each country's real functioning. For this purpose, the proximity and remoteness method and ten initial indicators were used for each helix until a consensus was reached. Subsequently, the SDGs were assigned to each helix.

Official sources use different measurement scales when presenting data, which could make analysis difficult. For this reason, the study will use the same scale to homogenize the data. In this case, the endecadary scale with 11 values of [0, 1] will be used. Thus, the value closest to 1 expresses an approach to sustainable development, and the value closest to 0 shows a move away from development. Table A2 shows the five helixes' analysis criteria, the 20 indicators, the concepts, and the SDGs' links. The results of the case study are presented below. Also, Table A3 details the results of QHIM indicators.

Firstly, Figure 2 presents the four indicators of political capital (PC). The results indicated that Chile leads in all indicators of helix 1. On the other hand, there was an alternation in second place between Peru (PC1), Brazil (PC2 and PC3), and Mexico (PC4). In general terms, Brazil, Colombia, Peru, and Mexico presented results below 0.50, which shows these countries' fragility in political capital. Consequently, low government regulatory capacity, corruption, political instability, and inadequate public services can be barriers to achieving the SDGs (3, 10, 11, 16, and 17).

Secondly, Figure 3 shows the human capital (HC) indicators. The results revealed that Brazil led in three indicators (HC1, HC2, and HC3) and Mexico in one indicator (HC4). Overall, all five countries had the best result in HC1, which refers to total R&D expenditure. However, the total score for helix 2 would be below 0.30 (except for Brazil with 0.32), which shows a weakness in the education helix. At the same time, a concern, since low investment in education will compromise the reach of SDGs 4 and 9.

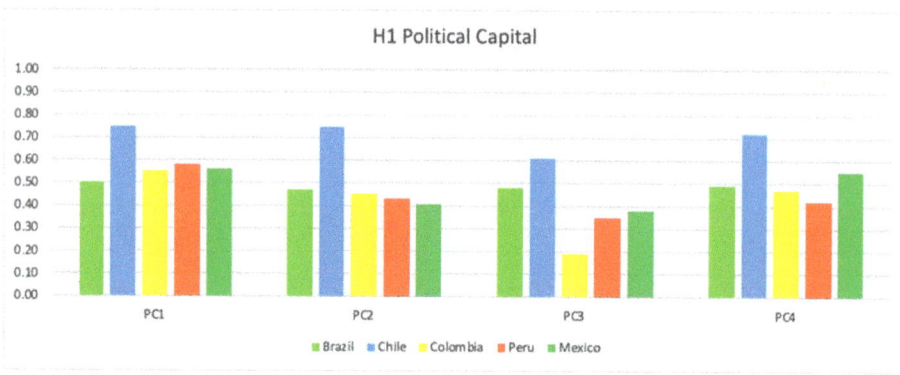

Figure 2. Political Capital (PC). Source: Own elaboration based on [45].

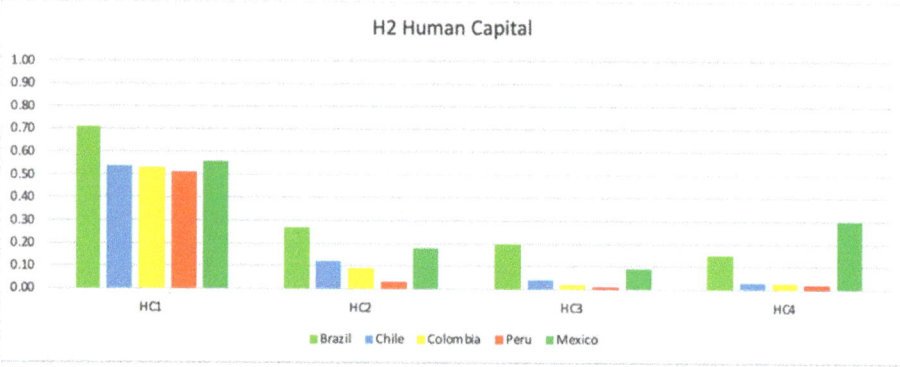

Figure 3. Human Capital (HC). Source: Own elaboration based on [45–48].

Thirdly, Figure 4 shows the indicators of helix 3, economic capital (EC). The results indicated that Colombia led in EC4, Mexico in EC2, Brazil in EC1, and Chile in EC3. In general, economic capital presented the worst result among all the helices. In general terms, the region has low foreign direct investment, high unemployment, a weak current account balance, and low purchasing power. As a consequence, it will negatively affect the fight against poverty (SDG1), hunger (SDG2), decent work (SDG8), industry, innovation and infrastructure (SDG9), and partnerships (SDG17).

Fourthly, Figure 5 presents the indicators of ecological capital (EN). Peru led in EN1, EN2, and EN4, and Brazil in EN3. Except for Mexico, the other four countries achieved a score above 0.60. In Mexico's case, the lowest ratings were in renewable energy (EN3) and population density (EN4), which impacted the final result. In helix 4, the countries analyzed show a small advance towards achieving the objectives (SDG 1, 2, 6, 7, 13, 14, and 15).

Fifthly, Figure 6 shows the social capital indicators (SC). Chile led in SC1, SC2, and SC4, and Mexico in SC3. All five countries presented total scores above 0.50, indicating some progress in gender development, human development, and poverty reduction, contributing to the SDGs (SDG 1, 2, 5, 10, 11, 12, and 15). However, the results point to the existence of gaps in the social sphere.

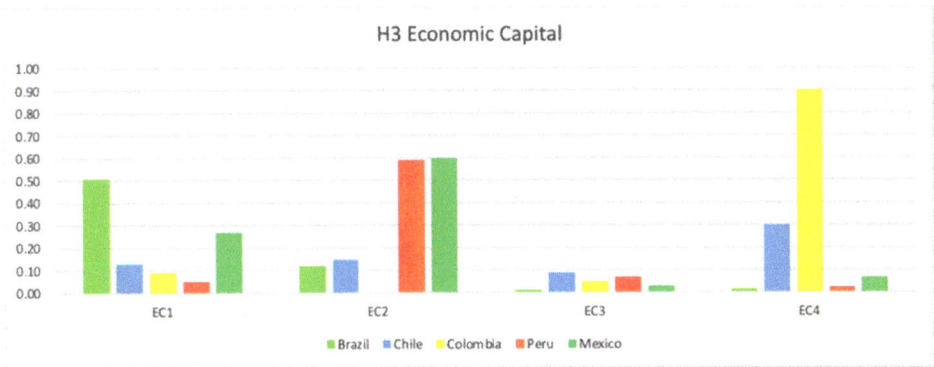

Figure 4. Economic Capital (EC). Source: Own elaboration based on [45].

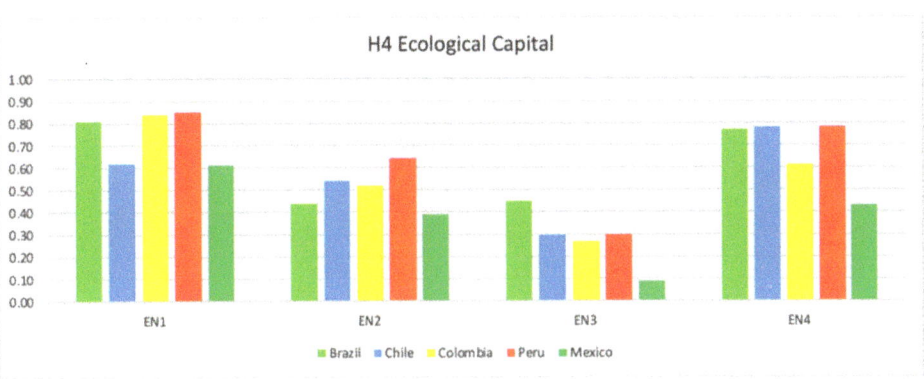

Figure 5. Ecological Capital (EN). Source: Own elaboration based on [45,49,50].

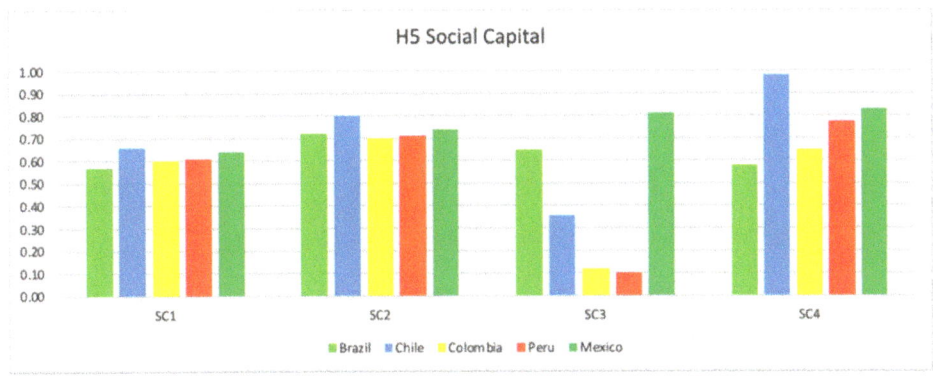

Figure 6. Social Capital (SC). Source: Own elaboration based on [45,51,52].

Figure 7 reveals the overall result of the five helices. Also, Table A4 shows the results of QHIM per each helix and country. The result of each helix is the average value of the four indicators per block. Social capital led the ranking (0.63), followed by ecological capital (0.55). Political capital would be in third position (0.51), followed by human capital (0.22) and economic capital (0.20). Chile led in H1, Brazil in H2, Colombia in H3, Peru in H4, and Mexico in H5.

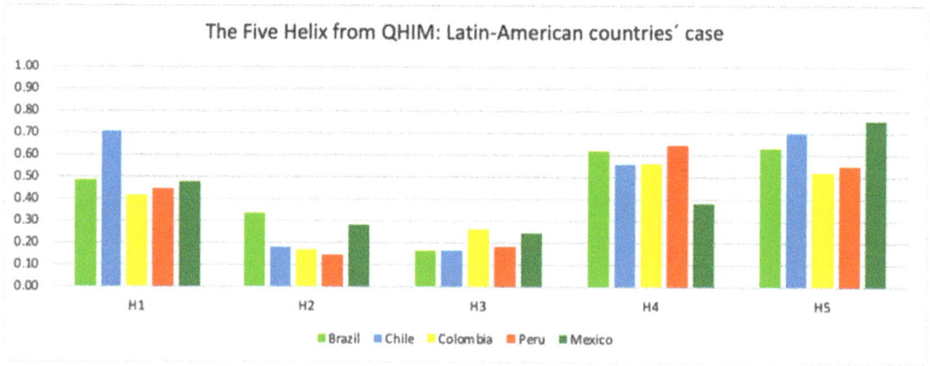

Figure 7. The Five Helix (H1–5) from QHIM: Latin-American countries' case. Source: Own elaboration based on [8,45–53].

Finally, Figure 8 shows the result of applying the QHIM through the case study in Latin America. The total value represents the average of the five helices for each country. Chile led the ranking with an overall score of 0.46. The second position would be Brazil (0.45), followed by Mexico (0.43), Peru (0.39), and Colombia (0.38).

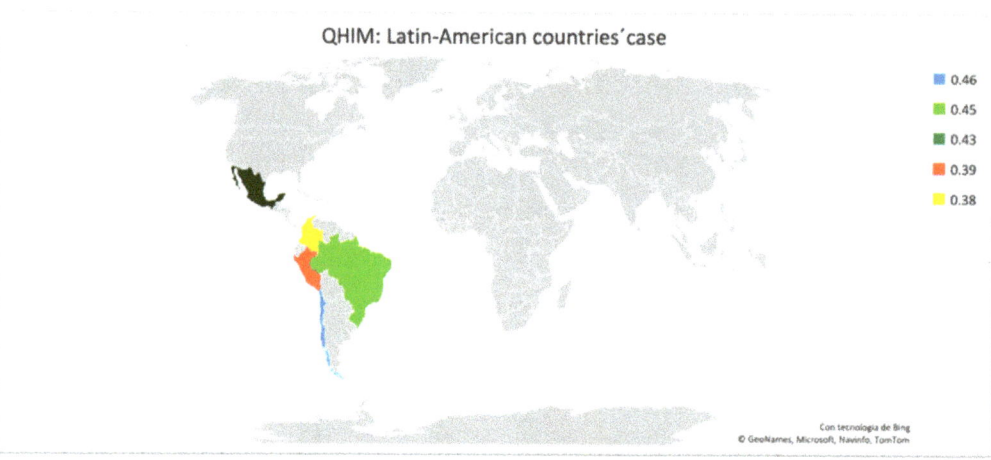

Figure 8. QHIM: Latin-American countries' case. Source: Own elaboration based on [8,45–53].

It should be remembered that the rating scale used was 11 values [0, 1]. Thus, the value closest to 1 expresses an approach to sustainable development, and the value closest to 0 shows a distance to sustainable development. Therefore, results show that there is still a long way to go for the countries analyzed towards sustainable development according to the indicators proposed by the QHIM. The main reason for this would be the low performance in the human and economic capital indicators.

Therefore, the study recommends that countries increase investment in education, incentives for research and development, fiscal balance, economic stimuli, foreign investment, and quality employment. The next subsection explains the use of the Forgotten Effects Theory.

2.3. Forgotten Effects Theory

This subsection explains the simulation algorithm and presents the process carried out to identify sustainable development's forgotten effects.

The "Forgotten Effects Theory" [54] is the mathematical model chosen to simulate this research. This algorithm was applied in several knowledge areas based on previous studies and presented reliable decision-making results [14,55]. However, the model has drawbacks associated with the selection of variables and the choice of experts. Useful tips for solving these problems are: first, it is necessary to know the research topic well and support the use of the variables scientifically. Secondly, it is essential to invite experts on the subject under investigation with time available to collaborate.

The process begins with the presence of a direct incidence relationship, defined by a cause-and-effect matrix defined by two sets of elements: $C = \{c_i / i = 1, 2, \ldots, n\}$ which act as causes; $E = \{e/j = 1, 2, \ldots, m\}$ which act as effects and a causality relationship $\widetilde{G}$ defined by the $n \times m$ dimension matrix: $\left[\widetilde{G}\right] = \left\{\mu_{c_i e_j} \in [0,1] / i = 1, 2, \ldots, n; j = 1, 2, \ldots, m\right\}$ being $\mu(c_i, e_j)$ of the values the characteristic function of belonging of each one of the elements of the matrix $\widetilde{G}$ (formed by the rows corresponding to the set's elements-causes-and the columns corresponding to the elements of the set-effects). The matrix $\widetilde{G}$, also named first-generation, is the result of cause-effect estimates. The assigned value belongs to the interval [0, 1], where zero means the lowest value, and the closer to 1, the higher the incidence rate.

The second step is to calculate the relationships between the causes, and the relationships between the effects, through two square auxiliary matrices. These two matrices include the possible effects derived from relating causes and effects to each other: $\left[\widetilde{C}\right] = \left\{\mu_{c_i c_j} \in [0,1] / i, j = 1, 2, \ldots, n\right\}$ and $\left[\widetilde{E}\right] = \left\{\mu_{e_i e_j} \in [0,1] / i, j = 1, 2, \ldots, m\right\}$.

The Matrix $\left[\widetilde{C}\right]$ shows the incidence relationships that can occur between causes, and the matrix $\left[\widetilde{E}\right]$ presents the incidence relationships that can occur between effects. Both matrices are reflexive: $\mu_{c_i c_j} = 1 \ \forall_{i=1,2,\ldots,n}$ and $\mu_{e_i e_b_j} = 1 \ \forall_{j=1,2,\ldots,m}$. Therefore, an element, either cause or effect, affects itself with the greatest presumption. Neither $\left[\widetilde{C}\right]$ nor $\left[\widetilde{E}\right]$ are symmetrical matrices, there is at least some pair of subscripts i, j so: $\mu_{c_i c_j} \neq \mu_{c_j c_i}$ and $\mu_{e_i e_j} \neq \mu_{e_j e_i}$.

The third step is to establish the direct and indirect incidences, through the maximum-minimum composition of the three matrices (1): $\left[\widetilde{C}\right] \circ \left[\widetilde{G}\right] \circ \left[\widetilde{E}\right] = \left[\widetilde{G}^*\right]$. The result is the matrix $\left[\widetilde{G}^*\right]$ that collects the incidences between causes and effects of second generation.

$$\left[\widetilde{G}^*\right] = \begin{array}{c} \\ c_1 \\ c_2 \\ \vdots \\ c_n \end{array} \begin{pmatrix} e_1 & e_2 & \cdots & e_m \\ \mu^*_{c_1 e_1} & \mu^*_{c_1 e_2} & \cdots & \mu^*_{c_1 e_m} \\ \mu^*_{c_2 e_1} & \mu^*_{c_2 e_2} & \cdots & \mu^*_{c_2 e_m} \\ \vdots & \vdots & \vdots & \vdots \\ \mu^*_{c_n e_1} & \mu^*_{c_n e_2} & \cdots & \mu^*_{c_n e_m} \end{pmatrix} \quad (1)$$

The fourth step is to calculate the degree to which some causal relationships were forgotten or overlooked (2): $\left[\widetilde{F}\right] = \left[\widetilde{G}^*\right] - \left[\widetilde{G}\right]$.

$$\left[\widetilde{F}\right] = \begin{array}{c} \\ c_1 \\ c_2 \\ \vdots \\ c_n \end{array} \begin{pmatrix} e_1 & \cdots & e_m \\ \mu^*_{c_1 e_1} - \mu_{c_1 e_1} & \cdots & \mu^*_{c_1 e_m} - \mu_{c_1 e_m} \\ \mu^*_{c_2 e_1} - \mu_{c_1 e_1} & \cdots & \mu^*_{c_2 e_m} - \mu_{c_2 e_m} \\ \vdots & \vdots & \vdots \\ \mu^*_{c_n e_1} - \mu_{c_n e_1} & \cdots & \mu^*_{c_n e_m} - \mu_{c_n e_m} \end{pmatrix} \quad (2)$$

With the result, it is possible to know the element that has been interposed between cause and effect. Figure 9 indicates the steps to follow.

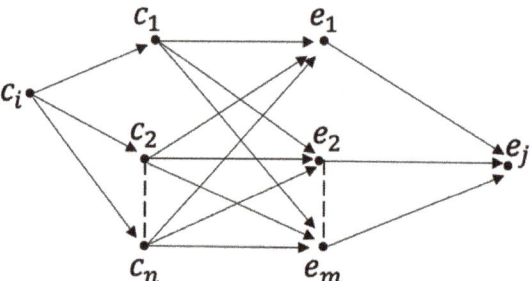

Figure 9. The max-min composition of the matrices.

Finally, the forgotten effects matrix shows that values closer to number 1 have a more significant forgotten effect. Therefore, some effects were not considered initially, and that can generate negative impacts.

The algorithm identifies an interposed element that enhances and accumulates the causal relationship's effects from its application. Therefore, the results allow predicting and acting more effectively on the causes, thus minimizing the effects.

2.4. Simulation Process

To proceed with the calculations, the software FuzzyLog© was used, which allows the elaboration and work with models based on the mathematics of uncertainty to recover the so-called forgotten effects in the causality relations. This program facilitates the values' insertion and automatically solves the incidence matrices' calculation, showing all the results directly in their different versions and variants in graphic and numerical form for their corresponding analysis. Researchers have validated this tool's effectiveness as robust, reliable, and easy to operate [13,56].

The research used the simulation process proposed by the authors [16], which consisted of four stages: (1) Analysis of a real-world problem, (2) Development and validation of the conceptual model, (3) Codification and verification of the model, and (4) Experimental development and simulation results.

The first stage of the simulation process corresponds to the five Latin American countries' case analysis presented in Section 2.2.

The second stage consisted of developing and validate the conceptual model, beginning with identifying the study variables. Two sets of interrelated elements have been proposed from the literature review that act as causes and effects. Three academic specialists on the subjects validate the 22 variables that are the study object in the simulation. They are professors and researchers in Brazil, Colombia, and Spain.

In this case, the set of causes represents the five innovation helixes and is presented as: $C = \{c_1, c_2, c_3, c_4, c_5\}$. Table 1 presents a set of causes.

Table 1. The five helices of innovation.

C	The Five Helices of Innovation
c_1	Political Capital
c_2	Human Capital
c_3	Economic Capital
c_4	Ecological Capital
c_5	Social Capital

Source: Own elaboration based on [10]. C/c: Cause.

The set of effects constitutes the SDGs and is presented as: $E = \{e_1, e_2, e_3, e_4, e_5, e_6, e_7, e_8, e_9, e_{10}, e_{11}, e_{12}, e_{13}, e_{14}, e_{15}, e_{16}, e_{17}\}$. Table 2 presents a set of effects.

Table 2. The 17 Sustainable Development Goals.

E	SDGs
e_1	1 No Poverty
e_2	2 Zero Hunger
e_3	3 Good Health and Well-being
e_4	4 Quality Education
e_5	5 Gender Equality
e_6	6 Clean Water and Sanitation
e_7	7 Affordable and Clean Energy
e_8	8 Decent Work and Economic Growth
e_9	9 Industry, Innovation and Infrastructure
e_{10}	10 Reducing Inequality
e_{11}	11 Sustainable Cities and Communities
e_{12}	12 Responsible Consumption and Production
e_{13}	13 Climate Action
e_{14}	14 Life Below Water
e_{15}	15 Life on Land
e_{16}	16 Peace, Justice, and Strong Institutions
e_{17}	17 Partnerships for the Goals

Source: Own elaboration based on [8]. E/e: effect.

The third stage was to code and verify the model proposed. All variables were inserted into the FuzzyLog© software provided by Anna María Gil-Lafuente, and a review of the data performed. Lastly, with the appropriate programming, the simulation was carried out.

Finally, in the experimental development stage, the specialists estimated the direct incidence between the two sets of causes and effects shown in the matrix $[\widetilde{M}]$. The assigned value belongs to the interval [0, 1], where zero means the lowest value, and the closer to 1, the higher the incidence rate. After collecting the assessments of each specialist, an average is calculated to obtain the consolidated outcome. Figure A1 shows the results.

Next, the same specialists evaluated the incidences between the causes and between the effects. The specialists sent the answers by e-mail in spreadsheet format. An average was then calculated to aggregate the values. As a result, the matrix between causes and the matrix between effects is generated, represented in Figures A2 and A3, respectively.

With the three matrices $[\widetilde{M}]$, $[\widetilde{A}]$, and $[\widetilde{B}]$, the accumulated effects matrix was calculated $[\widetilde{M}^*]$. Figure A4 presents the results of the matrix calculation $[\widetilde{A}] \circ [\widetilde{M}] \circ [\widetilde{B}] = [\widetilde{M}^*]$.

Finally, the Forgotten Effects Matrix was calculated: $[\widetilde{F}] = [\widetilde{M}^*] - [\widetilde{M}]$. Figure A5 shows the results.

The results of the Forgotten Effects Matrix indicated the effects not observed or forgotten during the assessment stage. The higher the value, the greater the degree of forgotten effect. Therefore, values closer to the number 1 deserve special attention from the decision-maker. The most relevant results of the simulation are presented below.

3. Results of Simulation Applying the Forgotten Effects Theory

This section presents the main results applying the Forgotten Effects Theory. The selection criterion used was to detail one result for each helix due to this publication's page limit. These five results presented are sufficient to validate the model. The incidences chosen should be between 0.8 (almost full incidence) and 0.9 (practically full incidence) [57]. Future studies may explore other application results.

Table 3 shows these cause-effect relationships that presented high incidences of 0.8 and 0.9 and recovered with the model's application.

Table 3. Cause-effect relationships.

Causes (Helices of Innovation)	Effects (SDGs)
H1 Political Capital	5 Gender Equality
H2 Human Capital	13 Climate Action
H3 Economic Capital	13 Climate Action
H4 Ecological Capital	2 Zero Hunger
H5 Social Capital	16 Peace, Justice, and Strong Institutions

Source: Own elaboration based on [54]. SDGs: sustainable development goals.

Firstly, Figure 10 shows the non-existence of a relationship between political capital (H1) and Gender Equality (SDG 5). However, it can be seen that the interposed element (SDG16) Peace, Justice, and Strong Institutions potentiated this relationship to 0.8. The figure also shows the path traveled with all incidents. Therefore, the result indicates that to achieve SDG5 political capital and strong institutions are needed to promote gender equality.

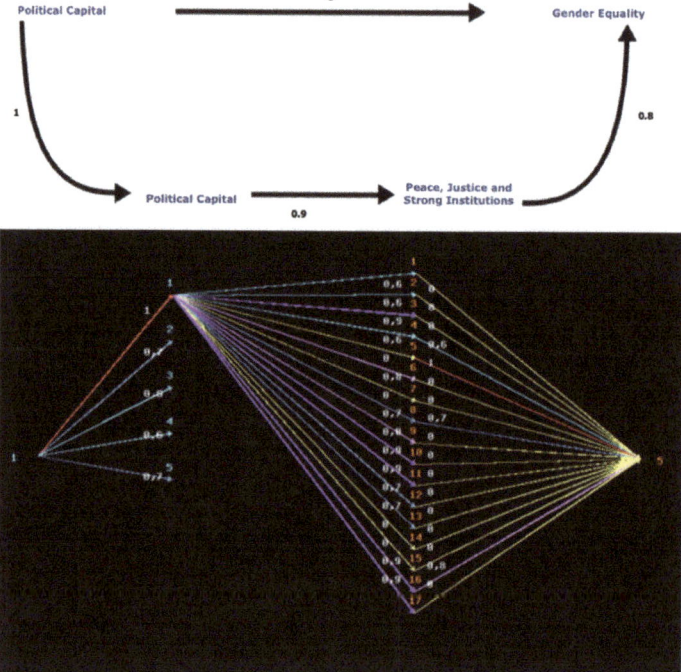

Figure 10. Relation between Political Capital (H1) and Gender Equality (SDG5).

Secondly, Figure 11 presents the relationship between the Human Capital (H2) and Climate Action (SDG13) variables. The result shows no direct relationship between the variables, but the interposed element (SDG9) Industry, Innovation, and Infrastructure potentiated this relationship to 0.9. Also, the figure presents all existing incidences. Therefore, the result shows the importance of H2 to reach the SDG13. In this case, investment in R&D strengthens the industry with sustainable production, reducing global warming.

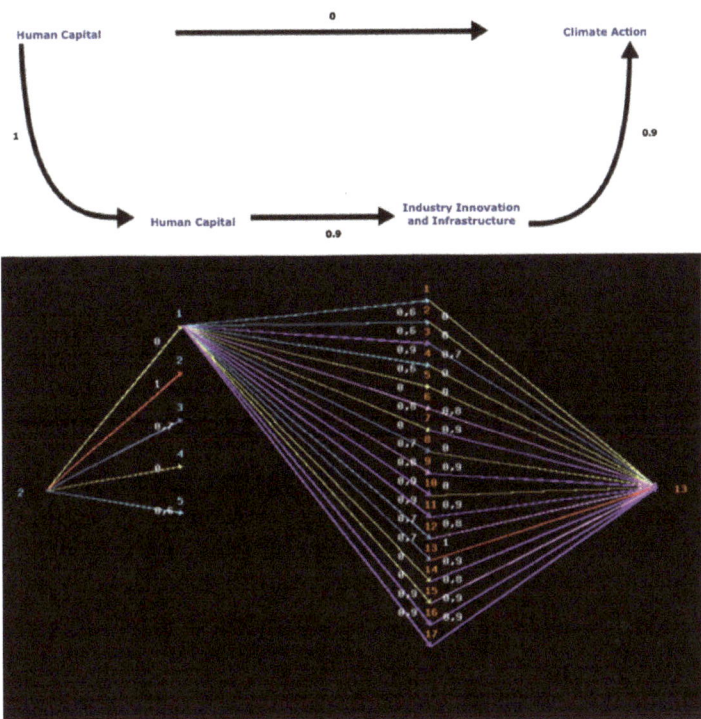

Figure 11. Relation between Human Capital (H2) and Climate Action (SDG 13).

Thirdly, Figure 12 shows the relationship between the Economic Capital (H3) and Climate Action (SDG 13) variables. The result shows no direct relationship between the variables, but the interposed element (SDG9) Industry, Innovation, and Infrastructure potentiated this relationship to 0.9. The figure also presents all existing incidences. Soon the result shows the importance of the H3 to reach the SDG13. In this case, the economic stimuli will increase the opportunities for SDG9 with sustainable solutions and thus allow to face climate change.

Fourthly, Figure 13 shows the relationship between the Ecological Capital (H4) and Zero Hunger (SDG2) variables. At first, this relationship did not exist, but the interposed elements (H3) Economic capital and (SDG1), No poverty, potentiated this relationship to 0.9. The figure also shows the path traveled with all incidents. Then, the result indicates that to reach SDG1, and it is necessary to involve the H4; for example, the use of clean energy will expand employment opportunities and, as a consequence, contribute to the reduction of poverty and hunger.

Fifthly, Figure 14 presents the relationship between the Social Capital (H5) and Peace, Justice, and Strong Institutions (SDG16) variables. The result showed no direct relationship between the variables, but the interposed element (SDG10) Reducing Inequality potentiated this relationship to 0.9. Also, the figure presents all existing incidences. Therefore, the result shows the importance of H5 to reach the SDG16. Therefore, social protection policies' adoption contributes to achieving greater equality, peace, and social justice progressively.

In summary, the results reinforce the existing links between the helices and the SDGs (Table A5). The algorithm's application allowed the identification of forgotten effects that can impact the scope of sustainable development. It is up to the decision-maker to use the simulation results or adjust the model and apply it in their country or company.

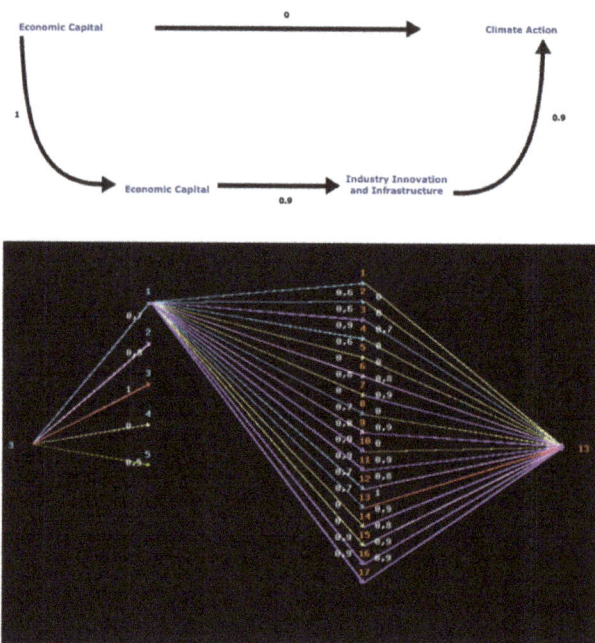

Figure 12. Relation between Economic Capital (H3) and Climate Action (SDG13).

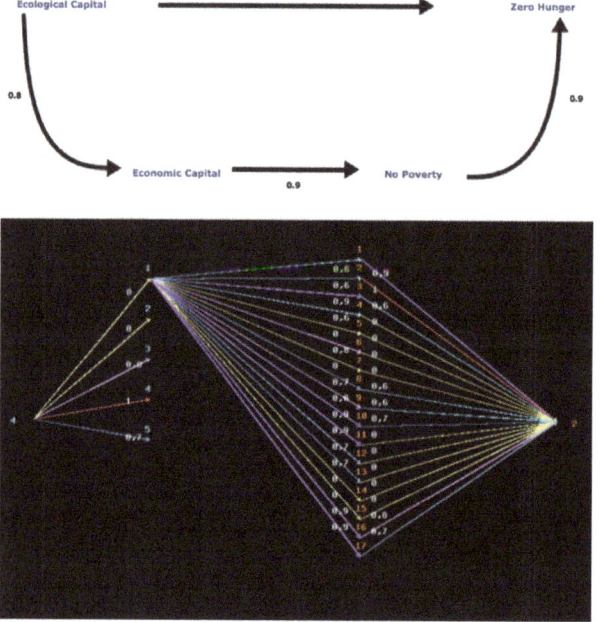

Figure 13. Relation between Ecological Capital (H4) and Zero Hunger (SDG2).

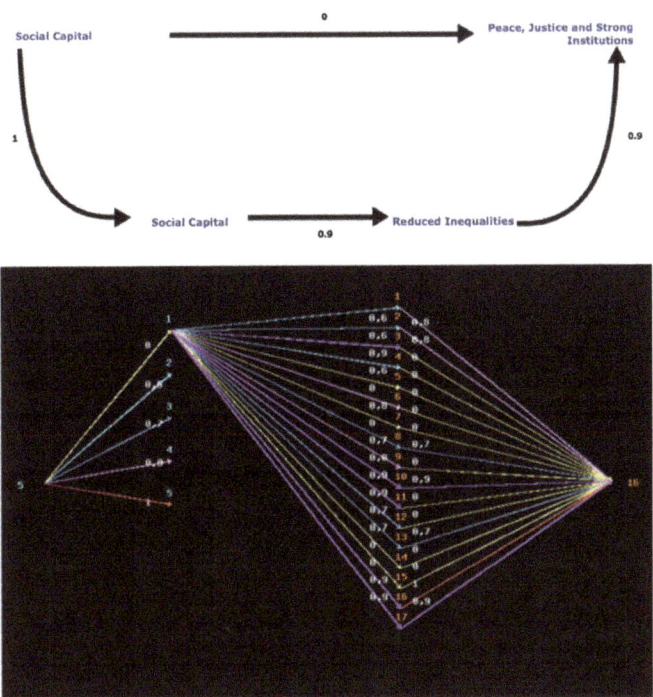

Figure 14. Relation between Social Capital (H5) and Peace, Justice, and Strong Institutions (SDG16).

4. Discussion of the Results

Applying the proposed QHIM model indicated that Chile was the country with the highest score, followed by Brazil, Mexico, Colombia, and Peru. In the ranking of the five helixes, social capital (H5) would rank first, followed by ecological (H4), political (H1), human (H2), and economic (H3) capital, respectively. Despite the progress made by these countries in recent years, the study identified opportunities for improvement in all helixes, which can support decision-making on strategy and prioritization of actions. In this sense, the study recommends that countries increase investment in education, incentives for research and development, fiscal balance, economic stimulus, foreign investment, and quality employment.

In response to other studies [37–39], the application of the QHIM provides a set of indicators with quality, accessible and timely data from official sources, which reduces uncertainty in decision making. In this way, the research contributes to a conceptual framework of indicators reducing the identified gap [37]. In line with another study [44], the QHIM can help implement sustainable development systems. With this model, it is possible to know each helix's result and its links with the SDGs and the country's overall performance. Also, the model makes it possible to identify the interrelationships between the systems.

However, it would be interesting in the future to compare the results with other methods, such as generations of Ordered Weighted Averaging (OWA) Operators [58] or Pythagorean Fuzzy Uncertain Environments [59]. As advantages, these methods allow to add weights to the variables, to deal with large amounts of data, and to prioritize the results. In this way, they are methods that facilitate management and decision making.

On the other hand, the case study confirmed that the SDGs' scope depends on several systems [10], so it is necessary to evaluate the five helices (social, ecological, political, human, and economic) analyze them in an integrated manner. Table A3 shows the main study results. These results reinforce the findings of other studies [3,36]. Countries should have a systemic vision since one helix will affect the others' performance and, consequently, sustainable development [8].

Nevertheless, they reinforce research [36,40] on understanding the correlation between the SDGs in decision making. Also, the application of the QHIM and the simulation conducted seek to reduce the gaps identified by other authors [8,41,42].

The research also reveals the importance of correctly identifying cause-effects by seeking harmony between systems. The application of a Fuzzy Logic algorithm identified the forgotten effects of sustainable development. It confirmed other authors' findings [37] on the uncertainty caused by the SDGs' interactions. The simulation also confirmed the results of the QHIM application from official sources.

Finally, the simulation corroborated the indications of other studies' results [14,42] by understanding this complexity, reducing uncertainty [16], and facilitating SDG-related decision making.

5. Conclusions

The research deepened the debate on sustainable development by relating the Quintuple Helix Innovation Model (QHIM) to the Sustainable Development Goals (SDGs). The literature review identified knowledge gaps in the implementation of the SDGs. For this reason, the study proposed 20 indicators related to the QHIM through a case analysis in Latin American countries to respond to the identified gaps. The study also applied a Fuzzy Logic algorithm to identify forgotten effects that may affect the SDGs' achievement, confirming the case study's findings. The results revealed the importance of correctly identifying cause-effects by seeking harmony between systems.

As theoretical contributions, the research advanced the frontier of knowledge by reducing the identified gaps, and at the same time, contributes to sustainable development with the proposed models. As practical contributions, the applied study offers governments, society, academia, and companies solutions adjusted to the problems identified, such as the lack of integration and systemic vision to achieve SDGs. Also, the research involved the participation of stakeholders in decision making.

Therefore, the present study is novel and useful for various stakeholders. A limitation of the research may be the number of variables used. Finally, the study results indicate promising research lines on sustainable development and decision making in uncertain environments applied to other countries. The study also opens up new research opportunities in prioritization models that facilitate decision-making.

Author Contributions: Conceptualization, L.B.-P. and I.D.l.V.; methodology, L.B.-P. and A.M.G.-L.; software, A.M.G.-L.; validation, L.B.-P., I.D.l.V., and A.M.G.-L.; formal analysis, I.D.l.V.; investigation, L.B.-P., and I.D.l.V.; resources, L.B.-P. and A.M.G.-L.; writing—original draft preparation, L.B.-P.; writing—review and editing, L.B.-P. and I.D.l.V.; supervision, A.M.G.-L. All authors have read and agreed to the published version of the manuscript.

Funding: This research received no external funding.

Institutional Review Board Statement: Not applicable.

Informed Consent Statement: Not applicable.

Data Availability Statement: Not applicable.

Acknowledgments: The authors wish to thank The Royal Academy of Economic and Financial Sciences and the University of Barcelona from Spain, and CENTRUM Católica Graduate Business School, Peru.

Conflicts of Interest: The authors declare no conflict of interest.

Appendix A

Table A1. Evolution of theoretical models.

Typologies	First Definition	Scope and Approaches of the Proposals
Triple Helix [20]	This model is a tool that applies to developed countries because it assumes that some activities are automatically related to economic growth. While this is true, the model focuses on universities but recognizes the other actors' dynamic interactions. It even proposes the creation of new intermediary organizations that are relevant for promoting knowledge generation processes.	This model seeks to establish a paradigm based on the growing interest in knowledge production among government, university, and business. In particular, it gives a central role to universities. It defines them as the actor that creates knowledge and plays a fundamental role in the relationship between companies and government policies. The aim is to identify the dynamic interactions in which innovative initiatives are generated.
Quadruple Helix Innovation Model [27]	This model describes the university-industry-government-public environment interactions within a knowledge economy. In this theory, each sector is represented by a knowledge subsystem (helix), which shows the four actors' overlapping interactions and seeks to boost innovation initiatives, giving a relevant social capital role.	The evolution of the model from Triple Helix to Quadruple Helix was due to incorporating an actor who changes the analysis perspective. The public environment (or social capital) allows us to examine society's behavior, understood as the public that interacts with other actors, from the concept of media-based democracy.
Quintuple Helix Innovation Model [10]	This model describes five knowledge subsystems (helices) and incorporates the environment as a key actor in the decision-making process. Here, know-how plays a predominant role since it allows the creation and transformation through innovation initiatives that activate the circulation of knowledge among the subsystems.	This model's model focuses on promoting innovation initiatives that seek to generate socio-ecological interactions from the circulation of knowledge from the subsystems (helices). This model's central value focuses on environmental impact and seeks to generate awareness of the responsibility of societies concerning this issue.
Smart Quintuple Helix Innovation System [18]	This theoretical model based on innovation is based on the interaction of five subsystems that exchange knowledge to generate and promote sustainable development. The subsystems are political capital, educational capital, economic capital, environmental capital, and social capital.	These theoretical models are evolutions of the triple helix incorporating relevant actors conceptualized as knowledge subsystems. It starts from the conception of a search for developing countries from the articulation of innovative initiatives. Each capital plays a relevant role and in which the environment is critical. Likewise, social capital and social networks, and media outlets have vital roles in generating opinion matrices and in which society can 'audit' the decisions of other actors.

Source: Own elaboration based on [10,18,20,27].

Table A2. Analysis criteria.

Helix	Indicators	Concepts	[1] SDGs
H1. Politics (Political Capital)	PC1. Regulatory Quality PC2. Control of Corruption PC3. Political Stability and Absence of violence PC4. Government Effectiveness	PC1. Reflects perceptions of the government's ability to formulate and implement sound policies and regulations that permit and promote private sector development. PC2. Reflects perceptions of the extent to which public power is exercised for private gain, including both petty and grand forms of corruption and capture of the state by elites and private interests. PC3. Measures perceptions of the likelihood of political instability and politically motivated violence, including terrorism. PC4. Reflects perceptions of the quality of public services, the quality of the civil service and the degree of its independence from political pressures, the quality of policy formulation and implementation, and the credibility of the government's commitment to such policies.	3, 10, 11, 16, and 17.

Table A2. Cont.

Helix	Indicators	Concepts	[1] SDGs
H2. Education (Human Capital)	HC1. GERD (GDP %) HC2. GERD (US$PPP) HC3. Scientific articles published HC4. Granted patents	HC1. Defined as the total expenditure (current and capital) on R&D carried out by all resident companies, research institutes, university, and government laboratories, etc., in a country. HC2. Total intramural expenditure on R&D performed during a specific reference period expressed in Purchasing Power Parity dollars. HC3. Scientific articles published in the Web of Science. HC4. Measures the number of granted patents.	4 and 9.
H3. Economy (Economic Capital)	EC1. Foreign Direct Investment EC2. Unemployment EC3. Current Account Balance EC4. Purchasing power parity (PPP)	EC1. Refers to direct investment equity flows in the reporting economy. EC2. Refers to the share of the labor force that is without work but available for and seeking employment. EC3. The current account balance is the sum of net exports of goods and services, net primary income, and net secondary income. Data are in current US$ dollars. EC4. PPP conversion factor is the number of units of a country's currency required to buy the same amount of goods and services in the domestic market as a U.S. dollar would buy in the United States.	1, 2, 8, 9, and 17.
H3. Economy (Economic Capital)	EC1. Foreign Direct Investment EC2. Unemployment EC3. Current Account Balance EC4. Purchasing power parity (PPP)	EC1. Refers to direct investment equity flows in the reporting economy. EC2. Refers to the share of the labor force that is without work but available for and seeking employment. EC3. The current account balance is the sum of net exports of goods and services, net primary income, and net secondary income. Data are in current US$ dollars. EC4. PPP conversion factor is the number of units of a country's currency required to buy the same amount of goods and services in the domestic market as a U.S. dollar would buy in the United States.	1, 2, 8, 9, and 17.
H4. Environment (Ecological Capital)	EN1. CO_2 Emissions EN2. Renewable water resources EN3. Renewable energy consumption EN4. Population density	EN1. Carbon dioxide emissions are those stemming from the burning of fossil fuels and the manufacture of cement. They include carbon dioxide produced during consumption of solid, liquid, and gas fuels and gas flaring. EN2. Total real renewable water resources in m^3 per inhabitant per year. EN3. Renewable energy consumption is the share of renewable energy in total final energy consumption. EN4. Population density is midyear population divided by land area in square kilometers.	1, 2, 6, 7, 13, 14, and 15.
H5. Society (Social Capital)	SC1. Gender Development SC2. Human Development Index SC3. International migrant stock SC4. Poverty headcount	SC1. It measures the human development costs of gender inequality. SC2. The Human Development Index is a summary measure of human development. It measures the average achievements in a country in three primary human development dimensions: a long and healthy life, access to knowledge, and a decent living standard. SC3. International migrant stock is the number of people born in a country other than those they live in. It also includes refugees. SC4. The poverty headcount ratio at $1.90 a day is the percentage of the population living on less than $1.90 a day at 2011 international prices.	1, 2, 5, 11, 10, 12, 13 and 15.

Source: [8,45–53]. [1] SDGs: 1 No Poverty, 2 Zero Hunger, 3 Good Health and Well-being, 4 Quality Education, 5 Gender Equality, 6 Clean Water and Sanitation, 7 Affordable and Clean Energy, 8 Decent Work and Economic Growth, 9 Industry, Innovation and Infrastructure, 10 Reducing Inequality, 11 Sustainable Cities and Communities, 12 Responsible Consumption and Production, 13 Climate Action, 14 Life Below Water, 15 Life on Land, 16 Peace, Justice, and Strong Institutions, and 17 Partnerships for the Goals.

Table A3. Results of QHIM indicators.

S	Indicators	Brazil	Chile	Colombia	Peru	Mexico
H1. Political Capital	PC1. Regulatory Quality	0.50	0.75	0.55	0.58	0.56
	PC2. Control of Corruption	0.47	0.75	0.45	0.43	0.41
	PC3. Political Stability and Absence of violence	0.48	0.61	0.19	0.35	0.38
	PC4. Government Effectiveness	0.49	0.72	0.47	0.42	0.55
H2. Human Capital	HC1. GERD (GDP %)	0.71	0.54	0.53	0.51	0.56
	HC2. GERD (US$PPP)	0.27	0.12	0.09	0.03	0.18
	HC3. Scientific articles published	0.20	0.04	0.02	0.01	0.09
	HC4. Granted patents	0.15	0.03	0.03	0.02	0.30
H3. Economic Capital	EC1. Foreign Direct Investment	0.51	0.13	0.09	0.05	0.27
	EC2. Unemployment	0.12	0.15	0.00	0.59	0.60
	EC3. Current Account Balance	0.01	0.09	0.05	0.07	0.03
	EC4. Purchasing power parity (PPP)	0.01	0.30	0.90	0.02	0.07
H4. Ecological Capital	EN1. CO_2 Emissions	0.81	0.62	0.84	0.85	0.61
	EN2. Renewable water resources	0.44	0.54	0.52	0.64	0.39
	EN3. Renewable energy consumption	0.45	0.30	0.27	0.30	0.09
	EN4. Population density	0.77	0.78	0.61	0.78	0.43
H5. Social Capital	SC1. Gender Development	0.57	0.66	0.60	0.61	0.64
	SC2. Human development Index	0.72	0.80	0.70	0.71	0.74
	SC3. International migrant stock	0.65	0.36	0.12	0.10	0.81
	SC4. Poverty headcount	0.58	0.98	0.65	0.77	0.83

Note: Official sources use different measurement scales when presenting the data, which could make analysis difficult. For this reason, the study used the same scale to homogenize the data. In this case, the endecadary scale with 11 values of [0, 1] was used. Thus, the value closest to 1 expresses an approach to sustainable development, and the value closest to 0 shows a move away from sustainable development. Source: Own elaboration based on [8,45–53].

Table A4. Results of QHIM per each helix and country.

	Brazil	Chile	Colombia	Peru	Mexico
H1. Political Capital	0.49	0.71	0.42	0.45	0.48
H2. Human Capital	0.33	0.18	0.17	0.14	0.28
H3. Economic Capital	0.16	0.17	0.26	0.18	0.24
H4. Ecological Capital	0.62	0.56	0.56	0.64	0.38
H5. Social Capital	0.63	0.70	0.52	0.55	0.76
TOTAL	0.45	0.46	0.38	0.39	0.43

Note: The result of each helix is the average value of the four indicators per block. The total value represents the average of the five helices for each country. Source: Own elaboration based on [8,45–53].

Table A5. Main study results.

	Case Analysis Applying the QHIM	SDGs (Effects)	Simulation Applying the Forgotten Effects Theory	SDGs (Forgotten Effects)
H1	Low government regulatory capacity, corruption, political instability, and inadequate public services.	3,10,11, 16 and 17	Political Capital (H1) and Gender Equality (SDG5).	16
H2	Low investment in education.	4 and 9.	Human Capital (H2) and Climate Action (SDG 13)	9
H3	Low foreign direct investment, high unemployment, weak current account balance, and low purchasing power.	1, 2, 8, 9 and 17.	Economic Capital (H3) and Climate Action (SDG13).	9

Table A5. Cont.

	Case Analysis Applying the QHIM	SDGs (Effects)	Simulation Applying the Forgotten Effects Theory	SDGs (Forgotten Effects)
H4	Little progress in environmental indicators. Better performance in CO_2 emissions. Lower performance in renewable energy consumption.	1, 2, 6, 7, 13, 14 and 15.	Ecological Capital (H4) y Zero Hunger (SDG2).	1
H5	Some progress in gender development, human development, and poverty reduction.	1, 2, 5, 10, 11, 12 and 15.	Social Capital (H5) and Peace, Justice, and Strong Institutions (SDG16).	10

Source: Own elaboration.

	E_1	E_2	E_3	E_4	E_5	E_6	E_7	E_8	E_9	E_{10}	E_{11}	E_{12}	E_{13}	E_{14}	E_{15}	E_{16}	E_{17}
C_1	0.6	0.6	0.9	0.6	0	0.8	0	0.7	0.8	0.9	0.9	0.7	0.7	0	0	0.9	0.9
C_2	0.4	0.4	0	0.9	0	0	0	0	0.9	0.5	0.8	0	0	0	0	0.5	0
C_3	0.9	0.9	0.6	0.7	0	0.7	0	0.9	0.9	0.6	0	0.7	0	0	0	0	0.9
C_4	0.7	0	0.6	0	0	0.9	0.9	0	0	0	0.7	0.6	0.9	0.9	0.9	0	0
C_5	0.9	0.9	0.7	0.7	0.9	0.8	0	0	0	0.9	0.9	0.9	0	0	0.9	0	0

Figure A1. Matrix with a valuation of direct incidences between causes (C) and effects (E).

	C_1	C_2	C_3	C_4	C_5
C_1	1	0.7	0.6	0.6	0.7
C_2	0	1	0.7	0	0.6
C_3	0.7	0.8	1	0	0.5
C_4	0	0	0.8	1	0.7
C_5	0	0.6	0.7	0.8	1

Figure A2. Matrix with a valuation of incidences among causes.

	E_1	E_2	E_3	E_4	E_5	E_6	E_7	E_8	E_9	E_{10}	E_{11}	E_{12}	E_{13}	E_{14}	E_{15}	E_{16}	E_{17}
E_1	1	0.9	0.6	0.7	0	0.7	0	0	0	0.8	0.6	0.5	0	0	0	0.8	0
E_2	0.8	1	0.8	0	0	0	0	0.5	0	0.8	0.7	0	0	0	0	0.8	0
E_3	0.7	0.6	1	0	0	0.8	0	0.7	0	0.7	0	0	0.7	0	0	0	0
E_4	0.6	0	0	1	0.6	0	0	0.8	0.7	0	0	0	0	0	0	0	0
E_5	0.8	0	0	0.8	1	0	0	0.8	0	0.7	0	0	0	0	0	0	0
E_6	0.7	0	0	0	0	1	0	0	0.8	0	0.9	0.7	0.8	0.9	0	0	0
E_7	0	0	0.8	0	0	0	1	0	0	0	0.8	0	0.9	0	0	0	0
E_8	0.7	0.6	0.7	0	0.7	0	0	1	0.7	0.8	0	0	0	0	0	0.7	0
E_9	0.6	0.6	0.5	0	0	0	0.5	1	0.6	0.7	0.7	0.9	0.6	0.6	0	0	
E_{10}	0.6	0.7	0.6	0	0	0	0	0.6	0	1	0	0	0	0	0	0.9	0
E_{11}	0	0	0.8	0	0	0.8	0.7	0	0	0	1	0.7	0.9	0.7	0.7	0	0
E_{12}	0	0	0	0	0	0	0.7	0.6	0.8	0	0.7	1	0.8	0.7	0.7	0	0
E_{13}	0	0	0	0	0	0	0	0	0	0	0.7	0	1	0.9	0.8	0.7	0
E_{14}	0	0	0.6	0	0	0	0	0	0	0	0	0	0.9	1	0.7	0	0
E_{15}	0	0	0.7	0	0	0.6	0	0	0	0	0	0	0.8	0.7	1	0	0
E_{16}	0.9	0.8	0.7	0.8	0.8	0.7	0.6	0.7	0.8	0.6	0.7	0	0.9	0.7	0.6	1	0.6
E_{17}	0.8	0.7	0	0	0	0	0.6	0.7	0.8	0	0.6	0.9	0	0	0.9	0	1

Figure A3. Matrix with a valuation of incidents between effects.

	E_1	E_2	E_3	E_4	E_5	E_6	E_7	E_8	E_9	E_{10}	E_{11}	E_{12}	E_{13}	E_{14}	E_{15}	E_{16}	E_{17}
C_1	0.9	0.8	0.9	0.8	0.8	0.8	0.7	0.7	0.8	0.9	0.9	0.7	0.9	0.8	0.7	0.9	0.9
C_2	0.7	0.7	0.8	0.9	0.7	0.8	0.7	0.8	0.9	0.7	0.8	0.7	0.9	0.7	0.7	0.7	0.7
C_3	0.9	0.9	0.8	0.8	0.7	0.8	0.7	0.9	0.9	0.8	0.8	0.7	0.9	0.7	0.7	0.9	0.9
C_4	0.8	0.8	0.8	0.7	0.7	0.9	0.9	0.8	0.8	0.8	0.7	0.9	0.9	0.9	0.9	0.8	0.8
C_5	0.9	0.9	0.8	0.8	0.9	0.8	0.8	0.8	0.8	0.9	0.9	0.9	0.9	0.8	0.9	0.9	0.7

Figure A4. Cumulative effects matrix.

	E_1	E_2	E_3	E_4	E_5	E_6	E_7	E_8	E_9	E_{10}	E_{11}	E_{12}	E_{13}	E_{14}	E_{15}	E_{16}	E_{17}
C_1	0.3	0.2	0	0.2	0.8	0	0.7	0	0	0	0	0	0.2	0.8	0.7	0	0
C_2	0.3	0.3	0.8	0	0.7	0.8	0.7	0.8	0	0.2	0	0.7	0.9	0.7	0.7	0.2	0.7
C_3	0	0	0.2	0.1	0.7	0.1	0.7	0	0	0.2	0.8	0	0.9	0.7	0.7	0.9	0
C_4	0.1	0.8	0.2	0.7	0.7	0	0	0.8	0.8	0.8	0.2	0.1	0	0	0	0.8	0.8
C_5	0	0	0.1	0.1	0	0	0.8	0.8	0.8	0	0	0	0.9	0.8	0	0.9	0.7

Figure A5. Forgotten Effects matrix.

References

1. Gil-Lafuente, A.M.; De Paula, L.B. Algorithm applied in the identification of stakeholders. *Kybernetes* **2013**, *42*, 674–685. [CrossRef]
2. Sachs, J.D. How Can We Achieve the SDGS? Strategic Policy Directions. *Dubai Policy Rev.* **2019**, *1*, 25–31. [CrossRef]
3. Cordova, M.F.; Celone, A. SDGs and Innovation in the Business Context Literature Review. *Sustainability* **2019**, *11*, 7043. [CrossRef]
4. De la Vega, I.; De Paula, L.B. Scientific mapping on the convergence of innovation and sustainability (innovability): 1990–2018. *Kybernetes* **2020**. Ahead-of-Publication. [CrossRef]
5. Ghosh, S.; Rajan, J. The business case for SDGs: An analysis of inclusive business models in emerging economies. *Int. J. Sustain. Dev. World Ecol.* **2019**, *26*, 344–353. [CrossRef]
6. Hernández, I.M.D.L.V.; De Paula, L.B. The quintuple helix innovation model and brain circulation in central, emerging and peripheral countries. *Kybernetes* **2019**, *49*, 2241–2262. [CrossRef]
7. World Commission on Environment and Development. *Our Common Future*; Oxford University Press: Oxford, UK, 1987.
8. Take Action for the Sustainable Development Goals. Available online: https://www.un.org/sustainabledevelopment/sustainable-development-goals/ (accessed on 10 October 2020).
9. Freeman, R.E.; Harrison, J.S.; Wicks, A.C.; Parmar, B.; Colle, S. *Stakeholder Theory: The State of the Art*; Cambridge University Press: Cambridge, UK; New York, NY, USA, 2010; ISBN 978-0-521-19081-7/978-0-521-13793-5.
10. Carayannis, E.G.; Barth, T.D.; Campbell, D.F. The Quintuple Helix innovation model: Global warming as a challenge and driver for innovation. *J. Innov. Entrep.* **2012**, *1*, 2. [CrossRef]
11. Zadeh, L.A. Fuzzy sets. *Inf. Control.* **1965**, *8*, 338–353. [CrossRef]
12. Mesa, F.R.B.; Merigó, J.M.; Gil Lafuente, A.M. Fuzzy decision making: A bibliometric-based review. *J. Intell. Fuzzy Syst.* **2017**, *32*, 2033–2050. [CrossRef]
13. Carles, M.-F.; Patricia, H.; Antonio, S.; Merigo, J.M. The Forgotten Effects: An Application in the Social Economy of Companies of the Balearic Islands. *Econ. Comput. Econ. Cybern. Stud. Res.* **2018**, *52*, 147–160. [CrossRef]
14. De Paula, L.B.; Gil-Lafuente, A.M.; Alvares, D.F. A contribution of fuzzy logic to sustainable tourism through a case analysis in Brazil. *J. Intell. Fuzzy Syst.* **2021**, *40*, 1851–1864. [CrossRef]
15. Bertrand, J.W.M.; Fransoo, J.C. Operations management research methodologies using quantitative modeling. *Int. J. Oper. Prod. Manag.* **2002**, *22*, 241–264. [CrossRef]
16. Harper, A.; Mustafee, N.; Yearworth, M. Facets of trust in simulation studies. *Eur. J. Oper. Res.* **2021**, *289*, 197–213. [CrossRef]
17. Yin, R. *Case Study Research, Design Methods*, 5th ed.; SAGE Publications Ltd.: Thousand Oaks, CA, USA, 2014; ISBN 978-1452242569.
18. Carayannis, E.G.; Campbell, D.F.J. *Smart Quintuple Helix Innovation Systems*; SpringerBriefs in Business; Springer International Publishing: Cham, Switzerland, 2019; ISBN 978-3-030-01516-9.
19. The Endless Transition: A "Triple Helix" of University-Industry-Government Relations. *Minerva* **1998**, *36*, 203–208. [CrossRef]
20. Etzkowitz, H. The norms of entrepreneurial science: Cognitive effects of the new university–industry linkages. *Res. Policy* **1998**, *27*, 823–833. [CrossRef]
21. Etzkowitz, H.D.; Leydesdorff, L. The dynamics of innovation: From National Systems and "Mode 2" to a Triple Helix of university–industry–government relations. *Res. Policy* **2000**, *29*, 109–123. [CrossRef]

22. Leydesdorff, L. The Triple Helix of University-Industry-Government Relations (February 2012). *SSRN Electron. J.* **2012**. [CrossRef]
23. Etzkowitz, H.; Zhou, C. Innovation incommensurability and the science park. *R&D Manag.* **2018**, *48*, 73–87. [CrossRef]
24. Etzkowitz, H. The evolution of the entrepreneurial university. *Int. J. Technol. Glob.* **2004**, *1*, 64. [CrossRef]
25. Etzkowitz, H.; Zhou, C. Introduction to special issue Building the entrepreneurial university: A global perspective. *Sci. Public Policy* **2008**, *35*, 627–635. [CrossRef]
26. Dzisah, J.; Etzkowitz, H. Triple helix circulation: The heart of innovation and development. *Int. J. Technol. Manag. Sustain. Dev.* **2008**, *7*, 101–115. [CrossRef]
27. Carayannis, E.G.; Campbell, D.F. 'Mode 3' and 'Quadruple Helix': Toward a 21st century fractal innovation ecosystem. *Int. J. Technol. Manag.* **2009**, *46*, 201. [CrossRef]
28. Carayannis, E.G.; Campbell, D.F. Triple Helix, Quadruple Helix and Quintuple Helix and How Do Knowledge, Innovation and the Environment Relate To Each Other? *Int. J. Soc. Ecol. Sustain. Dev.* **2010**, *1*, 41–69. [CrossRef]
29. Carayannis, E.G.; Campbell, D.F.J. Open Innovation Diplomacy and a 21st Century Fractal Research, Education and Innovation (FREIE) Ecosystem: Building on the Quadruple and Quintuple Helix Innovation Concepts and the "Mode 3" Knowledge Production System. *J. Knowl. Econ.* **2011**, *2*, 327–372. [CrossRef]
30. Mondini, G. Sustainability assessment: From brundtland report to sustainable development goals. *Valori Valutazioni* **2019**, *23*, 129–137.
31. Brown, J.A.; Clark, C.; Buono, A.F. The United Nations Global Compact: Engaging Implicit and Explicit CSR for Global Governance. *J. Bus. Ethic* **2016**, *147*, 721–734. [CrossRef]
32. Lomazzi, M.; Borisch, B.; Laaser, U. The Millennium Development Goals: Experiences, achievements and what's next. *Glob. Health Action* **2014**, *7*, 23695. [CrossRef]
33. Mitchell, D.; Allen, M.R.; Hall, J.W.; Muller, B.; Rajamani, L.; Le Quéré, C. The myriad challenges of the Paris Agreement. *Philos. Trans. R. Soc. A Math. Phys. Eng. Sci.* **2018**, *376*, 20180066. [CrossRef] [PubMed]
34. Fraisl, D.; Campbell, J.; See, L.; Wehn, U.; Wardlaw, J.; Gold, M.; Moorthy, I.; Arias, R.; Piera, J.; Oliver, J.L.; et al. Mapping citizen science contributions to the UN sustainable development goals. *Sustain. Sci.* **2020**, *15*, 1735–1751. [CrossRef]
35. Junior, R.M.; Fien, J.; Horne, R. Implementing the UN SDGs in Universities: Challenges, Opportunities, and Lessons Learned. *Sustain. J. Rec.* **2019**, *12*, 129–133. [CrossRef]
36. Pradhan, P.; Costa, L.; Rybski, D.; Lucht, W.; Kropp, J.P. A Systematic Study of Sustainable Development Goal (SDG) Interactions. *Earth's Future* **2017**, *5*, 1169–1179. [CrossRef]
37. Janoušková, S.; Hák, T.; Moldan, B. Global SDGs Assessments: Helping or Confusing Indicators? *Sustainability* **2018**, *10*, 1540. [CrossRef]
38. Yonehara, A.; Saito, O.; Hayashi, K.; Nagao, M.; Yanagisawa, R.; Matsuyama, K. The role of evaluation in achieving the SDGs. *Sustain. Sci.* **2017**, *12*, 969–973. [CrossRef]
39. Hák, T.; Janoušková, S.; Moldan, B. Sustainable Development Goals: A need for relevant indicators. *Ecol. Indic.* **2016**, *60*, 565–573. [CrossRef]
40. Morton, S.; Pencheon, D.; Squires, N. Sustainable Development Goals (SDGs), and their implementation. *Br. Med. Bull.* **2017**, *124*, 1–10. [CrossRef] [PubMed]
41. Nilsson, M.; Griggs, D.; Visbeck, M. Policy: Map the interactions between Sustainable Development Goals. *Nat. Cell Biol.* **2016**, *534*, 320–322. [CrossRef]
42. Collste, D.; Pedercini, M.; Cornell, S.E. Policy coherence to achieve the SDGs: Using integrated simulation models to assess effective policies. *Sustain. Sci.* **2017**, *12*, 921–931. [CrossRef] [PubMed]
43. Allen, C.; Metternicht, G.; Wiedmann, T. National pathways to the Sustainable Development Goals (SDGs): A comparative review of scenario modelling tools. *Environ. Sci. Policy* **2016**, *66*, 199–207. [CrossRef]
44. De la Vega, I. Estado y dinámicas de los sistemas tecnocientíficos: El caso de los países de la Alianza del Pacífico. *CLAD Mag. Reforma Democr.* **2018**, *70*, 29–60.
45. World Bank Open Data Countries. Available online: https://data.worldbank.org/%0Ahttps://data.worldbank.org/country (accessed on 10 October 2020).
46. OECD Gross Domestic Spending on R&D. Available online: https://data.oecd.org/rd/gross-domestic-spending-on-r-d.htm (accessed on 28 September 2020).
47. Web of Science Scientific Articles Published. Available online: http://login.webofknowledge.com/error/Error?Src=IP&Alias=WOK5&Error=IPError&Params=&PathInfo=%2F&RouterURL=http%3A%2F%2Fwww.webofknowledge.com%2F&Domain=.webofknowledge.com (accessed on 28 September 2020).
48. WIPO Statistics Database Total Patent Grants (Direct and PCT National Phase Entries). Total Count by Filling Office. Available online: https://www3.wipo.int/ipstats/index.htm?tab=patent (accessed on 28 September 2020).
49. FAO. Aquastat. Available online: http://www.fao.org/nr/water/aquastat/data/query/index.html?lang=en (accessed on 28 September 2020).
50. Our World in Data per Capita CO_2 Emissions. Available online: https://ourworldindata.org/grapher/co-emissions-per-capita?tab=chart (accessed on 28 September 2020).
51. UNDP Human Development Reports. Available online: http://hdr.undp.org/en/data# (accessed on 28 September 2020).

52. United Nations International Migrant Stock. Available online: https://www.un.org/en/development/desa/population/migration/data/estimates2/estimates19.asp (accessed on 28 September 2020).
53. UNESCO Institute for Statistics GERD in Purchasing Power Parities (PPPs). Available online: http://uis.unesco.org/en/glossary-term/gerd-purchasing-power-parities-ppps (accessed on 28 September 2020).
54. Kaufmann, A.; Gil-Aluja, J. *Modelos para la Investigación de Efectos Olvidados*; Editorial Milladoiro: Vigo, Spain, 1988; ISBN 84-404-3657-2.
55. Luciano, E.V.; Gil-Lafuente, A.M.; González, A.G.; Boria-Reverter, S. Forgotten effects of corporate social and environmental responsibility. *Kybernetes* **2013**, *42*, 736–753. [CrossRef]
56. Beatriz, F.-R.; Federico, G.-S. Study of the Competitiveness of the Michoacán Company and Variables that Affect it: Application of the Theory of Forgotten Effects. *Econ. Comput. Econ. Cybern. Stud. Res.* **2020**, *54*, 233–250. [CrossRef]
57. Linares-Mustarós, S.; Gil-Lafuente, A.M.; Coll, D.C.; Ferrer-Comalat, J.C. Premises for the Theory of Forgotten Effects. In *Advances in Intelligent Systems and Computing*; Springer: Berlin/Heidelberg, Germany, 2019; Volume 894, pp. 206–215.
58. Perez-Arellano, L.A.; Blanco-Mesa, F.; Leon-Castro, E.; Alfaro-Garcia, V. Bonferroni Prioritized Aggregation Operators Applied to Government Transparency. *Mathematics* **2020**, *9*, 24. [CrossRef]
59. Wang, L.; Garg, H.; Li, N. Pythagorean fuzzy interactive Hamacher power aggregation operators for assessment of express service quality with entropy weight. *Soft Comput.* **2021**, *25*, 973–993. [CrossRef]

Article

Fuzzy Techniques Applied to the Analysis of the Causes and Effects of Tourism Competitiveness

Martha B. Flores-Romero [1,†], Miriam E. Pérez-Romero [1,†], José Álvarez-García [2,*,†] and María de la Cruz del Río-Rama [3,†]

1. Faculty of Accounting and Management, Saint Nicholas and Hidalgo Michoacán State University (UMSNH), Morelia 58030, Mexico; betyf@umich.mx (M.B.F.-R.); miromero@umich.mx (M.E.P.-R.)
2. Financial Economy and Accounting Department, Faculty of Business, Finance and Tourism, University of Extremadura, 10071 Cáceres, Spain
3. Business Management and Marketing Department, Faculty of Business Sciences and Tourism, University of Vigo, 32004 Ourense, Spain; delrio@uvigo.es
* Correspondence: pepealvarez@unex.es
† These authors contributed equally to this work.

Abstract: The aim of this research is to identify and analyze the causes and effects of tourism competitiveness, as well as cause–effect relationships from the perspective of two groups of experts, which are decision makers versus academics/researchers, both from the tourism sector. The purpose is to respond to the question: do decision makers in the tourism sector share the same perspective as academics/researchers regarding the relationship between the causes and effects of tourism competitiveness? The methodology used is the theory of expertons, the theory of forgotten effects and the Hamming distance. It was found that in most cases, the groups of experts share perspective, since their differences are small or non-existent. However, in all the relationships analyzed (cause–effect, cause–cause, and effect–effect), academic experts reported the highest assessment. The greatest difference in opinion is identified in the evaluation of the "Environmental Commitment" and "Tourist Demand" relationship. Decision makers in the tourism sector are ignoring the growing inclination and sensitivity that tourists are adopting towards the environment. It is necessary for the tourism sector to develop and consolidate its commitment to caring for and preserving the environment, which is an element that contributes to a destination's competitiveness and has two main effects: tourism demand and customer satisfaction.

Keywords: tourist destination competitiveness; experton theory; forgotten effects theory; Hamming distance; decision making

Citation: Flores-Romero, M.B.; Pérez-Romero, M.E.; Álvarez-García, J.; del Río-Rama, M.d.l.C. Fuzzy Techniques Applied to the Analysis of the Causes and Effects of Tourism Competitiveness. *Mathematics* **2021**, *9*, 777. https://doi.org/10.3390/math9070777

Academic Editor: Jorge de Andres Sanchez

Received: 22 February 2021
Accepted: 29 March 2021
Published: 2 April 2021

Publisher's Note: MDPI stays neutral with regard to jurisdictional claims in published maps and institutional affiliations.

Copyright: © 2021 by the authors. Licensee MDPI, Basel, Switzerland. This article is an open access article distributed under the terms and conditions of the Creative Commons Attribution (CC BY) license (https://creativecommons.org/licenses/by/4.0/).

1. Introduction

Since the 80s and 90s, tourism has become a global economic phenomenon, which is a situation that has encouraged the search for competitive models that reveal what makes one destination more interesting than another [1]. Tourism is regarded as a sound alternative to achieve the economic development and social well-being of nations, but especially for developing or less developed ones [2]. However, tourism market trends and the life cycle of destinations bring about the need to undertake renovation processes [3]. Furthermore, the resources that position a destination as the most competitive today may not have any significance in the future [4].

The study of competitiveness became a topic of interest in tourism sector research in the 1990s, with the first researcher to conduct studies on tourism competitiveness being Poon [5]. Subsequently, to understand the role that competitiveness plays in tourism, researchers such as Ritchie and Crouch [6] defined a theoretical and conceptual framework to reveal how a tourist destination manages its competitiveness. The concept progressed from a perspective focused on tourist attractiveness to the strategic promotion of the tourism industry in a more holistic way, which considers different advantages of competitiveness [7].

Therefore, tourism competitiveness is presented as a key instrument to turn tourism into a factor of economic development [8–10]; it provides countries with the opportunity to maintain their position as leaders in the tourism activity [11] or to obtain a favorable position [12]. The continual interest of the tourism sector in competitiveness is due to the diversification that occurred in this activity; there are destinations competing to have more tourist arrivals or more tourism expenditure, which are indicators that reflect the economic prosperity of their residents [1].

Tourism competitiveness is defined as the ability of a destination to intensify tourism expenditure and gain more visitors while offering satisfying and memorable experiences in a profitable way, as well as improving the well-being of local residents and preserving the natural capital of the destination for future generations [6]. The concept proposed by Ritchie and Crouch [6] was adapted as the primary definition of tourism competitiveness [13], from which others arose such as the one proposed by Acerenza [14], who in a simple way defines tourism competitiveness as the ability of a destination to attract tourists. On the other hand, Dupeyras and MacCallum [15] conceptualize destination competitiveness as the ability of a place to take advantage of its attractiveness and offer quality and innovative tourist services, as well as to gain national and global market shares, while ensuring that the available resources that support tourism are used efficiently and sustainably.

In practice, tourism competitiveness is a construct in which various tangible and intangible factors participate, although it is only in a few, which are known as critical factors, where the greatest options for success or failure lie [16]. Tourism competitiveness brings about collective improvement, both in organizations and institutions, in favor of strengthening the tourism sector and focusing on increasing tourist flow and jobs [17]. It can also be said that destination competitiveness is the ability of the destination to conceive, integrate and provide tourist experiences, which include value-added goods and services that tourists consider substantial [18] and is characterized as a crucial element for the success of tourist destinations [19].

As a line of research, tourism competitiveness has had an interesting development in recent years, with one of its fields being the identification of the factors that affect it [2] and its relationship with variables such as tourism performance. Research papers in which this subject was addressed include those by Imali [20], Milicevic et al. [21], Hanafiah and Zulkifly [22]; Armenski et al. [23], Andradres and Dimanche [24], Amaya et al. [25], Cucculelli and Goffi [26], García and Siles [1], Decasper [27], Castellanos et al. [28], Leung and Baloglu [12], Goffi [19], Gandara et al. [17], Bolaky [29], Rodrigues and Carrasqueira [30] and Pascarella and Fontes [4], to name a few. Models have also been proposed to explain this phenomenon, including those developed by Poon [5], Hassan [31], Health [32], Ritchie and Crouch [6], Dwyer and Kim [33], Acerenza [14], Wei-Chiang [18], Alonso [16] and Jiménez and Aquino [34], among others.

Based on the research carried out to date on tourism competitiveness, it is possible to determine the causes that generate it and its effects on a tourist destination. However, a question arises: do decision makers in the tourism sector share the same perspective as academics/researchers regarding the relationship between the causes and effects of tourism competitiveness?

In this context, the aim of this paper is to compare the opinions held by two groups of experts, decision makers versus academics/researchers, both from the tourism sector, regarding the relationship between the causes and effects of tourism competitiveness. The methodology used to achieve these objectives is the experton and forgotten effects theories, and the Hamming distance.

This manuscript is structured in four sections. The first section establishes the bases of tourism competitiveness and its causes and effects, and sets out the objective of the research. The second section describes the materials and methods; the experton theory, the forgotten effects theory and the Hamming distance are discussed. In the third section, the results obtained are presented and discussed. In the fourth section, the conclusions of the paper are discussed.

2. Materials and Methods

Firstly, the causes and effects of tourism competitiveness were identified by reviewing the models proposed by Poon [5], Hassan [31], Health [32], Ritchie and Crouch [6], Dwyer and Kim [33], Acerenza [14], Wei-Chiang [18], Alonso [16], and Jiménez and Aquino [34]. The causes and effects found are shown in Tables 1 and 2.

At the same time, 8 experts were selected according to their academic and professional background; these people are directly immersed in the tourism sector, either offering a service or conducting research, as applicable. According to the purpose of this work, the government sector has been excluded. Two groups of experts were formed: the first group was made up of academics/researchers on the subject of tourism and competitiveness; the second group of experts was made up of people who work in the tourism sector. Next, both groups evaluated the cause–effect, cause–cause and effect–effect relationships based on the endecadary scale shown in Table 3, proposed by Kaufmann and Gil-Aluja [46]. It is worth mentioning that all the experts had the same weight in the evaluation.

Table 1. Causes of tourist competitiveness.

Causes		Definition
a.	Environmental commitment	The destination's commitment to the environment [32].
b.	Legacy resources	The endogenous resources of the territory itself, among which are the natural resources (physiography, flora, climate and fauna, among others) and cultural resources (history, customs, architecture, music and dances, among others). They are the main attractions of a destination [33].
c.	Resources created	Tourist infrastructure (accommodation, meals, transport, travel agencies and car rentals, among others), tourist services (medical services, security and gas stations, among others), special events, recreational activities, sports, leisure and entertainment, and souvenir shops [33].
d.	General infrastructure	The variety and quality of local transport services, drinking water supply, sanitation, communication systems, public facilities, electricity, financial services and telecommunications, among others [6,35].
e.	Quality of service	The degree of conformity of the attributes and characteristics of a service with respect to customer expectations [36].
f.	Accessibility	Means of access to a destination; airports and roads, mainly [33].
g.	Hospitality	The kindness and warmth of the local population [33].
h.	Destination management	The activities that can improve resource attractiveness, as well as strengthen the quality and effectiveness of support factors (infrastructure), such as destination marketing management; policies, planning and development of the destination; management organization; human resources development and resource management [33].
i.	Ties to the market	Business relationships, migration flows, culture and language, among others [33].
j.	Location	The geographical area in which the destination is located [37].
k.	Security	The conditions of political stability, crime, terrorism and disease, among others [33].
l.	Price	A relationship that indicates the amount of money needed to acquire a given amount of a good or service [38].
m.	The microenvironment	The capacities and resources of tourism companies, this also includes the stakeholders of a destination [33].
n.	The macro environment	The conditions of the economic, demographic, social, political and legal environment, mainly [33].
o.	Demand conditions	The understanding of the perception and preferences of tourism [33].

Source: Reproduced from [39], Journal of Intelligent & Fuzzy Systems: 2021.

Table 2. Effects of tourist competitiveness.

	Effects	Definition
A.	Sustainable development	The process of economic, human and environmental development that can be maintained without relying on external assistance [40].
B.	Tourist demand	A set of goods and services that tourists are willing to acquire in a certain destination [41].
C.	Customer satisfaction	Associated with the simple feeling of contentment, conditioned by a double human vision: utilitarian (to what extent the good of consumption or service fulfils the functions or tasks assigned to them) and hedonistic (activation of affective processes) [42].
D.	Customer loyalty	A favorable correspondence between the attitude of the individual towards an organization; in this case, the tourist destination and the behavior of purchasing the products and services thereof [43].
E.	Tourist spending	Total consumption expenses made by a visitor or on behalf of a visitor during their travel and tourist stay at the destination [41].
F.	Profitability	Return on investment [44].
G.	Prosperity	Favorable development, especially in the economic and social aspect.
H.	Arrival of tourist	Number of tourists arriving at a destination [41].
I.	Economic growth	Quantitative increase or expansion of income and value of goods and services produced in the economic system as measured through the growth rate of gross domestic product [45].

Source: Reproduced from [39], Journal of Intelligent & Fuzzy Systems: 2021.

Table 3. Endecadary scale.

Degree	Meaning
0	It has no incidence
0.1	Virtually no incidence
0.2	Almost without incidence
0.3	It has a very weak incidence
0.4	Has a weak incidence
0.5	Median incidence
0.6	Has a noticeable incidence
0.7	It has a lot of incidence
0.8	It has a strong incidence
0.9	It has a very strong incidence
1	It has the highest incidence

Source: Kaufmann and Gil-Aluja [46].

Based on the evaluations conducted by the experts, six expert tables were constructed, with three for each group. To construct each experton, first the absolute frequencies were calculated (number of experts suggesting each result), then the data were normalized through relative frequencies (division of the absolute frequencies by the total number of experts) and finally, the accumulated relative frequencies were obtained [47–49]. This is done level by level for the 11-point endecadary scale [49] (Table 3).

The forgotten effects theory was used to identify the variables and relationships that remain hidden or generate an indirect impact. This was done for each group of experts, for which the constructed expertons M (cause–effect relationship), A (cause–cause relationship) and B (effect–effect relationship) were used, which correspond to the direct incidence matrices. The matrices M, A and B were convoluted in the following way: $A \circ M = AM$

and $AM \circ B = M^*$ (M^* represents the accumulated effects matrix), then the forgotten effects matrix was calculated with the formula $O = M^* - M$.

Finally, the Hamming distance between expertons was used to compare the opinions that both groups of experts have regarding tourism competitiveness, particularly its causes, effects and the relationship between them. This last step follows a similar process to the Hamming distance between fuzzy subsets in the discrete domain. In this case, all expert evaluations are considered, except for the level α = 0, and the result is normalized by dividing by the number of n evaluations considered. The formulation of the Hamming distance between experts is as follows:

$$\text{Distance to the left} = d_I(A, B) = \frac{1}{2n} \sum_{j=1}^{n} |a_1(\alpha_j) - b_1(\alpha_j)| \qquad (1)$$

$$\text{Distance to the right} = d_D(A, B) = \frac{1}{2n} \sum_{j=1}^{n} |a_2(\alpha_j) - b_2(\alpha_j)| \qquad (2)$$

$$\text{Total distance} = d(A, B) = d_I(A, B) + d_D(A, B) \qquad (3)$$

It should be noted that only one of the distances was used (left or right), since the information was in individual data and not in intervals.

2.1. Theory of Expertons

There are phenomena in nature that humans evaluate through a subjective opinion and can hardly be classified according to whether or not a property is fulfilled [50]. In this regard, the theory of expertons suggests that to obtain realistic data of phenomena that cannot be measured directly, it is useful to have an aggregate set of the assessments given by experts [46,51].

The theory of expertons extends the probabilistic set concept [52] to uncertain environments that can be evaluated with interval numbers and allows for the analysis of group information considering all individual opinions, producing a single final result. Thus, it makes the information more robust because it is evaluated by several experts and the use of several experts in the analysis generally leads to better decisions [47].

An experton is defined as a generalization of a probabilistic set when the accumulated probabilities are replaced by intervals that decrease monotonically [53]. It arises as a result of a procedure of aggregation of different expert opinions regarding the degree of veracity of a statement and provides the percentage of experts who agree that the veracity of the statement is at least the given value [49].

The theory of expertons, which was developed by Kaufmann [51], is defined as follows:

Definition 1. *Let E be a referential set, finite or not; where r experts are required to express their subjective opinion about each element of E in the form of a confidence interval:*

$$\forall x \in E: \left[a_*^j(x), a_j^*(x)\right] \subset [0, 1] \qquad (4)$$

where $\subset$ is an inclusion set and j is the expert. A cumulative complementary law $F_*(a, x)$ can be established for all $a_*^j(x)$ and $F^*(a, x)$ for all $a_j^*(x)$, which is due to a statistic that indicates that for each $x \in E$, the lower limits are one way and the upper limits another way. By substituting this approach in Equation (4), the following is obtained:

$$\forall x \in E, \forall x \in [0, 1]: \tilde{A}(x) = [F_*(\alpha, x), F^*(\alpha, x)] \qquad (5)$$

The symbol $\tilde{A}$ that appears in Equation (5) recalls the nature of the concept.

So, the referential set E is the following experton:

$$\forall x \in E, \forall x \in [0, 1] : [F_*(\alpha, x), F^*(\alpha, x)] = 1 \qquad (6)$$

Additionally, an empty experton is given by:

$$\forall x \in E : [F_*(\alpha, x), F^*(\alpha, x)] = \begin{cases} 1, & \alpha = 0 \\ 0, & \alpha \neq 0 \end{cases} \qquad (7)$$

And it has the following properties:

$$\forall x \in E, \forall \alpha, \alpha' \in [0,1] : (\alpha < \alpha') \\ \implies ([F_*(\alpha', x), F^*(\alpha', x)] \subset_i [F_*(\alpha, x), F^*(\alpha', x)]) \qquad (8)$$

The expression $\subset_i$ that appears in Equation (8) is known as the inclusion interval. It can be expressed as follows:

$$(\alpha < \alpha') \implies ((F_*(\alpha', x) \leq F_*(\alpha, x)) \text{ and } (F^*(\alpha', x) \leq F^*(\alpha, x))) \qquad (9)$$

When a final consideration or interpretation of the phenomena is required, the experton can be reduced to a single representative value by reducing the entropy of the results. This can be obtained by calculating the mathematical expectation of the probabilistic set [53].

2.2. Forgotten Effects Theory

Any activity is subject to cause–effect incidents [54] and to the possibility of forgetting some causal relationships that are not explicit, obvious or visible [55]. In situations of uncertainty and volatility, there are variables that are not immediately detectable because they are hidden as a result of an accumulation of causes [46]. To identify the incidents that are not so evident between variables, but are fundamental for decision making, the theory of forgotten effects has proven to be an effective approach when seeking to maximize the information retrieved from the complex relationships between variables and to minimize errors that can occur in these processes [56].

The theory of forgotten effects is an extension of fuzzy logic applications [57]. This theory allows for an approach to the objective of globalizing the direct and indirect incidences between a set of causes and effects [46], since it suggests that all events, phenomena and facts that surround people are based on some type of system or subsystem. Therefore, they are subject to some type of cause–effect relationship, with the possibility that voluntarily or involuntarily some relationships are not directly perceived [58].

The theory of forgotten effects developed by Kaufmann and Gil-Aluja [46] is defined as follows:

Definition 2. *The existence of two sets, $A = \left\{\frac{a_i}{i} = 1, 2, \ldots, n\right\}$ and $B = \left\{\frac{b_j}{j} = 1, 2, \ldots, m\right\}$, is assumed. It is conjectured that an impact of a_i on b_j prevails if the value of the characteristic membership function of (a_i, b_j) is evaluated in a $[0, 1]$ range, that is:*

$$\forall (a_i, b_j) \Rightarrow \mu(a_i, b_j) \in [0,1] \qquad (10)$$

The set of pairs of elements evaluated is the direct incidence matrix (M), which shows the cause–effect relationship in different degrees caused by the corresponding set A (causes) and set B (effects), as shown below:

$$M = \begin{array}{c|ccccc} & b_1 & b_2 & b_3 & \cdots & b_m \\ \hline a_1 & \mu_{a_1b_1} & \mu_{a_1b_2} & \mu_{a_1b_3} & \cdots & \mu_{a_1b_m} \\ a_2 & \mu_{a_2b_1} & \mu_{a_2b_2} & \mu_{a_2b_3} & \cdots & \mu_{a_2b_m} \\ a_3 & \mu_{a_3b_1} & \mu_{a_3b_2} & \mu_{a_3b_3} & \cdots & \mu_{a_3b_m} \\ \vdots & \cdots & \cdots & \cdots & \cdots & \cdots \\ a_n & \mu_{a_nb_1} & \mu_{a_nb_2} & \mu_{a_nb_3} & \cdots & \mu_{a_nb_m} \end{array}$$

The next step is to proceed to detecting the forgotten effects. For this, it is assumed that there is a third set of elements, called C, expressed in the following way $C = \left\{\frac{C_k}{k} = 1, 2, \ldots, k\right\}$. This set consists of elements that are effects of set B and has an incidence matrix which is expressed as follows:

$$N = \begin{array}{c|ccccc} & c_1 & c_2 & c_3 & \cdots & c_k \\ \hline b_1 & \mu_{b_1c_1} & \mu_{b_1c_2} & \mu_{b_1c_3} & \cdots & \mu_{b_1c_k} \\ b_2 & \mu_{b_2c_1} & \mu_{b_2c_2} & \mu_{b_2c_3} & \cdots & \mu_{b_2c_k} \\ b_3 & \mu_{b_3c_1} & \mu_{b_3c_2} & \mu_{b_3c_3} & \cdots & \mu_{b_2c_k} \\ \vdots & \cdots & \cdots & \cdots & \cdots & \cdots \\ b_m & \mu_{b_mc_1} & \mu_{b_mc_2} & \mu_{b_mc_3} & \cdots & \mu_{b_mc_k} \end{array}$$

So far, there are two incidence matrices, M and N, and they both have element B in common. This relationship is expressed as:

$$M \subset AxB \; y \; N \subset BxC \tag{11}$$

Next, the max–min operator (convolution) is used to detect the relationship between sets A and C using B. As a result, a new incidence matrix is generated, which is expressed by:

$$\begin{aligned} M \circ N &= P \\ P &\subset AxC \end{aligned} \tag{12}$$

This new relationship is formulated in the following way:

$$\forall (a_i, c_z \in AxC) \tag{13}$$

$$\mu(a_i, c_z) M \circ N = \forall_{bj} \left(\mu_M(a_i, b_j) \wedge \mu_N(b_j, c_z) \right) \tag{14}$$

The matrix that results from the max–min operation is:

$$P = \begin{array}{c|ccccc} & c_1 & c_2 & c_3 & \cdots & c_k \\ \hline a_1 & \mu_{a_1c_1} & \mu_{a_1c_2} & \mu_{a_1c_3} & \cdots & \mu_{a_1c_k} \\ a_2 & \mu_{a_2c_1} & \mu_{a_2c_2} & \mu_{a_2c_3} & \cdots & \mu_{a_2c_k} \\ a_3 & \mu_{a_3c_1} & \mu_{a_3c_2} & \mu_{a_3c_3} & \cdots & \mu_{a_2c_k} \\ \vdots & \cdots & \cdots & \cdots & \cdots & \cdots \\ a_n & \mu_{a_nc_1} & \mu_{a_nc_2} & \mu_{a_nc_3} & \cdots & \mu_{a_nc_k} \end{array}$$

Matrix P defines the causal relationships between the elements of the A and C sets, at the intensity or degree that B allows for.

2.3. The Hamming Distance between Fuzzy Subsets

The distance measurement plays a vital role in pattern recognition, information fusion, decision making and other fields [59]. It is an important issue in fuzzy sets and their extensions [60,61]. There are several distance measures that have been introduced by researchers to solve problems in various fields; among these distances, the Euclid distance and the Hamming distance are widely used [62]. Specifically, the Hamming distance was developed by Richard Wesley Hamming in 1950 and is defined as follows [63]:

Definition 3. *The distance $D(x,y)$ between two x and y points is defined as the number of coordinates for which x and y are different. This function fulfils the three usual conditions for a metrics:*

$$D(x,y) = 0 \text{ if and only if } x = y \tag{15}$$

$$D(x,y) = D(y,x) > 0 \text{ if } x \neq y \tag{16}$$

$$D(z,y) + D(y,z) \geq D(x,z) \text{ (triangular inequality)} \tag{17}$$

The Hamming distance, like the Theory of Expertons, is considered a fuzzy numerical model [64]; it is known for its ability to calculate the difference between two sets or elements and is identified as one of the multi-criteria decision-making approaches. Also, when counting the number of specific differences between two permutations, it is a natural choice to measure the distance between assignments or pairings [65]. This approach helps to solve many problems related to biology, science and technology due to its ability to construct some related distance measures, in particular, similarity and proximity, which usually become a norm in several problems [66].

Before defining the Hamming distance between fuzzy subsets in the discrete domain, it is necessary to understand the concepts of fuzzy sets and subsets, which in the words of Zadeh [67] are defined below:

Definition 4. *Let X be a universe of analysis, then a fuzzy set A in X is defined as a set of pairs established in the following way:*

$$A = \{\langle x, \mu_A(x)\rangle : x \in X\} \tag{18}$$

where $\mu_A : X \to [0, 1]$ is the membership function that characterizes the universe of analysis A and $\mu_A(x)$ is the degree of membership of x in A.

Definition 5. *A fuzzy subset A of a universe of analysis X is characterized by a membership function $\mu_A : X \to [0, 1]$ which associates to each element x of X a number $\mu_A(x)$ in an interval $[0, 1]$; thus, $\mu_A(x)$ represents the degree of membership of x in A.*

Regarding the degree of membership $\mu_A(x)$ shown in the two previous definitions, it can be interpreted as the degree of compatibility of x with the concept represented by A or as the degree of possibility of x given A. In these cases, the function $\mu_A : X \to [0, 1]$ can be referred to as a compatibility function. It should be noted that the meaning attributed to a particular numerical value of the membership function is purely subjective in nature [57]. From the above, it is possible to state that the degree of non-membership of x in A is equal to $1 - \mu_A(x)$ [68].

Therefore, the Hamming distance of two fuzzy subsets is defined as follows [69]:

Definition 6. *Given two fuzzy subsets A and B with a reference set $X = \{x_1, x_2, \ldots, x_n\}$ and membership functions μ_A and μ_B, the Hamming distance is defined as:*

$$d(A,B) = \sum_{j=1}^{n} |\mu_A(x_j) - \mu_B(x_j)| \tag{19}$$

where μ_A and $\mu_B \in [0, 1]$.

In this case, the Hamming distance measures the relationship between variables in a study of facts and how they fit a profile. Finally, it calculates the distance between the extremes of the intervals [64].

3. Results

To compare the perspectives that decision makers in the tourism sector and tourism academics/researchers have regarding the causes and effects of tourism competitiveness, as well as the relationship between them, the following tools were used: experton, forgotten effects and Hamming distance between experts, with the last tool showing numerical differences of opinion. Each expert evaluated the cause–effect, cause–cause and effect–effect relationships; the expertons shown in Tables 4–9 were constructed based on these evaluations.

Figure 1 shows the Hamming distance between the two groups of experts regarding the evaluation of the relationship between the causes and effects of tourism competitiveness. The greatest distance, with a value of 0.19, occurs in the relationship between Environmental Commitment and Tourism Demand. The experton which was constructed based on the assessment of this relationship by the sector of academics/researchers was considerably higher than the one obtained by experts in the tourism sector.

In this regard, there is a growing acceptance of public and private stakeholders interested in tourism due to the assumption of a compatibility between the economic benefit and the minimization of the socio-cultural impacts on hosts and tourists with the protection of the natural environment; a situation that raises conflicting attitudes among the actors involved, which are favorable among administrators, researchers and environmental groups, but reluctant in the private sector [68]. It is visualized that this is one of the reasons why the group of academics/researchers reported a higher evaluation with respect to the experts of the tourism sector.

Table 4. Expertons made up of the cause–effect valuation carried out by the group of experts from the tourism sector (Matrix M1).

Cause–Effect	A	B	C	D	E	F	G	H	I
a	0.75	0.40	0.38	0.53	0.28	0.50	0.58	0.58	0.40
b	0.50	0.63	0.50	0.60	0.45	0.55	0.48	0.73	0.50
c	0.53	0.63	0.55	0.55	0.58	0.50	0.55	0.65	0.53
d	0.70	0.70	0.78	0.68	0.70	0.73	0.75	0.78	0.70
e	0.38	0.85	0.93	0.93	0.75	0.83	0.70	0.85	0.58
f	0.35	0.63	0.65	0.60	0.55	0.55	0.60	0.83	0.63
g	0.25	0.75	0.88	0.83	0.70	0.50	0.48	0.83	0.48
h	0.38	0.83	0.55	0.68	0.75	0.60	0.55	0.75	0.50
i	0.50	0.65	0.63	0.58	0.50	0.58	0.78	0.70	0.68
j	0.45	0.70	0.75	0.73	0.50	0.75	0.70	0.78	0.78
k	0.45	0.83	0.90	0.90	0.80	0.83	0.90	0.93	0.88
l	0.33	0.80	0.78	0.78	0.90	0.88	0.80	0.75	0.80
m	0.65	0.68	0.68	0.65	0.60	0.73	0.65	0.65	0.65
n	0.60	0.70	0.48	0.50	0.53	0.70	0.65	0.60	0.68
o	0.55	0.93	0.68	0.58	0.78	0.70	0.63	0.80	0.43

Source: own elaboration.

Table 5. Expertons formed from the cause–effect evaluation carried out by the group of academic experts/researchers (Matrix M2).

Cause–Effect	A	B	C	D	E	F	G	H	I
a	0.83	0.78	0.68	0.53	0.63	0.48	0.53	0.63	0.53
b	0.70	0.73	0.63	0.65	0.68	0.65	0.70	0.83	0.60
c	0.60	0.63	0.60	0.55	0.65	0.65	0.65	0.75	0.58
d	0.60	0.70	0.83	0.78	0.75	0.75	0.78	0.75	0.78
e	0.65	0.83	0.90	0.90	0.90	0.85	0.80	0.88	0.73
f	0.65	0.68	0.68	0.60	0.58	0.58	0.50	0.68	0.58
g	0.58	0.78	0.88	0.85	0.80	0.63	0.55	0.80	0.55
h	0.63	0.80	0.75	0.75	0.78	0.65	0.63	0.73	0.53
i	0.60	0.65	0.63	0.65	0.68	0.55	0.68	0.68	0.58
j	0.58	0.70	0.78	0.78	0.65	0.75	0.73	0.85	0.78
k	0.73	0.88	0.90	0.90	0.80	0.83	0.83	0.83	0.85
l	0.60	0.68	0.73	0.75	0.73	0.78	0.75	0.75	0.73
m	0.53	0.68	0.63	0.70	0.68	0.75	0.68	0.68	0.58
n	0.55	0.60	0.58	0.55	0.58	0.68	0.60	0.58	0.60
o	0.58	0.75	0.78	0.75	0.75	0.80	0.70	0.73	0.65

Source: own elaboration.

Table 6. Expertons made up of the cause–cause valuation carried out by the group of experts from the tourism sector (Matrix A1).

Cause–Cause	a	b	c	d	e	f	g	h	i	j	k	l	m	n	o
a	1.00	0.53	0.55	0.58	0.55	0.33	0.20	0.35	0.43	0.40	0.28	0.30	0.35	0.48	0.45
b	0.53	1.00	0.55	0.63	0.48	0.60	0.35	0.60	0.55	0.38	0.43	0.53	0.68	0.43	0.45
c	0.58	0.43	1.00	0.68	0.68	0.70	0.58	0.63	0.58	0.45	0.58	0.60	0.68	0.55	0.45
d	0.45	0.40	0.50	1.00	0.45	0.68	0.25	0.50	0.40	0.28	0.65	0.63	0.60	0.38	0.50
e	0.45	0.38	0.58	0.65	1.00	0.60	0.88	0.48	0.63	0.43	0.73	0.73	0.70	0.55	0.68
f	0.40	0.30	0.55	0.55	0.63	1.00	0.55	0.48	0.48	0.63	0.63	0.53	0.65	0.60	0.65
g	0.25	0.30	0.55	0.50	0.88	0.60	1.00	0.55	0.58	0.30	0.55	0.58	0.68	0.45	0.70
h	0.23	0.45	0.65	0.53	0.78	0.60	0.40	1.00	0.70	0.65	0.75	0.65	0.60	0.55	0.60
i	0.38	0.40	0.55	0.55	0.73	0.55	0.75	0.78	1.00	0.48	0.60	0.58	0.58	0.48	0.65
j	0.33	0.60	0.68	0.65	0.60	0.83	0.35	0.68	0.70	1.00	0.60	0.68	0.58	0.53	0.65
k	0.28	0.35	0.75	0.58	0.85	0.58	0.60	0.85	0.88	0.63	1.00	0.70	0.75	0.65	0.68
l	0.25	0.33	0.75	0.80	0.80	0.73	0.40	0.80	0.70	0.58	0.65	1.00	0.65	0.55	0.63
m	0.35	0.38	0.58	0.50	0.70	0.58	0.55	0.68	0.60	0.45	0.60	0.58	1.00	0.73	0.50
n	0.43	0.35	0.65	0.73	0.60	0.70	0.53	0.58	0.63	0.60	0.83	0.68	0.85	1.00	0.63
o	0.45	0.40	0.63	0.73	0.73	0.68	0.60	0.78	0.63	0.63	0.73	0.73	0.63	0.53	1.00

Source: own elaboration.

Table 7. Expertons made up of the cause–cause valuation carried out by the group of academic experts/researchers (Matrix A2).

Cause–Cause	a	b	c	d	e	f	g	h	i	j	k	l	m	n	o
a	1.00	0.55	0.63	0.55	0.65	0.53	0.55	0.50	0.53	0.68	0.60	0.55	0.58	0.55	0.55
b	0.50	1.00	0.55	0.58	0.55	0.70	0.60	0.68	0.50	0.65	0.60	0.55	0.65	0.58	0.60
c	0.68	0.60	1.00	0.63	0.58	0.73	0.50	0.68	0.55	0.65	0.63	0.63	0.63	0.63	0.60
d	0.70	0.70	0.83	1.00	0.78	0.85	0.70	0.83	0.83	0.80	0.83	0.85	0.78	0.83	0.83
e	0.68	0.65	0.75	0.78	1.00	0.83	0.83	0.75	0.83	0.70	0.88	0.80	0.80	0.70	0.80
f	0.63	0.50	0.68	0.70	0.75	1.00	0.70	0.63	0.70	0.83	0.83	0.60	0.73	0.68	0.78
g	0.55	0.55	0.68	0.70	0.85	0.68	1.00	0.55	0.63	0.63	0.75	0.68	0.75	0.60	0.78
h	0.58	0.60	0.65	0.75	0.83	0.70	0.68	1.00	0.78	0.88	0.83	0.65	0.75	0.58	0.73
i	0.53	0.50	0.65	0.73	0.73	0.65	0.73	0.70	1.00	0.70	0.75	0.65	0.63	0.55	0.78
j	0.48	0.53	0.70	0.70	0.68	0.80	0.60	0.75	0.78	1.00	0.78	0.73	0.63	0.60	0.75
k	0.53	0.53	0.75	0.85	0.85	0.83	0.65	0.80	0.85	0.88	1.00	0.80	0.83	0.60	0.80
l	0.48	0.60	0.68	0.68	0.78	0.70	0.48	0.70	0.75	0.75	0.78	1.00	0.70	0.55	0.80
m	0.53	0.55	0.65	0.63	0.68	0.70	0.60	0.63	0.68	0.63	0.63	0.63	1.00	0.68	0.63
n	0.43	0.38	0.60	0.70	0.65	0.68	0.40	0.45	0.50	0.43	0.50	0.48	0.78	1.00	0.50
o	0.45	0.45	0.63	0.73	0.70	0.78	0.58	0.83	0.73	0.80	0.80	0.78	0.68	0.65	1.00

Source: own elaboration.

Table 8. Expertons made up of the effect–effect valuation carried out by the group of experts from the tourism sector (Matrix B1).

Effect–Effect	A	B	C	D	E	F	G	H	I
A	1.00	0.63	0.70	0.60	0.60	0.53	0.65	0.53	0.53
B	0.65	1.00	0.65	0.68	0.85	0.73	0.73	0.95	0.83
C	0.50	0.55	1.00	0.93	0.65	0.68	0.58	0.80	0.55
D	0.28	0.53	0.80	1.00	0.48	0.50	0.10	0.88	0.45
E	0.60	0.68	0.55	0.60	1.00	0.68	0.68	0.58	0.68
F	0.50	0.68	0.55	0.58	0.55	1.00	0.65	0.73	0.73
G	0.60	0.73	0.68	0.70	0.70	0.70	1.00	0.75	0.63
H	0.48	0.90	0.75	0.70	0.68	0.68	0.63	1.00	0.73
I	0.70	0.65	0.70	0.73	0.78	0.95	0.85	0.75	1.00

Source: own elaboration.

In Figure 1, it can also be seen that there are relationships in which the distance is zero in the evaluation of both groups of experts. This means that both groups agree on those evaluations; this is found in the relationships of Location–Tourist Demand, Hospitality–Customer Satisfaction and Location–Economic Growth. Regarding the Hospitality–Customer Satisfaction relationship, previous studies show the relationship between them, such as those developed by Oliver [70], who proposed that satisfaction is deduced from the guest's perception of the attention given. Alves and Barcellos [71] indicated that experiences in hospitality and tourist services are the main product of the sector, with an impact and influence on its competitiveness.

Table 9. Expertons made up of the effect–effect assessment carried out by the group of academic experts/researchers (Matrix B2).

Effect–Effect	A	B	C	D	E	F	G	H	I
A	1.00	0.65	0.65	0.50	0.58	0.60	0.58	0.58	0.55
B	0.60	1.00	0.60	0.55	0.53	0.43	0.50	0.48	0.38
C	0.65	0.68	1.00	0.85	0.68	0.70	0.65	0.78	0.63
D	0.78	0.83	0.85	1.00	0.73	0.83	0.80	0.93	0.80
E	0.80	0.80	0.58	0.78	1.00	0.68	0.60	0.73	0.65
F	0.53	0.78	0.63	0.73	0.68	1.00	0.58	0.78	0.60
G	0.53	0.65	0.68	0.60	0.58	0.50	1.00	0.60	0.65
H	0.70	0.83	0.88	0.85	0.75	0.73	0.63	1.00	0.70
I	0.50	0.75	0.58	0.53	0.45	0.78	0.45	0.55	1.00

Source: own elaboration.

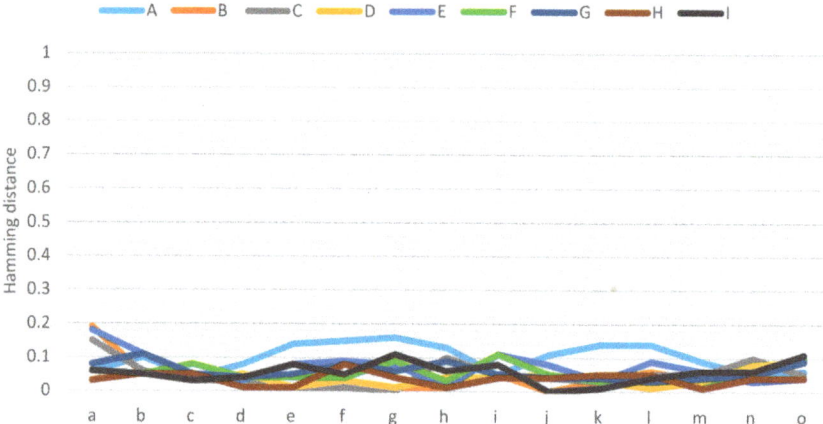

Figure 1. Hamming distance between expert groups in cause–effect relationships. Source: own elaboration.

Figure 2 shows the Hamming distance between the two groups of experts regarding the evaluation of the relationship between causes and the cause of tourism competitiveness. The greatest distance, with a value of 0.26, occurs in the relationship found between General Infrastructure and Location, and again in this relationship, academic experts were the ones who gave a higher evaluation compared to the evaluations made by the tourism sector experts. In this matrix, there is a diagonal line of the relationships in which the distance is zero. However, this occurs because the evaluation of the cause–cause relationship produces a value of one when it is the same.

Figure 3 shows the Hamming distance between the two groups of experts regarding the evaluation of the relationship between effect–effect of tourism competitiveness. The greatest distance, with a value of 0.25, occurs in the relationship that is found between Customer Loyalty and Sustainable Development. This time, it was also academic experts who evaluated the relationship higher. Note that in this matrix there is also a diagonal line of the relationships in which the distance is zero, which is for the same reasons as in the previous figure.

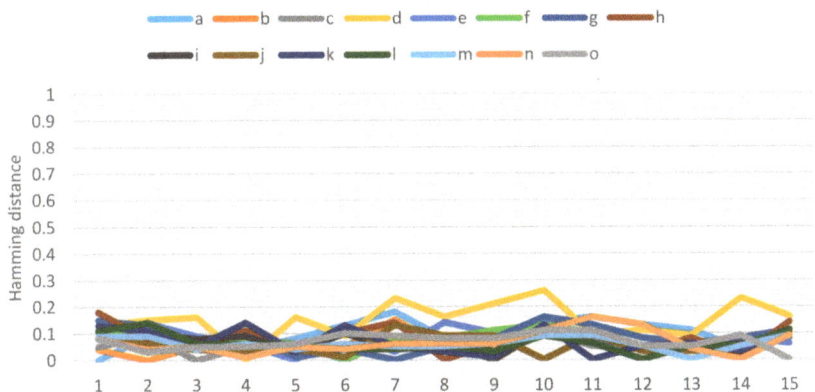

Figure 2. Hamming distance between experts in cause–cause relationships. Source: own elaboration.

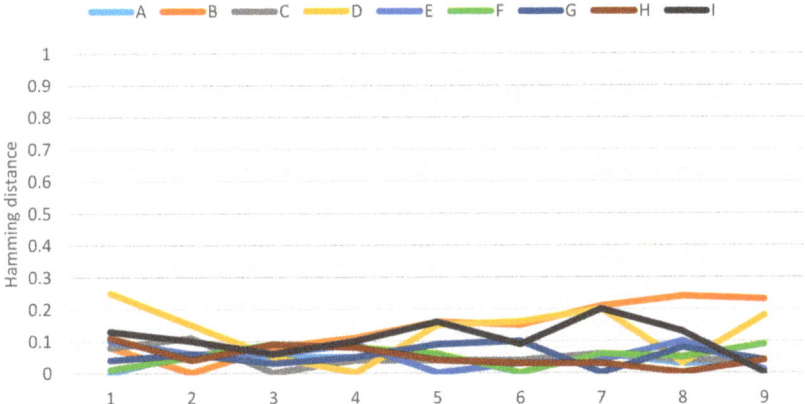

Figure 3. Hamming distance between experts in effect–effect relationships. Source: own elaboration.

From the information shown in Tables 4–9, the forgotten effects were obtained by each group of experts and with these results. The Hamming distance of the forgotten effects was calculated, which is shown in Figure 4. The greatest distance, with a value of 0.33, occurs in the relationship between Environmental Commitment (cause) and Customer Satisfaction (effect). In this case, the forgotten effect by the academic sector was zero, while the forgotten effect by the tourism sector was 0.33, which indicates that the academic sector has clearly identified the impact of the environmental commitment that a tourist destination has on customer satisfaction, which is a situation in which people who work in the tourism sector have not yet recognized.

Figure 4 shows 30 relationships whose Hamming distance is zero. In this situation, all the effects and all the causes are present except for Accessibility and The Macroenvironment. Cumulatively, the Hamming distance of the effect that shows the highest value is the "Sustainable Development" effect with a value of 1.78, while for the Hamming distance of the causes, the causes with the highest values are "Environmental Commitment" and "Accessibility", with both causes having a value of 1.18.

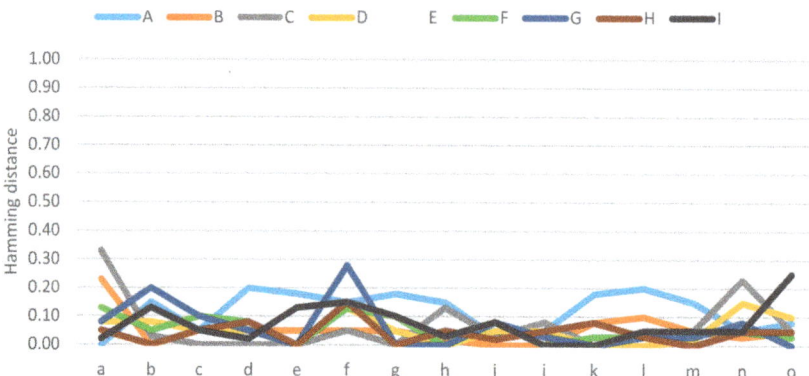

Figure 4. Hamming distance in the forgotten effects. Source: own elaboration.

4. Conclusions

This research paper began by discussing the importance of the tourism sector in the economy, as well as the need to develop competitiveness in tourist destinations. An analysis from the perspective of two expert groups (decision makers from the tourism sector and academics/researchers) regarding the relationship between the causes and effects of tourism competitiveness using fuzzy techniques was presented, such as the Theory of Expertons, the Theory of Forgotten Effects and the Hamming distance between expertons.

The Theory of Expertons enabled us to group all of the experts' opinions of each group into a single group result. The Forgotten Effects Theory helped to identify, for each group of experts, the variables and relationships that remained hidden or whose impact was indirect on cause–effect relationships. Finally, the Hamming distance between experts helped to detect differences of opinion between the perspective of a group of decision makers in the tourism sector versus the perspective of a group of academics/researchers regarding the relationship between the causes and effects of tourism competitiveness.

The results obtained showed that the experts' perspectives regarding the cause–effect relationship have a distance that ranges from 0.00 to 0.19, with the value of 0.19 being the one corresponding to the relationship of Environmental Commitment (cause) and Tourist Demand (effect). Regarding the cause–cause relationship, a distance was found in a range of 0.00 to 0.26, with the value of 0.26 referring to the General Infrastructure and Location relationship. Regarding the effect–effect relationship, the resulting distance ranges from 0.00 to 0.25, which corresponds to the relationship between Customer Loyalty and Sustainable Development. Finally, the distances found in the Forgotten Effects fluctuate between 0.00 and 0.33, with the value of 0.33 present in the Environmental Commitment (cause) and Customer Satisfaction (effect) relationship. In all of the aforementioned relationships, academic experts gave the highest evaluation.

According to the experience of the group of academics/researchers, Environmental Commitment (cause) has a significant impact on Tourism Demand (effect) and on Customer Satisfaction (effect). When applying Forgotten Effects on these relationships, zero values were obtained, which indicates that academics/researchers have clearly defined these cause–effect relationships. On the other hand, decision makers in the tourism sector have a Forgotten Effects value of 0.23 in the Environmental Commitment–Tourism Demand relationship and the Forgotten Effects in the Environmental Commitment–Customer Satisfaction relationship is 0.33. In both cases, decision makers omitted the indirect impact that causes sustainable development. In the aforementioned cause–effect relationship, decision makers in the tourism sector do not share the same perspective as academics/researchers.

In most cases, decision makers in the tourism sector share the same perspective as academics regarding the relationship between the different causes and effects of tourism competitiveness. However, they are ignoring the growing inclination and sensitivity that the tourist is adopting towards the environment. The tourism sector must adopt an attitude of responsibility, protection and respect for the environment, pay special attention to the implementation of actions that minimize the environmental impact that the activity itself causes, in addition to joining the programs that are present in each tourist destination that pursue a sustainable use of resources. It is necessary for the tourism sector to develop and consolidate its commitment to caring for and preserving the environment, which is an element that contributes to the competitiveness of the destination and has two main effects: tourism demand and customer satisfaction.

With the results of this work, the tourism sector can benefit from knowing the elements of a destination, which beyond the attractiveness itself, today interest the tourist and to whom they must direct their actions based on the effect(s) they want to achieve. On the other hand, the benefit for academics/researchers is that based on these results, new works and lines of research are made visible that allow for finding a way to support the business sector to understand the changes that are experienced in the tourist environment and in the interest of the tourist. A joint work between academics/researchers and the tourism sector could close the perception gaps that exist and generate results for both.

The results of this research are a starting point for future research and for companies that are part of the tourism sector to make decisions and take actions that contribute to making the tourist destination a competitive place. Future lines of research could address the reasons that limit decision makers in the tourism sector to develop an environmental commitment. On the other hand, it would be worth expanding the studies on the subject by using other tools such as the Pichart algorithm and different distance measures, as well as including interval values, in order to support the results obtained. One of the limitations of this study lies in the number of experts that made up each group, as to identify a pattern of behavior or opinion more easily, a larger group is required in terms of quantity and in terms of the role played in tourism.

Author Contributions: Conceptualization, Investigation, Methodology, Formal Analysis, Writing—Original Draft, Preparation and Writing—Review & Editing, M.B.F.-R., M.E.P.-R., J.Á.-G., and M.d.l.C.d.R.-R. All authors have read and agreed to the published version of the manuscript.

Funding: This research received no external funding.

Institutional Review Board Statement: Not applicable.

Informed Consent Statement: Not applicable.

Data Availability Statement: Not applicable.

Acknowledgments: The authors are also very grateful to the anonymous referees for their comments and suggestions which have considerably improved this paper.

Conflicts of Interest: The authors declare no conflict of interest.

References

1. García, A.; Siles, D. Cómo mejorar la competitividad turística de un destino: Análisis del mediterráneo español y recomendaciones a los gestores de los destinos. *Revista de Análisis Turístico* **2015**, *19*, 1–11.
2. Sánchez, M. Análisis cuantitativo del impacto económico de la competitividad en destinos turísticos internacionales. *Revista de Economía Mundial* **2012**, *32*, 103–125.
3. Rebollo, J.F.V.; Castiñeira, C.J.B. Renovación y reestructuración de los destinos turísticos consolidados del litoral: Las prácticas recreativas en la evolución del espacio turístico. *Boletín de la Asociación de Geógrafos Españoles* **2010**, *53*, 329–353.
4. Pascarella, R.; Fontes, J.R. Competitividad de los destinos turísticos, modelo de evaluación basado en las capacidades dinámicas y sus implicancias en las políticas públicas. *Estudios y Perspectivas en Turismo* **2010**, *19*, 1–17.
5. Poon, A. *Tourism, Technology and Competitive Strategies*; CAB International: Oxford, UK, 1993.
6. Ritchie, J.R.B.; Crouch, G.I. *The Competitive Destination: A Sustainable Tourism Perspective*; CABI Publishing: Wallingford, UK, 2003.
7. Kim, N. Tourism Destination Competitiveness, Globalization and Strategic Development from a Development Economics Perspective. Ph.D. Thesis, University of Illinois at Urbana—Champaign, Champaign, IL, USA, 2012.

8. Sánchez-Cañizares, S.M.; López-Guzmán, T. Gastronomy as a tourism resource: Profile of the culinary tourist. *Curr. Issues Tour.* **2012**, *15*, 229–245. [CrossRef]
9. Miguel, A.E.; Solís, N.; Torres, J.C. El impacto territorial del turismo en el desarrollo sostenible: El caso de las regiones de México 2000–2010. *Rev. Tur. Patrim. Cult.* **2014**, *12*, 357–368. [CrossRef]
10. Pulido, J.I.; Sánchez, M. Competitividad versus crecimiento en destinos turísticos. Un análisis mediante técnicas multivariantes. *Cuad. Econ.* **2020**, *22*, 159–182.
11. Pérez, R.M.I. Diagnóstico de la calidad y competitividad del sector turístico en México. *Cuad. Tur.* **2011**, 121–143. Available online: https://revistas.um.es/turismo/article/view/147261 (accessed on 21 June 2020).
12. Leung, S.Y.; Baloglu, S. Tourism competitiveness of Asia Pacific Destinations. *Tour. Anal.* **2013**, *18*, 371–384. [CrossRef]
13. Francoise, D.; Du Plessis, E. A review on tourism destination competitiveness. *J. Hosp. Tour. Manag.* **2020**, *45*, 256–265.
14. Acerenza, M. *Competitividad de los Destinos Turísticos*; Trillas: Mexico City, Mexico, 2009.
15. Dupeyras, A.; MacCallum, N. *Indicators for Measuring Competitiveness in Tourism: A Guidance Document. OECD Tourism Papers*; OECD Publishing: Paris, France, 2013.
16. Alonso, V.H. Factores críticos de éxito y evaluación de la competitividad de destinos turísticos. *Estud. Perspect. Tur.* **2020**, *19*, 201–220.
17. Gandara, J.M.; Chim-Miki, A.F.; Domareski, T.C.; Biz, A.A. La competitividad turística de Foz Do Iguacu según los determinantes del Integrative Model de Dwyer & Kim: Analizando la estrategia de construcción del futuro. *Cuad. Tur.* **2013**, *31*, 105–128.
18. Wei-Chiang, H. Global competitiveness measurement for the tourism sector. *Curr. Issues Tour.* **2009**, *12*, 105–132.
19. Goffi, G. A model of tourism destination competitiveness: The case of the Italian destinations of excellence. *Anu. Tur. Soc.* **2013**, *14*, 121–147.
20. Imali, F. Tourism Competitiveness by Shift-Share Analysis to way-forward Destination Management: A case study for Sri Lanka. *J. Tour. Serv.* **2020**, *11*, 88–102.
21. Milicevic, S.; Petrovic, J.; Kostic, M.; Lakicevic, M. Tourism product in the function of improving destination competitiveness: Case of Vrnjacka Banja, Serbia. *Qual. Access Success* **2020**, *21*, 133–138.
22. Hanafiah, M.H.; Zulkifly, M.I. Tourism destination competitiveness and tourism performance: A secondary data approach. *Compet. Rev. Int. Bus. J.* **2019**, *29*, 592–621. [CrossRef]
23. Armenski, T.; Dwyer, K.; Pavlukovic, V. Destination competitiveness: Public and private sector tourism management in Serbia. *J. Travel Res.* **2017**, *57*, 384–398. [CrossRef]
24. Andrades, L.; Dimanche, F. Destination competitiveness and tourism development in Russia: Issues and challenges. *Tour. Manag.* **2017**, *62*, 360–376. [CrossRef]
25. Amaya-Molinar, C.M.; Sosa-Ferreira, A.P.; Ochoa-Llamas, I.; Moncada-Jiménez, P. The perception of destination competitiveness by tourists. *Investigaciones Turísticas* **2017**, *14*, 1–20. [CrossRef]
26. Cucculelli, M.; Goffi, G. Does sustainability enhance tourism destination competitiveness? Evidence from Italian Destinations of Excellence. *J. Cleaner Prod.* **2016**, *III*, 370–382. [CrossRef]
27. Decasper, S.M. Competitividad y desarrollo sostenible en el turismo. *Anais Bras. Estud. Turísticos ABET* **2015**, *5*, 47–58.
28. Castellanos, C.A.; Hernández, Y.; Castellanos, J.R.; Campos, L.M. La competitividad del destino turístico Villa Clara, Cuba. Identificación de sus factores determinantes mediante análisis estructural (MIC-MAC). *Estud. Perspect. Tur.* **2014**, *23*, 250–277.
29. Bolaky, B. Tourism competitiveness in the Caribbean. *Rev. CEPAL* **2011**, *104*, 55–76. [CrossRef]
30. Rodrigues, L.; Carrasqueira, H. Análisis del desempeño competitivo de los destinos turísticos balnearios. El caso de Algarbe versus el Sur de España. *Estud. Perspect. Tur.* **2011**, *20*, 855–875.
31. Hassan, S. Determinants of Market Competitiveness in an Environmentally Sustainable Tourism Industry. *J. Travel Res.* **2000**, *38*, 239–245. [CrossRef]
32. Heath, E. Towards a Model to Enhance Destination Competitiveness: A Southern African Perspective. *J. Hospitality Tour. Manag.* **2003**, *10*, 124.
33. Dywer, L.; Kim, C. Destination competitiveness: A models and determinants. *Curr. Issues Tour.* **2003**, *6*, 369–414. [CrossRef]
34. Jiménez, P.; Aquino, F.K. Propuesta de un modelo de competitividad de destinos turísticos. *Estud. Perspect. Tur.* **2012**, *21*, 977–995.
35. Crouch, G.; Ritchie, J.R. Tourism, competitiveness and societal prosperity. *J. Bus. Res.* **1999**, *44*, 137–152. [CrossRef]
36. Morillo, M.C.; Morillo, M.C. Satisfacción del usuario y calidad del servicio en alojamientos turísticos del estado Mérida, Venezuela. *Rev. Cienc. Soc.* **2016**, *21*, 111–131.
37. Ramírez, B.R.; López, L. *Espacio, Paisaje, Región, Territorio y Lugar: La Diversidad en el Pensamiento Contemporáneo*; Universidad Nacional Autónoma de México: Mexico City, Mexico, 2015.
38. Sainz, A.A. Claves para gestionar precio, producto y marca. In *Conceptos Clave en la Gestion del Producto*; Belío Galindo, J.L., Ed.; Wolters Kluwer España, S.A.: Madrid, Spain, 2007.
39. Pérez-Romero, M.E.; Flores-Romero, M.B.; Alfaro-García, V.G. Tourism and destination competitiveness an exploratory analysis applying the forgotten effects theory. *J. Intell. Fuzzy Syst.* **2021**, *40*, 1795–1804. [CrossRef]
40. Pujadas, C. ¿Desarrollo sostenible o sustentable? *Adn Agua Medioambiente* **2011**, *5*, 1–28.
41. DATATUR. Glosario: Demanda Turística, Gasto Turístico, Llegada de Turistas. 2019. Available online: https://www.datatur.sectur.537gob.mx/SitePages/Glosario.aspx (accessed on 20 July 2020).
42. Morales, V.; Hernández, A. Calidad y Satisfacción en Los Servicios: Conceptualización. *Revista Digital* **2004**, *1*, 1–6.

43. Mesén, V. Fidelización de clientes: Concepto y perspectiva contable. *Tec Empresarial* **2011**, *5*, 29–35.
44. Contreras-Salluca, N.P.; Díaz-Correa, E.D. Estructura financiera y rentabilidad: Origen, teoría y definiciones. *Rev. Científica Valor Contab.* **2015**, *2*, 35–44. [CrossRef]
45. Enríquez, I. Las teorías del crecimiento económico: Notas críticas para incursionar en un debate inconcluso. *Rev. Latinoam. Desarro. Económico* **2016**, *25*, 73–125. [CrossRef]
46. Kaufmann, A.; Gil-Aluja, J. *Models for the Research of Forgotten Effects*; Milladoiro: Santiago de Compostela, Spain, 1988. (In Spanish)
47. Merigó, J.M.; Casanovas, M.; Yang, J.B. Group decision making with expertons and uncertain generalized probabilistic weighted aggregation operators. *Eur. J. Oper. Res.* **2014**, *235*, 215–224. [CrossRef]
48. Luna, K.; Tinto, J.; Sarmiento, W.; Cisneros, D. Tratamiento de impagos bajo el enfoque de la incertidumbre con la aplicación de redes neuronales. *Rev. Cienc. Pedagógicas Innovación* **2017**, *5*, 61–70.
49. Linares-Mustarós, S.; Ferrer-Comalat, J.C.; Corominas-Coll, D.; Merigó, J.M. The ordered weighted average in the theory of expertons. *Int. J. Intell. Syst.* **2018**, *34*, 1–21. [CrossRef]
50. López, C.; Linares-Mustarós, S.; Viñas, J. The use of fuzzy mathematical tools for local public services outsourcing according to typology. *J. Intell. Fuzzy Syst.* **2020**, *38*, 5379–5389. [CrossRef]
51. Kaufmann, A. Theory of expertons and fuzzy logic. *Fuzzy Sets Syst.* **1988**, *28*, 295–304. [CrossRef]
52. Hirota, K. Concept of probabilistic sets. *Fuzzy Sets Syst.* **1981**, *5*, 31–46. [CrossRef]
53. Alfaro-García, V.G.; Gil-Lafuente, A.M.; Alfaro, G.G. A fuzzy methodology for innovation management measurement. *Kybernetes* **2017**, *46*, 50–66. [CrossRef]
54. Gil-Aluja, J. *Fuzzy Sets in the Management of Uncertainty*; Springer: London, UK, 2004.
55. Barcellos, L.; Gil-Lafuente, A.M. Algorithms applied in the sustainable management of human resources. *Fuzzy Econ. Rev.* **2010**, *15*, 39–52.
56. Gil-Lafuente, A.M. *Fuzzy Logic in Financial Analysis*; Springer: Berlin/Heidelberg, Germany, 2005.
57. Zadeh, L.A. Fuzzy sets. *Inf. Control* **1965**, *8*, 338–353. [CrossRef]
58. Saldaña, C.X.; Guamán, G.A. Análisis financiero basado en la técnica Fuzzy Logic, como instrumento para la toma de decisiones en la empresa Italimentos Cia. Ltd.a. *Rev. Econ. Política* **2019**, *XV*, 1–19.
59. Zhou, Q.; Mo, H.; Deng, Y. A new divergence measure of Pythagorean Fuzzy Sets Based on Belief Function and Its Applications in Medical Diagnosis. *Mathematics* **2020**, *8*, 142. [CrossRef]
60. Liu, X. Entropy, distance measure and similarity measure of fuzzy sets and their relations. *Fuzzy Sets Syst.* **1992**, *52*, 305–318.
61. Zeng, W.Y.; Guo, P. Normalized distance, similarity measure, inclusion measure and entropy of interval-valued fuzzy sets and their relationship. *Inf. Sci.* **2008**, *178*, 1334–1342. [CrossRef]
62. Dai, S.; Bi, L.; Hu, B. Distance Measures between the Interval-Valued Complex Fuzzy Sets. *Mathematics* **2019**, *7*, 549. [CrossRef]
63. Hamming, R.W. Error detecting and error correcting codes. *Bell Syst. Tech. J.* **1950**, *29*, 147–160. [CrossRef]
64. Pinto-López, I.N.; Gil-Lafuente, A.M.; Flores, G.S. Pichat´s Algorithm for the Sustainable Regional Analysis Management: Case Study of Mexico. In *Complex Systems: Solutions and Challenges in Economics, Management and Engineering*; Berger-Vachon, C., Gil Lafuente, A.M., Kacprzyk, J., Kondratenko, Y., Merigó, J.M., Morabito, C.F., Eds.; Springer: London, UK, 2018.
65. Arza, F.; Pérez, A.; Irurozki, E.; Ceberio, J. Kernels of Mallows Models under de Hamming Distance for solving the Quadratic Assignment Problem. *Swarm Evol. Comput.* **2020**, *59*, 1–13. [CrossRef]
66. Marinov, E.; Szmidt, E.; Kacprzyk, J.; Tcvetkov, R. A modified weighted Hausdorff distance between intuitionistic fuzzy sets. In Proceedings of the 6th IEEE International Conference on Intelligent System, Sofia, Bulgaria, 6–8 September 2012.
67. Zadeh, L.A. Fuzzy sets as a basis for a theory of possibility. *Fuzzy Sets Syst.* **1978**, *1*, 3–28. [CrossRef]
68. Valenzuela, M. La sostenibilidad ambiental del sector hotelero español. Una contribución al turismo sostenible entre el interés empresarial y el compromiso ambiental. *ARBOR Cienc. Pensam. Cultura* **2017**, *193*, 1–18. [CrossRef]
69. Grzegorzewski, P. Distances between intuitionistic fuzzy sets and/or interval-valued fuzzy sets based on the Hausdorff metric. *Fuzzy Sets Syst.* **2004**, *148*, 319–328. [CrossRef]
70. Oliver, R.L. *Satisfaction: A Behavioral Perspective on the Consumer*; Taylor & Francis Group: New York, NY, USA, 2015.
71. Alves, C.A.; Barcellos, R. Hospitalidad, emociones y experiencias en los servicios turísticos. *Estud. Perspect. Tur.* **2019**, *28*, 290–311.

Article

Model of Evaluation and Selection of Expert Group Members for Smart Cities, Green Transportation and Mobility: From Safe Times to Pandemic Times

Miroslav Kelemen [1], Volodymyr Polishchuk [2,*], Beáta Gavurová [3], Róbert Rozenberg [1], Juraj Bartok [4,5], Ladislav Gaál [4], Martin Gera [5] and Martin Kelemen, Jr. [1]

[1] Faculty of Aeronautics, Technical University of Kosice, 041 21 Kosice, Slovakia; miroslav.kelemen@tuke.sk (M.K.); robert.rozenberg@tuke.sk (R.R.); martin.kelemen@tuke.sk (M.K.J.)
[2] Faculty of Information Technologies, Uzhhorod National University, 88000 Uzhhorod, Ukraine
[3] Research and Innovation Centre Bioinformatics, USP TECHNICOM, Faculty of Mining, Ecology, Process Control and Geotechnology, Technical University of Košice, 040 01 Kosice, Slovakia; beata.gavurova@tuke.sk
[4] MicroStep-MIS, Čavojského 1, 841 04 Bratislava, Slovakia; juraj.bartok@fmph.uniba.sk (J.B.); ladislav.gaal@microstep-mis.com (L.G.)
[5] Department of Astronomy, Physics of the Earth, and Meteorology, Comenius University in Bratislava, Mlynská dolina, 842 48 Bratislava, Slovakia; martin.gera@fmph.uniba.sk
* Correspondence: volodymyr.polishchuk@uzhnu.edu.ua; Tel.: +380-664-207-484

Citation: Kelemen, M.; Polishchuk, V.; Gavurová, B.; Rozenberg, R.; Bartok, J.; Gaál, L.; Gera, M.; Kelemen, M., Jr. Model of Evaluation and Selection of Expert Group Members for Smart Cities, Green Transportation and Mobility: From Safe Times to Pandemic Times. *Mathematics* **2021**, *9*, 1287. https://doi.org/10.3390/math9111287

Academic Editor: Jorge de Andres Sanchez

Received: 3 May 2021
Accepted: 1 June 2021
Published: 3 June 2021

Publisher's Note: MDPI stays neutral with regard to jurisdictional claims in published maps and institutional affiliations.

Copyright: © 2021 by the authors. Licensee MDPI, Basel, Switzerland. This article is an open access article distributed under the terms and conditions of the Creative Commons Attribution (CC BY) license (https://creativecommons.org/licenses/by/4.0/).

Abstract: This paper presents the development of technologies to support the decision-making of local government executives and smart city concept managers in selecting and evaluating the competencies of new members for advisory groups for solving problems that are implemented in safe times in individual areas or in crises, such as pandemics. The reason for developing effective urban transformation strategies and for the transparent selection of independent experts (non-politicians) for policymaking, decision-making, and implementation teams is not only the heterogeneity of smart city dimensions together with the necessary complexity and systems approach, but also the nature of the capacities and tools needed for smart city concepts. The innovative hybrid competency assessment model is based on fuzzy logic and a network for neuro-fuzzy assessment. It is a technological model for evaluating the competencies of specialists, taking into account the influence of human factors on the processes of personnel selection and system management. An innovative web platform named "Smart City Concept Personnel Selection" has been designed, which can be adapted to various users of municipalities or regional institutions for the transparent selection of qualified personnel for effective decision-making and the use of public funds during safe times or emergencies, such as the COVID-19 pandemic.

Keywords: expert group; neuro-fuzzy assessment; evaluation of specialists; smart city; assessment risk; smart transport; mobility; transparent selection; public financial resources; recovery plan

1. Introduction

The success or failure of an organization is directly related to how its human resources are used and maintained. Due to increased competition and globalization and rapid technological improvements, global markets require high-quality and professional human resources. This can only be achieved by hiring potentially adequate staff. This also applies to the management of a smart city, which seeks to improve traditional methods of recruiting staff to achieve ideal solutions. By achieving ideal solutions, a smart city, as a community–government–business system, will improve resource management, which will lead to achieving the goals of the smart city concept and increasing capital for its development.

To implement projects within the concept of a smart city in order to create new value for services and customer satisfaction, it is necessary to implement the following tasks:

form a project team, generate service ideas, test the service ideas, select a development concept, design and develop the concept, and test the services and commercialize and improve the quality of the services. In this study, we will focus on the task of evaluating and selecting members of a smart city and green transportation and mobility expert group. Here, it is necessary to select a group of specialists (experts) to evaluate and recommend projects aimed at reducing congestion in the smart city and improving the environment. The target function of any project is to use minimal resources and deliver the maximal effect. Appropriate specialists are needed for a systematic understanding of the implementation processes of the presented projects. They must have both general and special competencies. The superstructure of the evaluation criteria is entrusted to a systems analyst, depending on the target needs of evaluation.

The main goal of the study was to develop technologies to support the decision-making processes of municipal managers and managers of smart city concepts in transparently evaluating the competencies of new specialists (experts, non-politicians) and selecting them for special or innovative tasks within advisory commissions at the local or regional level.

Innovative software in the form of the Smart City Concept Personnel Selection web platform was designed as a tool for this purpose. This tool was approved within the research on the evaluation and selection of an advisory group of experts for an intelligent city and ecological transport and mobility as members of the Commission for Transport and Construction of the City of Košice with 250,000 inhabitants. The innovative web platform "Smart City Concept Personnel Selection" can be adapted to various users of municipalities or regional institutions for the transparent selection of qualified personnel for effective decision-making and use of public funds during safe times or emergencies, such as a global pandemic.

The study provides an answer to a basic research question: what tool can be used to support the right and coordinated decisions of municipal leaders and responsible smart city concept managers in the transparent selection of independent experts for advisory commissions to make beneficial decisions for the public and use public resources effectively to implement measures in times of safety but also in emergencies, such as the COVID-19 pandemic, which endanger people's lives?

Overview of Domestic and Foreign Research Studies

The reason for developing effective urban transformation strategies and transparent selection of independent experts for policy teams for decision-making and implementation is not only due to the heterogeneity of smart city dimensions together with the necessary complexity and systems approach, but also the nature of the capacities and tools needed for smart city concepts. Their development is complex due to the diverse structure of the systems within which the decision-making processes take place, as well as the risks arising from the challenging quantification of some of the assumptions that the decision-making processes are based on. There is a lack of studies aimed at developing a methodological platform for the decision-making processes influenced by the pandemic (as in the work of Abebe et al. [1]; Gavril et al. [2]; Rathore and Khanna [3]; Wójcik and Ioannou [4]; or Waiho et al. [5]). We have seen the emergence of numerous research studies whose research trajectories have been linked to policymaking, but which have been largely compensatory (e.g., Åslund [6]; Hudson [7]; ILO, International Labour Office [8]; Juergensen et al. [9]; UNCTAD, United Nations Conference on Trade and Development [10,11]).

The relevance of this study is evidenced by significant global research, scientific publications, and the COVID-19 pandemic. At present, many publications and conferences are being devoted to the smart city concept, and startups and innovative projects are being developed to improve and stimulate the development of such cities [12,13].

While obtaining and processing intellectual knowledge on the concept of the smart city, there is a problem of formalizing the opinions of experts on the object of study. There are no general approaches to transforming experienced human expert knowledge into a knowledge base.

Hybrid models of multiple-criteria decision-making (MCDM) are rapidly emerging as alternative methods of information modeling [14,15]. Fuzzy or hybrid decision-making methods are widely used in many areas that require effective information management when evaluating alternative decisions and making optimal decisions [16,17]. The human experience uses fuzzy systems and, on their basis, a fuzzy initial estimate is obtained [18,19]. Current approaches and advantages of using fuzzy mathematics in decision support systems are investigated in [20,21]. For example, the applied application of fuzzy mathematics in various fields has been studied in [22,23]. In [24], a narrowly specialized approach to evaluating startup project developers using fuzzy networks is presented. In [25], an innovative fuzzy model for assessing expert knowledge is proposed to increase the security of decision-making. Unfortunately, the last two works are not suitable for evaluation by experts in different fields.

Many researchers around the world are working on selecting the best choice of specialists for vacant positions. In [26], Gungor et al. considered quantitative and qualitative factors using a vague analytical hierarchical process to solve the problem of personnel selection. Some researchers have used an object-oriented programming model and a machine learning method to solve the problem of recruitment for the project team [27,28]. Zhang and Liu [29] combined fuzzy numbers with a gray relational approach to select software engineers. In [30], Afshari used a fuzzy integral to recruit staff when the recruitment criteria were interdependent. In [31], a fuzzy number is used to assess the weight of each criterion and the performance of the evaluated specialists. Heidary Dahooie et al. [32] developed a system of competencies with five criteria to select the best IT expert. Researchers in [33] used vague linguistic terms to express the opinions of experts to solve the problem of recruitment. Ozdemir and Nalbant [34] evaluated the applicants using the method of fuzzy analytic hierarchy, based on five criteria and their pairwise comparison. As a result, a ranking number of applicants was built. In [35], a method of multi-criteria decision-making for the selection of suitable people for vacant positions is developed, taking into account quantitative and qualitative factors. In [36], a hybrid gray model of multiple-criteria decision-making methods for personnel selection is proposed, which allows for eliminating the vagueness of the input information. Only [37,38] solve the task of supporting decision-making for the management of a smart city regarding the evaluation and selection of experts. The aim of [37] was to develop a methodology for assessing and selecting instructor pilots for smart city UAM Urban Air Mobility, based on the principles of fuzzy mathematics and using different information criteria and competency models. A feature of [38] is the general fuzzy model of evaluation and selection of a group of experts (teams) for smart urban transport and mobility. Such experts are selected for various tasks of smart city management.

However, there is no comprehensive study assessing the competencies of advisory commissions' experts and their selection to solve special or innovative tasks within the functioning and smart city concept in safe and pandemic times.

This paper is systematized as follows: in Section 2, we define the formal problem statement and fuzzy, hybrid, and neuro-fuzzy models for assessing the competence of smart city specialists. The criteria were based on the experts' knowledge, their skills, and at least 15 years of practical experience in municipality management and personnel selection processes. In Section 3, we outline the results of the experiment. The model algorithm was used to create a web application to support municipal management decisions for the above-mentioned agenda, from safe times to a pandemic. In Section 4, we discuss the results of the hybrid fuzzy model and the SW developed in the study. In Section 5, we conclude the paper and present the main results. We also discuss ideas for future work and improvements.

2. Materials and Methods

2.1. Formal Problem Statement

In the study of the smart city as a complex system, there is a need to understand and influence the controllability of the processes to achieve its goals. In addition to many different factors, the controllability of the smart city system depends on the specialists [39]. Such entities may include the following:

- Decision makers (DM) at a certain stage of the smart city system development and at some point in time;
- Experts/system analysts who provide information on the performance indicators of the smart city system;
- Managers who are directly responsible for achieving the goals in the concept of the smart city system;
- Members of the City Council commission as experts who make management decisions within their competences, etc.

The subject of management is an expert (specialist) who is endowed with certain competencies and powers that allow for implementing their will in the form of management decisions, which management teams are required to perform.

Depending on the applied task, it is necessary to evaluate managers and their competencies. The level of controllability of processes in the complex systems of a smart city depends on the competence of the specialists. In the future, without reducing the generality, the subjects of management of such complex systems will be named specialists.

Adequate solutions of the problem of evaluation and selection of smart city specialists, taking into account their competencies, directly affect the achievement of goals by the functioning systems of the smart city.

To assess the competencies of specialists, there are tasks through which we can clearly understand what the ideal combination of knowledge and practical skills should be, as well as the ways of thinking; professional, ideological, and civic qualities; and moral and ethical values, which determine a person's ability to successfully pursue professional activities in this system of functioning. Examples of such tasks for a smart city include the evaluation and selection of instructor pilots for Smart City Urban Air Mobility to operate drones for the delivery of parcels or for the prospective operation of drone taxis (VTOL—Vertical Take Off and Landing); assessment of experts' knowledge with the implementation of IT tools; assessment of the competencies of infectious disease experts for staffing situations and others. Such tasks simulate linear processes of assessing the competencies of specialists.

On the other hand, there are tasks where, when innovative goals are achieved, a new task is created. It does not clearly articulate and investigate what properties the subject must have to in order to achieve the goal of the system. Therefore, to build an information model of criteria for assessing the competencies of specialists, managers use practical experience and intuitive knowledge. At the same time, managers predict that these competencies of specialists can achieve the target needs of the system of functioning of the smart city. Examples of such tasks include evaluating a team of experts to select a project aimed at reducing congestion in a smart city and improving the environment; evaluation and selection of an expert group for smart city and green transportation and mobility, as members of the Transport and Construction Commission; assessment of the competencies of a group of specialists for innovation, development, grant work and implementation of new research. In such problems, we cannot talk about the linearity of the processes of assessing the competencies of specialists.

Given the above, the task of assessing the competencies of specialists can be classified as modeling linear and nonlinear processes.

To adequately determine the impact of human factors on the controllability of the processes of smart city systems, we propose assessing the competencies of the system managers using the following approaches. Additionally, taking into account any type of data, there is a need and relevance to present fuzzy knowledge on the application of fuzzy sets to build information models. Information modeling of the presentation of fuzzy knowledge will provide an opportunity to adequately approach the evaluation of alternative decisions, while increasing the degree of validity of decision-making. When modeling linear processes, use the apparatus of fuzzy sets [19,22]. This will reveal the uncertainties of information and works with different scales of input data. Additionally, to assess the competencies of specialists in innovative tasks, when modeling nonlinear processes, we propose to use neuro-fuzzy models [24]. Fuzzy neural networks have the following advantages: the ability to work with fuzzy and high-quality information; the possibility of using expert knowledge in the form of fuzzy inference rules. Rationality and adequacy of the use of neuro-fuzzy models are proved in [19–23].

The model of technology for assessing the competencies of specialists in the system of functioning of the smart city, taking into account the influence of human factors on the processes of control, is presented as follows:

$$\{E, K, LP \mid Y\}, \tag{1}$$

where E is a set of specialists in some subject area in the smart city system; K is the information model of the criteria (groups of criteria) for assessing the competencies of the specialists, taking into account the human impact on the control processes of the smart city system; $LP = \{M_1; M_2; M_3\}$ is a type of model for assessing the competencies of experts. As part of our study, we offer the following models: M_1—a fuzzy model for the assessment of the competence of specialists, taking into account different models of competencies for linear assessment tasks; M_2—a hybrid fuzzy model that takes into account the experience of managers; M_3—a neuro-fuzzy network for modeling nonlinear evaluation processes.

As a result, we obtain an initial assessment of Y competencies of specialists, on the basis of which we make decisions on the selection of specialists. The algorithm for selecting the model for assessing the competencies of smart city specialists, taking into account the linearity of the assessment processes, is given as follows in Figure 1.

Let a set of specialists be given, $E = \{e_1, e_2, \ldots, e_n\}$, for a smart city. Specialists need to be evaluated according to different methods of competencies, which, in turn, consist of evaluation indicators (criteria). After that, specialists need to be organized according to a certain rule to select the most competent ones or determine their assessments. Different methods of assessing the knowledge, skills, abilities, or psychophysiological properties of specialists are used in the smart city system for different target tasks. For example, within this study, consider the following methods: m_1—evaluation of ways of thinking; m_2—assessment of theoretical knowledge; m_3—assessment of practical knowledge; m_4—assessment of knowledge in the theory of pedagogy, psychology, and communicative competence; m_5—assessment of narrowly specialized skills. For the fuzzy model of assessment of the competence of specialists, we use the methods m_1–m_4, and for the neuro-fuzzy network of assessment—m_5.

Take the set of criteria $K_d = \{K_{d1}, K_{d2}, \ldots, K_{dm}\}$, $d = \overline{1, \omega}$ for the corresponding methods of competence m_d. Each criterion is a question to which the answer must be chosen that is close to the truth. The answers to the question are denoted by Z_{djk}, $d = \overline{1, \omega}$, $j = \overline{1, m}$ and $k = \overline{1, l}$, where d is the evaluation method, j is the number of the criterion (question) in the corresponding model, and k is the number of the answer to the question. According to each criterion, the specialist chooses one of the answer options, which is assigned the appropriate score b_{djk} or some linguistic term, such as L = {H; HC; C; B}: H—"low level of the indicator"; HC—"level of the indicator below average"; C—"average level of the indicator"; B—"high level of the indicator".

Next, based on the experience of the authors and significant experiments on the evaluation of specialists in various fields, we present a block diagram of the technology for the evaluation and selection of specialists of the smart city, shown in Figure 2. According to this scheme, software support was developed in the form of a web platform, where there is a choice for special knowledge of specialists and involvement of managers' opinions.

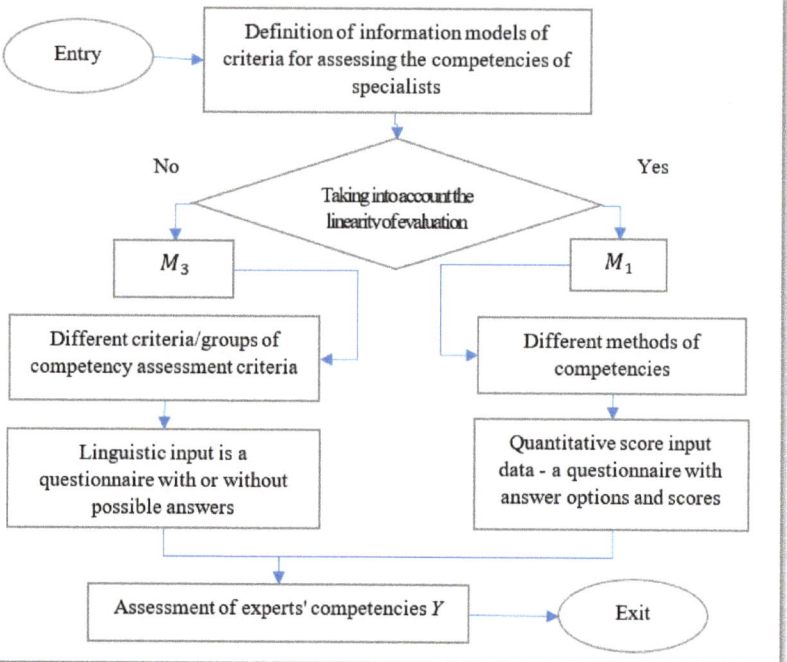

Figure 1. Algorithm for choosing a model for assessing the competencies of smart city specialists, taking into account the linearity of assessment processes: M_1—fuzzy model for assessment of the competence of specialists; M_3—neuro-fuzzy network.

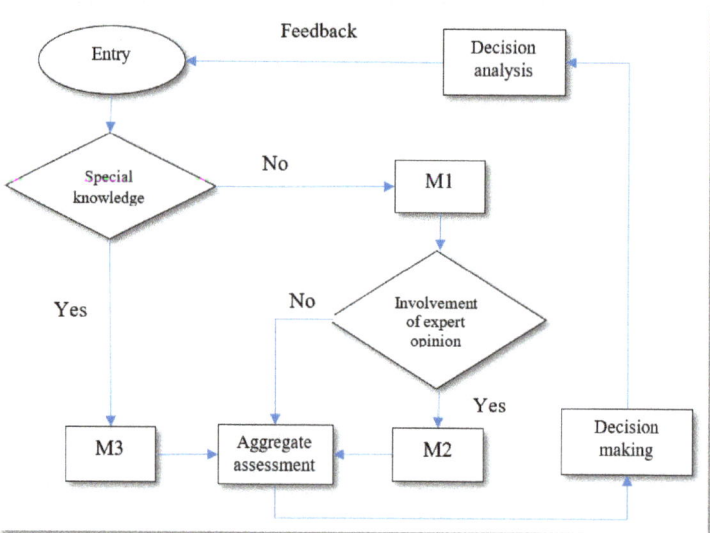

Figure 2. Block diagram of the technology for the assessment and selection of specialists of the smart city: M_1—fuzzy model for assessment of competence of specialists; M_2—hybrid fuzzy model; M_3—neuro-fuzzy network.

The choice of evaluation model depends on the specialist's need for special knowledge, skills, and abilities, the ability to attract expert opinion on the evaluated professionals, their competence, ability, and potential to adequately perform the task to achieve the target needs of the system. For this purpose, three models are proposed, which will give an aggregate assessment of specialists of the smart city.

The choice of the method M_1—a fuzzy model for the assessment of the competence of specialists, taking into account different models of competency tasks—follows from the fact that knowledge about the subject area is incomplete, subjective, and not always reliable. The use of accurate methods does not allow taking into account the verbal inaccuracy and subjectivity of expert information, which in turn imposes restrictions on the quality of knowledge for decision-making.

To apply the knowledge and experience of smart city managers in the evaluation process, the M_2 model is proposed—a hybrid fuzzy model that takes into account the experience of managers, which is able to derive a normalized assessment of the quality of selection of specialists. Its feature is that it combines different types of data in one structure.

The correctness of the choice of the method M_3—neuro-fuzzy network, for the studied problems—the authors prove in [24]. M_3 is based on Takagi–Sugeno–Kang (TSK) and determined that among neural networks, the TSK network gives more accurate results than the Adaptive-Network-Based Fuzzy Inference System (ANFIS). In addition, the change in the number of rules in the training sample does not have a significant effect on the forecasting results.

After that, the decision-maker analyzes and selects the decision, followed by approval of the decision. If over time the solution does not meet the target needs of the system, or new conditions appear, then there is feedback and evaluation occur first.

2.2. Fuzzy, Hybrid, and Neuro-Fuzzy Models for Assessing the Competence of Smart City Specialists

We present a fuzzy model for assessing the competence of specialists, M_1 [37,38].

At the first stage, it is necessary to select evaluation models and calculate the number of points scored.

To reveal the quality of the evaluated specialists in a specific applied task, it is necessary to adequately build methods and criteria of competencies. After that, we proceed to the survey of experts on the relevant models of competencies $g(M_d) = \{s_{d1}, s_{d2}, \ldots, s_{dn}\}$ and summarize the number of points scored.

Next, we phase the input data. Since the input data are presented in the form of questionnaires, with score points that are subjective in nature, there is a need to reveal the uncertainty of the input data [40]. As a result, the obtained numerical variables $\{s_{d1}, s_{d2}, \ldots, s_{dn}\}$ take different numerical values, and to compare them, the values need to be normalized. For this, we propose using the modeling of fuzzy knowledge using the apparatus of fuzzy sets and membership functions [41]. The procedure for transferring fuzzy data does not depend on the type of membership functions. Therefore, in fuzzy set theory, it is assumed that experts may have different representations of the types of membership functions and the base sets on which they are defined, and this does not affect the final result [42]. For example, the membership function of a harmonic S-spline is given by formula (2):

$$\partial_{di}(s_{di}, a, b) = \begin{cases} 0, & s_{di} < a; \\ \frac{1}{2} + \frac{1}{2}\cos\left(\frac{s_{di}-b}{b-a} \cdot \pi\right), & a \leq s_{di} \leq b; \\ 1, & s_{di} > b. \end{cases}, \; (i = \overline{1,n}; d = \overline{1,\omega}). \tag{2}$$

where a is the convolution of the sum of the minimum points, b is the convolution of the sum of the maximum points of the grading scale according to the criteria in the methods m_d, and s_{di} is the convolution of the sum of points of the i-th specialist in the competency model m_d. Thus, the obtained input data will be normalized and comparative.

In addition, fuzzification of the "desired values" is necessary [43]. For each competency model, the decision-maker (manager) has his own considerations, which should be the "desired values"—that is, the sum of the scores for each method m_d. We denote them by the respective vector $T = (t_1, t_2, \ldots, t_\omega)$ of the models m_d, $(d = \overline{1, \omega})$. To compare the "desired values", we calculate the value of the constructed membership function of the quadratic S-spline:

$$\delta_d(t_d, a, b) = \begin{cases} 0, & t_d \leq a; \\ 2\left(\frac{t_d-a}{b-a}\right)^2, & a < t_d \leq \frac{a+b}{2}; \\ 1 - 2\left(\frac{b-t_d}{b-a}\right)^2, & \frac{a+b}{2} < t_d < b; \\ 1, & t_d \geq b. \end{cases} \quad (d = \overline{1, \omega}). \tag{3}$$

For the model m_1, the DM chooses one of the values of the thinking strategies of the specialist, or a combination thereof, according to the proposed characteristic function [37]:

$$\delta_1 = \begin{cases} 0.2 & \text{if} & \text{synthesiser,} \\ 0.3 & \text{if} & \text{synthesiser/idealist,} \\ 0.4 & \text{if} & \text{idealist,} \\ 0.5 & \text{if} & \text{idealist/pragmatist,} \\ 0.6 & \text{if} & \text{pragmatist,} \\ 0.7 & \text{if} & \text{pragmatist/analyst,} \\ 0.8 & \text{if} & \text{analyst,} \\ 0.9 & \text{if} & \text{analyst/realist,} \\ 1 & \text{if} & \text{realist,} \end{cases} \tag{4}$$

Next, we polarize the input data and "desired values". To do this, define a set of quantities that are a relative estimate of the proximity of the elements μ_{di} to the corresponding element of the "desired values" δ_d [44]:

$$z_{di} = 1 - \frac{|\delta_d - \mu_{di}|}{\max\left\{\delta_d - \min_i \mu_{di}; \max_i \mu_{di} - \delta_d\right\}}, \quad d = \overline{1, \omega}; \, i = \overline{1, n}. \tag{5}$$

The matrix $Z = (z_{di})$, defined in this way, characterizes the relative estimates of the proximity of the corresponding specialist e_i to the "desired values" δ_d by the methods of assessing competencies.

If we have more than one method of assessing competencies, then they will have different levels of importance. In this regard, the DM sets the weight for each method of assessing the competence of specialists $\{p_1, p_2, \ldots, p_\omega\}$, for example, from the interval [12,21]. If there is no need to set weights, then we accept them as equally important. For further calculations, we carry out their rationing [37]:

$$w_d = \frac{p_d}{\sum_{d=1}^{\omega} p_d}, \quad d = \overline{1, \omega}; \, w_d \in [0, 1]. \tag{6}$$

where the condition $\sum_{d=1}^{\omega} w_d = 1$ is met.

To defuzzify the data, an aggregate estimate is constructed using the convolution model. For example, take a weighted average convolution:

$$\mu(e_i) = \sum_{d=1}^{\omega} w_d \cdot z_{di}, \, i = \overline{1, n}. \tag{7}$$

Based on the obtained estimates $\mu(e_i)$, we can build a ranking of specialists in terms of competency in the assessment methods—for example, to select a team: $A = $

$\{\mu(e_1), \mu(e_2), \ldots, \mu(e_n)\}$. Depending on the needs, a combination of the most competent specialists is chosen.

Next, consider a hybrid fuzzy model—M_2.

When choosing a specialist, there is a need to analyze their competency, work experience, and other additional indicators that are not taken into account in the methods of assessing competency. In this case, we offer a formalized toolkit that will allow managers of the municipality of a smart city to take into account their own opinions about the specialists in question.

Suppose that we have the calculated aggregate assessments of all specialists (7) assessed according to a fuzzy model. We present a hybrid fuzzy evaluation model in the following formal form:

$$M(\mu(e); E) \to \varphi(e). \quad (8)$$

where $\mu(e)$ is the aggregate assessment of the specialist e based on the methods of assessing competencies; E—expert assessment of the specialists by the managers of the smart city. At the output of the model, we have $\varphi(e)$—a normalized assessment of the quality of the selection of specialists.

We will introduce the concept of "risk trend" in the selection of specialists. To do so, we project the estimates of $\mu(e)$ on the "risk trend" of the evaluation and selection of specialists as well as considering the dependence in the form of the S-linear membership function:

$$\mu(e_i) = \begin{cases} 0, & P_i < \tau_1; \\ \frac{P_i - \tau_1}{\tau_2 - \tau_1}, & \tau_1 \leq P_i \leq \tau_2; \\ 1, & P_i > \tau_1. \end{cases} \quad i = \overline{1, n}. \quad (9)$$

where τ_1, τ_2 are numerical values that we consider in the percentage scale—$\tau_1 = 0, \tau_2 = 100$. For example, when it comes to risk, 100% is associated with the most critical risk. Since the values of the membership function are known $\mu(e_i)$ and are known numerical values, we express the following from formula (9) P_i: $P_i = \mu(e_i) \cdot 100$. The obtained value of P characterizes the assessment of the projection of the "risk trend" of specialist selection.

Let the reasoning of smart city managers regarding the evaluated specialists be presented in the form of linguistic variables—for example, G {the specialist is very well suited to perform the task}; H {the specialist suitable for the task}; S {the specialist is poorly suited to perform the task}. To adequately interpret the dependence of the aggregate assessment of the specialist on the projection of the "risk trend" of the selection and consideration of the opinions of the managers of the smart city, we construct the following membership function:

$$\varphi(e_i) = \begin{cases} 0, & P_i < 0; \\ (\mu(e_i))^k, & 0 \leq P_i \leq 100; \\ 1, & P_i > 100. \end{cases} \quad (10)$$

where k is the degree of compliance of the specialist to perform tasks. For example, let us say that one specialist is very well suited to perform the task $k = 1/3$, one specialist is suitable for the task $k = 1$, and one specialist is poorly suited to perform the task $k = 5/3$.

The obtained value of $\varphi(e_i)$ is an aggregate assessment of the quality of the selection of specialists.

Thus, the presented fuzzy hybrid model for assessing the competence of specialists, which is able to derive a normalized assessment of the quality of the selection of specialists, uses the analysis of the reasoning of smart city managers, reveals the vagueness of the input assessments, and increases the validity of further management decisions based on the results.

To assess narrowly specialized skills, we offer a neuro-fuzzy assessment network [24,40].

Suppose that a set of specialists $E = \{e_1, e_2, \ldots, e_n\}$ is given for the smart city, which we will evaluate using many indicators (criteria) grouped into S groups. Specialists need to be organized according to a certain rule and derive a linguistic rating of Y.

Each criterion for evaluating specialists is expertly evaluated using one of the terms—for example, the next term set of linguistic variables L = {H; HC; C; B}, where H is "low level"; HC—"below average"; C—"average level of the indicator"; B—"high level of the indicator" [32]. Additionally, for each assessment, the specialist receives a "coefficient of confidence" d [24] in assigning an assessment, from the interval [0; 1].

Obtaining an aggregate assessment of the competencies of specialists can be represented in the form of a four-layer neuro-fuzzy network, based on Takagi–Sugeno–Kang (TSK) [20]; Figure 3.

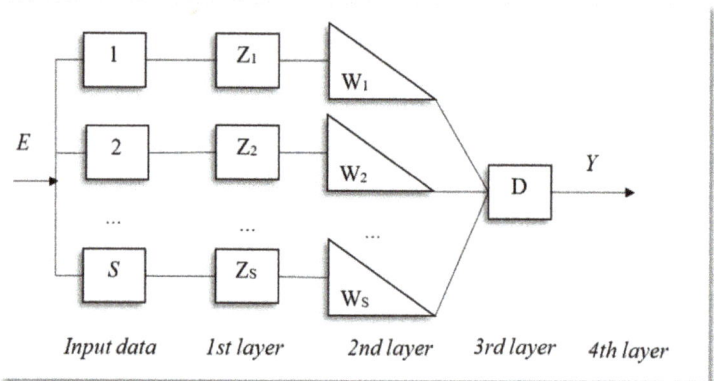

Figure 3. The structure of the neural fuzzy network, where E—a set of specialists; S—number of groups of criteria; Z—neurons; W—functions of postsynaptic potential; D—data defuzzification; Y—output assessment of competencies of specialists.

Next, we consider in more detail what happens on each layer of the neural network.

First layer: Fuzzification of input data

In the neurons of the first layer, the operation of fuzzification is performed—that is, the construction of the membership rule to obtain a normalized estimate of the input data $\left(L_{lji}; d_{lji}\right)$, where l is a group of evaluation criteria $l = \overline{1,S}$, j is the number of criteria groups $j = \overline{1,m}$, and i is the number of the evaluated specialist $i = \overline{1,n}$. Let the term set of linguistic variables L = {H; HC; C; B} be represented in some numerical interval $[a_1; a_5]$, where H ∈ $[a_1; a_2]$, HC ∈ $[a_2; a_3]$, C ∈ $[a_3; a_4]$ and B ∈ $[a_4; a_5]$. The values of the partitioning of the intervals can be determined in the process of learning the neural network.

Next, we present the rule of belonging by modeling fuzzy knowledge. The membership function is chosen on the basis of "x is greater than"; for example, this type is a linear S-shaped membership function:

$$\mu\left(O_{lji}\right) = \begin{cases} 0, & O_{lji} \leq a_1; \\ \frac{O_{lji}-a_1}{a_5-a_1}, & a_1 < O_{lji} \leq a_5; \\ 1, & O_{lji} \geq a_5. \end{cases} \text{ where } O_{lji} = \begin{cases} a_2 \cdot d_{lji}, & \text{if } L_{lji} \in H; \\ a_3 \cdot d_{lji}, & \text{if } L_{lji} \in HC; \\ a_4 \cdot d_{lji}, & \text{if } L_{lji} \in C; \\ a_5 \cdot d_{lji}, & \text{if } L_{lji} \in B. \end{cases} \quad (11)$$

The membership function constructed in this way means that the obtained value of $\mu\left(O_{lji}\right)$ will be 1 if the specialist has high scores.

Thus, in the neurons of the first layer, we reveal the subjectivity of the opinions of specialists and move from linguistic assessments and confidence in their assignment to normalized comparable data [24].

Second layer: Aggregation of values of activation conditions

The second layer aggregates the functions of postsynaptic potential by groups of evaluation criteria. The second layer contains the number of neurons that corresponds to the number of groups of criteria.

Let the municipality management set the synaptic weights $\alpha_{11}, \ldots, \alpha_{1j_1}, \alpha_{21}, \ldots, \alpha_{2j_2}, \ldots, \alpha_{s1}, \ldots, \alpha_{sj_s}$ from the interval $[1; \beta]$ for each criterion. Input signals with synaptic weights form the values of the level of excitation of neurons $Z_{1i}, Z_{2i}, \ldots, Z_{Si}$:

$$Z_{1i} = \frac{1}{\alpha_{11} + \alpha_{12} + \cdots + \alpha_{1j_1}} \cdot (\mu(O_{11i}) \cdot \alpha_{11} + \mu(O_{12i}) \cdot \alpha_{12} + \cdots + \mu(O_{1j_1i}) \cdot \alpha_{1j_1}),$$
$$Z_{2i} = \frac{1}{\alpha_{21} + \alpha_{22} + \cdots + \alpha_{2j_2}} \cdot (\mu(O_{21i}) \cdot \alpha_{21} + \mu(O_{22i}) \cdot \alpha_{22} + \cdots + \mu(O_{2j_2i}) \cdot \alpha_{2j_2}), \quad (12)$$
$$\ldots\ldots\ldots\ldots\ldots\ldots\ldots\ldots\ldots\ldots\ldots\ldots\ldots\ldots\ldots\ldots\ldots\ldots\ldots\ldots$$
$$Z_{Si} = \frac{1}{\alpha_{S1} + \alpha_{S2} + \cdots + \alpha_{Sj_S}} \cdot (\mu(O_{S1i}) \cdot \alpha_{S1} + \mu(O_{S2i}) \cdot \alpha_{S2} + \cdots + \mu(O_{Sj_Si}) \cdot \alpha_{Sj_S}), \quad i = \overline{1,n}.$$

The output neurons of the second layer $Z_1, Z_2, \ldots, Z_S$ will be normalized, because the calculations use the relative importance of the synaptic weights of the criteria [40].

Third layer: Aggregation of values of weights of groups of criteria

In the third layer, the neurons of the second layer are adjusted for the importance of a particular group of evaluation criteria. In this case, for each group of criteria, the management of the municipality has its own considerations regarding the synaptic weights $\alpha_1, \alpha_2, \ldots, \alpha_S$ according to the groups of criteria: the most important effect (IE), significant effect (SE), medium effect (ME), insignificant effect (BE), and little or no effect (LE).

Linguistic variables correspond to some interval—for example, IE = 10, SE = 8, ME = 6, BE = 4, and LE = 2. Values can be changed by system analysts or in the learning process, if there is sufficient information.

We calculate the functions of the postsynaptic potential of the neurons of the third layer as follows:

$$W_{1i} = \frac{\alpha_1}{\alpha_1 + \alpha_2 + \cdots + \alpha_S} \cdot Z_{1i}; \; W_{2i} = \frac{\alpha_2}{\alpha_1 + \alpha_2 + \cdots + \alpha_S} \cdot Z_{2i}; \ldots; \; W_{Si} = \frac{\alpha_S}{\alpha_1 + \alpha_2 + \cdots + \alpha_S} \cdot Z_{Si}. \quad (13)$$

Fourth layer: Output layer

In the fourth layer, the data $D = Z(e_i)$ undergo defuzzification and the levels of decision-making on the linguistic interpretation of the competencies of specialists are compared. To do this, use the following activation function in the original neuron:

$$Z(e_i) = W_{1i} + W_{2i} + \cdots + W_{Si}, \; i = \overline{1,n}. \quad (14)$$

The levels of competence of specialists are determined as a result of training in a neuro-fuzzy network.

All the obtained aggregate estimates, $\mu(e)$, $\varphi(e)$, $Z(e)$, according to the given models, can be compared with the initial variable, $Y = \{y_1, y_2, \ldots, y_t\}$.

The scale of the original variable Y is proposed as follows: $[0.87; 1]$—y_1; $[0.67; 0.87]$—y_2; $[0.37; 0.67]$—y_3; $[0.21; 0.37]$—y_4; $[0; 0.21]$—y_5.

- y_1 = "specialist rating is high". The highest level of specialist rating; very high ability to respond in a timely manner and solve current or strategic problems in the implementation of tasks.
- y_2 = "specialist rating is above average". High level of specialist rating; able to respond in a timely manner and solve current or strategic problems in the implementation of tasks, but negative changes in circumstances may reduce this ability.
- y_3 = "specialist rating is average". Speculative level of specialist rating; there is a possibility of developing different types of risks in the implementation of tasks.
- y_4 = "specialist rating is low". According to the rating, it is not possible to fulfill the tasks set on time; the ability to fulfill the obligations depends entirely on the favorable situation.
- y_5 = "specialist rating is very low". Very high probability of non-fulfillment of tasks by a specialist; unable to work on tasks.

Thus, a fuzzy model for assessing the competence of specialists, a hybrid fuzzy model, and a neuro-fuzzy network have been developed as components of decision support technologies for smart city managers in assessing the competencies of specialists and choosing them to solve special or innovative tasks.

2.3. The Information Model for Evaluation and Selection of the Expert Group as Members of the Transport and Construction Commission

Here, the criteria for assessing narrowly specialized skills (m_5) are presented, which allow for assessing specialists as members of the Transport and Construction Commission of the city of Košice using M_3, the neuro-fuzzy assessment network.

The City Councils of Košice have 14 commissions [45]: the Commission for the Protection of the Public Interest in the Performance of the Functions of Public Officials; Legislative and Legal Commission; Culture Commission; Commission of National Minorities; Finance Commission; Commission for Education; Commission for Sport and Active Recreation; Commission for Regional Development and Tourism; Transport and Construction Commission; Commission for the Environment, Public Order and Health; Social and Housing Commission; Property Commission; Church Commission; Temporary Commission for the Organization of Relations between the City of Košice and EEI Ltd. (canceled).

The city commissions are made up of deputies of the City Parliament and experts, without a parliamentary mandate. The deputies of the City Parliament are elected by the citizens, who cannot influence in which city commission the deputies will work, but can help in the selection of experts without a parliamentary mandate [46].

In our study, we present an information model for the evaluation and selection of the expert group members (who are not deputies of the City Council) of the Transport and Construction Commission to evaluate and oversee the implementation of innovative projects. The Transport and Construction Commission consists of nine politicians (commission chair and eight city deputies) and eleven experts (non-politicians). The work of the Commission follows [46]:

(a) Assesses and gives recommendations on drafts of the city's zoning plans, its amendments and the Košice Development Program, as well as on the proposals of city-wide concepts for the development of individual areas of city life and urban areas from the point of view of spatial development, construction, and transport;
(b) Comments on the concepts, the design of preparation, and the method of implementation of construction projects in the city, which significantly affect the lives of the city's people from the point of view of transport;
(c) Expresses its opinion on fundamental conceptual proposals for the development of urban public transport and passenger, bicycle, and pedestrian transport in the city;
(d) Comments on changes in the public transport tariffs in the city of Košice;
(e) Comments on proposals for the preparation and implementation of transport constructions in the city and on decisive constructions of a city-wide characteristic;
(f) Comments on the concept and solutions of urban static transport;
(g) Takes a position on activities in the field of administration and maintenance of roads in the city pursuant to the Road Act;
(h) Gives opinions on individual points of City Council meetings, which are substantively related to the issues of construction and transport, as well as on points that are related to the actual activity of the commission.

Under the new Program for Economic Development and Social Development for 2022–2027, which is part of the Integrated Territorial Strategy for Sustainable Urban Development of the Functional Area of Košice [47], certain specific objectives are planned with regard to transport, such as the following:

Priority no. 1: green area—SMART Integrated Transport System (EUR 9.615 million); low-emission vehicle fleet for public passenger transport (EUR 164.197 million); modernized transport infrastructure for public passenger transport (EUR 190.955 million); comprehensive network of transport cycle routes in the UMR (EUR 24.57 million);

Priority no. 2: quality public services—safety of the population (EUR 27.683 million, of which public lighting in the city requires EUR 24.12 million and safe pedestrian crossings require EUR 2.0 million); active leisure (EUR 28.827 million).

For these intentions, the great importance of selecting quality experts (external analysts, members of commissions—non-deputies) can already be seen in order to effectively use public funds, not only "according to political and populist proposals of regional and city politicians".

Thus, for the information model, we present the following set of evaluation criteria for experts of the Transport and Construction Commission of the city of Košice, which is classified into three groups.

The first group of criteria K_1 includes professional competence and experience in spatial development, construction, and transport. For this group, we offer the following indicators and answer options:

K_{11}—Possession of professional knowledge, skills, and abilities in the transport and/or construction industry, which is confirmed by education: 1. Bachelor's degree in technical/managerial direction; 2. Master's degree in technical/managerial direction; 3. Master's degree in technical/managerial orientation and availability of various advanced training certificates; 4. Available scientific degree.

K_{12}—Successful experience in working on various tasks related to the concept of transport in the city, control over the implementation of transport solutions, reduction in environmental impact, and traffic assessment to reduce accidents: 1. Absent; 2. Available for one or two tasks; 3. Available for two or three tasks; 4. Available for all tasks.

K_{13}—Successful experience of working on investment, innovative, or scientific projects, programs and development strategies aimed at transport and construction: 1. Absent; 2. Project executor; 3. Expert in the project; 4. Project manager.

K_{14}—Ability to anticipate risks related to the implementation of transport solutions with reduced environmental impact and accidents: 1. Available; 2. Average; 3. High; 4. Very high.

K_{15}—Understanding of pricing of passenger transportation services on public transport: 1. Available; 2. Project participant on this topic; 3. Bachelor's degree in economics; 4. Master's degree or Ph.D. in economics.

The second group of criteria K_2 includes the professional activity and sociality of experts.

K_{21}—Communication ability, openness and ability to cooperate with different partners: 1. Insignificant; 2. Average; 3. Significant; 4. Intensive.

K_{22}—Availability of publications on the issues of analysis, development, innovations in the field of transport and the concept of smart city: 1. Absent; 2. Available publication; 3. Available articles of a scientific nature; 4. Numerous articles.

K_{23}—Teamwork skills: 1. No experience; 2. Experience of project work in a team; 3. Middle manager; 4. Senior manager.

K_{24}—Ability to manage people, set tasks, delegate authority, and desire to compete in the struggle for supremacy and authority: 1. Absent; 2. Available experience in people management; 3. The middle manager or teaching experience; 4. Senior manager.

K_{25}—Focus on winning and maintaining their own reputation, recognition, and the achievement of goals and respect among people: 1. Low dynamics; 2. Average dynamics; 3. High dynamics; 4. Very high dynamics.

The third group of criteria K_3 includes the creativity and psycho-physiology of experts.

K_{31}—Openness to new ideas, constant movement forward, growth and focus on innovative changes to achieve the most effective result: 1. Insignificant; 2. Average; 3. Significant; 4. Intensive.

K_{32}—The correct assessment of their strengths and weaknesses, the constant development of professional and personal qualities, the desire to solve complex professional problems for self-development, and accumulation of knowledge and experience: 1. Available; 2. Average; 3. High; 4. Very high.

K_{33}—Ability to acquire knowledge and its implementation in practice: 1. Available; 2. Average; 3. High; 4. Very high.

K_{34}—Efficiency and systematic thinking: 1. Available; 2. Average; 3. High; 4. Very high.

K_{35}—Stress resistance and emotional balance: 1. Available; 2. Average; 3. High; 4. Very high.

3. Results

We tested the study on the example of the evaluation of experts for smart city and green transportation and mobility as members of the Transport and Construction Commission of the city of Košice, using a neural network. For the evaluation, we had five candidate experts for the position of member of the Transport and Construction Commission [46]. The input data were the answers of the applicants to the questions given in the form of the term set of linguistic variables, L = {H; HC; C; B}, proposed in the study, and estimates of the "coefficient of confidence" d; see Table 1.

Table 1. Input data.

Group of Criteria	The Name of the Criterion	Weight Criterion	e_1		e_2		e_3		e_4		e_5	
K_1	K_{11}	10	HC	0.5	HC	0.6	C	0.9	C	0.9	B	0.8
	K_{12}	8	HC	0.9	HC	0.6	HC	0.8	C	0.8	C	0.7
	K_{13}	10	C	0.7	HC	0.5	C	0.8	C	0.7	B	0.9
	K_{14}	9	C	0.8	HC	0.8	C	0.7	C	0.9	HC	0.8
	K_{15}	8	HC	0.6	B	0.8	B	0.9	B	0.9	C	0.9
K_2	K_{21}	9	C	0.8	C	0.8	C	0.9	B	0.8	B	0.9
	K_{22}	7	B	0.7	C	0.9	C	0.7	C	0.7	B	0.8
	K_{23}	8	C	0.8	C	0.8	B	0.8	B	0.9	C	0.9
	K_{24}	7	HC	0.9	HC	0.8	B	0.6	B	0.9	B	0.8
	K_{25}	7	C	0.9	C	0.9	C	0.7	C	0.6	C	0.9
K_3	K_{31}	8	C	0.8	HC	0.8	C	0.8	HC	0.8	C	0.9
	K_{32}	8	HC	0.9	HC	0.9	B	0.9	HC	0.8	C	0.9
	K_{33}	8	C	0.7	HC	0.8	C	0.6	C	0.9	B	0.8
	K_{34}	9	HC	0.7	B	0.9	B	0.7	B	0.8	C	0.9
	K_{35}	7	B	0.8	HC	0.8	C	0.7	C	0.8	B	0.8

We next show the process of obtaining an aggregate assessment of the competencies of the specialists using the proposed four-layer neural network. To fuzzify the input data, we defined the membership function in the numerical interval [0; 10], where H ∈ [0; 2], HC ∈ [2; 5], C ∈ [5; 8] and B ∈ [8; 10].

The value of $\mu\left(O_{lji}\right)$ was calculated using formula (11); Figure 4.

Next, to aggregate the values of the activation conditions, we calculated the value of the level of excitation of neurons using formula (12). To aggregate the values of the weights of the groups of criteria, the manager of the municipality had his own considerations about the synaptic weights according to the groups of criteria: K_1—the most important effect ($\alpha_1 = 10$); K_2—significant effect ($\alpha_3 = 8$); K_3—average effect ($\alpha_2 = 6$). Then, we calculated the functions of the postsynaptic potential of the neurons of the third layer using formula (13). The results of the calculations are shown in Figure 5.

Next, we defuzzified the data and compared the levels of decision-making on the linguistic interpretation of the competencies of specialists using the activation function (14): $Z(e_1) = 0.5424$; $Z(e_2) = 0.5048$; $Z(e_3) = 0.6524$; $Z(e_4) = 0.687$; $Z(e_5) = 0.7371$. We built a ranking of the specialists using the obtained values: e_5; e_4; e_3; e_1; e_2. We compared the quantitative result with the original variable Y and found that the specialists e_5 and e_4 had a "specialist rating above average" and all the others had a "specialist rating—average".

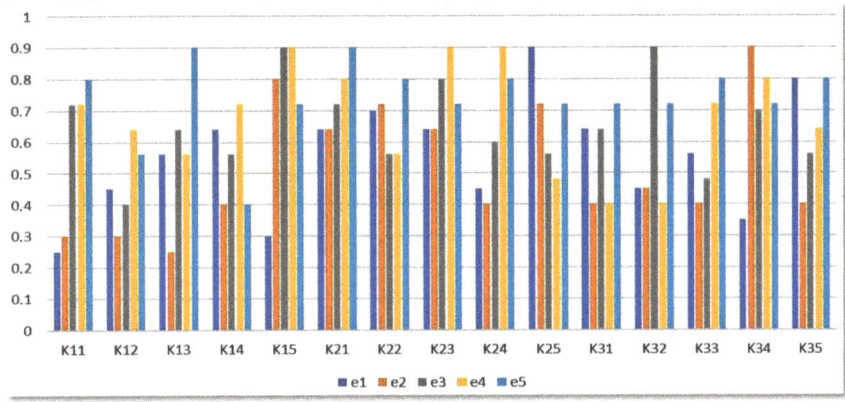

Figure 4. Fuzzification of input data.

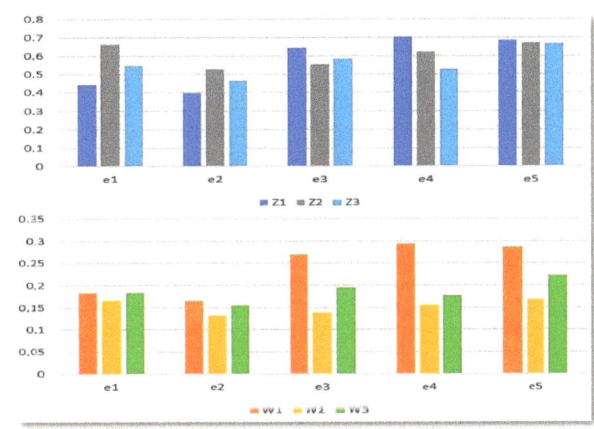

Figure 5. Values of neuronal excitation level (Z_1; Z_2; Z_3) and postsynaptic potential function (W_1; W_2; W_3).

As part of the study, an innovative web platform named Smart City Concept Personnel Selection [48] was developed on the basis of the proposed technology for the evaluation and selection of specialists, as shown in Figure 6.

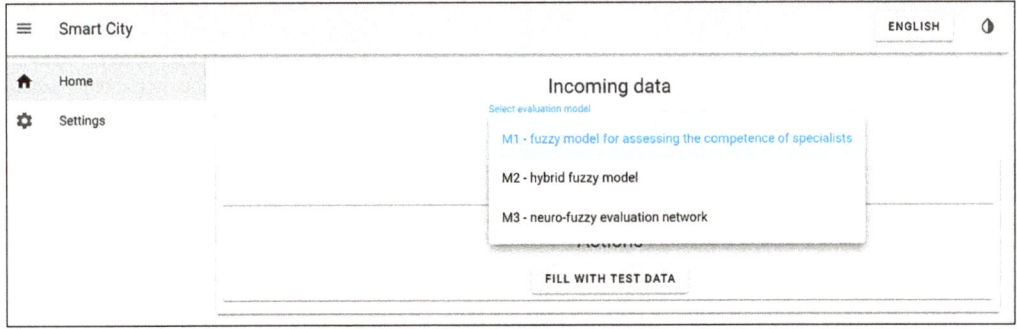

Figure 6. Head screen of the Smart City Concept Personnel Selection web platform.

All of the important components on which the technology of decision support for assessing the competencies of specialists and their selection is based were placed in the settings (Figure 7).

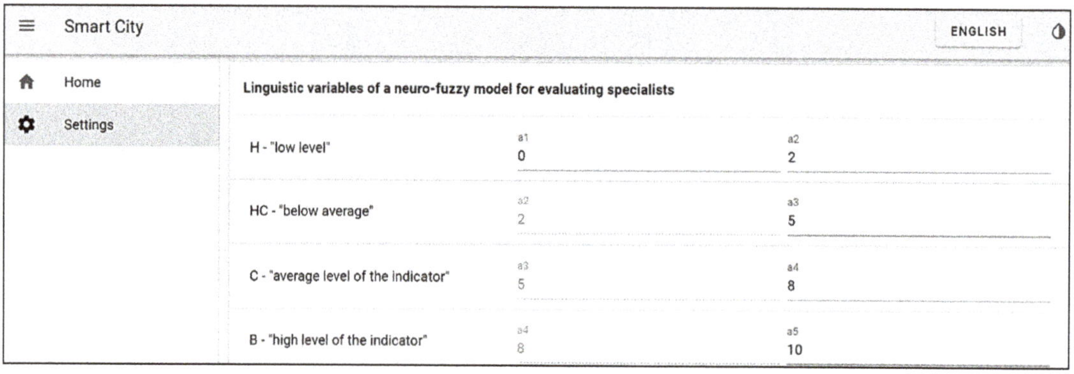

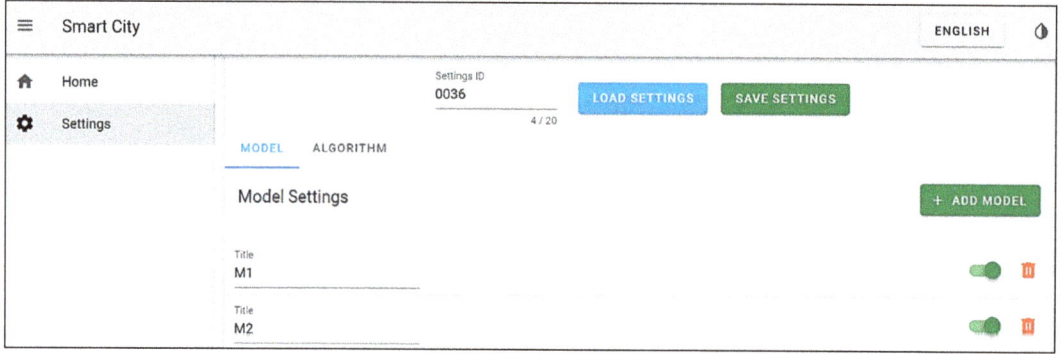

Figure 7. Smart City Concept Personnel Selection web platform setup screen.

In addition, in the setup, it is easy for system analysts to build information evaluation models for various experts, such as members of other commissions. Thus, using the identifier (ID), it is possible to store and protect the adjusted information models on the server for the assessment of various experts and various users (Figure 8).

Figure 8. ID for the evaluation information model members of the Transport and Construction Commission.

After selecting the model for assessing the competencies of smart city specialists and entering the input data, we proceeded to the calculations. Based on the initial assessment of the competencies of specialists and the level of rating, further decisions could be made.

The developed technology and web platform is a useful innovative tool for managers in the concept of a smart city when assessing the competencies of specialists and choosing them to solve special or innovative tasks, which reveals the vagueness of input assessments, increases the validity of further management solutions, and uses the analysis of reasoning, experience, and knowledge of managers.

4. Discussion

The result of the study is technology providing decision support for the managers of smart cities for the estimation of the competencies of experts and their selection for special or innovative tasks. For this purpose, a fuzzy model for assessing the competencies of specialists, a hybrid fuzzy model, and a neuro-fuzzy network were developed. The models are able to assess the levels of competencies of specialists and derive their ranking based on many subjects of management (specialists) of a subject area in the system of a smart city, information models of criteria (groups of criteria) for assessing the competencies of specialists that take into account human impact on the controllability of the smart city system, and a kind of competency assessment model. At the same time, the tools of intellectual analysis of knowledge, a system approach, processing of fuzzy data, and a neuro-fuzzy network are used. The models developed in the work reveal the vagueness of the input estimates and increase the degree of validity of further management decisions by the managers of the municipality regarding the selection of specialists to perform innovative tasks. The output of the models is an assessment of the competencies of specialists and their rating.

Depending on the requirements for the choice of specialists, information methods are proposed that allow to obtain a set of input data for the work of the developed models, namely evaluation of ways of thinking; assessment of theoretical knowledge; assessment of practical knowledge; assessment of knowledge in the theory of pedagogy, psychology and communicative competence; and assessment of narrowly specialized skills. For the proposed information methods, it is necessary to define a set of evaluation criteria for specialists, which will allow assessing their competencies depending on the tasks assigned to them.

The paper proposes an information model for evaluation and selection of an expert group as members of the Transport and Construction Commission. Innovative software support in the form of a web platform has been constructed, which implements the developed methods, algorithms, and information models for application by smart city managers in the evaluation and selection of specialists. The toolkit allows smart city analysts to change decision-making levels, synaptic weights, and the degree of suitability of the specialist to perform tasks in the context of reasoning by municipal managers and to set different sets of criteria for evaluating specialists within the proposed information methods. All of this allows for maximizing the adaptation process and supports decision-making for specific professionals within the functioning of the smart city system. In addition, the study was tested on the example of the evaluation of experts for smart city and green transportation and mobility as members of the Transport and Construction Commission of the city of Košice, using a neuro-fuzzy network.

The advantages of technology to support decision-making by smart city managers in the selection of specialists (management entities) stem from the advantages of the developed models in particular. The model of assessing the competence of specialists is based on different methods of competence assessment of specialists, which uses not only the skills of specialists but also different qualitative properties, and can be used for different inputs and different methods of competencies and criteria, taking into account the wishes of the manager and importance assessment competency methods. The advantage of the fuzzy hybrid model is its ability to take into account the own opinions of municipal managers regarding the specialists in question—if necessary, taking into account other additional indicators. The advantages of a neuro-fuzzy assessment network are the objectivity of expert assessments using input linguistic variables and the "coefficient of confidence" of the specialist's reasoning regarding their assignment; it is based on a neuro-fuzzy network, which has the ability to change the settings of synaptic scales; upon receipt of experimental data, it is possible to conduct training of the neural network.

The disadvantages of this technology for supporting decision-making include the fact that it is necessary to adequately select membership functions for the criteria of competency models. The membership function in a fuzzy network corresponds to the stage of rough debugging, and this process depends on the partitioning of the interval $[a_1; a_5]$, which requires the sampling of reliable experimental data. Additionally, the use of different types of convolutions can lead to ambiguity of the final results.

The mathematical substantiation of the models has already been described in more detail by the authors in [24,25,37,38]. Determining the effectiveness of the developed models M_1—fuzzy model, for assessment of the competence of specialists, and M_2—hybrid fuzzy model, was performed by the number of comparison operations. The number of operations of comparison of alternatives decreases, in comparison with methods using the technology of pairwise comparisons (analytic hierarchy process), by 100*(n − 3)/(n − 1) percent or in (n − 1)/2 times (n—number of evaluation specialists) [22,23]. Thus, at n = 4 by 33.3%, at n = 5 by 50%, at n = 11 by 80%, etc.

The proposed model M_3—neuro-fuzzy network was tested for different data sets [24,25] and compared with known, widely used artificial neural networks and teaching methods, forming a knowledge base by generating new production rules that do not contradict the rules of the knowledge base of the system, based on the analysis of experimental data. The method of teaching corresponds to a simplified method of fuzzy inference, but it differs in that the knowledge base is not fixed, but supplemented by the receipt of experimental data [24]. The consistency of the new production rule is guaranteed by the procedure of replenishment of the knowledge base.

The rationality of the research results is proved by the advantages of the developed technology. The reliability of the obtained results is ensured by the correct use of the apparatus of fuzzy sets, system approach, and neural-fuzzy networks, which is confirmed by the research results.

5. Conclusions

In this study, we researched the actual task of developing technology to support decision-making by the managers of smart cities in assessing the competencies of specialists (management entities) and their selection to solve special or innovative tasks. The following results were obtained:

- For the first time, a model of technology for assessing the competencies of specialists in the system of functioning of a smart city was constructed that takes into account the influence of human factors on the processes of controllability of the system. This technology uses a fuzzy model for assessing the competence of specialists to take into account different models of competencies for linear assessment tasks, a hybrid fuzzy model that takes into account the experience of managers, and a neuro-fuzzy network in modeling nonlinear assessment processes.
- For the first time, a hybrid fuzzy model for assessing the competencies of specialists in the smart city system was developed. The peculiarity of the model is that it is able to derive a normalized assessment of the quality of selection of specialists; uses the analysis of reasoning, experience, and knowledge of smart city managers; reveals the vagueness of input assessments; and increases the validity of further management decisions based on results.
- A fuzzy model of assessment of the competence of specialists and a neuro-fuzzy network were further developed. The fuzzy model is adapted to evaluate different specialists on different indicators, including both quantitative and qualitative properties. The neuro-fuzzy network is adapted to assess the narrowly specialized skills of specialists. For each group of criteria, scales in the form of a linguistically effective assessment were used.
- For the first time, an information model was presented for evaluation and selection of an expert group as members of the Transport and Construction Commission. To do this, we proposed 15 evaluation criteria (divided into three groups) of narrowly specialized skills to evaluate members of the Transport and Construction Commission using the neuro-fuzzy network.
- An example evaluation of five candidate experts for smart city and green transportation and mobility, as potential members of the Transport and Construction Commission of the city of Košice, was conducted using real data.
- An innovative web platform was constructed, which implements the developed methods, algorithms, and information models for application by smart city managers in the evaluation and selection of specialists.

The individual components of the smart city concept will form future work on the separate application areas for innovative modeling and evaluation of measures with an impact on public funds. As we have emphasized, the innovative web platform "Smart City Concept Personnel Selection" can be adapted to different users of municipalities or regional institutions for the transparent selection of qualified personnel for effective decision-making and use of public funds. The innovative web platform will respect local conditions and national law regulations and standards. The importance of innovative solutions to support transparent decision-making and the efficient use of public resources is growing in the context of current European Union support for the development and implementation of country-specific recovery and resilience plans, primarily for the recovery of member states and regions in 2021–2027 after the global COVID-19 pandemic.

Among other things, we also see further research in ablation experiments aimed at extending scientific validation of developed technological models to support the decisions of smart city managers, with this method. According to other researchers, we respect that although an ablation study is of course usually not sufficient to conclude the contributions of different modules, when coupled with other empirical and statistical methods, it can provide valuable insights to practitioners and researchers [49].

Author Contributions: Conceptualization, M.K., V.P. and B.G.; methodology, M.K. and V.P.; resources, B.G., V.P. and M.K.J.; data curation, V.P. and M.K.J.; investigation and formal analysis, M.K., V.P., B.G. and R.R.; supervision, M.K.; project administration, V.P.; writing—original draft preparation, M.K., V.P. and B.G.; writing—review and editing, M.K., V.P., B.G., R.R., J.B.; L.G. and M.G.; funding acquisition, M.K. All authors have read and agreed to the published version of the manuscript.

Funding: This work was supported and funded by the Slovak Research and Development Agency under the contract no. PP-COVID-20-0002.

Institutional Review Board Statement: Not applicable.

Informed Consent Statement: Not applicable.

Data Availability Statement: Data supporting the reported results can be found at the following links: https://www.kosice.sk/mesto/samosprava/mestske-zastupitelstvo/2018-2022/komisie (accessed on 25 April 2021), https://www.kosice.sk/clanok/komisia-dopravy-a-vystavby-2018-2022 (accessed on 25 April 2021), https://static.kosice.sk/files/manual/umr/Vstupna_sprava_UMR_Kosice_a_funk_oblast.pdf (accessed on 25 April 2021).

Conflicts of Interest: The authors declare no conflict of interest.

References

1. Abebe, G.; Bundervoet, T.; Wieser, C. *Monitoring COVID-19 Impacts on Firms in Ethiopia: Results from a High-Frequency Phone Survey of Firms*; World Bank: Washington, DC, USA, 2020. [CrossRef]
2. Gavril, S.; Coca, O.; Creanga, D.E.; Mursa, G.C.; Mihai, C. The impact of the crisis generated by Covid-19 on the population concerns. A comparative study at the level of the European Union. *Transform. Bus. Econ.* **2020**, *19*, 703–719.
3. Rathore, U.; Khanna, S. From Slowdown to Lockdown: Effects of the COVID-19 Crisis on Small Firms in India (31 May 2020). Available online: https://ssrn.com/abstract=3615339 (accessed on 25 April 2021). [CrossRef]
4. Wójcik, D.; Ioannou, S. COVID-19 and finance: Market developments so far and potential impacts on the financial sector and centres. *Geogr. COVID 19 Pandemic* **2020**, *111*, 387–400. [CrossRef]
5. Waiho, K.; Fazhan, H.; Dahlianis Ishak, S.; Azman Kasan, N.; Jung Liew, J.; Norainy, H.; Ikhwanuddin, M. Potential impacts of COVID-19 on the aquaculture sector of Malaysia and its coping strategies. *Aquac. Rep.* **2020**, *18*, 100450. [CrossRef]
6. Åslund, A. Responses to the COVID-19 crisis in Russia, Ukraine, and Belarus. *Eurasian Geogr. Econ.* **2020**, 1–14. [CrossRef]
7. Hudson, D.S. (Ed.) *Chapter 5 the Economic, Social and Environmental Impacts of COVID 19*; Goodfellow Publishers: Oxford, UK, 2020.
8. International Labour Office ILO. *Prevention and Mitigation of COVID-19 at Work for Small and Medium-Sized Enterprises. Action Checklist and Follow up*; International Labour Office (ILO): Geneva, Switzerland, 2020.
9. Juergensen, J.; Guimón, J.; Narula, R. European SMEs amidst the COVID-19 crisis: Assessing impact and policy responses. *J. Ind. Bus. Econ.* **2020**, *47*, 499–510. [CrossRef]
10. UNCTAD. *United Nations Conference on Trade and Development. Maximizing Sustainable Agri-Food Supply Chain Opportunities to Redress COVID-19 in Developing Countries*; UNCTAD: Geneva, Switzerland, 2020; p. 17. Available online: https://unctad.org/system/files/official-document/ditctabinf2020d9_en.pdf (accessed on 25 April 2021).

11. UNCTAD. United Nations Conference on Trade and Development. Impact of the COVID-19 Pandemic on Trade and Development: Transitioning to a New Normal. UNCTAD. 2020. Available online: https://unctad.org/system/files/official-document/osg2020d1_en.pdf (accessed on 25 April 2021).
12. Kamolov, S.G.; Korneeva, A.M. Future technologies for smart cities. *Bull. Mosc. Reg. State Univ. Ser. Econ.* **2018**, *2*, 100–114. [CrossRef]
13. Finger, M. Smart city—Hype and/or realit. *IGLUS Q.* **2018**, *4*, 2–6.
14. Shen, K.Y.; Zavadskas, E.K.; Tzeng, G.H. Updated discussions on "Hybrid multiple criteria decision-making methods: A review of applications for sustainability issues". *Econ. Res. Ekon. Istraz.* **2018**, *31*, 1437–1452. [CrossRef]
15. Zavadskas, E.K.; Govindan, K.; Antucheviciene, J.; Turskis, Z. Hybrid multiple criteria decision-making methods: A review of applications for sustainability issues. *Econ. Res. Ekon. Istraz.* **2016**, *29*, 857–887. [CrossRef]
16. Mardani, A.; Zavadskas, E.K.; Khalifah, Z.; Jusoh, A.; Nor, K. Multiple criteria decision-making techniques in transportation systems: A systematic review of the state of the art literature. *Transport* **2016**, *31*, 359–385. [CrossRef]
17. Kotsovsky, V.; Batyuk, A. Representational capabilities and learning of bithreshold neural networks. In Proceedings of the International Scientific Conference Intellectual Systems of Decision Making and Problem of Computational Intelligence (ISDMCI-2020), Kherson, Ukraine, 25–29 May 2020; Springer: Cham, Switzerland, 2020; pp. 499–514.
18. Zade, L. *The Concept of a Linguistic Variable and Its Application to Approximate Decision Making (Ponyatiye Lingvisticheskoy Peremennoy i Yego Primeneniye k Prinyatiyu Priblizhennykh Resheniy)*. 1976, pp. 1–167. Available online: https://scask.ru/a_book_zade.php (accessed on 25 April 2021).
19. Rotshteyn, O.P. *Intelligent Identification Technologies: Fuzzy Sets, Genetic Algorithms, Neural Networks (Intelektualni Tekhnolohiyi Identyfikatsiyi: Nechitki Mnozhyny, Henetychni Alhorytmy, Neyronni Merezhi)*; UNIVERSUM: Vinnytsya, Ukraine, 1999; pp. 1–320.
20. Snytyuk, V. *Prognostication. Models. Methods. Algorithms: Textbook (Prohnozuvannya. Modeli. Metody. Alhorytmy: Navch. Posib)*; Maklaut: Kiev, Ukraine, 2008; pp. 1–364.
21. Subbotin, S.O. *Submission and Processing of Knowledge in Artificial Intelligence Systems and Decision Support (Podannya ta Obrobka znan u Systemakh Shtuchnoho Intelektu ta Pidtrymky Pryynyattya Rishen)*; ZNTU: Zaporizhzhya, Ukraine, 2008; pp. 1–341.
22. Zaychenko, Y.P. *Fuzzy Models and Methods in Intelligent Systems: Tutorial (Nechetkiye Modeli i Metody v Intellektualnykh Sistemakh: Navchalny Posibnyk)*; Slovo: Kiev, Ukraine, 2008; pp. 1–344.
23. Zgurovsky, M.; Zaychenko, Y. *Big Data: Conceptual Analysis and Applications*; Springer: New York, NY, USA, 2020.
24. Polishchuk, V.; Kelemen, M. Information model of evaluation and output rating of start-up projects development teams. In Proceedings of the Second International Workshop on Computer Modeling and Intelligent Systems (CMIS-2019), Zaporozhye, Ukraine, 15–19 April 2019; Volume 2353, pp. 674–688.
25. Polishchuk, V.; Liakh, I. Enhancing the safety of assessment of expert knowledge in fuzzy conditions global environment. In *Collective Monograph: Management Mechanisms and Development Strategies of Economic Entities in Conditions of Institutional Transformations of the Global Environment*; Bezpartochnyi, M., Ed.; ISMA University Landmark: Riga, Latvia, 2019; Volume 2, pp. 234–143.
26. Gungor, Z.; Serhadlioglu, G.; Kesen, S.E. A fuzzy AHP approach to personnel selection problem. *Appl. Soft Comput.* **2009**, *9*, 641–646. [CrossRef]
27. Fan, Z.P.; Feng, B.; Jiang, Z.Z.; Fu, N. A method for member selection of R&D teams using the individual and collaborative information. *Expert Syst. Appl.* **2009**, *36*, 8313–8323.
28. Feng, B.; Jiang, Z.Z.; Fan, Z.P.; Fu, N. A method for member selection of cross-functional teams using the individual and collaborative performances. *Eur. J. Oper. Res.* **2010**, *203*, 652–661. [CrossRef]
29. Zhang, S.F.; Liu, S.Y. A GRA-based intuitionistic fuzzy multi-criteria group decision making method for personnel selection. *Expert Syst. Appl.* **2011**, *38*, 11401–11405. [CrossRef]
30. Afshari, A.R.; Yusuff, R.M.; Derayatifar, A.R. Linguistic extension of fuzzy integral for group personnel selection problem. *Arab. J. Sci. Eng.* **2013**, *38*, 2901–2910. [CrossRef]
31. Sang, X.; Liu, X.; Qin, J. An analytical solution to fuzzy TOPSIS and its application in personnel selection for knowledge-intensive enterprise. *Appl. Soft Comput.* **2015**, *30*, 190–204. [CrossRef]
32. Heidary Dahooie, J.; Beheshti Jazan Abadi, E.; Vanaki, A.S.; Firoozfar, H.R. Competency-based IT personnel selection using a hybrid SWARA and ARAS-G methodology. *Hum. Factors Ergon. Manuf. Serv. Ind.* **2018**, *28*, 5–16. [CrossRef]
33. Yalçın, N.; Yapıcı Pehlivan, N. Application of the fuzzy CODAS method based on fuzzy envelopes for hesitant fuzzy linguistic term sets: A case study on a personnel selection problem. *Symmetry* **2019**, *11*, 493. [CrossRef]
34. Ozdemir, Y.; Nalbant, K.G. Personnel selection for promotion using an integrated consistent fuzzy preference relation—Fuzzy analytic hierarchy process methodology: A real case study. *Asian. J. Interdiscip. Res.* **2020**, *3*, 219–236. [CrossRef]
35. Chen, C.-T.; Hung, W.-Z. A two-phase model for personnel selection based on multi-type fuzzy information. *Mathematics* **2020**, *8*, 1703. [CrossRef]
36. Ulutaş, A.; Popovic, G.; Stanujkic, D.; Karabasevic, D.; Zavadskas, E.K.; Turskis, Z. A new hybrid MCDM model for personnel selection based on a novel grey PIPRECIA and grey OCRA methods. *Mathematics* **2020**, *8*, 1698. [CrossRef]
37. Rozenberg, R.; Szabo, S.; Polishchuk, V.; Némethová, H.; Jevčák, J.; Choma, L.; Kelemen, M., Jr.; Vuković, D. Information model for evaluation and selection of instructor pilots for smart city urban air mobility. In Proceedings of the 5th International Conference on Smart and Sustainable Technologies (SpliTech 2020), Split, Croatia, 23–26 September 2020.

38. Kelemen, M.; Polishchuk, V.; Liakh, I.; Némethová, H.; Jevčák, J.; Choma, L.; Kelemen, M., Jr.; Vuković, D. Municipality management and model of evaluation and selection of the expert group members for smart city transportation and mobility including UAV/UAS. In Proceedings of the 5th International Conference on Smart and Sustainable Technologies (SpliTech 2020), Split, Croatia, 23–26 September 2020.
39. Polishchuk, V.; Malyar, M.; Kelemen, M.; Polishchuk, A. The technology for determining the level of process control in complex systems. In Proceedings of the 7th International Conference "Information Technology and Interactions" (IT&I-2020): Workshops Proceedings, Kyiv, Ukraine, 2–3 December 2020; Technical University of Aachen: Aachen, Germany, 2020; pp. 138–148. Available online: http://ceur-ws.org/Vol-2845/Paper_14.pdf (accessed on 25 April 2021).
40. Kelemen, M.; Polishchuk, V.; Gavurová, B.; Szabo, S.; Rozenberg, R.; Gera, M.; Kozuba, J.; Andoga, R.; Divoková, A.; Blišťan, P. Fuzzy model for quantitative assessment of environmental start-up projects in air transport. *Int. J. Environ. Res. Public Health* **2019**, *16*, 3585. [CrossRef]
41. Voloshyn, O.F.; Polishchuk, V.V.; Malyar, M.M.; Sharkadi, M.M. Informatsiyne modelyuvannya nechitkykh znan. *RIU* **2018**, *4*, 84–95. [CrossRef]
42. Kelemen, M.; Polishchuk, V.; Gavurová, B.; Andoga, R.; Szabo, S.; Yang, W.; Christodoulakis, J.; Gera, M.; Kozuba, J.; Kaľavský, P.; et al. Educational model for evaluation of airport NIS security for safe and sustainable air transport. *Sustainability* **2020**, *12*, 6352. [CrossRef]
43. Polishchuk, V.; Kelemen, M.; Gavurová, B.; Varotsos, C.; Andoga, R.; Gera, M.; Christodoulakis, J.; Soušek, R.; Kozuba, J.; Blišťan, P.; et al. A fuzzy model of risk assessment for environmental start-up projects in the air transport sector. *Int. J. Environ. Res. Public Health* **2019**, *16*, 3573. [CrossRef]
44. Polishchuk, V. Fuzzy method for evaluating commercial projects of different origin. *J. Autom. Inf. Sci.* **2018**, *50*, 60–73. [CrossRef]
45. Commission of the Municipal Council in Košice. 2021. Available online: https://www.kosice.sk/mesto/samosprava/mestske-zastupitelstvo/2018-2022/komisie (accessed on 25 April 2021).
46. Transport and Construction Commission 2018–2022. 2018. Available online: https://www.kosice.sk/clanok/komisia-dopravy-a-vystavby-2018-2022 (accessed on 25 April 2021).
47. Entry Message for Processing PHRSR/IUS UMR—Functional Area of the City of Košice. 2021. Available online: https://static.kosice.sk/files/manual/umr/Vstupna_sprava_UMR_Kosice_a_funk_oblast.pdf (accessed on 25 April 2021).
48. Smart City Concept Personnel Selection. 2021. Available online: https://smartcity-2.herokuapp.com (accessed on 24 May 2021).
49. Meyes, R.; Lu, M.; Waubert de Puiseau, C.; Meisen, T. Ablation Studies in Artificial Neural Networks. *arXiv* **2019**, arXiv:1901.08644.

 *mathematics*

Article

Evaluation and Selection of the Quality Methods for Manufacturing Process Reliability Improvement—Intuitionistic Fuzzy Sets and Genetic Algorithm Approach

Ranka Gojković [1], Goran Đurić [2], Danijela Tadić [3], Snežana Nestić [3] and Aleksandar Aleksić [3,*]

1. Faculty of Mechanical Engineering, University of East Sarajevo, 71123 East Sarajevo, Bosnia and Herzegovina; ranka.gojkovic@ues.rs.ba
2. Faculty of Mechanical Engineering, University of Belgrade, 11120 Belgrade, Serbia; gdjuric@mas.bg.ac.rs
3. Faculty of Engineering, University of Kragujevac, 34000 Kragujevac, Serbia; galovic@kg.ac.rs (D.T.); s.nestic@kg.ac.rs (S.N.)
* Correspondence: aaleksic@kg.ac.rs

Abstract: The aim of this research is to propose a hybrid decision-making model for evaluation and selection of quality methods whose application leads to improved reliability of manufacturing in the process industry. Evaluation of failures and determination of their priorities are based on failure mode and effect analysis (FMEA), which is a widely used framework in practice combining with triangular intuitionistic fuzzy numbers (TIFNs). The all-existing uncertainties in the relative importance of the risk factors (RFs), their values, applicability of the quality methods, as well as implementation costs are described by pre-defined linguistic terms which are modeled by the TIFNs. The selection of quality methods is stated as the rubber knapsack problem which is decomposed into subproblems with a certain number of solution elements. The solution of this problem is found by using genetic algorithm (GA). The model is verified through the case study with the real-life data originating from a significant number of organizations from one region. It is shown that the proposed model is highly suitable as a decision-making tool for improving the manufacturing process reliability in small and medium enterprises (SMEs) of process industry.

Keywords: selection of quality methods; manufacturing process; intuitionistic fuzzy sets; genetic algorithm

Citation: Gojković, R.; Đurić, G.; Tadić, D.; Nestić, S.; Aleksić, A. Evaluation and Selection of the Quality Methods for Manufacturing Process Reliability Improvement—Intuitionistic Fuzzy Sets and Genetic Algorithm Approach. *Mathematics* **2021**, *9*, 1531. https://doi.org/10.3390/math9131531

Academic Editors: Jorge de Andres Sanchez and Laura González-Vila Puchades

Received: 29 May 2021
Accepted: 24 June 2021
Published: 29 June 2021

Publisher's Note: MDPI stays neutral with regard to jurisdictional claims in published maps and institutional affiliations.

Copyright: © 2021 by the authors. Licensee MDPI, Basel, Switzerland. This article is an open access article distributed under the terms and conditions of the Creative Commons Attribution (CC BY) license (https://creativecommons.org/licenses/by/4.0/).

1. Introduction

Nowadays, the problem of increasing reliability of manufacturing process, which represents the core of any industrial enterprise, affects the realization of business goals so that it is an interesting field of research for both researchers and practitioners. According to the results of the best practice, and literature sources [1], it can be said that failures have the greatest impact on the reliability of the manufacturing process. Although there is a significant number of production and organization concepts, such as world class manufacturing (WCM), just-in-time (JIT), the scope of this paper is set to the lean manufacturing and in compliance with that, failures leading to lean waste are identified. According to the lean concept [2], there are seven types of waste found in any process: transportation, inventory, motion, waiting, overprocessing, overproduction, and defect. Later, Liker [3] introduced waste related to underutilization of labor creativity, and it is denoted as Unused employee creativity. Reducing or eliminating the impact of identified failures can be performed by implementing various quality methods.

Quality methods are extremely important, because without reliable and complete information, it is practically impossible to undertake effective measures aimed at improving manufacturing processes [4,5].

There are many quality methods that are defined in the literature [6,7]. Application of these quality methods can lead to the simultaneous elimination or reduction of the impact of one or more identified failures of the manufacturing process. Respecting the limited resources (money, time, etc.), it can be concluded that it is almost impossible to implement a number of quality methods at the same time. In practice, the basic task of the reliability manager is how to choose a set of quality methods whose application of the considered problem can be effectively solved in the shortest possible time. Many authors suggest that the choice of quality methods should be based on the priority of failures [8–11]. Experiences of best practice show that decision makers (DMs), when choosing quality methods, respect many criteria besides the priority of failures. In this research, the authors suggest that in determining the set of quality methods, besides the mentioned, it is necessary to consider applicability of methods and implementation costs due to limited financial resources of small and medium enterprises in the manufacturing industry. Prioritization can be seen as a problem in itself. According to [12], the priority assessment under an uncertain environment is based on the failure mode and effect analysis (FMEA) which is combined with fuzzy sets theory [13], as in this research. In other words, evaluation of failures is performed by respecting to risk factors (RFs): severity, occurrence, and detection. According to suggestions of Liu [14], in this research, assumptions were introduced: (i) RFs have not equal the relative importance, and (ii) the relative importance of RFs and their values can be adequately described in linguistic terms, and (iii) the priority of failures is based on the rank of risk priority number (RPN) which is calculated as the product of these three RFs with respects to their weights.

Respecting the nature of human thinking, it can be said that DMs hardly give an accurate evaluation of uncertain data in practical decision problems. Hence, DMs express their assessments easier when using natural language words instead of the precise numbers. The development of theories of mathematics, such as the theory of intuitive fuzzy sets (IFSs) [15] allows vagueness to be represented fairly quantitatively. The advantages of using IFSs are: (1) transient stages during decision making can be rendered by intuitionistic indices, and (2) it is possible to foresee the best and worst results.

As it is known, in the manufacturing process management literature, it is almost impossible to find papers where the problem of choosing the quality methods set is considered as a combinatorial optimization problem. Since, in this research, the treated problem is stated as a rubber knapsack problem which represents the version of the classic 0–1 knapsack problem (0–1 KP) with the goal of finding, in a set of items of given values and weights, the subset of items with the highest total value, subject to a total weight constraint [16,17]. This can be marked as one of this paper's novelties. Finding the optimal solution of the considered problem is based on using the genetic algorithm (GA) [18]. By using GAs, perform a search in complex, and large landscapes, and provide near-optimal solutions for objective function an optimization problem. In the literature, GA is a widely used metaheuristic method for solving difficult nondeterministic polynomial time (NP) problems from different domains [19–21], as in this research.

Motivation for this research comes from the fact that there are almost no literature sources that treat the selection of quality methods where uncertainties are given by exact ways.

The wider objective of this research may be interpreted as: (1) identification of failures in manufacturing process in SMEs of the process industry, (2) assessment identified failures based on FMEA, (3) modelling of the existing uncertainties are performed by using triangular intuitionistic fuzzy number (TIFN) [22,23], (4) the priority of failure is determined by applying the RPN with TIFNs with respects to RF weights, (6) selection of quality methods is stated as KP, and obtaining an adequate set of quality methods is based on the application of GA. The authors believe that the solution obtained in this way is significantly less burdened by the subjective attitudes of DMs, and therefore the effectiveness of solving the problem of reliability of the manufacturing process is significantly better.

The paper is organized in the following way: In Section 2, there is a literature review. Methodology is presented in Section 3. The proposed Algorithm is presented in Section 4. The case study is presented in Section 5. Conclusion and discussion are given in Section 6.

2. Literature Review

In the literature, there are no proposed procedures, rules, or recommendations on how to choose quality methods whose application improves the reliability of the manufacturing process. With the respect to the best industrial practice, it can be said that the choice of quality methods is always based on the knowledge and experience of reliability managers. In this way, the chosen set of quality methods is significantly burdened by the subjective attitudes of DMs. In order to increase the accuracy of the solution, in this research, the treated problem is stated as a discrete optimization problem and the solution is given by exact ways.

The classical KP can be defined as filling knapsack with a given set of objects with associated values, and space requirements associated with them although these problems have a simple structure, but they are known to be NP-hard. The KP has very important applications in the financial and industry domains, such as resource distribution, investment decision-making, items shipment, budget controlling, production planning [17], etc. There are many variants of KPs which are presented in the research literature.

In this research, the considered problem should be stated as KP which does not have a fixed value constraint. The knapsack constraint is not a specific value but a function of the number of solution elements. In the relevant literature, this version of KP is more complex than 0–1 KP and is relatively rarely investigated. The solution to this problem can be solved in different ways. In this paper, the solution to the considered problem is found by decomposing the given problem into subproblems with a certain number of solution elements. Each of the subproblems described above is a 2-dimensional KP. Each obtained solution can be further decomposed into several possible subversions due to the fact that many quality methods have the same applicability and/or implementation costs, so they are equivalent from the point of view of solution optimality.

Many researchers suggest applying GA for solving KPs [24,25]. Initially, GA generates randomly a population consisting of representative individuals (chromosomes) over which genetic operators of mutation, crossbreeding, and selection are successively applied. Fitness function is defined through the goal function of the considered problem. Based on the value of fitness function, a decision is made whether a representative individual remains in the population or not. In this way, a randomly selected population is transformed into a new population.

In classical KPs problems, all variables are described by precise numbers. In real life problems that exist in a changing environment, it is almost impossible to use precise measurement scales. Since, DMs better express their assessments of the relative importance and values of variables by using linguistic expressions. The development of theories of mathematics, such as the theory of fuzzy sets [13], has enabled these linguistic terms to be presented quantitatively in a sufficiently good way. In the literature, there are a large number of papers in which the relative importance of RFs and its values are modeled by: (i) type 1 fuzzy sets [26–28], (ii) the interval type 2 fuzzy numbers [10,29,30], and (iii) intuitionistic fuzzy sets [31–33]. The intuitionistic fuzzy set using two characteristic membership functions expressing the degree of membership and the degree of non-membership of elements of the universal set. It can cope with the presence of vagueness and hesitancy originating from imprecise [15]. It may be suggested that the natural language words can be adequately quantitatively described by using intuitionistic fuzzy sets.

Many authors suggest that the assessment of the relative importance of RFs should be set as a fuzzy group decision making problem. Aggregation DMs opinions into a single rating can be obtained by applying different operators, for instance: the intuitionistic fuzzy weighted average operator [34], intuitionistic fuzzy analytic hierarchy process [35,36], the utilized methods [37], fuzzy geometric mean [38], fuzzy averaging operator [39]. In this paper, all DMs originate from SMEs in which the same economic activity is realized, so to determine the aggregated value of the weight of risk RFs it is adequate to apply a fuzzy averaging operator as in [39].

3. Methodology

They base their assessments on knowledge and experience as well as data from records.

3.1. Basic Definitions of Intuitionistic Fuzzy Sets

Definition 1. *An intuitionistic fuzzy set A in the universe of discourse X is defined with the form* [15]:

$$\widetilde{A} = (x, \mu_{\widetilde{A}}(x), \vartheta_{\widetilde{A}}(x) | x \in X), \tag{1}$$

where:

The numbers $\mu_{\widetilde{A}}(x) \to [0,1]$ and $\vartheta_{\widetilde{A}}(x) \to [0,1]$ denote the membership degree and non-membership degree.

With the condition
$0 \leq \mu_{\widetilde{A}}(x) + \vartheta_{\widetilde{A}}(x) \leq 1, \forall x \in X$
$\pi_{\widetilde{A}}(x) = 1 - \mu_{\widetilde{A}}(x) - \vartheta_{\widetilde{A}}(x)$
$0 \leq \pi_{\widetilde{A}}(x) \leq 1, \forall x \in X$

The value of $\pi_{\widetilde{A}}(x)$ is called the degree of indeterminacy (or hesitation). The smaller $\pi_{\widetilde{A}}(x)$, more certain $\widetilde{A}$.

Definition 2. *An IFS* $\widetilde{A} = (x, \mu_{\widetilde{A}}(x), \vartheta_{\widetilde{A}}(x) | x \in X)$ *of the real line is called an intuitionistic fuzzy number (IFN) whose membership function and non-membership function are defined as follows* [40]:

$$\mu_{\widetilde{A}}(x) = \begin{cases} \frac{x-a}{b-a} \cdot \mu_{\widetilde{A}}(x) & if \quad a \leq x < b \\ \mu_{\widetilde{A}}(x) & if \quad x = b \\ \frac{c-x}{c-b} \cdot \mu_{\widetilde{A}}(x) & if \quad b < x \leq c \\ 0 & else \end{cases} \tag{2}$$

and

$$\vartheta_{\widetilde{A}}(x) = \begin{cases} \frac{b-x+(x-a) \cdot \vartheta_{\widetilde{A}}(x)}{b-a} & if \quad a \leq x < b \\ \vartheta_{\widetilde{A}}(x) & if \quad x = b \\ \frac{x-b+(c-x) \cdot \vartheta_{\widetilde{A}}(x)}{c-b} & if \quad b < x \leq c \\ 1 & else \end{cases} \tag{3}$$

where a, b, c are real numbers, and $a \leq b \leq c$.

TIFN can be denoted as

$$\widetilde{A} = ([a, b, c]; \mu_{\widetilde{A}}(x), \vartheta_{\widetilde{A}}(x)), \tag{4}$$

Definition 3. *In compliance with the Definition 2, let* $\widetilde{A} = ([a_1, b_1, c_1]; \mu_{\widetilde{A}}(x), \vartheta_{\widetilde{A}}(x))$ *and* $\widetilde{B} = ([a_2, b_2, c_2]; \mu_{\widetilde{B}}(x), \vartheta_{\widetilde{B}}(x))$ *be two positive TIFNs. Additionally, λ is the real number. The operations of these TIFNs are given by* [41]:

$$\widetilde{A} + \widetilde{B} = ([a_1 + a_2, b_1 + b_2, c_1 + c_2]; \min(\mu_{\widetilde{A}}(x), \mu_{\widetilde{B}}(x)), \max(\vartheta_{\widetilde{A}}(x), \vartheta_{\widetilde{B}}(x))), \tag{5}$$

$$\widetilde{A} - \widetilde{B} = ([a_1 - c_2, b_1 - b, c_1 - a_2]; \min(\mu_{\widetilde{A}}(x), \mu_{\widetilde{B}}(x)), \max(\vartheta_{\widetilde{A}}(x), \vartheta_{\widetilde{B}}(x))), \tag{6}$$

$$\widetilde{A} \cdot \widetilde{B} = ([a_1 \cdot a_2, b_1 \cdot b_2, c_1 \cdot c_2]; \min(\mu_{\widetilde{A}}(x), \mu_{\widetilde{B}}(x)), \max(\vartheta_{\widetilde{A}}(x), \vartheta_{\widetilde{B}}(x))), \tag{7}$$

$$\lambda \cdot \widetilde{A} = ([\lambda \cdot a_1, \lambda \cdot b_1, \lambda \cdot c_1]; \mu_{\widetilde{A}}(x), \vartheta_{\widetilde{A}}(x)), \tag{8}$$

$$\widetilde{A}^{\lambda} = ([a_1^{\lambda}, b_1^{\lambda}, c_1^{\lambda}]; \mu_{\widetilde{A}}^{\lambda}(x), (1 - (1 - \vartheta_{\widetilde{A}}(x))^{\lambda})), \tag{9}$$

$$(\widetilde{A})^{-1} = \left(\left[\frac{1}{d_1}, \frac{1}{c_1}, \frac{1}{b_1}, \frac{1}{a_1}\right]; \mu_{\widetilde{A}}(x), \vartheta_{\widetilde{A}}(x)\right), \tag{10}$$

Definition 4. *A usual defuzzification method can be defined as mapping IFNs into scale value and taking the median* [42]*, so that:*

$$a = \frac{1}{12} \cdot (a_1 + 4 \cdot b_1 + c_1) \cdot \left(1 - \vartheta_{\widetilde{A}}(x) + \mu_{\widetilde{A}}(x)\right) \quad (11)$$

Definition 5. *Let* $\widetilde{A} = \left([a_1, b_1, c_1]; \mu_{\widetilde{A}}(x), \vartheta_{\widetilde{A}}(x)\right)$ *and* $\widetilde{B} = \left([a_2, b_2, c_2]; \mu_{\widetilde{B}}(x), \vartheta_{\widetilde{B}}(x)\right)$ *be two positive TIFNs. The Hamming distance* [43] *is:*

$$d_H = \frac{1}{3} \cdot (|a_1 - a_2| + |b_1 - b_2| + |c_1 - c_2|) + \max\left(\left|\mu_{\widetilde{A}}(x) - \mu_{\widetilde{B}}(x)\right|, \left|\left(\vartheta_{\widetilde{A}}(x) - \vartheta_{\widetilde{B}}(x)\right)\right|\right) \quad (12)$$

3.2. Definition of a Finite Set of Decision Makers

In this manuscript, the term DM means the FMEA team of each company. It is common for the FMEA team at the level of each company to be composed of: production manager, FMEA leader, and quality manager. It should be emphasized that they make the decision by consensus. DMs can be presented by the set $e = \{1, \ldots, e, \ldots E\}$ The total number of considered DMs is marked with E, DM index $e = 1, \ldots, E$.

3.3. Choice of Appropriate Linguistic Variables for Describing the Relative Importance of RFs

Defining a set of RFs against which failures are evaluated can be formally presented as a set $\{1, \ldots, k, \ldots, K\}$. Total number of RFs is denoted as K, and k, $k = 1, \ldots, K$ is index of RF. In conventional FMEA [44] three RFs are defined: severity of consequence ($k = 1$), frequency of failure ($k = 2$), and possibility of detecting failure ($k = 3$), as in this research.

In compliance with the evidence from literature [10,14], the RFs may have different relative importance. The fuzzy rating of the relative importance of RFs are based on the pre-defined linguistic expressions and their corresponding TIFNs are presented in the Table 1.

Table 1. The relative importance of RFs.

	TIFNs
low importance (L)	$([1, 1, 3.5]; 0.8, 0.1)$
medium importance (M)	$([1, 3.5, 5]; 0.6, 0.3)$
high importance (H)	$([2.5, 5, 5]; 0.7, 0.2)$

The domains of these TIFNs are defined into real line into interval [1–5]. The value 1 indicates the lowest and the value 5 the highest relative importance of the considered RFs. The overlap of TIFNs describing the relative importance of RFs is large. This indicates a lack of knowledge of DMs about the importance of the considered criteria in SMEs of the manufacturing sector.

3.4. Choice of Appropriate Linguistic Variables for Describing the RF Values, Aplicability of Quality Methods, and Implementation Costs of Quality Methods

In the production process, numerous failures can occur, which can be formally represented by the set of indices $\{1, \ldots, i, \ldots, I\}$, where I presents the total number of failures, and the index of each failure is denoted as i, $I = 1, \ldots, I$. In this research, failures that are identified at the level of each observed SME are considered.

Analysis and reduction of the identified failures can be performed by applying numerous quality methods that can be formally represented by the set of indices $\{1, \ldots, m, \ldots, M\}$. The total number of quality methods is denoted as M, and m, $m = 1, \ldots, M$ is index of quality method. In this research, quality methods are selected according to [6].

The evaluation of the value of RFs, $\widetilde{v}_{ik}$ is performed by DM at the level of each SME. Applicability of quality method m for failure analysis i, $\widetilde{v}_{mi}$, $I = 1, \ldots, I$, are assessed by quality manager at the level of each considered SME and presented in Table 2.

Table 2. The RF values and degree of belief that quality methods are applicable.

	TIFNs
very low value (VLV)	([1, 1, 2.5]; 0.65, 0.3)
low value (LV)	([2, 3.5, 5]; 0.7, 0.25)
medium value (MV)	([3.5, 5, 6.5]; 0.5, 0.45)
high value (HV)	([6, 7.5, 9]; 0.75, 0.2)
very high value (VHV)	([7.5, 9, 9]; 0.8, 0.15)

The domains of these TIFNs are defined in the common measurement scale [1–9]. The value 1 indicates the lowest and the value 9 indicates the highest values of RFs.

Implementation costs of considered quality methods, $\tilde{c}_m$ were evaluated by the quality manager and presented in Table 3.

Table 3. Implementation costs.

	DESCRIPTION	TIFNs
Extremely low cost (C1)	Implementing quality methods requires almost no cost	([0, 0.1, 0.25]; 0.7, 0.25)
Very low value (C2)	The implementation of the quality method does not require equipment	([0.1, 0.3, 0.6]; 0.65, 0.3)
Medium value (C3)	Implementing the method requires standard equipment	([0.3, 0.55, 0.7]; 0.55, 0.4)
High value (C4)	Implementation of quality methods requires the use of computer equipment	([0.5, 0.75, 0.9]; 0.6, 0.3)
Very high value (C5)	Implementation of quality methods requires expensive specialized equipment	([0.7, 0.95, 1]; 0.65, 0.3)

The domains of these TIFNs are defined in the common measurement scale [0, 1]. The value 0 indicates the lowest and the value 1 indicates the highest values of implementation costs.

4. The Proposed Algorithm

Step 1. The relative importance of RF k, $k = 1, \ldots, K$ is assessed by each DM e, $e = 1, \ldots, E$:

$$\widetilde{W}_k^e \qquad (13)$$

Step 2. The aggregated relative importance of RF k, $\widetilde{W}_k$, $k = 1, \ldots, K$, is given by using fuzzy averaging operator:

$$\widetilde{W}_k = \frac{1}{E} \cdot \widetilde{W}_k^e \qquad (14)$$

Step 3. The representative scalar of TIFN, $\widetilde{W}_k$, W_k, $k = 1, \ldots, K$ is given by using the defuzzification procedure [42].

Step 4. Construct the weights vector of RFs, $[\omega_k]_{1 \times K}$. The element of weights vector of RFs, ω_k given by using the linear normalization procedure, so that:

$$\omega_k = \frac{W_k}{\sum_{k=1,\ldots,K} W_k} \qquad (15)$$

Step 5. The value of each RF, $k = 1, \ldots, K$ for each failure i, $I = 1, \ldots, I$ is assessed by DM and could be presented by TIFN, $\tilde{v}_{ik}$.

Step 6. Determine the priority index for each failure i, $I = 1, \ldots, I$ by using fuzzy geometric mean:

$$\widetilde{RPN}_i = \prod_{k=1,\ldots,K} (\tilde{v}_{ik})^{\omega_k} \qquad (16)$$

Step 7. Degree of beliefs that quality methods are applicable to the analysis of identified failure, $\tilde{v}_{mi}$ and costs of implementation of quality methods, $\tilde{c}_m$ are assessed by DM.

Step 8. Determine the applicability of the quality method, $\tilde{z}_{mi}$, to eliminate failure i, $m = 1, \ldots, M$; $I = 1, \ldots, I$:

$$\tilde{z}_{mi} = \tilde{v}_{mi} \cdot \widetilde{RPN}_i \qquad (17)$$

Step 9. The normalized value of applicability of the quality method m at the level of failure i, $m = 1, ..., M$; $I = 1, ..., I$, is:

$$\tilde{r}_{mi} = \frac{\tilde{z}_{mi}}{\tilde{z}^*} \qquad (18)$$

where, $\tilde{z}^*$ is the maximum applicability of the method, so that:

$$\tilde{z}^* = VHV \prod_{k=1,...,K} (VHV)^{\omega_k} \qquad (19)$$

Step 10. The total applicability of the method m, $m = 1, ..., M$ is obtained according to the expression:

$$\tilde{r}_m = \max_{i=1,...,I} (\tilde{r}_{mi}) \qquad (20)$$

Ranking of uncertain values, $\tilde{r}_{mi}$ is performed according to crisp values r_{mi}, so that:

$$r_m = \max_{i=1,...,I} defuzz\, (\tilde{r}_{mi}) \qquad (21)$$

Step 11. Let us set the KP problem:
The fitness function:

$$\max_{j=1,...,J} \sum d(\tilde{r}_m, \tilde{c}_m)_j, \ j \in \{1, ..., m, ..., M\} \qquad (22)$$

where, $d\,(\tilde{r}_m, \tilde{c}_m)$ is calculated as the Hamming distance between two TIFNs [43].
The objective to:

$$\frac{1}{J-1} \cdot \sum_{j=1,...,J} (d(\tilde{c}_m, \tilde{c}))^2 \leq \frac{1}{M-1} \cdot \sum_{m=1,...,M} (d(\tilde{c}_m, \tilde{c}))^2 \qquad (23)$$

where:

$$\tilde{c} = \frac{1}{M} \cdot \sum_{m=1,...,M} \tilde{c}_m$$

And

$$d(\tilde{r}_m, \tilde{c}_m)_j = \begin{cases} 1 & \text{if object } j \text{ is selected} \\ 0 & \text{otherwise} \end{cases} \qquad (24)$$

Step 12. The near optimal solution of the treated KP problem is found by using GA. The encoding schemes are differentiated according to the problem domain. The well-known encoding schemes are binary, octal, hexadecimal, value-based, and tree. Binary encoding is the used encoding scheme in this paper. Each chromosome is represented using a binary string. In binary encoding, every chromosome is 0 or 1. In knapsack problem, binary encoding is used in GA to show the presence of items, 1 for the presence of an item and 0 for the absence of an item.

The initial parameter setting with a population of 100 individuals was gradually reduced to 30 without loss in solution quality, also a further increase in the number of interactions over 1000 was not significant.

The parameters of the applied GA are: generational GA, roulette wheel parent selection, elitism 0.05%, number of individuals in the population 30, selection 0.95, mutation 0.02, and number of iterations 1000. Furthermore, a part of the code of the implemented GA is given, which provides the condition that the chromosome has a precisely determined number of units. In this way, the generation of correct units is ensured without the need for subsequent rejection of defective ones. The following part of the code is given below:

```
public string Generate(int NumberOfAllMeasures, int NumberOfMeasures)
{
string ret = new String('0', NumberOfAllMeasures);
int[] measuresRB = new int[NumberOfAllMeasures];
double[] measuresRand = new double[b NumberOfAllMeasures];
for (int i = 1; i <= NumberOfAllMeasures; i++)
{
measuresRB[i-1] = i-1;
measuresRand[i-1] = random.NextDouble();
}
Array.Sort(measuresRand, measuresRB);
char[] ch = ret.ToCharArray();
for (int i = 1; i <= NumberOfMeasures; i++)
{
ch[measuresRB[i-1]] = '1';
}
return new string(ch);
```

5. Case Study

In this research, the research is realized in 24 manufacturing SMEs located in Bosnia and Herzegovina (B&H). Respecting the official data, it can be said that: (i) account of considered SMEs for more than 15% of total gross domestic product (GDP), (ii) about 20% of total employment in B&H, and (iii) almost 90% of exports, such as they have a significant impact on economic growth in B&H. Experiences of good practice show that one of the most important problems of an operational management team is to maintain the reliability of the manufacturing process over a long period of time. In this way, the defined business goals of the company can be realized to a high degree. Improving the reliability of the manufacturing process is realized through the accurate identification of potential failures and through the selection of adequate quality methods by which the identified failures are reduced and/or eliminated.

As is known in the literature, there are no procedures or recommendations on how to identify failures in manufacturing processes or how to choose quality methods, which leads to increased reliability of the manufacturing process. Experiences of best practice in the process industry show that failure identification is always based on evidence data and experience and knowledge of decision makers, as in this research. The set of quality methods that can potentially be applied is defined according to the recommendations from the relevant literature.

Assessment of the relative importance of RFs, their values, applicability of treated quality methods, as well as the costs of implementation were obtained using the interview method. Questionnaires were sent to the FMEA team asking them to express all estimates using the given pre-defied linguistic expressions. They also returned the completed questionnaires by e-mail.

The Illustration of the Proposed Model

Fuzzy assessment of the relative importance of RFs is performed by DM at the level of each SME (Step 1 of the proposed Algorithm). This is presented in Table 4.

Table 4. Fuzzy rating of the relative importance of RFs.

	E=1	E=2	E=3	E=4	E=5	E=6	E=7	E=8	E=9	E=10	E=11	E=12	E=13	E=14	E=15	E=16	E=17	E=18	E=19	E=20	E=21	E=22	E=23	E=24
K=3	H	H	M	H	H	M	H	H	H	M	H	H	H	M	H	H	H	M	H	H	H	H	H	M
K=2	H	M	M	M	M	M	M	H	H	L	M	M	M	M	M	M	M	L	H	M	H	H	M	M
K=1	L	H	H	M	H	H	L	M	M	L	L	L	L	L	M	M	L	M	M	L	M	M	L	H

The proposed aggregation and defuzzification procedures (Step 2 to Step 3 of the proposed Algorithm) is illustrated for example ($k = 1$):

$$\widetilde{W}_1 = \frac{1}{24} \cdot (18 \cdot H + 6 \cdot M) = \frac{1}{24} \cdot ([51, 111, 120]; 0.6, 0.3) = ([2.13, 4.63, 5]; 0.6, 0.3)$$

The precise number of TIF $\widetilde{W}_1$, W_1 is:

$$W_1 = \frac{1}{12} \cdot (2.13 + 4 \cdot 4.63 + 5) \cdot (1 - 0.3 + 0.6) = 2.776$$

Similarity, the relative importance of the rest RFs are given:

$$\widetilde{W}_2 = ([1.375, 3.666, 4.875]; 0.6, 0.3) \text{ and } W_2 = 2.266$$
$$\widetilde{W}_3 = ([1.312, 2.771, 4.375]; 0.6, 0.3) \text{ and } W_3 = 1.817$$

The weights vector (Step 4 of the proposed Algorithm) is:

$$\omega_k = \frac{W_k}{\sum_{k=1,\ldots,K} W_k}$$

$$[\ 0.40, \quad 0.33, \quad 0.26\ ]$$

The assessment values of RFs for SME ($e = 1$) are given in a Table 5 (Step 5 of the proposed Algorithm):

Table 5. Fuzzy rating of the relative importance of RFs.

Lean Waste	Failures	S	O	D	Lean Waste	Failures	S	O	D
Overprocessing	Inadequate level of automation (I = 1)	HV	HV	VLV	Overproduction	Inadequate use of automation (I = 23)	VLV	VLV	HV
	Inadequate processing modes (I = 2)	MV	LV	MV		Poor assessment of market demands (I = 24)	MV	LV	MV
	Worker error (I = 3)	MV	MV	MV		Poor application of just in case logic (I = 25)	HV	MV	MV
	Product design requires too many processing steps (I = 4)	LV	VLV	VLV		Low knowledge and skills of employees (I = 26)	LV	MV	VLV
	Too many processing processes, too many iterations (I = 5)	MV	LV	VLV	Motion	Poor workplace ergonomics (I = 27)	VLV	HV	VLV
	Customer needs are not clear (I = 6)	VHV	LV	HV		Large distance between operations (I = 28)	MV	MV	VLV
Inventory	Imbalance of material flow (I = 7)	LV	LV	VLV		Frequent hand movements (I = 29)	HV	MV	LV
	The unreliability of suppliers (I = 8)	LV	VLV	VLV		Multiple taking the same piece (I = 30)	MV	HV	LV
	Excessive supply of raw materials (I = 9)	VHV	VLV	VLV		Employees have to move to get information (I = 31)	MV	MV	MV
	Poor communication (I = 10)	LV	LV	LV		Manual work to compensate for some shortcomings in the production process (I = 32)	MV	MV	MV
	Protection of the company from risk and unexpected event (I = 11)	MV	VLV	LV		Inexperience of the operator (I = 33)	HV	MV	LV

Table 5. Cont.

Lean Waste	Failures	S	O	D	Lean Waste	Failures	S	O	D
Transportation	Not understanding the process flow (I = 12)	VLV	VLV	LV	Waiting	Waiting for material between operations (I = 34)	LV	LV	VLV
	Inadequate layout of technological equipment (I = 13)	MV	LV	VLV		Interruption of the machine or system (I = 35)	MV	MV	VLV
	Large storage space (I = 14)	LV	VLV	VLV		Lack of work (I = 36)	HV	MV	VLV
	Communication failure (I = 15)	VLV	MV	LV		Waiting for the information needed to continue the process (I = 37)	MV	VLV	MV
	Using old layouts (I = 16)	VLV	VLV	VLV		Imbalance with subsequent processes (I = 38)	MV	MV	MV
Defect	Insufficient knowledge and skills of workers (I = 17)	MV	MV	MV		Long preparatory-final time (I = 39)	MV	VLV	MV
	Inaccuracies in the documentation (I = 18)	VHV	LV	LV	Unused employee creativity	Narrowly defined jobs (I = 40)	MV	LV	VLV
	Insufficient process control (I = 19)	HV	MV	MV		Not involving workers in creating new ideas (I = 41)	LV	MV	MV
	Design-construction omissions (I = 20)	MV	VLV	LV		Employees do not work in the appropriate position (I = 42)	MV	LV	LV
	Inadequate state of technical and technological equipment (I = 21)	MV	VLV	MV		Not involving all employees and their knowledge and skills in business and production processes (I = 43)	MV	MV	MV
Overproduction	Imbalance of production lines (I = 22)	MV	LV	LV		Workers' absenteeism (I = 44)	MV	MV	MV

The proposed procedure (Step 6 of the proposed Algorithm) is illustrated for the failure (I = 1):

$$\widetilde{RPN}_1 = ([6, 7.5, 9]; 0.75, 0.2)^{0.40} \cdot ([6, 7.5, 9]; 0.75, 0.2)^{0.33} \cdot ([1, 1, 2.5]; 0.65, 0.3)^{0.26}$$

$$\widetilde{RPN}_1 = ([2.07, 2.26, 2.43]; 0.89, 0.09) \cdot ([1.81, 1.95, 2.07]; 0.91, 0.07) \cdot ([1, 1, 1.27]; 0.89, 0.09) = ([3.73, 4.40, 6.41]; 0.89, 0.09)$$

RPN values for all other failures were calculated in a similar way and shown in Appendix A.

According to the proposed procedure (Step 7 of the proposed Algorithm), the degree of beliefs is calculated.

Applicability of methods with respects to priority failures as well as calculation of normalized values (Step 8 to Step 9 of the proposed Algorithm) is illustrated in the following example:

$$\widetilde{z}_{61} = \widetilde{v}_{61} \cdot \widetilde{RPN}_1 = ([6, 7.5, 9]; 0.75, 0.2) \cdot ([3.73, 4.40, 6.41]; 0.89, 0.09)$$
$$= ([27.41, 47.08, 66.70]; 0.75, 0.20)$$

Let the maximum applicability of the method be given by using the expression (4.7):

$$\tilde{z}^* = ([7.5, 9, 9]; 0.8, 0.15) \prod_{k=1,\ldots,3} ([7.5, 9, 9]; 0.8, 0.15)^{0.4} \cdot ([7.5, 9, 9]; 0.8, 0.15)^{0.33}$$
$$\cdot ([7.5, 9, 9]; 0.8, 0.15)^{0.26} = ([55.12, 79.24, 79.24]; 0.8, 0.15)$$

So that, the normalized assessment of the applicability of the method ($m = 6$) at the level of failure ($i = 1$) is:

$$\tilde{r}_{61} = \frac{([27.41, 47.08, 66.70]; 0.75, 0.20)}{([55.12, 79.24, 79.24]; 0.8, 0.15)} = ([0.35, 0.59, 1.21]; 0.75, 0.20)$$

To apply the proposed procedure (Step 10 of the proposed Algorithm) it is necessary to determine representative scalars, so that:

$$r_{61} = \frac{1}{12} \cdot (0.35 + 4 \cdot 0.59 + 1.21) \cdot (1 - 0.2 + 0.75) = 0.506$$

Ranking of uncertain values, $\tilde{r}_{61}$ is performed according to crisp values r_{61}, so that:

$$r_6 = \max_{i=1,\ldots,I} defuzz \begin{pmatrix} 0.178; 0.169; 0.413; 0.076; 0.120; 0.506; 0.105; 0.078; 0.230; 0.137; 0.274; 0.066; 0.258; 0.080; 0.101; 0.043; 0.413; \\ 0.418; 0.486; 0.243; 0.268; 0.156; 0.080; 0.204; 0.487; 0.117; 0.089; 0.133; 0.208; 0.191; 0.191; 0.204; 0.227; 0.289; \\ 0.340; 0.149; 0.199; 0.129; 0.120; 0.167; 0.156; 0.191; 0.191 \end{pmatrix} = 0.506$$

Therefore:

$$\tilde{r}_{61} = ([0.35, 0.59, 1.21]; 0.55, 0.4)$$

In the same way, each quality method m, $m = 1, \ldots, M$ is accompanied by the normalized total applicability shown in Table 6. The same table shows the other input data.

Table 6. Input data.

	$\tilde{r}_m$	$\tilde{c}_m$	$d(\tilde{r}_m, \tilde{c}_m)$	$(d(\tilde{c}_m, \tilde{c}))^2$
M = 1	([0.34, 0.59, 1.21]; 0.5, 0.45)	C3	0.2467	0.0215
M = 2	([0.34, 0.59, 1.21]; 0.5, 0.45)	C3	0.2467	0.0215
M = 3	([0.09, 0.22, 0.62]; 0.5, 0.45)	C2	0.1867	0.0187
M = 4	([0.27, 0.47, 1.12]; 0.5, 0.45)	C2	0.4367	0.0187
M = 5	([0.27, 0.47, 1.12]; 0.5, 0.45)	C2	0.4367	0.0187
M = 6	([0.34, 0.59, 1.21]; 0.5, 0.45)	C2	0.5300	0.0187
M = 7	([0.43, 0.71, 1.28]; 0.5, 0.45)	C2	0.6233	0.0187
M = 8	([0.01, 0.02, 0.16]; 0.5, 0.45)	C2	0.4200	0.0187
M = 9	([0.01, 0.02, 0.16]; 0.5, 0.45)	C2	0.4200	0.0187
M = 10	([0.09, 0.18, 0.58]; 0.5, 0.45)	C2	0.2000	0.0187
M = 11	([0.06, 0.10, 0.41]; 0.5, 0.45)	C2	0.2933	0.0187
M = 12	([0.17, 0.34, 0.86]; 0.5, 0.45)	C2	0.2733	0.0187
M = 13	([0.22, 0.42, 1.03]; 0.5, 0.45)	C4	0.3967	0.1995
M = 14	([0.06, 0.10, 0.41]; 0.5, 0.45)	C2	0.2933	0.0187
M = 15	([0.10, 0.16, 0.57]; 0.5, 0.45)	C2	0.1867	0.0187
M = 16	([0.06, 0.10, 0.41]; 0.5, 0.45)	C2	0.2933	0.0187
M = 17	([0.10, 0.16, 0.57]; 0.5, 0.45)	C2	0.1867	0.0187
M = 18	([0.10, 0.16, 0.57]; 0.5, 0.45)	C1	0.3600	0.1627
M = 19	([0.43, 0.71, 1.28]; 0.5, 0.45)	C2	0.6233	0.0187
M = 20	([0.34, 0.59, 1.21]; 0.5, 0.45)	C2	0.5300	0.0187
M = 21	([0.43, 0.71, 1.28]; 0.5, 0.45)	C2	0.6233	0.0187
M = 21	([0.43, 0.711.28]; 0.5, 0.45)	C2	0.233	0.0187
M = 22	([0.34, 0.59, 1.21]; 0.5, 0.45)	C2	0.5300	0.0187
M = 23	([0.34, 0.59, 1.21]; 0.5, 0.45)	C5	0.4867	0.3762
M = 24	([0.12, 0.19, 0.57]; 0.5, 0.45)	C2	0.2033	0.0187
M = 25	([0.01, 0.02, 0.16]; 0.5, 0.45)	C2	0.4200	0.0187

Table 6. Cont.

	$\tilde{r}_m$	$\tilde{c}_m$	$d(\tilde{r}_m, \tilde{c}_m)$	$(d(\tilde{c}_m, \tilde{c}))^2$
M = 26	([0.34, 0.59, 1.21]; 0.5, 0.45)	C4	0.3600	0.1995
M = 27	([0.34, 0.59, 1.21]; 0.5, 0.45)	C2	0.5300	0.0187
M = 28	([0.20, 0.39, 0.92]; 0.5, 0.45)	C4	0.3767	0.1995
M = 29	([0.43, 0.71, 1.28]; 0.5, 0.45)	C2	0.6233	0.0187
M = 30	([0.01, 0.02, 0.16]; 0.5, 0.45)	C2	0.4200	0.0187
M = 31	([0.01, 0.02, 0.16]; 0.5, 0.45)	C2	0.4200	0.0187
M = 32	([0.01, 0.02, 0.16]; 0.5, 0.45)	C2	0.4200	0.0187
M = 33	([0.06, 0.10, 0.41]; 0.5, 0.45)	C3	0.3767	0.0215
M = 34	([0.01, 0.02, 0.16]; 0.5, 0.45)	C2	0.4200	0.0187
M = 35	([0.43, 0.71, 1.28]; 0.5, 0.45)	C2	0.6233	0.0187
M = 36	([0.34, 0.59, 1.21]; 0.5, 0.45)	C2	0.5300	0.0187
M = 37	([0.01, 0.02, 0.16]; 0.5, 0.45)	C2	0.4200	0.0187
M = 38	([0.34, 0.59, 1.21]; 0.5, 0.45)	C2	0.5300	0.0187
M = 39	([0.34, 0.59, 1.21]; 0.5, 0.45)	C2	0.5300	0.0187
M = 40	([0.34, 0.59, 1.21]; 0.5, 0.45)	C2	0.5300	0.0187
M = 41	([0.34, 0.59, 1.21]; 0.5, 0.45)	C2	0.5300	0.0187
M = 42	([0.34, 0.59, 1.21]; 0.5, 0.45)	C2	0.5300	0.0187
M = 43	([0.43, 0.71, 1.28]; 0.5, 0.45)	C2	0.6233	0.0187
M = 44	([0.34, 0.59, 1.21]; 0.5, 0.45)	C2	0.5300	0.0187
M = 45	([0.35, 0.61, 1.19]; 0.5, 0.45)	C2	0.5333	0.0187
M = 46	([0.43, 0.71, 1.28]; 0.5, 0.45)	C2	0.6233	0.0187
M = 47	([0.34, 0.59, 1.21]; 0.5, 0.45)	C2	0.5300	0.0187
M = 48	([0.34, 0.59, 1.21]; 0.5, 0.45)	C2	0.5300	0.0187
M = 49	([0.34, 0.59, 1.21]; 0.5, 0.45)	C1	0.8467	0.1627

Determine the mean value of implementation costs:

$$\tilde{c} = \frac{1}{49} \cdot \sum_{m=1,\ldots,49} \tilde{c}_m = ([0.14, 0.35, 0.62]; 0.55, 0.40)$$

And the variance of implementation costs:

$$\frac{1}{49-1} \sum_{m=1,\ldots,49} d(d(\tilde{c}_m, \tilde{c}))^2 - \frac{1}{48} \cdot 2.1117 = 0.044$$

Assumed that the reliability manager should select 10 quality methods. This assumption was introduced based on best practice experience. Under this assumption, the KP problem (Step 11 of the proposed Algorithm) is:

The objective function

$$\max_{j=1,\ldots,10} d(\tilde{r}_m, \tilde{c}_m)_j, j \in \{1, \ldots, m, \ldots, M\}$$

The objective to:

$$\frac{1}{9} \cdot \sum_{j=1,\ldots 10} (d(\tilde{c}_m, \tilde{c}))^2 \leq 0.04$$

GA (Step 12 of the proposed Algorithm) was applied to find a near-optimal solution. The stop criterion is defined to have a number of iterations equal to 1000.

It has been shown that about 300 iterations are achieved near the optimal solution (see Figure 1).

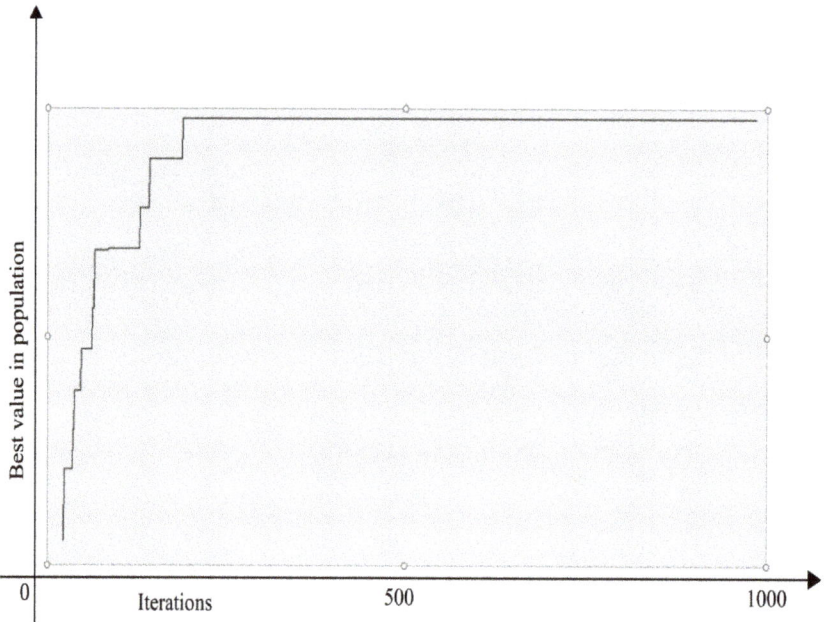

Figure 1. Value of fitness function by iterations.

For the near-optimal solution, the solution obtained in the last iteration was adopted:

By applying GA, the quality methods that the reliability manager in the considered company should implement in order to eliminate failures, or increase the reliability of the manufacturing process is:

solution 1: $\{m = 7, m = 19, m = 21, m = 29, m = 35, m = 38, m = 43, m = 45, m = 46, m = 49\}$

solution 2: $\{m = 7, m = 19, m = 21, m = 22, m = 29, m = 35, m = 43, m = 45, m = 46, m = 49\}$

It is worth to mention that the treated problem could be solved by the branch and bound algorithm [45] but it has a significant limitation since it is applicable to the lower size problems. In the presented case study, the output of the GA and branch and bound algorithm does not indicate output results deviation. As the number of quality methods in practice is increasing, the application of GA seems to be more adequate due to the nonpalatability of the branch and bound algorithm.

Based on the results of the survey, it can be concluded that most quality tools can be successfully applied to the analysis and identification of failures, which will ultimately lead to the elimination of failures, and thus lean waste. The obtained results show 10 methods that need to be applied in order to finally eliminate failures and increase the reliability of the manufacturing process. These methods can be implemented simultaneously. The period of implementation and education of employees for the implementation of these methods is not long. Another important fact is that the implementation of the above methods is not faced with the challenge of reorganization. The organizational culture of different SMEs will affect the effectiveness of the application of methods, so the effectiveness of the application should be observed at the level of each individual enterprise. In SMEs that have implemented the total quality management (TQM) and quality system according to the ISO 9001 standard, most of these tools are already in use and do not require additional investment and additional costs for the application of these methods.

This new problem-solving approach sets the standard for those who are just applying the problem-solving approach, as well as for those who are interested in continuously improving existing problem-solving methodologies.

6. Conclusions

As continuous changes that occur both in the company and in the environment impact the overall business performance, improving the reliability of the manufacturing process is one of the most important problems of operational management. This might be very significant since a reliable manufacturing process further positively impacts the stability of other business processes and enables competitive advantages in the long run. Experiences of best practice show that improving the reliability of the manufacturing process depends on the knowledge and experience of the FMEA team to identify failures as well as the knowledge of quality managers to choose quality methods by which identified failures can be eliminated or reduced leading to improved reliability of the manufacturing process.

In this research, identification of failures that can occur in manufacturing processes of considered SMES, is performed by FMEAs and based on their knowledge and experience as well as on best practice experience. The identified failures are assessed with respects to the RFs which are defined according to the FMEA framework. The set of possible quality methods are defined according to literature sources [6].

It is assumed that it is closer to the human way of thinking that the existing uncertainties into the relative importance of RFs, their values, application possibilities, as well as application costs of quality methods can be described better by using pre-defined linguistic expressions. These linguistic terms are modeled by the TIFNs. It can be said that the use of TIFNs does not require complex mathematical calculations and at the same time, linguistic terms are quantitatively described in a sufficiently good way.

Respecting to the results of the best practice, it can be said that the choice of quality methods depends on several variables. In this research, it is assumed that quality managers simultaneously consider the overall applicability of the method and implementation costs. The overall applicability of quality methods at the level of each failure is calculated as the product of the estimated degree of belief that the application of the quality method can lead to the reduction or elimination of failures and fuzzy weighted RPN with TIFNs associated with each failure. Determination weights of failures is sated as a fuzzy group decision making problem. In the literature, there are almost no papers in which procedures have been developed in which the choice of the quality method is made in an exact way. In this manuscript, the treated problem is stated as KP whose: (i) fitness function is defined as the distance between total applicability and implementation costs, and (ii) constraint is defined as a function of the number of solution elements. Near optimal solution with respect to fitness function and given constraints, simultaneously, may be efficiently achieved by small consumption of computation resources through the application of the GA.

The main advantage in theoretical terms of the presented model that combines FMEA, intuitive fuzzy sets [15], KP, and GA are: (i) that DMs can express their estimates using natural language words which in a good enough way can be quantitatively described by TIFNs, (ii) selection of a set of quality methods whose application can effectively increase the reliability of the manufacturing process while minimizing the consumption of financial resources. In this way, prioritizing quality methods is not significantly burdened by subjective attitudes of the quality manager, which would be one of the main advantages of this way of defining the improvement strategy.

The proposed model can be quickly and easily adjusted due to changes in the number of failures, number of quality methods, as well as their values.

The general limitations of the model are the need for a well-structured list of failures that can be realized in the manufacturing process in the production of SMEs.

Future research could include the extension of the proposed model in terms of: (1) increasing the number of variables on which fitness function depends and (3) applying other metaheuristic methods and comparing the obtained results.

Author Contributions: Conceptualization, D.T. and A.A.; investigation, R.G. and S.N.; methodology, D.T. and R.G., validation, G.Đ., S.N., A.A. and R.G.; visualization, G.Đ. and S.N.; writing—original draft, R.G.; A.A. and D.T. All authors have read and agreed to the published version of the manuscript.

Funding: This research received no external funding.

Institutional Review Board Statement: Not applicable.

Informed Consent Statement: Not applicable.

Data Availability Statement: Not applicable.

Conflicts of Interest: The authors declare that they have no competing interest.

Appendix A

$\widetilde{RPN}_1 = ([3.73, 4.40, 6.41]; 0.89, 0.09)$	$\widetilde{RPN}_{16} = ([1, 1, 2.5]; 0.84, 0.13)$	$\widetilde{RPN}_{31} = ([3.5, 5, 6.5]; 0.76, 0.21)$
$\widetilde{RPN}_2 = ([2.91, 4.44, 5.96]; 0.76, 0.21)$	$\widetilde{RPN}_{17} = ([3.5, 5, 6.5]; 0.76, 0.21)$	$\widetilde{RPN}_{32} = ([3.5, 5, 6.5]; 0.76, 0.21)$
$\widetilde{RPN}_3 = ([3.5, 5, 6.5]; 0.76, 0.21)$	$\widetilde{RPN}_{18} = ([3.41, 5.13, 6.34]; 0.91, 0.06)$	$\widetilde{RPN}_{33} = ([3.75, 5.36, 6.92]; 0.8, 0.09)$
$\widetilde{RPN}_4 = ([1.32, 1.66, 3.31]; 0.87, 0.11)$	$\widetilde{RPN}_{19} = ([4.35, 5.89, 7.42]; 0.8, 0.18)$	$\widetilde{RPN}_{34} = ([1.66, 2.51, 4.16]; 0.87, 0.11)$
$\widetilde{RPN}_5 = ([2.09, 2.9, 4.63]; 0.76, 0.21)$	$\widetilde{RPN}_{20} = ([2, 267, 4.42]; 0.76, 0.07)$	$\widetilde{RPN}_{35} = ([2.51, 3.26, 5.05]; 0.76, 0.21)$
$\widetilde{RPN}_6 = ([4.57, 6.28, 7.41]; 0.89, 0.09)$	$\widetilde{RPN}_{21} = ([2.31, 2.94, 4.74]; 0.76, 0.21)$	$\widetilde{RPN}_{36} = ([3.12, 3.85, 5.76]; 0.8, 0.18)$
$\widetilde{RPN}_7 = ([1.66, 2.51, 4.16]; 0.87, 0.11)$	$\widetilde{RPN}_{22} = ([2.51, 4.04, 5.56]; 0.76, 0.21)$	$\widetilde{RPN}_{37} = ([2.31, 2.94, 4.74]; 0.76, 0.21)$
$\widetilde{RPN}_8 = ([1.32, 1.66, 3.31]; 0.87, 0.09)$	$\widetilde{RPN}_{23} = ([1.61, 1.71, 3.51]; 0.84, 0.13)$	$\widetilde{RPN}_{38} = ([3.5, 5, 6.5]; 0.76, 0.21)$
$\widetilde{RPN}_9 = ([2.26, 2.43, 4.20]; 0.87, 0.06)$	$\widetilde{RPN}_{24} = ([2.91, 4.44, 5.96]; 0.76, 0.21)$	$\widetilde{RPN}_{39} = ([2.31, 2.94, 4.74]; 0.76, 0.21)$
$\widetilde{RPN}_{10} = ([2, 3.5, 5]; 0.87, 0.07)$	$\widetilde{RPN}_{25} = ([4.35, 5.89, 7.42]; 0.80, 0.18)$	$\widetilde{RPN}_{40} = ([2.09, 2.9, 4.63]; 0.76, 0.21)$
$\widetilde{RPN}_{11} = ([2, 2.67, 4.42]; 0.76, 0.21)$	$\widetilde{RPN}_{26} = ([2, 2.83, 4.54]; 0.8, 0.18)$	$\widetilde{RPN}_{41} = ([2.79, 4.33, 5.85]; 0.80, 0.18)$
$\widetilde{RPN}_{12} = ([1.2, 1.39, 3.00]; 0.84, 0.13)$	$\widetilde{RPN}_{27} = ([1.81, 1.95, 3.82]; 0.84, 0.13)$	$\widetilde{RPN}_{42} = ([2.51, 4.04, 5.56]; 0.76, 0.21)$
$\widetilde{RPN}_{13} = ([2.09, 2.9, 4.63]; 0.76, 0.21)$	$\widetilde{RPN}_{28} = ([2.51, 3.26, 5.05]; 0.76, 0.21)$	$\widetilde{RPN}_{43} = ([3.5, 5, 6.5]; 0.76, 0.21)$
$\widetilde{RPN}_{14} = ([1.32, 1.66, 3.31]; 0.87, 0.11)$	$\widetilde{RPN}_{29} = ([3.75, 5.36, 6.92]; 0.80, 0.18)$	$\widetilde{RPN}_{44} = ([3.5, 5, 6.5]; 0.76, 0.21)$
$\widetilde{RPN}_{15} = ([1.82, 2.37, 4.12]; 0.8, 0.18)$	$\widetilde{RPN}_{30} = ([3.61, 5.2, 6.75]; 0.76, 0.07)$	

References

1. Karaulova, T.; Kostina, M.; Sahno, J. Framework of Reliability Estimation for Manufacturing Processes. *Mechanics* **2013**, *18*, 713–720. [CrossRef]
2. Ohno, T. *Toyota Production System: Beyond Large-Scale Production*, 3rd ed.; CRC Press: New York, NY, USA, 1988.
3. Liker, J. *The Toyota Way: 14 Management Principles from the World's Greatest Manufacturer*; McGraw-Hill: New York, NY, USA, 2004.
4. Starzyńska, B.; Hamrol, A. Excellence toolbox: Decision support system for quality tools and techniques selection and application. *Total. Qual. Manag. Bus. Excel.* **2013**, *24*, 577–595. [CrossRef]
5. Anand, G.; Ward, P.T.; Tatikonda, M.V. Role of explicit and tacit knowledge in Six Sigma projects: An empirical examination of differential project success. *J. Oper. Manag.* **2009**, *28*, 303–315. [CrossRef]
6. Tague, N.R. *The Quality Toolbox*; ASQ Quality Press: Milwaukee, WI, USA, 2005; Volume 600.
7. Hagemeyer, C.; Gershenson, J.K.; Johnson, D.M. Classification and application of problem solving quality tools: A manufacturing case study. *The TQM Magazine* **2006**, *18*, 455–483. [CrossRef]
8. Arunagiri, P.; Gnanavelbabu, A. Identification of Major Lean Production Waste in Automobile Industries using Weighted Average Method. *Procedia Eng.* **2014**, *97*, 2167–2175. [CrossRef]
9. Gnanavelbabu, A.; Arunagiri, P. Ranking of MUDA using AHP and Fuzzy AHP algorithm. *Mater. Today: Proc.* **2018**, *5*, 13406–13412. [CrossRef]
10. Aleksic, A.; Ristic, M.R.; Komatina, N.; Tadic, D. Advanced risk assessment in reverse supply chain processes: A case study in Republic of Serbia. *Adv. Prod. Eng. Manag.* **2019**, *14*, 421–434. [CrossRef]
11. Nestic, S.; Lampón, J.F.; Aleksic, A.; Cabanelas, P.; Tadic, D. Ranking manufacturing processes from the quality management perspective in the automotive industry. *Expert Syst.* **2019**, *36*, 12451. [CrossRef]
12. Huang, J.; You, J.-X.; Liu, H.-C.; Song, M.-S. Failure mode and effect analysis improvement: A systematic literature review and future research agenda. *Reliab. Eng. Syst. Saf.* **2020**, *199*, 106885. [CrossRef]
13. Zadeh, L.A. The concept of a linguistic variable and its application to approximate reasoning. *Part III Inf. Sci.* **1975**, *9*, 43–80. [CrossRef]
14. Liu, H.C. FMEA Using Uncertainty Theories and MCDM Methods. In *FMEA Using Uncertainty Theories and MCDM Methods*; Springer: Singapore, 2016; pp. 13–27.
15. Atanassov, K. *Intuitionistic Fuzzy Sets: Theory and Applications*; Physica-Verlag: Wyrzburg, Germany, 1999.

16. Mathews, G.B. On the Partition of Numbers. *Proc. Lond. Math. Soc.* **1896**, *s1-28*, 486–490. [CrossRef]
17. Kellerer, H.; Pferschy, U. Improved dynamic programming in connection with an FPTAS for the knapsack prob-lem. *J. Comb. Optim.* **2004**, *8*, 5–11. [CrossRef]
18. Shanmugam, G.; Ganesan, P.; Vanathi, D.P. Meta heuristic algorithms for vehicle routing problem with stochastic demands. *J. Comput. Sci.* **2011**, *7*, 533. [CrossRef]
19. Senvar, O.; Turanoglu, E.; Kahraman, C. Usage of metaheuristics in engineering: A literature review. In *Me-ta-Heuristics Optimization Algorithms in Engineering, Business, Economics, and Finance*; IGI Global: Hershey, PA, USA, 2013; pp. 484–528.
20. Lu, S.; Pei, J.; Liu, X.; Qian, X.; Mladenovic, N.; Pardalos, P.M. Less is more: Variable neighborhood search for inte-grated production and assembly in smart manufacturing. *J. Sched.* **2019**, *23*, 649–664. [CrossRef]
21. Tadić, D.; Đorđević, A.; Aleksić, A.; Nestić, S. Selection of recycling centre locations by using the interval type-2 fuzzy sets and two-objective genetic algorithm. *Waste Manag. Res.* **2019**, *37*, 26–37. [CrossRef] [PubMed]
22. Tian, Z.-P.; Wang, J.-Q.; Zhang, H.-Y. An integrated approach for failure mode and effects analysis based on fuzzy best-worst, relative entropy, and VIKOR methods. *Appl. Soft Comput.* **2018**, *72*, 636–646. [CrossRef]
23. Mirghafoori, S.H.; Izadi, M.R.; Daei, A. Analysis of the barriers affecting the quality of electronic services of libraries by VIKOR, FMEA and entropy combined approach in an intuitionistic-fuzzy environment. *J. Intell. Fuzzy Syst.* **2018**, *34*, 2441–2451. [CrossRef]
24. Spillman, R. Solving large knapsack problems with a genetic algorithm. In Proceedings of the 1995 IEEE International Conference on Systems, Man and Cybernetics, Intelligent Systems for the 21st Century, Vancouver, BC, Canada, 22–25 October 1995; Volume 1, pp. 632–637.
25. Ezugwu, A.E.; Akutsah, F.; Olusanya, M.O.; Adewumi, A.O. Enhanced intelligent water drops algorithm for multi-depot vehicle routing problem. *PLoS ONE* **2018**, *13*, e0193751. [CrossRef]
26. Fattahi, R.; Khalilzadeh, M. Risk evaluation using a novel hybrid method based on FMEA, extended MULTIMOORA, and AHP methods under fuzzy environment. *Saf. Sci.* **2018**, *102*, 290–300. [CrossRef]
27. Panchal, D.; Mangla, S.K.; Tyagi, M.; Ram, M. Risk analysis for clean and sustainable production in a urea fertiliz-er industry. *Int. J. Qual. Reliab. Manag.* **2018**, *35*, 1459–1476. [CrossRef]
28. Zimmermann, H.J. *Fuzzy Set Theory—And Its Applications*; Springer Science & Business Media: Berlin/Heidelberg, Germany, 2011.
29. Wu, Q.; Zhou, L.; Chen, Y.; Chen, H. An integrated approach to green supplier selection based on the interval type-2 fuzzy best-worst and extended VIKOR methods. *Inf. Sci.* **2019**, *502*, 394–417. [CrossRef]
30. Mendel, J.M. Type-2 Fuzzy sets. In *Uncertain Rule-Based Fuzzy Systems*; Springer: Cham, Switzerland, 2017; pp. 259–306.
31. Can, G.F. An intuitionistic approach based on failure mode and effect analysis for prioritizing corrective and preven-tive strategies. *Hum. Factors Ergon. Manuf.* **2018**, *28*, 130–147. [CrossRef]
32. Liu, H.-C.; You, J.-X.; Duan, C.-Y. An integrated approach for failure mode and effect analysis under interval-valued intuitionistic fuzzy environment. *Int. J. Prod. Econ.* **2019**, *207*, 163–172. [CrossRef]
33. Tooranloo, H.S.; Ayatollah, A.S.; Alboghobish, S. Evaluating knowledge management failure factors using intui-tionistic fuzzy FMEA approach. *Knowl. Inf. Syst.* **2018**, *57*, 183–205. [CrossRef]
34. Wan, S.P.; Wang, Q.Y.; Dong, J.Y. The extended VIKOR method for multi-attribute group decision making with triangular intuitionistic fuzzy numbers. *Knowl.-Based Syst.* **2013**, *52*, 65–77. [CrossRef]
35. Xu, Z.; Liao, H. Intuitionistic Fuzzy Analytic Hierarchy Process. *IEEE Trans. Fuzzy Syst.* **2014**, *22*, 749–761. [CrossRef]
36. Dutta, B.; Guha, D. Preference programming approach for solving intuitionistic fuzzy AHP. *Int. J. Comput. Intell. Syst.* **2015**, *8*, 977–991. [CrossRef]
37. Ervural, B.C.; Oner, S.C.; Coban, V.; Kahraman, C. A novel Multiple Attribute Group Decision Making methodology based on Intuitionistic Fuzzy TOPSIS. In Proceedings of the 2015 IEEE International Conference on Fuzzy Systems (FUZZ-IEEE), Istanbul, Turkey, 2–5 August 2015; pp. 1–6.
38. Wu, J.; Cao, Q.-W. Same families of geometric aggregation operators with intuitionistic trapezoidal fuzzy numbers. *Appl. Math. Model.* **2013**, *37*, 318–327. [CrossRef]
39. Wan, S.-P. Power average operators of trapezoidal intuitionistic fuzzy numbers and application to multi-attribute group decision making. *Appl. Math. Model.* **2013**, *37*, 4112–4126. [CrossRef]
40. Hao, Y.; Chen, X.; Wang, X. A ranking method for multiple attribute decision-making problems based on the possibility degrees of trapezoidal intuitionistic fuzzy numbers. *Int. J. Intell. Syst.* **2019**, *34*, 24–38. [CrossRef]
41. Wang, J.Q.; Zhang, Z. Multi-criteria decision-making method with incomplete certain information based on intuitionistic fuzzy number. *Control. Decis.* **2009**, *24*, 226–230.
42. Gupta, P.; Mehlawat, M.K.; Grover, N. Intuitionistic fuzzy multi-attribute group decision-making with an application to plant location selection based on a new extended VIKOR method. *Inf. Sci.* **2016**, *370*, 184–203. [CrossRef]
43. Grzegorzewski, P. Distances between intuitionistic fuzzy sets and/or interval-valued fuzzy sets based on the Hausdorff metric. *Fuzzy Sets Syst.* **2004**, *148*, 319–328. [CrossRef]
44. Stamatis, D.H. *Risk Management Using Failure Mode and Effect Analysis (FMEA)*; Quality Press: Milwaukee, WI, USA, 2019.
45. Shih, W. A branch and bound method for the multiconstraint zero-one knapsack problem. *J. Oper. Res. Soc.* **1979**, *30*, 369–378. [CrossRef]

Article

Fuzzy Set Qualitative Comparative Analysis on the Adoption of Environmental Practices: Exploring Technological- and Human-Resource-Based Contributions

Lucía Muñoz-Pascual [1,*], Carla Curado [2] and Jesús Galende [1]

[1] Departamento de Administración y Economía de la Empresa, Facultad de Economía y Empresa, Instituto Multidisciplinar de Empresa (IME), Campus Miguel de Unamuno s/n, Universidad de Salamanca, Edificio FES, 37007 Salamanca, Spain; jgalende@usal.es

[2] Advance/CSG ISEG, University of Lisbon, Rua do Quelhas, 6, 1200-781 Lisbon, Portugal; ccurado@iseg.ulisboa.pt

* Correspondence: luciamp@usal.es

Citation: Muñoz-Pascual, L.; Curado, C.; Galende, J. Fuzzy Set Qualitative Comparative Analysis on the Adoption of Environmental Practices: Exploring Technological- and Human-Resource-Based Contributions. *Mathematics* **2021**, *9*, 1553. https://doi.org/10.3390/math9131553

Academic Editors: Laura González-Vila Puchades and Jorge de Andres Sanchez

Received: 7 June 2021
Accepted: 29 June 2021
Published: 1 July 2021

Publisher's Note: MDPI stays neutral with regard to jurisdictional claims in published maps and institutional affiliations.

Copyright: © 2021 by the authors. Licensee MDPI, Basel, Switzerland. This article is an open access article distributed under the terms and conditions of the Creative Commons Attribution (CC BY) license (https:// creativecommons.org/licenses/by/ 4.0/).

Abstract: Our main objective was to analyze which paths can lead to the adoption of environmental practices (PRAC) in firms, for which we developed three original alternative research models. Model 1 involves five sources for the adoption of environmental practices: human resource costs, organizational learning capability, firm size, manager educational level and manager experience. Model 2 adopts five sources for PRAC: human resource costs, information technology support, firm size, manager educational level and manager experience. Finally, Model 3 adopts six sources for PRAC: human resource costs, organizational learning capability, information technology support, firm size, manager educational level and manager experience. Therefore, Model 1 uses the organizational learning capability for PRAC, Model 2 uses the information technology support for PRAC and Model 3 uses both organizational learning capability and information technology support for PRAC. We used a fuzzy set qualitative comparative analysis on 349 small- and medium-sized Portuguese firms in twelve industrial sectors. The results show that organizational learning capability (OLC) and information technology support (ITS) are important sources for the development of PRAC. In this line, the three research models show that there are different pathways that lead to PRAC. These research models also show pathways that lead to the absence of PRAC. Therefore, the qualitative findings show the relevancy of OLC and ITS to PRAC. In addition, our findings indicate that, by focusing on variables such as OLC, a firm can find more paths that lead to PRAC. Additionally, with the combination of OLC and ITS, it must be taken into account that only developing ITS without OLC is riskier when obtaining PRAC.

Keywords: fuzzy set qualitative comparative analysis; adoption of environmental practices; human resource costs; organizational learning capability; information technology support; size; education level; experience

1. Introduction

The aim of this study was to answer four research questions: (i) Can organizational learning capability (OLC) help to implement the adoption of environmental practices in SMEs? (ii) Can the increased use of ITS in SMEs improve the adoption of environmental practices? (iii) What exactly are the pathways and components of human resources and technology that lead to the adoption of environmental practices by SMEs? (iv) What characteristics of managers and types of firm according to size help to take paths towards the adoption of environmental practices in SMEs?

Since the 2005 World Summit on Social Development, organizations have been expected to design their strategies in accordance with the three main sustainable development goals—that is, economic development, social inclusion and environmental protection— which reflect the three pillars of the triple bottom line approach [1,2]. Therefore, sustainable management models are needed for the application of environmental practices that also help manage the other two pillars, namely economic and social.

Consequently, the adoption of environmental practices represents a means through which organizations can promote full sustainable development [3]. At the same time, current sustainability challenges can become a source of inspiration for the adoption and improvement of new practices in firms and with it improve competitiveness and contribute to building a better society [4,5]. However, this triple bottom line approach is not a dominant business model in the world [6,7] as firms focus their efforts mainly on obtaining economic benefits.

Our study presents three comparative and integrative research models for PRAC. In addition, our research examines differences in PRAC between SMEs with OLC and SMEs with ITS. Our research models also analyze other characteristics of firms (size) and managers (educational level and experience) for PRAC. Therefore, our study aims to find the paths that lead firms to PRAC or its absence, supported by a triple bottom line approach.

In this study, we used a qualitative method, fuzzy set qualitative comparative analysis (fsQCA), to identify the pathways and alternative configurations for PRAC or its absence. We used fsQCA instead of other methodologies because we aimed to identify the essential and necessary conditions of the configurations that lead to the result variable or its absence [8]—that is, to PRAC or its absence. Therefore, this study completely covers the PRAC pathways and applies the most appropriate qualitative methodology to meet the research objectives.

This study focuses its attention on sustainability in SMEs because they constitute more than 95% of firms worldwide. That is why conducting a study of this type and taking into account the size of the firms is very important, because PRAC, ITS and OLC can be very different between firms depending on their sizes. Therefore, this study also addressed sustainable pathways for three types of SMEs (SMEs with OLC, SMEs with ITS and SMEs with OLC and ITS) and several managers (educational level and experience) and firm characteristics (size).

This study shows that OLC and ITS are beneficial for PRAC and to boost the competitive advantage of SMEs. A firm that adopts new practices of learning for its employees and new technologies with a culture of sustainability can obtain several pathways for PRAC or its absence.

This study makes several contributions.

First, the findings significantly extend the knowledge of the adoption of environmental practices.

Second, there are seven alternative causal configurations that lead to PRAC with OLC and only three for its absence.

Third, there are six alternative causal configurations that lead to PRAC with ITS and only four for its absence.

Fourth, there are eight alternative causal configurations that lead to PRAC with OLC and ITS and seven for its absence. Therefore, there are more alternative causal configurations that lead to PRAC than to its absence in the three research models.

Finally, the factor of human resource variables as OLC is very important for PRAC, because with OLC, there are seven pathways for PRAC, and with ITS, there are six pathways for PRAC.

The article is structured as follows: Section 2 shows the theory and the constructs. In Section 3, the formulation of the hypotheses, the methodology, the sample and the measures are shown. Section 4 shows the main results of the investigation and the paths for PRAC and its absence. Finally, Section 5 shows the discussion and the conclusions.

2. Three Research Models to Explore the Adoption of Environmental Practices in SMEs

In light of the global crisis caused by COVID-19, one of the most important future challenges that can help improve the lives of people, firms and society as a whole is to find a real balance between the implementation of new technological systems and the care, improvement and learning of human capacities. For this, firms must make a significant investment in the management and maintenance of the factors derived from two key pillars, namely social and economic factors, both fundamental for the adoption of environmental practices and the development of business sustainability supported by the triple bottom line approach [2,9].

In this sense, our research model raises human factors such as OLC, the characteristics of the manager as social factors and technological and business factors as economic factors to achieve the adoption of environmental practices.

Therefore, it is necessary to continue advancing in research that helps to provide new mechanisms and pathways that interrelate human resource management, new technologies and business sustainability.

In addition, this is justified through the triple bottom line approach, where sustainability and the adoption of environmental practices should be sought through three approaches: economic, social and environmental. Here, we rely on this approach to assess which variables or factors can lead to the adoption of environmental practices [2,10].

We thus propose the following research questions:

- Can organizational learning capability (OLC) help to implement the adoption of environmental practices in SMEs?
- Can increased use of ITS in SMEs improve the adoption of environmental practices?
- What exactly are the pathways and components of human resources and technology that lead to the adoption of environmental practices by SMEs?
- What characteristics of managers and types of firm according to size help to take paths towards the adoption of environmental practices in SMEs?

In this sense, our research models try to be comprehensive and comparative research models for the adoption of environmental practices via the contributing factors or variables of the three pillars of the triple bottom line approach. Therefore, the OLC and human factors of managers comprise social factors, and technology and business variables, such as company size, comprise the most economical factors of the approach [2,10].

2.1. Triple Bottom Line Approach

The triple bottom line approach is a term related to sustainable business that refers to the impact that the activity of a firm has in three dimensions: social, economic and environmental. The concept was coined by John Elkington in 1994 and further developed in his book *The Triple Bottom Line: Does it All Add Up*, published in 2004 [11]. The evidence of performance in relation to the triple bottom line is manifested in the reports of sustainability or corporate social responsibility. Until 2009, their preparation and publication continued to be voluntary and evolving throughout the world.

Ideally, an organization with good performance in triple bottom line terms would be able to maximize its economic benefit and environmental responsibility, as well as minimizing or eliminating its negative externalities and emphasizing the social responsibility of the organization towards groups of interest, not just to shareholders. In this case, a stakeholder refers to anyone who is influenced, directly or indirectly, by the firm's actions. Therefore, the triple bottom line approach facilitates the performance of a business entity as a vehicle for coordinating sustainable interests.

2.2. Economic Development

Economic development is a dimension that deals with the outcomes and flow of money. It may involve looking at income or expenses, business climate factors, employment and business diversity factors and investing in technology. Specific examples include personal income, human resource costs, rotation of establishments, job growth, percentage of firms in each sector, technology, etc.

In this article, we propose two internal dimensions of a firm for economic development: human resource costs and information technology support.

2.2.1. Human Resource Costs

A firm must invest in the management of its human resources, which will help to improve the performance of its employees and will more easily build a culture of values [12].

The literature has distinguished three main dimensions of human resources: intellectual, emotional and social. These are linked and interrelated to the improvement of organizational capacities [13]. Continued investment in employee job improvements can help firms implement PRAC [14]. Additionally, each dimension of human resources can enhance a firm's resources. For example, within the intellectual dimension, the firm can invest in improving knowledge; within the emotional dimension, the firm can invest in improving motivation; and within the social dimension, the firm can invest in improving employee relationships. All of these resources together are a source of PRAC [15,16].

2.2.2. Information Technology Support

This concept refers to the application of technology to introduce and implement PRAC [17]. Technology is a key element for the development of new knowledge and the implementation of new practices [18] by facilitating the rapid collection, storage and exchange of knowledge [19–22]. Well-developed technology integrates new mechanisms and practices [17] that can eliminate barriers to communication among departments in an organization [23,24]. Firms that are at the forefront of ITS have more mechanisms to adopt new sustainable practices. Therefore, ITS is a fundamental component in highly sustainable firms since they can reduce costs throughout their production [25].

ITS has non-reproducibility, non-substitutability and appropriability factors that help firms to obtain a competitive advantage [26]. The literature indicates that firms that have a high investment in ITS tend to have a good PRAC [27,28], as technology provides the necessary mechanisms to convert inputs into sustainable outputs [25,29]. Therefore, this study presents ITS as a significant component of PRAC. Previous studies indicated that ITS must be evaluated at the organizational level to determine its effects [27]. Therefore, ITS is considered a clear precedent for the creation and implementation of PRAC in firms [28,30].

2.3. Social Development

Social development refers to the social dimensions of a firm, community or region and can include measures of education, learning, equity and access to social resources, health and well-being, quality of life and social capital. The following list provides some examples of possible variables: female participation rate in the labor force, median income, relative poverty, percentage of employees with a post-secondary degree or certificate, organizational learning ability, interorganizational awareness share, etc. In this article, we propose an internal dimension of firms: organizational learning capacity. Organizational learning capacity is a relevant tool for SMEs because it grants access to social resources. In this way, SMEs with a high level of organizational learning ability can collect and report information on what is happening inside and outside the firm.

Organizational Learning Capability

Learning capability refers to those factors that facilitate organizational learning [29,30]. This is a concept related to the renewal of dynamic and continuous knowledge. Knowledge renewal is undoubtedly the greatest source of ability to learn and explore new knowledge while exploiting already known knowledge [31]. Organizational learning is a broad concept and occurs both within and outside of the organization itself [32].

Lichtenthaler [33] classified organizational learning into three processes: exploratory, exploitative and transformative. All three processes have positive effects on PRAC. Organizational learning requires organizations to plan, visualize and transact. According to Chiva et al. [34], OLC was born from experimentation, risk-taking, interaction with the external environment, dialogue and participatory decision-making.

Experimentation is the search for new ideas by firms, whereby they are curious to know or make changes in work processes. Risk-taking shows tolerance for ambiguity, uncertainty and mistakes that lead to organizational learning. Interaction with the environment shows the breadth of the firm's relationships with its environment. Dialogue is the collective search for assumptions, processes and certainties. Finally, decision making relates to the collaboration and participation that employees have in the process [35]

2.4. Environment Development

Environmental dimensions represent the adoption of natural resources and reflect the potential influences on their viability. They could incorporate air and water quality, energy consumption, natural resources, solid and toxic waste and land use/land cover. Ideally, organizations should track long-term trends for each of the environmental dimensions to help identify the impacts that a project, policy or product will have in an area, market or community. In this context, we focus on the complete concept of the adoption of environmental practices.

Adoption of Environmental Practices

The reinforcement of environmental regulations worldwide in recent years and the COVID-19 crisis have motivated countries and firms to seek the adoption of environmental management practices [35,36]. Investment in human resources, organizational learning and technology are essential for economic and social development [37]. In addition, a business and social culture that leads to PRAC is relevant, which can be promoted and driven by suppliers, employees, clients and other public and private stakeholders [38].

PRAC can be an important advancement for organizations as they can introduce more efficient consumption and recycling methods and even help to reduce total costs [39]. Aragon-Correa et al. [40] showed that an active environmental strategy requires changes in operational routines and methods. Chan and Hawkins [41] indicated that the adoption of environmental practices helps to improve safety standards and working conditions. In addition, firms that enforce PRAC will have greater economic and fiscal advantages because they can benefit from public aid [42].

2.5. Firm Size and Managers' Characteristics

2.5.1. Firm Size

According to the European Union, the definition of SMEs is shown in Annex I of Commission Regulation (EU) No. 651/2014 of June 17, 2014. The document shows the types of firms based on the number of employees (small: <50 employees; medium: >50 employees) [43]. A large or medium size can help with the PRAC because there is the possibility of obtaining economies of scale, lower risk and the possibility of better and higher performance. Therefore, SMEs can have access to a larger battery of information and capabilities that allow PRAC more easily [44]. In contrast, small firms may be able to facilitate PRAC quickly by having fewer intermediaries or hierarchies. There could be better communication and coordination for PRAC. Furthermore, a smaller firm usually has more informal links to support the adoption of PRAC [44].

2.5.2. Manager Education Level

This paper distinguishes the educational competencies of managers as they could play an essential role in PRAC (secondary, undergraduate and graduate). Vila et al. [45] indicated that there are significant effects of these competencies on the possibility that managers act as drivers in PRAC. These competencies are directly related to the level of education as they are based on being alert to new opportunities and the ability to present ideas or reports, mobilize the skills of others and propose new ideas and solutions as well as the ability to use computers and the Internet. Therefore, the high technological level of a firm is directly linked to the education of the manager and his level of awareness about investing in ITS and PRAC. The technological level creates effects of power in the PRAC and reduces the risk of resistance of the administrators to the new systems [46]. Ultimately, introducing a business process system without a supportive learning environment could have drastic consequences [47].

However, it is important to test this in our research models because, in turn, a higher level of manager education can mean that the manager has a greater knowledge of certain environmental issues related to ethics when it comes to doing things well or not achieving the adoption of environmental practices in the firm. For example, a manager may be educated on how to implement a certain quota for pollution in his firm, but this knowledge can also be used to circumvent the ethical mechanisms to comply with said quota if using the black market to buy and sell pollution quotas. Therefore, our research models aim to show whether a manager's level of education is a factor that leads to the adoption of environmental practices or their absence [48].

Both the size of a firm and the characteristics of the manager (level of education and experience) are relevant factors included in our research models and, therefore, in our qualitative analyses, because they are factors that, a priori, can affect the adoption of environmental practices in firms in a predefined way. In this sense, large firms are expected to adopt environmental practices in a very different way than small ones do, and the same is expected in the case of manager characteristics—that is, a manager with a high level of education and experience will apply environmental practices to his firm in a very different way than a manager with a lower level of education and experience would.

2.5.3. Manager Experience

The willingness of firms to adopt new ideas or practices can be analyzed through the manager's experience (junior: <2 years; intermediate: 2 to 5 years; senior: >5 years). A manager's experience shows their intentions according to their knowledge and know-how when introducing new tools that improve the environment and culture of their organization. For example, the manager can contribute to an improvement in communication and creativity that translates into the adoption of new practices within the firm [48]. Kumar and Saqib [49] found a positive relationship between experience, measured through age, and the performance of research and development (R and D) and the adoption of new practices. Kuemmerle [48] analyzed the relationship between the innovative results of R&D laboratories, measured through interviews with their managers and by obtaining patents, and their experience, showing a positive relationship. Gumbau [50] found that the length of time that a manager has been with a firm has a positive influence on the level of resources invested in new practices.

However, other studies such as those by Molero and Buesa [51] revealed that firms with less experienced and younger managers have a more active attitude in the implementation of new technologies and environmental management practices.

3. Hypotheses and Methods

3.1. Hypotheses: Alternative Configurations

The theory of configuration indicates the sufficient and necessary conditions to lead to a result (PRAC). Here, there is equifinality if more than one pathway leads to the same result (PRAC). These configurations are important because through various asymmetric paths, we are able to reach a result and we do not remain in simple traditional bivariate interactions.

Therefore, given the complexity of the involved phenomena, we propose that parallel nonlinear configurations of conditions lead to PRAC (and its absence) in several ways [52].

For Model 1:

Hypotheses 1 (H1). *There are alternative configurations leading to PRAC considering the contribution of HRC.*

Hypotheses 2 (H2). *There are alternative configurations leading to the absence of PRAC considering the contribution of HRC.*

For Model 2:

Hypotheses 3 (H3). *There are alternative configurations leading to PRAC considering the contribution of ITS.*

Hypotheses 4 (H4). *There are alternative configurations leading to the absence of PRAC considering the contribution of ITS.*

For Model 3:

Hypotheses 5 (H5). *There are alternative configurations leading to PRAC considering the contribution of HRC and ITS.*

Hypotheses 6 (H6). *There are alternative configurations leading to the absence of PRAC considering the contribution of HRC and ITS.*

Our paper shows the configurations that managers and firms must select to achieve PRAC. In this way, different pathways constituting the related paths between HRC, OLC, ITS and PRAC can be found. Furthermore, the paper includes managerial experience and educational level together with the size of the firm as factors that can also facilitate PRAC [53].

The three comparative and integrative research models are shown in Figures 1–3.

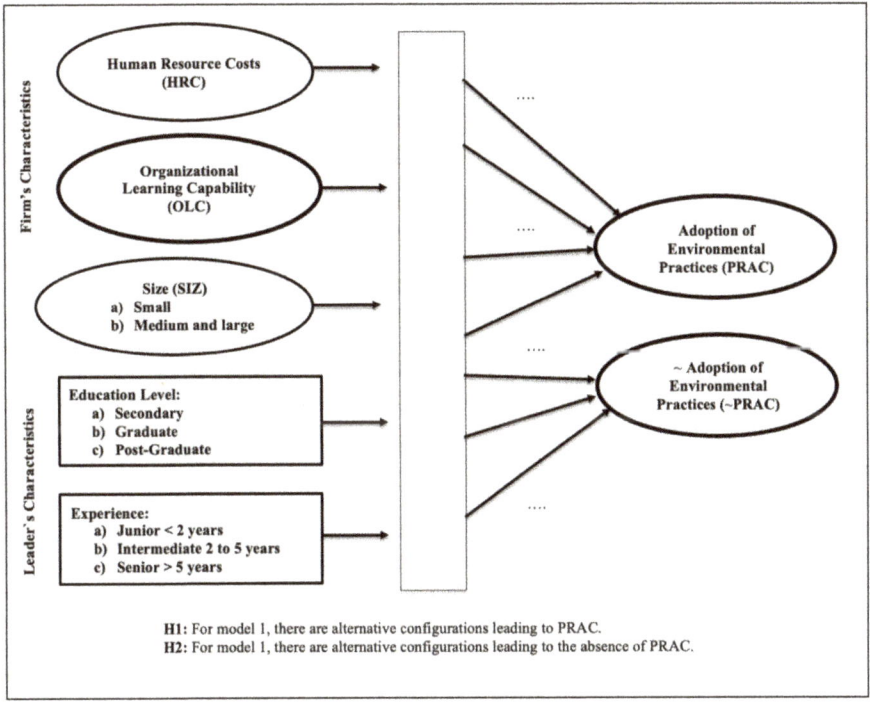

Figure 1. Research Model 1 for adoption of environmental practices (with OLC).

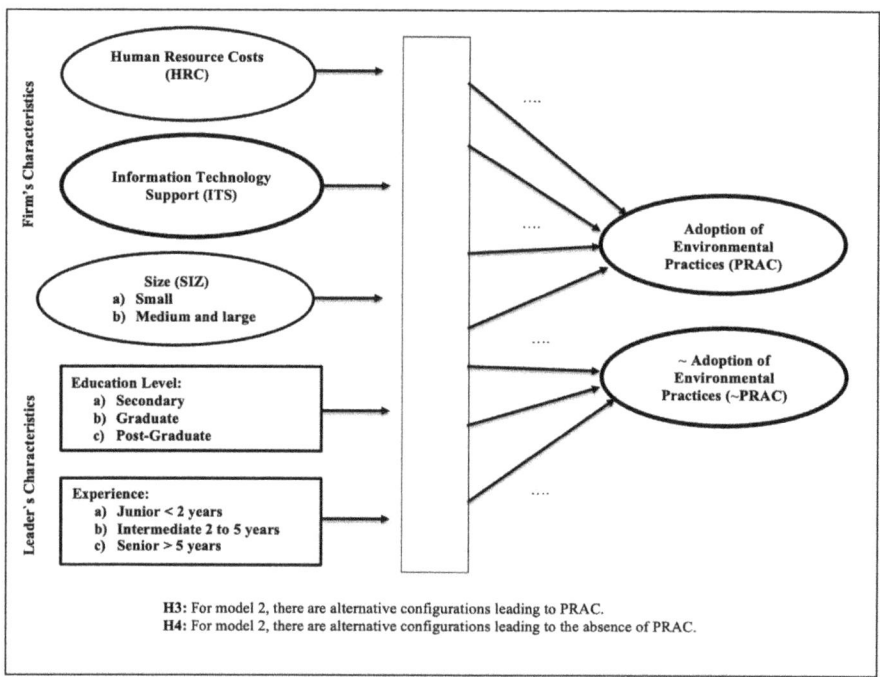

Figure 2. Research Model 2 for adoption of environmental practices (with ITS).

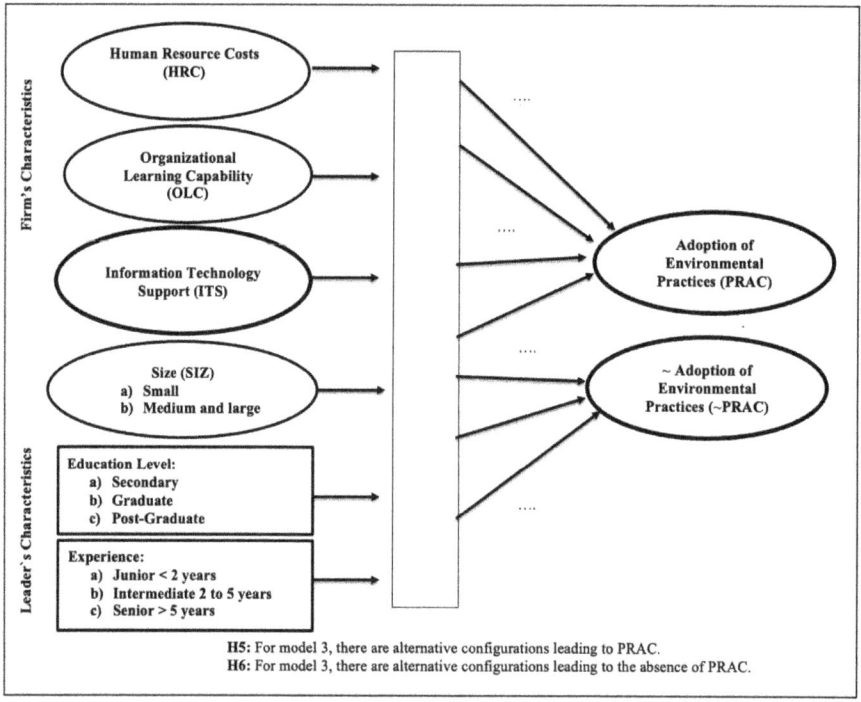

Figure 3. Research integrative Model 3 for adoption of environmental practices.

3.2. Methodology: Qualitative Methods

We used a qualitative method to test PRAC (hypotheses H1 and H2 in Model 1; hypotheses H3 and H4 in Model 2; hypotheses H5 and H6 in Model 3). Osabutey and Jin [54] showed that traditional quantitative methods have limitations when it comes to explaining complex interactions between variables. In addition, other recent studies applied qualitative methods. For example, Oyemomi et al. [55] used fsQCA. FsQCA is a qualitative method that identifies the essential and necessary conditions of the configurations that will lead to the outcome variable or its absence [56]. Therefore, this study completely covers all pathways for PRAC.

In this paper, we used fuzzy set QCA (fsQCA) from [57] to evaluate these hypotheses. There are many applications of quantitative methods in management and business [37] but few applications of qualitative methods and mixed methods.

In similar fields to management, fsQCA has been applied instead of quantitative methods (PLS, SEM and others) and also as a complementary method to quantitative methods [58,59]. Any correlational method in general assumes symmetrical relations between variables and measures the net effect of each variable on the assessed output. On the other hand, fsQCA allows discovery of the combinatorial effects of variables on the output as well as accepting that these interactions could be asymmetrical [58].

Therefore, in this paper, we applied fsQCA and found the logical implication that combining the presence/absence of input variables provides better output results. Consequently, consistency and coverage measures inform the relevance of the discovered logical implications.

3.3. Sample and Measurement Assessment

An online survey was conducted in Qualtrics® and sent to 6846 innovative SMEs in Portugal within twelve industrial sectors: manufacturing, energy supply and gas; water supply and pollution; edifice; trade and repair of vehicles; transportation and storage; catering; information and communication; accommodation; scientific activities; administrative activities; health activities; and other services. In total, 385 responses were obtained. After applying a rigorous cleaning process [60], the final sample constituted 349 firms. Therefore, the response rate was 5.1%. The sample included firms from the twelve industrial sectors existing in the population.

The questionnaire was originally written in English and was later translated into Portuguese by an expert translator for sending to firms. It was then back-translated into English. In this way, errors related to language interpretation were minimized. In addition, to carry out the questionnaire, five academics and managers who are experts in the field were contacted to show them a pilot questionnaire and subsequently launch the final questionnaire. Finally, the firms were called by telephone to inform them of the study and the questionnaires were then sent to them. It is a simple and quick survey to answer that takes an average of 20 min to answer, and the people surveyed were the CEOs in each firm. Our goal was for CEOs to respond to the questionnaire because they are the top decision-makers and know all the tools and information necessary to try to achieve economic and environmental development in the firm [61].

The characteristics of the sample are detailed below. The majority of CEOs surveyed were women (56.4%), and more than 76.5% had undergraduate or graduate degrees. Their average age was 43.6 years and more than 77.4% had more than five years of seniority in their firms. In relation to the characteristics of the firm, more than 92.4% had more than 10 years of experience and the majority (65.9%) had 50 employees or fewer. Of them, 63.6% were public limited companies and 36.4% were collective firms. On the other hand, to test the non-response bias, a trend extrapolation test was used to compare the responses that took the longest to arrive with those that arrived first. Our study considers that the responses that took the longest to arrive are those that arrived in the second phase of the study and after carrying out a reminder. These responses may be very similar to those of those firms that never responded since, had it not been for the reminder, they might never have responded [62]. After conducting a one-way analysis of variance (ANOVA), we could see that there were no significant differences between CEOs who responded early and those who responded late in terms of company size (number of employees in the firm) and age.

Therefore, the sample was representative of the population.

3.4. Variables and Measurements

HRC measures the total expenditure on human resources per year as a proportion of the firm's total invoicing [63].

All other measurements used Likert-type scales [64]. The ranges were from 7 (strongly agree) to 1 (strongly disagree). To measure OLC, we used a scale validated by Alegre and Chiva [65]. This scale has five dimensions: risk-taking, experimentation, participatory decision-making, interaction with the external environment and dialogue. To measure ITS, a validated scale by Lee and Choi [18] was used, and finally, PRAC was measured with the scale by Molina-Azorín et al. [66]. Therefore, we had the construct OLC with five dimensions and the two constructs ITS and PRAC with one dimension.

The level of education of the manager (EL) was measured taking into account what type of education they have: high school, undergraduate and graduate. The experience (EXP) of the manager was measured according to the number of years in the firm: junior, <2 years; intermediate, 2 to 5 years; senior, >5 years. Finally, the size of the firm (SIZ) was measured according to the number of employees: small: <50 employees; medium: >50 employees [67,68].

A confirmatory factor analysis (CFA) was used in AMOS® to study the validity of the measurement. All of the items showed good levels; therefore, there is a good validity in the measurement.

Table 1 summarizes the variables' descriptions and the CFA.

Table 1. Variables' descriptions and CFA.

Constructs	Mean	SD	CFA
Organizational Learning Capability (OLC) [65] (V.E = 70.09%); (α = 0.95)			
Experimentation (OLC-E)			
OLC-E1. In my organization, people receive support and encouragement when presenting new ideas	4.91	1.65	0.86
OLC-E2. In my organization, initiative receives a favorable response, so people feel encouraged to generate new ideas	4.85	1.58	0.86
Risk-taking (OLC-R)			
OLC-R3. In my organization, people are encouraged to take risks	4.54	1.62	0.85
OLC-R4. In my organization, people often venture into unknown territory	3.79	1.58	0.84
Interaction with the external environment (OLC-I)			
OLC-I5. In my organization, it is part of the work of all staff to collect, bring back and report information about what is going on outside the firm	4.21	1.78	0.83
OLC-I6. In my organization, there are systems and procedures for receiving, collating and sharing information from outside the firm	3.90	1.73	0.83
OLC-I7. In my organization, people are encouraged to interact with the environment: competitors, customers, technological institutes, universities, suppliers, etc.	4.48	1.69	0.82
Dialogue (OLC-D)			
OLC-D8. In my organization, employees are encouraged to communicate	5.07	1.59	0.80
OLC-D9. In my organization, there is free and open communication within my work group	5.38	1.47	0.77
OLC-D10. In my organization, managers facilitate communication	5.26	1.60	0.77
OLC-D11. In my organization, cross-functional teamwork is a common practice	4.84	1.70	0.73
Participative decision-making (OLC-P)			
OLC-P12. Managers in my organization frequently involve employees in important decisions	4.64	1.76	0.67
OLC-P13. In my organization, policies are significantly influenced by the employees' views	4.33	1.67	0.63
OLC-P14. In my organization, people feel involved in main firm decisions	4.28	1.72	0.58
Information technology support (ITS) [18]			
ITS1. My organization provides information technology support for collaborative works regardless of time and place	3.91	1.99	0.70
ITS2. My organization provides information technology support for communication among organization members	5.01	1.69	0.86
ITS3. My organization provides information technology support for searching for and accessing necessary information	5.13	1.63	0.90
ITS4. My organization provides information technology support for simulation and prediction	4.34	1.79	0.71
ITS5. My organization provides information technology support for systematic storing	4.90	1.68	0.80
Adoption of environmental practices (PRAC) [66] (V.E = 66.89%); (α = 0.87)			
PRAC1. My organization buys ecological products	4.27	1.69	0.86
PRAC2. My organization has reduced the use of cleaning products that are harmful to the environment	4.76	1.71	0.88
PRAC3. My organization implements energy-saving practices	5.39	1.44	0.82
PRAC4. My organization implements water-saving practices	5.21	1.48	0.77
PRAC5. My organization implements the selective collection of solid residues	5.79	1.45	0.72

The survey was designed to reduce common method bias (CMB) [69]. We used Harman's single factor test to evaluate the existence of CMB. No evidence of CMB existed.

3.5. Fuzzy Set Qualitative Comparative Analysis (fsQCA)

As demonstrated in this study, by using fsQCA [52], we can obtain more than one configuration or causal pathway that leads to PRAC. In this way, we can find a set of alternative causal configurations that a firm can use to arrive at PRAC. In our study, the causal pathways in fsQCA that could lead to PRAC corresponded to combinations of the following variables: HRC, OLC, ITS, EL, EXP and SIZ. Since OLC has five dimensions (OLC-E, OLC-R, OLC-I, OLC-D and OLC-P), it was measured using the fsQCA "fuzzyand" function. This function corresponds to the mathematical logical operation in Boolean algebra called "intercept." Therefore, when OLC appears in a causal configuration, it means that this condition is a five-dimensional cumulative condition. The end result will be PRAC.

Calibration

FsQCA is a methodological tool that uses calibrated data to transform categorical, demographic and Likert scale variables into transformed conditions with values ranging from zero to one. The process of ranking conditions from full membership to non-membership is known as calibration. In order to transform the variables measured with the Likert scale (OLC, ITS and PRAC) into a fuzzy set, the mean values of the items must be calculated [70]. Our measurement scale was a seven-point scale, and so we identified total non-membership, the crossover point and total membership as two, four and six, respectively. According to Woodside et al. [71] the cut-off values were adjusted according to the number of elements of each variable and their statistics. EXP and EL are categorical variables that we calibrated at three levels (1, 0.5, 0), and SIZ is a binary variable that did not need calibration (it adopts the value of zero or one) (Table 2).

Table 2. Descriptive statistics and calibrations of outcome and causal conditions.

Outcome and Causal Conditions	Descriptive Statistics	Calibration Cuts
PRAC	$\mu = 5.08$; $\sigma = 1.28$; min = 1.00; max = 7.00	(6.8; 5.3; 2.6) *
HRC	$\mu = 0.15$; $\sigma = 0.10$; min = 0.00; max = 0.72	(0.31; 0.13; 0.025) *
OLC-E	$\mu = 4.89$; $\sigma = 1.57$; min = 1.00; max = 7.00	(7; 5.5; 3) *
OLC-R	$\mu = 4.17$; $\sigma = 1.49$; min = 1.00; max = 7.00	(6.5; 4.4; 2) *
OLC-I	$\mu = 4.22$; $\sigma = 1.49$; min = 1.00; max = 7.00	(6.2; 4.3; 2) *
OLC-D	$\mu = 5.14$; $\sigma = 1.44$; min = 1.00; max = 7.00	(7;5;4.3) *
OLC-P	$\mu = 4.41$; $\sigma = 1.62$; min = 1.00; max = 7.00	(6.4; 4.8; 1.8) *
ITS	$\mu = 4.67$; $\sigma = 1.46$; min = 1.00; max = 7.00	(6.5; 4.9; 1.9) *
EL (manager education level)	Secondary = 23.43% Graduation = 54.23% Post-graduation = 22.34%	Secondary = 0 Graduation = 0.5 Post-graduation = 1
EXP (# years at the firm)	<2 = 7.36% 2 to 5 = 15.26% >5 = 77.38%	<2 = 0 2 to 5 = 0.5 >5 = 1
SIZ (# employees of the firm)	<50 employees = 65.94% ≥ 50 and ≤ 250 employees = 34.06%	<50 employees = 0; ≥ 50 and ≤ 250 employees = 1;

μ = average; σ = standard deviation; min = minimum; max = maximum; * = (0.95; 0.50; 0.05).

4. Results

4.1. fsQCA

The present study used fsQCA [52,55,72] to assess the conditions of need and sufficiency. The need for a condition was based on seeing the impact on the achievement of PRAC (a consistency score greater than 0.90) [73]. The sufficiency of a condition was based on seeing the extent of its relationship with PRAC (configurations of various conditions that lead to PRAC). Acceptable solutions must respect the coverage level within the range of limits from 0.25 to 0.90 [72] and the coherence threshold of 0.75 [52,72].

In Model 1 (PRAC = f(HRC, OLC, EL, EXP, SIZ)), seven alternative causal configurations exist that lead to PRAC, and three alternative causal configurations exist for the absence of PRAC. Therefore, in Model 1, both H1 and H2 are supported.

In Model 2 (PRAC = f(HRC, ITS, EL, EXP, SIZ)), six alternative causal configurations exist that lead to PRAC, and four alternative causal configurations exist for the absence of PRAC. Therefore, in Model 2, both H3 and H4 are supported.

Finally, in Model 3 (PRAC = f(HRC, ITS, OLC, EL, EXP, SIZ)), eight alternative causal configurations exist that lead to PRAC, and seven alternative causal configurations exist for the absence of PRAC. Therefore, in Model 3, both H5 and H6 are supported.

Therefore, these results offer managers seven different options to reach PRAC for Model 1, six different options to reach PRAC adoption for Model 2 and eight different options to reach PRAC for Model 3. In short, all the hypotheses proposed in our research models are supported.

The results show that firms that invest heavily in establishing learning systems for their employees and achieve a high OLC are more likely to be able to perform PRAC because they will have more alternative paths that lead to it. Specifically, seven causal configurations lead to PRAC compared to only three that lead to its absence according to Model 1. This is a clear sign that PRAC in firms can be achieved through the implementation of a culture of learning where firms implement mechanisms that contribute to the development of knowledge and ethical values to obtain good adoption and implementation of PRAC. On the other hand, from the results obtained from Model 2, it is clear that technology is also a very necessary tool in firms for PRAC. In this case, Model 2 shows that there are six paths that lead to PRAC with technology, versus only four that lead to its absence.

If we interpret the results of both models separately, we can indicate that technology is important for PRAC in firms, but the factors derived from the learning capacity are essential, since with OLC, we can find more ways that lead to achieving PRAC (seven in Model 1) and less in its absence (three in Model 1) than with a model in which there is only a strong investment in technological and non-human factors, where we see that there is one less path that leads to PRAC (six in Model 2) and one more path than in Model 1 that can lead to its absence (four in Model 2). This means that only with the implementation of technology can firms achieve PRAC through different, alternative routes, but there are also a high number of ways that can lead to its absence—that is, there is a probability that the expected result can be achieved as much as other results that are not expected. However, with investment in OLC, it can occur because there are also causal configurations that lead us to the absence of PRAC. The probability of this being the case is lower because in this scenario, there are more paths that lead us to the expected result—that is, to the adoption of PRAC.

Finally, it should be noted that the results show that joint investment in OLC and ITS is optimal for the PRAC, since by integrating factors of both types, we add one more path that leads to the desired result with respect to Model 1. However, it is important to point out that Model 3, where OLC and ITS are included together to reach PRAC, is also the research model that presents more paths towards PRAC absence. From this, it can be interpreted that technology is an adequate tool to achieve the result, but always with adequate human control so as to not incur precisely the opposite effect.

4.2. Causal Configurations

Table 3 shows the reported intermediate solution related to Model 1. This solution shows the seven causal configurations that lead to PRAC. It can be seen that there are four causal configurations with four conditions and three configurations of three conditions. Table 3 shows that HRC, OLC and SIZ are the three relevant core conditions for PRAC. HRC is present in one causal configuration that leads to PRAC, whereas OLC is present in three and SIZ in two. Model 1 also shows that there are three causal configurations for the absence of PRAC (Table 4). Here, HRC, EL, EXP and SIZ are the core conditions for the absence of PRAC. However, OLC is not present in these solutions.

Table 3. Intermediate solutions for PRAC with OLC. Model 1(a): PRAC = f(HRC, OLC, EL, EXP, SIZ). Overall solution coverage: 0.641398. Overall solution consistency: 0.770815.

Configurations	HRC	OLC	EL	EXP	SIZ	Coverage		Consistency
						Raw	Unique	
1	○	●			○	0.308783	0.028830	0.906541
2	●	●		●		0.327869	0.027125	0.898529
3	●		○	○	○	0.317438	0.059153	0.815325
4			○	●	●	0.294517	0.006982	0.804077
5	○	○		●	●	0.359214	0.015025	0.798292
6		●	●	●		0.306512	0.000376	0.877476
7	○		●	●	●	0.293203	0.000721	0.803997

Note: Black circles (●) indicate the presence of a condition, and open circles (○) indicate its absence. Large circles indicate core conditions (present in both the parsimonious and intermediate solutions), and small ones identify peripheral conditions (present only in the intermediate solution). Blank spaces indicate that the condition does not contribute to the configuration. PRAC = adoption of environmental practices; HRC = human resource cost; OLC = organizational learning capability; ITS = information technology support; EL = education level; EXP = experience; SIZ = firm size. In the three research models PRAC correspond to a function with combinations of the following variables. Research model 1: HRC, OLC, EL, EXP and SIZ. Research model 2: HRC ITS, EL, EXP and SIZ. Research model 3: HRC, ITS, OLC, EL, EXP and SIZ. However, in research model 1 (PRAC = f(HRC, OLC, EL, EXP, SIZ)) and research model 3 (PRAC = f(HRC, ITS, OLC, EL, EXP, SIZ)), OLC has five dimensions (OLC-E, OLC-R, OLC-I, OLC-D and OLC-P), it is measure using the fsQCA "fuzzyand" function. It is a mathematical logical operation in Boolean algebra called "intercept." Therefore, when OLC appears in a causal configuration, it means that this condition is a five-dimensional cumulative condition.

Table 4. Intermediate solutions for ~PRAC with OLC. Model 1(b): ~PRAC = f(HRC, OLC, EL, EXP, SIZ). Overall solution coverage: 0.393352. Overall solution consistency: 0.806792.

Configurations	HRC	OLC	EL	EXP	SIZ	Coverage		Consistency
						Raw	Unique	
1	○	○	●	●	○	0.339403	0.294387	0.811081
2	○	○	●	○	●	0.063939	0.010719	0.836640
3	●	○	●	○	○	0.085222	0.026853	0.811686

Note: Black circles (●) indicate the presence of a condition, and open circles (○) indicate its absence. Large circles indicate core conditions (present in both the parsimonious and intermediate solutions), and small ones identify peripheral conditions (present only in the intermediate solution). Blank spaces indicate that the condition does not contribute to the configuration. PRAC = adoption of environmental practices; HRC = human resource cost; OLC = organizational learning capability; ITS = information technology support; EL = education level; EXP = experience; SIZ = firm size. In the three research models PRAC correspond to a function with combinations of the following variables. Research model 1: HRC, OLC, EL, EXP and SIZ. Research model 2: HRC ITS, EL, EXP and SIZ. Research model 3: HRC, ITS, OLC, EL, EXP and SIZ. However, in research model 1 (PRAC = f(HRC, OLC, EL, EXP, SIZ)) and research model 3 (PRAC = f(HRC, ITS, OLC, EL, EXP, SIZ)), OLC has five dimensions (OLC-E, OLC-R, OLC-I, OLC-D and OLC-P), it is measure using the fsQCA "fuzzyand" function. It is a mathematical logical operation in Boolean algebra called "intercept." Therefore, when OLC appears in a causal configuration, it means that this condition is a five-dimensional cumulative condition.

The intermediate solution in Model 2 (Table 5) shows six configurations that lead to PRAC. ITS and SIZ are the core conditions for PRAC. ITS is present in five causal configurations and SIZ in one. Model 2 also shows four causal configurations for the absence of PRAC (Table 6). Here, HCR, EL and SIZ are the core conditions for the absence of PRAC. However, ITS is not present in these solutions. The literature indicates that firms that have a high investment in ITS tend to have a good PRAC [27,28], since technology provides the necessary mechanisms to convert inputs into sustainable outputs [25,29].

Table 5. Intermediate solutions for PRAC with ITS. Model 2(a): PRAC = f(HRC, ITS, EL, EXP, SIZ). Overall solution coverage: 0.786787. Overall solution consistency: 0.737200.

Configurations	HRC	ITS	EL	EXP	SIZ	Coverage		Consistency
						Raw	Unique	
1		•		•		0.653200	0.076208	0.755129
2		•	○		○	0.392448	0.007527	0.864343
3	○			•	•	0.385734	0.054554	0.792995
4		•	•		•	0.378529	0.005893	0.832660
5	○	•			○	0.447573	0.005584	0.825554
6	○	•	•			0.389092	0.000000	0.836647

Note: Black circles (•) indicate the presence of a condition, and open circles (○) indicate its absence. Large circles indicate core conditions (present in both the parsimonious and intermediate solutions), and small ones identify peripheral conditions (present only in the intermediate solution). Blank spaces indicate that the condition does not contribute to the configuration. PRAC = adoption of environmental practices; HRC = human resource cost; OLC = organizational learning capability; ITS = information technology support; EL = education level; EXP = experience; SIZ = firm size. In the three research models PRAC correspond to a function with combinations of the following variables. Research model 1: HRC, OLC, EL, EXP and SIZ. Research model 2: HRC ITS, EL, EXP and SIZ. Research model 3: HRC, ITS, OLC, EL, EXP and SIZ. However, in research model 1 (PRAC = f(HRC, OLC, EL, EXP, SIZ)) and research model 3 (PRAC = f(HRC, ITS, OLC, EL, EXP, SIZ)), OLC has five dimensions (OLC-E, OLC-R, OLC-I, OLC-D and OLC-P), it is measure using the fsQCA "fuzzyand" function. It is a mathematical logical operation in Boolean algebra called "intercept." Therefore, when OLC appears in a causal configuration, it means that this condition is a five-dimensional cumulative condition.

Table 6. Intermediate solutions for~PRAC with ITS. Model 2(b): ~PRAC = f(HRC, ITS, EL, EXP, SIZ). Overall solution coverage: 0.634377. Overall solution consistency: 0.797445.

Configurations	HRC	ITS	OLC	EL	EXP	SIZ	Coverage		Consistency
							Raw	Unique	
1	○	•				○	0.447573	0.045468	0.825554
2		•	○	○		○	0.357278	0.006751	0.878233
3	○		○	•		•	0.359214	0.043247	0.798292
4		•	○	•		•	0.339853	0.015138	0.846580
5	•	•		○	•		0.361423	0.010868	0.855247
6		•	•	•	•		0.286312	0.013407	0.893125
7	○	•		○	•	○	0.216127	0.005891	0.922435
8	•	•		○	•	•	0.220276	0.002983	0.900171

Note: Black circles (•) indicate the presence of a condition, and open circles (○) indicate its absence. Large circles indicate core conditions (present in both the parsimonious and intermediate solutions), and small ones identify peripheral conditions (present only in the intermediate solution). Blank spaces indicate that the condition does not contribute to the configuration. PRAC = adoption of environmental practices; HRC = human resource cost; OLC = organizational learning capability; ITS = information technology support; EL = education level; EXP = experience; SIZ = firm size. In the three research models PRAC correspond to a function with combinations of the following variables. Research model 1: HRC, OLC, EL, EXP and SIZ. Research model 2: HRC ITS, EL, EXP and SIZ. Research model 3: HRC, ITS, OLC, EL, EXP and SIZ. However, in research model 1 (PRAC = f(HRC, OLC, EL, EXP, SIZ)) and research model 3 (PRAC = f(HRC, ITS, OLC, EL, EXP, SIZ)), OLC has five dimensions (OLC-E, OLC-R, OLC-I, OLC-D and OLC-P), it is measure using the fsQCA "fuzzyand" function. It is a mathematical logical operation in Boolean algebra called "intercept." Therefore, when OLC appears in a causal configuration, it means that this condition is a five-dimensional cumulative condition.

Finally, the intermediate solution in Model 3 (Table 7) shows eight configurations that lead to PRAC. Here, ITS, OLC and SIZ are the core conditions. ITS is present in five causal configurations, OLC in three and SIZ in one. In addition, there are seven causal configurations for the absence of PRAC (Table 8). Here, HRC, OLC, EL and SIZ are the core conditions. However, ITS is not present in these solutions.

Table 7. Intermediate solutions for PRAC with OLC and ITS. Model 3(a): PRAC = f(HRC, ITS, OLC, EL, EXP, SIZ). Overall solution coverage: 0.721535. Overall solution consistency: 0.771006.

Configurations	HRC	ITS	EL	EXP	SIZ	Coverage		Consistency
						Raw	Unique	
1		o		•	•	0.462298	0.069423	0.794464
2	o	o	•		o	0.321358	0.085458	0.852616
3	•	o	•	o		0.100179	0.013094	0.938528
4	•	o	o		•	0.368816	0.029335	0.829176

Note: Black circles (•) indicate the presence of a condition, and open circles (o) indicate its absence. Large circles indicate core conditions (present in both the parsimonious and intermediate solutions), and small ones identify peripheral conditions (present only in the intermediate solution). Blank spaces indicate that the condition does not contribute to the configuration. PRAC = adoption of environmental practices; HRC = human resource cost; OLC = organizational learning capability; ITS = information technology support; EL = education level; EXP = experience; SIZ = firm size. In the three research models PRAC correspond to a function with combinations of the following variables. Research model 1: HRC, OLC, EL, EXP and SIZ. Research model 2: HRC ITS, EL, EXP and SIZ. Research model 3: HRC, ITS, OLC, EL, EXP and SIZ. However, in research model 1 (PRAC = f(HRC, OLC, EL, EXP, SIZ)) and research model 3 (PRAC = f(HRC, ITS, OLC, EL, EXP, SIZ)), OLC has five dimensions (OLC-E, OLC-R, OLC-I, OLC-D and OLC-P), it is measure using the fsQCA "fuzzyand" function. It is a mathematical logical operation in Boolean algebra called "intercept." Therefore, when OLC appears in a causal configuration, it means that this condition is a five-dimensional cumulative condition.

Table 8. Intermediate solutions for ~PRAC with OLC and ITS. Model 3(b): ~PRAC = f(HRC, ITS, OLC, EL, EXP, SIZ). Overall solution coverage: 0.631863. Overall solution consistency: 0.796658.

Configurations	HRC	ITS	OLC	EL	EXP	SIZ	Coverage		Consistency
							Raw	Unique	
1		o	o		•	•	0.448225	0.062615	0.805560
2	o	o	o	•		o	0.316626	0.077496	0.859679
3	•	o	o	•	o		0.097161	0.008904	0.947500
4	•	o	o	o		•	0.363947	0.026436	0.830786
5	•	o		o	•	•	0.316766	0.003879	0.839240
6	o	o	•	o	•	o	0.142530	0.005234	0.843909
7	o	•	o	•	o	•	0.056089	0.004986	0.819520

Note: Black circles (•) indicate the presence of a condition, and open circles (o) indicate its absence. Large circles indicate core conditions (present in both the parsimonious and intermediate solutions), and small ones identify peripheral conditions (present only in the intermediate solution). Blank spaces indicate that the condition does not contribute to the configuration. PRAC = adoption of environmental practices; HRC = human resource cost; OLC = organizational learning capability; ITS = information technology support; EL = education level; EXP = experience; SIZ = firm size. In the three research models PRAC correspond to a function with combinations of the following variables. Research model 1: HRC, OLC, EL, EXP and SIZ. Research model 2: HRC ITS, EL, EXP and SIZ. Research model 3: HRC, ITS, OLC, EL, EXP and SIZ. However, in research model 1 (PRAC = f(HRC, OLC, EL, EXP, SIZ)) and research model 3 (PRAC = f(HRC, ITS, OLC, EL, EXP, SIZ)), OLC has five dimensions (OLC-E, OLC-R, OLC-I, OLC-D and OLC-P), it is measure using the fsQCA "fuzzyand" function. It is a mathematical logical operation in Boolean algebra called "intercept." Therefore, when OLC appears in a causal configuration, it means that this condition is a five-dimensional cumulative condition.

Therefore, the results show that the assumptions of fsQCA [53] are fulfilled. It can be seen how more than one configuration of causal conditions leads to PRAC, and thus, the existence of alternative combinations of causal conditions leading to PRAC is confirmed. Furthermore, since these alternative causal configurations can produce the same result, the principle of equifinality is confirmed. The solutions present coverage and coherence values that respect the limits set by the literature [72].

Furthermore, our research models show that there are no pathways leading to PRAC when ITS or OLC are not included in the model as a causal condition. In this way, our work reveals the key role that technology and human resources management have in achieving PRAC. The increasing introduction, improvement and application of new technologies in firms help, in a clear and simple way, to improve and apply more environmental practices within firms. New technologies, greater investment in human resources, a culture of organizational learning and other firm and manager characteristics such as education, experience or firm size can actually encourage the adoption of environmental practices [2].

5. Discussion and Conclusions

Few empirical studies have shown the different pathways for PRAC [8,37,53]. This paper contributes to the theoretical and empirical research with a relevant and original qualitative method, fsQCA. The fsQCA in our research models shows new relations, pathways and alternative configurations for PRAC that other quantitative tools do not discover. Therefore, fsQCA offers alternative complex configurations of conditions for PRAC. Our research models show that both sources are relevant: technological- and human-resources-based. Contrary to quantitative approaches that only show single linear models of PRAC in SMEs [73], our results show several pathways for PRAC and its absence. They are relevant findings: the former points SMEs towards the ways to reach PRAC and the latter warns SMEs about the pathways that do not lead to PRAC.

Given the results from our research models, all hypotheses are supported. In Model 1(a), OLC is a core condition for PRAC in three pathways. It is the main core condition in this model for PRAC. However, in Model 1(b), the absence of OLC is a core condition in two of three pathways for the absence of PRAC. Similar considerations can be made about ITS in Model 2(a and b). In Model 2(a), ITS is a core condition for PRAC in five pathways of six. However, in Model 2(b), the absence of ITS is a core condition in all pathways for the absence of PRAC, totaling four pathways.

It is notable that OLC and ITS are complete substitutes (configuration 1 in Table 3 and configuration 5 in Table 5). Such results reveal similar substitutable contributions of the technological- and human-resources-based sources to PRAC. Yet, SMEs have other options that lead them to PRAC. Regarding Tables 4 and 6, the evidence shows that the absence of either OLC or ITS is associated with not adopting environmental practices.

Finally, in the full and integrative Model 3(a), ITS is a core condition for PRAC. ITS is present in five pathways for PRAC, and OLC is present in three pathways for PRAC, for eight total pathways. It is a relevant finding because technology is the most important factor leading to PRAC. Model 3(b) also shows that the absence of ITS is a core condition in six pathways for the absence of PRAC.

Therefore, our separate research models show that there are more ways to reach PRAC if we take into account human learning factors within the organization, such as OLC, than if we only adopt technological factors, such as ITS. In this sense, with Model 1 (OLC), seven paths lead us to PRAC, and with Model 2 (ITS), six paths lead us to PRAC. It is also relevant to note that the model that incorporates OLC without ITS (Model 1) offers firms fewer avenues in which they face the absence of PRAC, as there are only three. On the other hand, the model that incorporates ITS without OLC (Model 2) offers firms one more way to avoid PRAC; that is to say, in this case, four routes can condition the absence of PRAC, one more than in the previous case. Therefore, if we analyze these results separately, we could advise firms that, mainly, the optimum component that can shape and be present in different combinations to arrive at PRAC is the OLC. Additionally, we may interpret it such that depending on technology involves a higher risk of not achieving PRAC.

On the other hand, the main conclusions drawn from the complete and integrative research model shown in this paper to obtain PRAC show that it is desirable for firms to dedicate their efforts to joint investment in personnel management practices and the adoption of new technologies for obtaining PRAC. As can be seen in Model 3, when the two sources (OLC and ITS) are integrated, more paths are obtained to reach PRAC than when a firm decides to focus more on one source than on another to obtain PRAC. Our third model makes it clear that the adoption of information technologies such as key training, storage and communication can help firms to be more efficient when implementing PRAC. Regarding the lack of PRAC (Table 8), it is worth noting that all the configurations present the lack of OLC or ITS or both. Such findings seem to validate a contrario the relevancy of such conditions to achieve PRAC.

Finally, there seems to be a special contribution of ITS to PRAC, since there is no configuration leading to PRAC in the absence of ITS (Tables 5 and 7). On the contrary, it is possible to reach PRAC with the absence of OLC (configurations 3, 4 and 5 in Table 3, and configurations 2, 3 and 4 in Table 7). Such findings give strong support to arguments in the literature on the technological basis of PRAC [74].

In addition, in the three research models, other conditions such as EL, EXP and SIZ are relevant for obtaining PRAC. For example, in Model 1, SIZ is a central condition to reach PRAC in two causal alternatives. EL and EXP are important as well, but they are not central conditions. In addition, it must also be taken into account that the conditions related to the manager and the size of the firm, if not managed properly, can lead to other combinations or paths heading towards the absence of PRAC. In Model 2, SIZ is also a central condition for obtaining PRAC in at least one path. The same happens in Model 3.

In summary, our findings indicate that by focusing on variables such as OLC, a firm can find more paths that lead to PRAC. Additionally, with a joint focus on OLC and ITS, it must be taken into account that only investing in ITS and not in OLC is riskier when trying to implement PRAC.

The main goal of this paper was to find complementary pathways and conclusions about PRAC by applying fsQCA [73–75]. Our study shows that fsQCA is a strong tool that allows discernment of how factors are combined to generate (or not) PRAC. Moreover, fsQCA allows the important roles of OLC and ITS in inducing PRAC to be checked. Likewise, fsQCA discovers asymmetrical relations between conditions.

For managers, the findings indicate that to adopt PRAC in firms, it is necessary to consider a TBL approach and factors derived from it. States should focus on new formulas and regulations that help firms to be able to adopt PRAC through a clear commitment to their human resources and technology. In addition, several practical recommendations can be derived from our study:

1. Investment in human resources for the adoption of PRAC:
    * Training in sustainability for the application of PRAC.
    * Management of motivation and values of employees oriented towards PRAC.
    * Foster a corporate culture of sustainability.
    * Creation of internal rules for correct application of PRAC.

2. Organizational learning capacity for PRAC:
    * Foster a culture of trial and error—that is, of experimentation.
    * Possibility of taking risks when implementing new sustainable practices.
    * Promotion of teamwork and communication.
    * Let employees give ideas and make decisions regarding PRAC.

3. Information technology support for the adoption of PRAC:
    * Implementation of environmental information collection systems to determine the needs internal and external to the firm.

4. Adoption of environmental practices:
    * Code of ethics for all members of the firm that facilitates the adoption of PRAC.
    * Implementation of energy-saving practices.
    * Implementation of reverse logistics practices.
    * Diversification of products towards economic products.
    * Training on social and environmental needs.

We are aware that this study has some limitations. This paper focused on a limited population segment: 349 Portuguese SMEs certified as innovative in various industry categories (manufacturing, power and gas supply; water supply and pollution; building; vehicle trade and repair; transport and storage; catering; information and communication; housing; scientific activities; administrative activities; health activities; other services). Future studies should focus on other segments but also on other economic agents (large businesses, transnational corporations, etc.). Although our study to analyze the representativeness of the sample in the population was based on an analysis of variance (ANOVA) between firms that responded early and those that responded late, future studies could perform further analyses of representativeness, for example, between how the firms of one industrial sector and another behave among the twelve that were represented in this study to ensure that there are no significant differences between the responses of the different industrial sectors. In addition, our study focused only on three factors that can affect the adoption of environmental practices in firms: company size and the level of education and experience of the manager. Future studies may include more business factors such as the type of firm according to its legal and commercial form and more characteristics of the manager, such as political ideology, among others, that may affect the adoption of environmental practices. Another constraint is that this research is confined to Portugal. The conclusions may have been slightly different if the survey had a wider geographical extension or if it were answered in another country, so the use of an international database may allow improvement of the conclusions that we have extracted in this work. Although the response rate was representative, it can be considered slightly low (5.1%). Another issue that could be addressed in future research is the sustainability of ITS. It is certain that ITS leads to PRAC, but the development of technology can also affect PRAC negatively. Technology requires intensive computation resources with large energy consumption. Therefore, the findings of this research should be interpreted with the above considerations and new studies. Finally, let us point out that the use of alternative analytical tools to fsQCA based on fuzzy sets, such as fuzzy correlation indexes and fuzzy multiple criteria decision-making, and other quantitative methods, such as SEM, etc., should be considered in future research.

Author Contributions: Conceptualization, L.M.-P., J.G. and C.C.; methodology, L.M.-P., J.G. and C.C.; software, L.M.-P.; validation, L.M.-P., J.G. and C.C.; formal analysis, L.M.-P.; investigation, L.M.-P., J.G. and C.C.; data curation, L.M.-P.; writing—original draft preparation, L.M.-P.; writing—review and editing, L.M.-P., J.G. and C.C.; visualization, L.M.-P., J.G. and C.C.; supervision, J.G. and C.C.; project administration, J.G. and C.C.; funding acquisition, L.M.-P., J.G. and C.C. All authors have read and agreed to the published version of the manuscript.

Funding: The authors are grateful for the economic support provided by FCT (Fundação para a Ciência e Tecnologia Portugal) grant number UIDB/04521/2020, by "Agencia Estatal de Investigación" (AEI) of the Ministerio de Ciencia e Innovación del Gobierno de España grant number PID2019-107546GA-I00, by Consejería de Educación of the Junta de Castilla y León and the "Fondo Europeo de Desarrollo Regional" (FEDER) grant number SA106P20, and by Junta de Castilla y León and the European Regional Development Fund for the financial support to the Research Unit of Excellence "Economic Management for Sustainability" (GECOS) grant number CLU-2019-03.

Institutional Review Board Statement: Not applicable.

Informed Consent Statement: Not applicable.

Data Availability Statement: Not applicable.

Acknowledgments: Authors acknowledge helpful comments of anonymous reviewers.

Conflicts of Interest: The authors declare no conflict of interest.

References

1. Cohen, B.; Smith, B.; Mitchell, R. Toward a sustainable conceptualization of dependent variables in entrepreneurship research. *Bus. Strategy Environ.* **2008**, *17*, 107–119. [CrossRef]
2. Glavas, A.; Mish, J. Resources and Capabilities of Triple Bottom Line Firms: Going Over Old or Breaking New Ground? *J. Bus. Ethics* **2015**, *127*, 623–642. [CrossRef]
3. Kennedy, E.B.; Marting, T.A. Biomimicry: Streamlining the Front End of Innovation for Environmentally Sustainable Products. *Res. Technol. Manag.* **2016**, *59*, 40–47. [CrossRef]
4. Wagner, M. Innovation and competitive advantages from the integration of strategic aspects with social and environmental management in European firms. *Bus. Strategy Environ.* **2009**, *18*, 291–306. [CrossRef]
5. Hall, J.; Wagner, M. Integrating Sustainability into Firms' Processes: Performance Effects and the Moderating Role of Business Models and Innovation. *Bus. Strategy Environ.* **2012**, *21*, 183–196. [CrossRef]
6. Gasbarro, F.; Annunziata, E.; Rizzi, F.; Frey, M. The Interplay between Sustainable Entrepreneurs and Public Authorities: Evidence from Sustainable Energy Transitions. *Organ. Environ.* **2017**, *30*, 226–252. [CrossRef]
7. Vollenbroek, F.A. Sustainable development and the challenge of innovation. *J. Clean. Prod.* **2002**, *10*, 215–223. [CrossRef]

8. Muñoz-Pascual, L.; Curado, C.; Galende, J. The Triple Bottom Line on Sustainable Product Innovation Performance in SMEs: A Mixed Methods Approach. *Sustainability* **2019**, *11*, 1689. [CrossRef]
9. Argote, L. *Organizational Learning: Creating, Retaining and Transferring Knowledge*, 2nd ed.; Springer: Media, PA, USA, 2013.
10. Lichtenthaler, U. Absorptive Capacity, Environmental Turbulence, and the Complementarity of Organizational Learning Processes. *Acad. Manag. J.* **2009**, *52*, 822–846. [CrossRef]
11. Henriques, A.; Richardson, J. *The Triple Bottom Line, Does It All Add Up?: Assessing the Sustainability of Business and CSR*; Earthscan Publications Ltd.: London, UK, 2004; ISBN 9781844070152.
12. Roig-Tierno, N.; Kraus, S.; Cruz, S. The relation between coopetition and innovation/entrepreneurship. *Rev. Manag. Sci.* **2018**, *12*, 379–383. [CrossRef]
13. Celma, D.; Martínez-Garcia, E.; Coenders, G. Corporate Social Responsibility in Human Resource Management: An analysis of common practices and their determinants in Spain. *Corp. Soc. Responsib. Environ. Manag.* **2014**, *21*, 82–99. [CrossRef]
14. Gratton, L.; Ghoshal, S. Managing Personal Human Capital: New Ethos for the 'Volunteer'employee. *Eur. Manag. J.* **2003**, *21*, 1–10. [CrossRef]
15. Barney, J.B. Looking inside for competitive advantage. *Acad. Manag. Perspect.* **1995**, *9*, 49–61. [CrossRef]
16. Davidescu, A.A.; Apostu, S.-A.; Paul, A.; Casuneanu, I. Work Flexibility, Job Satisfaction, and Job Performance among Romanian Employees—Implications for Sustainable Human Resource Management. *Sustainability* **2020**, *12*, 6086. [CrossRef]
17. Gold, A.H.; Malhotra, A.; Segars, A.H. Knowledge Management: An Organizational Capabilities Perspective. *J. Manag. Inf. Syst.* **2001**, *18*, 185–214. [CrossRef]
18. Lee, H.; Choi, B. Knowledge Management Enablers, Processes, and Organizational Performance: An Integrative View and Empirical Examination. *J. Manag. Inf. Syst.* **2003**, *20*, 179–228. [CrossRef]
19. Roberts, J. From Know-how to Show-how? Questioning the Role of Information and Communication Technologies in Knowledge Transfer. *Technol. Anal. Strateg. Manag.* **2000**, *12*, 429–443. [CrossRef]
20. Eggers, F.; Hatak, I.; Kraus, S.; Niemand, T. Technologies That Support Marketing and Market Development in SMEs-Evidence from Social Networks. *J. Small Bus. Manag.* **2017**, *55*, 270–302. [CrossRef]
21. Richter, C.; Kraus, S.; Brem, A.; Durst, S.; Giselbrecht, C. Digital entrepreneurship: Innovative business models for the sharing economy. *Creat. Innov. Manag.* **2017**, *26*, 300–310. [CrossRef]
22. Kraus, S.; Roig-Tierno, N.; Bouncken, R.B. Digital innovation and venturing: An introduction into the digitalization of entrepreneurship. *Rev. Manag. Sci.* **2019**, *13*, 519–528. [CrossRef]
23. Riggins, F.J.; Rhee, H. Developing the learning network using extranets. *Int. J. Electron. Commer.* **1999**, *4*, 65–83. [CrossRef]
24. Medina-Molina, C.; Rey-Moreno, M.; Felício, J.A.; Romano Paguillo, I. Participation in crowdfunding among users of collaborative platforms: The role of innovativeness and social capital. *Rev. Manag. Sci.* **2019**, *13*, 529–543. [CrossRef]
25. Song, W.; Wang, G.Z.; Ma, X. Environmental innovation practices and green product innovation performance: A perspective from organizational climate. *Sustain. Dev.* **2020**, *28*, 224–234. [CrossRef]
26. Wade, M.; Hulland, J. Review: The Resource-Based View and Information Systems Research: Review, Extension, and Suggestions for Future Research. *MIS Q.* **2004**, *28*, 107. [CrossRef]
27. Kohli, R.; Grover, V. Business value of IT: An essay on expanding research directions to keep up with the times. *J. Assoc. Inf. Syst.* **2008**, *9*, 22–39. [CrossRef]
28. Kim, G.; Shin, B.; Kim, K.K.; Lee, H.G. IT capabilities, process-oriented dynamic capabilities, and firm financial performance. *J. Assoc. Inf. Syst.* **2011**, *12*, 487–587. [CrossRef]
29. Orlikowski, W.J.; Iacono, C.S. Research commentary: Desperately seeking the "IT" in IT research—A call to theorizing the IT artifact. *Inf. Syst. Res.* **2001**, *12*, 121–134. [CrossRef]
30. Vrchota, J.; Řehoř, P.; Maříková, M.; Pech, M. Critical Success Factors of the Project Management in Relation to Industry 4.0 for Sustainability of Projects. *Sustainability* **2020**, *13*, 281. [CrossRef]
31. Jaw, B.-S.; Liu, W. Promoting organizational learning and self-renewal in Taiwanese companies: The role of HRM. *Hum. Resour. Manag.* **2003**, *42*, 223–241. [CrossRef]
32. Argote, L.; Ingram, P.; Levine, J.M.; Moreland, R.L. Knowledge Transfer in Organizations: Learning from the Experience of Others. *Organ. Behav. Hum. Decis. Processes.* **2000**, *82*, 1–8. [CrossRef]
33. Lichtenthaler, U. Determinants of absorptive capacity: The value of technology and market orientation for external knowledge acquisition. *J. Bus. Ind. Mark.* **2016**, *31*, 600–610. [CrossRef]
34. Chiva, R.; Alegre, J.; Lapiedra, R. Measuring Organizational Learning Capability among the Workforce. *Int. J. Manpow.* **2007**, *28*, 224–242. [CrossRef]
35. Saitua-Iribar, A.; Corral-Lage, J.; Peña-Miguel, N. Improving Knowledge about the Sustainable Development Goals through a Collaborative Learning Methodology and Serious Game. *Sustainability* **2020**, *12*, 6169. [CrossRef]
36. Gavronski, I.; Ferrer, G.; Paiva, E.L. ISO 14001 certification in Brazil: Motivations and benefits. *J. Clean. Prod.* **2008**, *16*, 87–94. [CrossRef]
37. Kraus, S.; Ribeiro-Soriano, D.; Schussler, M. Fuzzy-set Qualitative Comparative Analysis (fsQCA) in Entrepreneurship and Innovation Research—The Rise of a Method. *Int. Entrep. Manag. J.* **2018**, *14*, 15–33. [CrossRef]
38. Fernández-Viñé, M.B.; Gómez-Navarro, T.; Capuz-Rizo, S.F. Eco-efficiency in the SMEs of Venezuela. Current status and future perspectives. *J. Clean. Prod.* **2010**, *18*, 736–746. [CrossRef]

39. Ramanathan, R.; Black, A.; Nath, P.; Muyldermans, L. Impact of environmental regulations on innovation and performance in the UK industrial sector. *Manag. Decis.* **2010**, *48*, 1493–1513. [CrossRef]
40. Aragón-Correa, J.A.; Hurtado-Torres, N.; Sharma, S.; García-Morales, V.J. Environmental strategy and performance in small firms: A resource-based perspective. *J. Environ. Manag.* **2008**, *86*, 88–103. [CrossRef]
41. Chan, E.S.W.; Hawkins, R. Attitude towards EMSs in an international hotel: An exploratory case study. *Int. J. Hosp. Manag.* **2010**, *29*, 641–651. [CrossRef]
42. Mohamed, S.T. The impact of ISO 14000 on developing world businesses. *Renew. Energy* **2001**, *23*, 579–584. [CrossRef]
43. Camisón-Zornoza, C.; Lapiedra-Alcamí, R.; Segarra-Ciprés, M.; Boronat-Navarro, M. A Meta-analysis of Innovation and Organizational Size. *Organ. Stud.* **2004**, *25*, 331–361. [CrossRef]
44. Rothwell, R.; Dodgson, M. Innovation and Size of Firm. In *The Handbook of Industrial Innovation*; Dodgson, M., Rothwell, R., Eds.; Edward Elgar Publishing: Cheltenham, UK, 1994; pp. 310–324.
45. Vila, L.E.; Pérez, P.J.; Coll-Serrano, V. Innovation at the workplace: Do profesional competencies matter? *J. Bus. Res.* **2014**, *67*, 752–757. [CrossRef]
46. Wang, M.H.; Yang, T.Y.; Liu, P.C. The impact of knowledge sharing and projects complexity on team creativity: An example of information systems development. *J. Bus. Res.* **2010**, *12*, 73–102.
47. Bassellier, G.; Benbasat, I.; Reich, B.H. The Influence of Business Managers' IT Competence on Championing IT. *Inf. Syst. Res.* **2003**, *14*, 317–336. [CrossRef]
48. Kuemmerle, W. Optimal scale for research and development in foreign environments—An investigation into size and performance of research and development laboratories abroad. *Res. Policy* **1998**, *27*, 111–126. [CrossRef]
49. Kumar, N.; Saqib, M. Firm size, opportunities for adaptation and in-house R & D activity in developing countries: The case of Indian manufacturing. *Res. Policy* **1996**, *25*, 713–722. [CrossRef]
50. Gumbau, M. Análisis microeconómico de los determinantes de la innovación: Aplicación a las empresas industriales españolas. *Rev. Española Econ.* **1997**, *14*, 41–66.
51. Molero, J.; Buesa, M. Patterns of technological change among Spanish innovative firms: The case of the Madrid region. *Res. Policy* **1996**, *25*, 647–663. [CrossRef]
52. Fiss, P.C. Building Better Causal Theories: A Fuzzy Set Approach to Typologies in Organization Research. *Acad. Manag. J.* **2011**, *54*, 393–420. [CrossRef]
53. Drummond, S.; O'Driscoll, M.P.; Brough, P.; Kalliath, T.; Siu, O.L.; Timms, C.; Riley, D.; Sit, C.; Lo, D. The relationship of social support with well-being outcomes via work–family conflict: Moderating effects of gender, dependants and nationality. *Hum. Relat.* **2017**, *70*, 544–565. [CrossRef]
54. Osabutey, E.L.C.; Jin, Z. Factors influencing technology and knowledge transfer: Configurational recipes for Sub-Saharan Africa. *J. Bus. Res.* **2016**, *69*, 5390–5395. [CrossRef]
55. Oyemomi, O.; Liu, S.; Neaga, I.; Alkhuraiji, A. How knowledge sharing and business process contribute to organizational performance: Using the fsQCA approach. *J. Bus. Res.* **2016**, *69*, 5222–5227. [CrossRef]
56. Curado, C.; Muñoz-Pascual, L.; Galende, J. Antecedents to innovation performance in SMEs: A mixed methods approach. *J. Bus. Res.* **2018**, *89*, 206–215. [CrossRef]
57. Ragin, C.C. Using qualitative comparative analysis to study causal complexity. *Health Serv. Res.* **1999**, *34*, 1225–1239.
58. Leischnig, A.; Henneberg, S.C.; Thornton, S.C. Net versus combinatory effects of firm and industry antecedents of sales growth. *J. Bus. Res.* **2016**, *69*, 3576–3583. [CrossRef]
59. Kaya, B.; Abubakar, A.M.; Behravesh, E.; Yildiz, H.; Mert, I.S. Antecedents of innovative performance: Findings from PLS-SEM and fuzzy sets (fsQCA). *J. Bus. Res.* **2020**, *114*, 278–289. [CrossRef]
60. Hair, J.; Anderson, R.; Tatham, R.; Black, W. *Multivariate Data Analysis*; Prentice-Hall: Saddle River, NJ, USA, 2005.
61. Muñoz-Pascual, L.; Galende, J. The impact of knowledge and motivation management on creativity: Employees of innovative Spanish companies. *Empl. Relat.* **2017**, *39*, 732–752. [CrossRef]
62. Armstrong, J.S.; Overton, T.S. Estimating nonresponse bias in mail surveys. *J. Mark Res.* **1977**, *14*, 396–402. [CrossRef]
63. Celma, D.; Martínez-Garcia, E.; Raya, J.M. Socially responsible HR practices and their effects on employees' wellbeing: Empirical evidence from Catalonia, Spain. *Eur. Res. Manag. Bus. Econ.* **2018**, *24*, 82–89. [CrossRef]
64. Westland, J.C. Data collection, control, and sample size. In *Structural Equation Models: From Paths to Networks*; Springer: Cham, Switzerland, 2015; pp. 83–115.
65. Alegre, J.; Chiva, R. Assessing the impact of organizational learning capability on product innovation performance: An empirical test. *Technovation* **2008**, *28*, 315–326. [CrossRef]
66. Molina-Azorín, J.F.; Claver-Cortés, E.; Pereira-Moliner, J.; Tarí, J.J. Environmental practices and firm performance: An empirical analysis in the Spanish hotel industry. *J. Clean. Prod.* **2009**, *17*, 516–524. [CrossRef]
67. Blau, J.R.; McKinley, W. Ideas, Complexity, and Innovation. *Adm. Sci. Q.* **1979**, *24*, 200. [CrossRef]
68. Graves, S.B.; Langowitz, N.S. Innovative productivity and returns to scale in the pharmaceutical industry. *Strateg. Manag. J.* **1993**, *14*, 593–605. [CrossRef]
69. Podsakoff, P.M.; MacKenzie, S.B.; Lee, Y.; Podsakoff, N.P. Common method biases in behavioral research: A critical review of the literature and recommended remedies. *J. Appl. Psychol.* **2003**, *88*, 879–903. [CrossRef]

70. Woodside, A.; Hsu, S.-Y.; Marshall, R. General theory of cultures' consequences on international tourism behavior. *J. Bus. Res.* **2011**, *64*, 785–799. [CrossRef]
71. Woodside, A.; Prentice, C.; Larsen, A. Revisiting Problem Gamblers' Harsh Gaze on Casino Services: Applying Complexity Theory to Identify Exceptional Customers. *Psychol. Mark.* **2015**, *32*, 65–77. [CrossRef]
72. Ragin, C. *Redesigning Social Inquiry: Fuzzy Sets and Beyond*; University of Chicago Press: Chicago, IL, USA, 2008.
73. Arias-Oliva, M.; de Andrés-Sánchez, J.; Pelegrín-Borondo, J. Fuzzy Set Qualitative Comparative Analysis of Factors In-fluencing the Use of Cryptocurrencies in Spanish Households. *Mathematics* **2021**, *9*, 324. [CrossRef]
74. Barcellos-Paula, L.; De la Vega, I.; Gil-Lafuente, A.M. The Quintuple Helix of Innovation Model and the SDGs: Lat-in-American Countries' Case and Its Forgotten Effects. *Mathematics* **2021**, *9*, 416. [CrossRef]
75. Flores-Romero, M.B.; Pérez-Romero, M.E.; Álvarez-García, J.; del Río-Rama, M.d.l.C. Fuzzy Techniques Applied to the Analysis of the Causes and Effects of Tourism Competitiveness. *Mathematics* **2021**, *9*, 777. [CrossRef]

Article

A Comparative Ranking Model among Mexican Universities Using Pattern Recognition

Daniel Edahi Urueta [1,*], Pedro Lara [2], Miguel Ángel Gutiérrez [2], Sergio Gerardo de-los-Cobos [2], Eric Alfredo Rincón [3] and Román Anselmo Mora [3]

1. Posgrado en Ciencias y Tecnologías de la Información, Departamento de Ingeniería Eléctrica, Universidad Autónoma Metropolitana-Iztapalapa, Av. San Rafael Atlixco 186, Col. Vicentina, Del. Iztapalapa, Ciudad de México C.P. 09340, Mexico
2. Departamento de Ingeniería Eléctrica, Universidad Autónoma Metropolitana-Iztapalapa, Av. San Rafael Atlixco 186, Col. Vicentina, Del. Iztapalapa, Ciudad de México C.P. 09340, Mexico; plara@xanum.uam.mx (P.L.); gamma@xanum.uam.mx (M.Á.G.); cobos@xanum.uam.mx (S.G.d.-l.-C.)
3. Departamento de Sistemas, Universidad Autónoma Metropolitana-Azcapotzalco, Av. San Pablo 180, Colonia Reynosa Tamaulipas, Ciudad de México C.P. 02700, Mexico; rigaeral@correo.azc.uam.mx (E.A.R.); mgra@correo.azc.uam.mx (R.A.M.)
* Correspondence: crownirv@hotmail.com; Tel.: +52-77-7330-5448

Citation: Urueta, D.E.; Lara, P.; Gutiérrez, M.Á.; de-los-Cobos, S.G.; Rincón, E.A.; Mora, R.A. A Comparative Ranking Model among Mexican Universities Using Pattern Recognition. *Mathematics* **2021**, *9*, 1615. https://doi.org/10.3390/math9141615

Academic Editor: Jorge de Andres Sanchez

Received: 6 June 2021
Accepted: 5 July 2021
Published: 8 July 2021

Publisher's Note: MDPI stays neutral with regard to jurisdictional claims in published maps and institutional affiliations.

Copyright: © 2021 by the authors. Licensee MDPI, Basel, Switzerland. This article is an open access article distributed under the terms and conditions of the Creative Commons Attribution (CC BY) license (https://creativecommons.org/licenses/by/4.0/).

Abstract: The evaluation of quality in higher education is today a matter of great importance in most countries because the allocation of resources should be in accordance with the quality of universities. Due to this, there are numerous initiatives to create instruments and evaluation tools that can offer a quality comparison among institutions and countries, the results of these efforts used to be called international rankings. These rankings include some that are "reputational" or subjective, based on opinion polls applied to groups that, which is estimated, can issue authorized views. There are also "objective" rankings, based on performance indicators, which are calculated from a certain set of empirical data; however, on many occasions these indicators are sponsored by universities with the desire to appear among the best universities and emphasize some characteristics more than others, which makes them untrustworthy and very variable between each other. In this sense, we considered the Comparative Study of Mexican Universities (CSMU), a database of statistical information on education and research of Mexican higher education institutions, this database allows users to be responsible for establishing comparisons and relationships that may exist among existing information items, or building indicators based on their own needs and analysis perspectives (Márquez, 2010). This work develops an unsupervised alternative model of ranking among universities using pattern recognition, specifically clustering techniques, which are based on public access data. The results of the CSMU database are obtained by analyzing 60 universities as a first iteration, but to present the final results UNAM is excluded.

Keywords: university ranking; unsupervised pattern recognition; clustering techniques

1. Introduction

The quality of higher learning and research in a university is usually measured by prestige or by publicity, even opinion surveys directed at certain audiences are frequent: general [1], academic [2] or alumni and students [3]. These surveys show results that are consistent with each other, and on many occasions are based on the prestige of an institution, current advertising on social networks and classic mass media (radio, television and print). For this reason, it is common for them to have a bias from a government or a university that is using those studies to self-advertise. An ideal characteristic to avoid this type of bias is the use of unsupervised classification systems, which allow finding the "natural" groups of a set of items to classify them according to their inherent properties, because the groups formed are due to the closeness in their attributes.

All Mexican universities have characteristics which are comparable to each other, such as: full-time professors, number of members in the National System of Researchers (SNI) and number of articles published in journals in different international indexes, among others. In this article, a classification of the 60 largest universities in Mexico was made using the obtained information from the comparative study of Mexican universities carried out by the National Autonomous University of Mexico (UNAM), which is based on the collection, organization and analysis of information obtained from official sources and recognized databases (SEP, CONACYT, INDAUTOR, IMPI, WoS, Scopus, among others). This database is available at www.execum.unam.mx (accessed on 8 February 2019). This information was divided into two groups: higher learning and research. In the first one registered students (technical professional, bachelor's degree, specialty, master's degree and doctorate) were taken into account, as well as the study degree and type of contract that professors have. On the research side, papers in different indexed journals (SCI, Scopus, CONACYT journal quality index) and patents were also considered. Despite the fact that there is a work related to this database in literature [4], it is limited to analyzing a single year. For the present study the available data was used ranging from 2009 to 2017, this range allows to analyze what is the tendency of the Mexican university system, observing the transitions of some universities among different groups over the years.

In this study three well-known classification techniques: *k*-means, Gaussian mixture method (GMM), and spectral clustering were used to analyze the database. Likewise, principal component analysis (PCA) was used, which is a fast and flexible unsupervised method for reducing dimensionality in data [5].

Just as there are different opinion polls, where some emphasize which are the best elements, some others highlight which are the bad elements, in the same way the classifying algorithms will emphasize either the good or bad characteristics.

This article is divided as follows: Section 1 describes the considered classification techniques in this paper. In Section 2 the database and its attributes used are described. Section 3 describes the proposed matrix model (higher learning and research axis and generated sectors). Section 4 includes the application of the model to the case of 60 Mexican universities. Section 5 shows the results obtained and finally conclusions are presented.

2. Technical Classification by Clustering

Clustering algorithms are methods that divide a set of data into groups in such a way that members of the same group are more similar to each other than members of different groups [6].

2.1. k-means Algorithm

Given a data set, the objective of this algorithm is to *set k* groups to classify them, where *k* represents the number of groups previously specified by the analyst or by some method to select the ideal number of classes. When *k*-means classifies the objects, the objects within the same group are as similar as possible, while the objects in different groups are as different as possible; each group is represented by the center or middle of the data points that belong to the group [7]. The basic pseudocode is:

Begin
1. Randomly choose k cluster centers
2. While points stop changing assignment to centroids

    assign each data point to the nearest cluster center.
    Set the new cluster centroids based on the average (mean) position of each centroid point.
3. End While
End

Formally, let us consider that n observations must be partitioned in c groups. Let x_i and μ_c be the i-th observation, $1 \leq i \leq n$, and the mean of group $1 \leq c \leq k$, respectively.

The goal of k-means is to minimize the sum of the squared error over all groups denoted by $J(C)$; Thus, the objective function is stated as:

$$J(C) = \sum_{c=1}^{k} \sum_{x_i \in c} ||x_i - \mu_c||_2. \quad (1)$$

Minimizing this objective function is an NP-Hard problem, even for $k = 2$ [8]. Therefore, k-means, is a greedy algorithm, this means, that it builds up a solution choosing the best option at every step so it can be expected to converge to a local minimum. k-means starts with an initial partition with k groups and assigns observations to groups to reduce the squared error. Since the squared error tends to decrease with an increase in the number of k groups (with $J(C) = 0$ when $k = n$), it is minimized for a fixed number of groups [9].

A k-means algorithm requires some user-specified parameters such as the number of clusters: typically, k-means runs independently for different values of k and the partition that appears the most meaningful to the human expert is selected. Besides, different initializations can lead to different final clusters because k-means can only converge to local minima. Another user-specified parameter is the metric, while it is true that the most used metric for computing the distance between points and cluster centers is the Euclidean distance, which is why k-means is limited to linear cluster boundaries; however, some other metrics such as the Mahalanobis distance metric has been used to detect hyperellipsoidal clusters [10]. Moreover, it has a limitation about the number of observations, because k-means assumes that each group has roughly the same cardinality.

Due in k-means there is no assurance that it will lead to the global best solution, k-means run for multiple starting guesses, and it improves the result in each step. However, k-means is used because it is broadly easy and fast to code and implement it.

2.2. Gaussian Mixture Model

A Gaussian mixture model is a probabilistic model that assumes all the data points are generated from a mixture of a finite number of Gaussian distributions with unknown parameters. A GMM can be seen as a k-means generalization which incorporates information about the covariance structure of the data, as well as the centers of the latent Gaussians [11]. GMM attempts to find a mixture of multi-dimensional Gaussian probability distributions that best model any input dataset. The pseudocode is:

Begin
1. Choose starting guesses for the location and shape
2. While the convergence is not reached:

> For each point, find weights encoding the probability of membership in each cluster. For each cluster, update its location, normalization, and shape based on all data points, making use of the weights.

3. End While
End

Formally, let k and n be the number of clusters and the total number of observations, respectively. Let μ_c, Σ_c and π_c be the mean, covariance, and the mixing probability of cluster c, $1 \leq c \leq k$. For GMM, μ_c is the center of cluster c, Σ_c which represents its width and π_c defines how large or small the Gaussian function will be.

Then the probability that x_i, $1 \leq i \leq n$ is in the cluster c is given by:

$$\gamma_i^c = \frac{\pi_c \mathcal{N}(x_i \mid \mu_c, \Sigma_c)}{\sum_{c=1}^{k} \pi_c \mathcal{N}(x_i \mid \mu_c, \Sigma_c)} \quad (2)$$

where $\mathcal{N}(x \mid \mu, \Sigma)$ describes the multivariable Gaussian.

γ_i^c gives the probability that x_i is in cluster c, divided by the sum of the probabilities that x_i is in cluster c', for all $1 \leq c' \leq k$, so if x_i is very close to a Gaussian c, it will have high values of γ_i^c and relatively low values for any other case.

As a second step, for each cluster c: the total weight m_c is calculated (which can be considered as the fraction of points assigned to group c) and π_c, μ_c and Σ_c are updated using γ_i^c with:

$$m_c = \sum_{i=1}^{n} \gamma_i^c \tag{3}$$

$$\pi_c = \frac{m_c}{m} \tag{4}$$

$$\mu_c = \frac{1}{m_c} \sum_{i=1}^{n} \gamma_i^c x_i \tag{5}$$

$$\Sigma_c = \frac{1}{m_c} \sum_{i=1}^{n} \gamma_i^c (x_i - \mu_c)^T (x_i - \mu_c) \tag{6}$$

Finally, the first and second steps are repeated until convergence is reached [12]. The result of this is that each cluster is associated not with a hard-edged sphere, but with a smooth Gaussian model. Although GMM is categorized as a clustering algorithm, it is technically a generative probabilistic model describing the distribution of the data; due to this property, there are two important limitations with GMM: the first one is about its computation complexity because it is necessary to calculate the distributions, and whereby the algorithm can fail if the dimensionality of the problem is too high; the second limitation is that in many instances, the number of groups is unknown and it may be necessary to experiment with a number of different groups in order to find the most suitable.

2.3. Spectral Clustering

Spectral clustering is a technique whose goal is to cluster data that is connected, but not necessarily clustered within convex boundaries, so it has no limitations on the shape of data and can detect linearly non-separable patterns. The basic idea is to construct a weighted graph from the initial dataset where each node represents a pattern, and each weighted edge simply considers the similarity between two patterns [13]. In this context, this clustering problem can be seen as a graph cut problem, which can be tackled by means of the spectral graph theory. The core of this theory is the eigenvalue decomposition of the Laplacian matrix of the weighted graph obtained from data. The pseudocode is:

Begin
1. Compute A, the $n \times n$ affinity matrix
2. Get the eigensystem of A:

Compute the first k eigenvectors of its Laplacian matrix to define a feature vector for each object:
Set U = $n \times k$ matrix containing the normalized eigenvectors of the k largest eigenvalues of A in its columns

3. Apply k-means on the row space of U to find the k cluster
End

Formally, let n be the number of data points to be grouped and $W = [w_{i,j}]_{n \times n}$ the weight matrix where each $w_{i,j}$ is the similarity between x_i and x_j data points. So, a clustering problem can be formulated into the minimum cut problem, i.e.,

$$q^* = \arg \min_{q \in \{-1, 1\}^n} \sum_{i, j=1}^{n} w_{i, j} (q_i - q_j)^2 = q^T L q \tag{7}$$

where $q = (q_1, q_2, \ldots, q_n)$ is a vector for binary memberships and if we express a partition (A, B) as the vector q_i, each q_i can be 1 if $i \in A$ or -1 $i \in B$. L is the Laplacian matrix, defined as $L = D - W$, where $D = [d_{i,i}]_{n \times n}$ is a diagonal matrix with each element $d_{i,i} = \delta_{i,j} \sum_{j=1}^{n} w_{i,j}$.

For grouping into several classes, the objective function can be defined as:

$$J_{norm\_mc}(q) = \sum_{z=1}^{k} \sum_{z' \neq z} \frac{C_{z,z'}(q)}{D_z(q)} \quad (8)$$

where k is the number of clusters, $q \in \{1, 2, \ldots, k\}^n$, $C_{z,z'} = \sum_{i,j=1}^{n} \delta(q_i, z)\delta(q_j, z')w_{i,j}$ and $D_z = \sum_{i=1}^{n} \sum_{j=1}^{n} \delta(q_i, z)w_{i,j}$. However, efficiently finding the solution that minimizes the above equation is quite difficult. Therefore, a common strategy is to first get the smallest v eigen-vectors of the Laplacian matrix L (excluding the one with zero eigen-value), and project the data points in the low-dimensional space spanning the v eigen-vectors. Then, a standard clustering algorithm, such as k-means, is applied to the cluster data points in this low-dimensional space [14].

In short terms, spectral clustering is based on two main steps: first embedding the data points in a space in which clusters are more "obvious" (using the eigenvectors of a Gram matrix), and then applying a classical clustering algorithm such as k-means [15]. The affinity matrix M is formed using a kernel such as the Gaussian kernel. To obtain m clusters, the first m principal eigenvectors of M are computed, and k-means is applied on the unit-norm coordinates. So, if we consider a data set which consists of n data points, the time complexity of spectral clustering is $O(n^3)$, which makes it prohibitive for large-scale data application [16]. Moreover, there is evidence [13] that spectral clustering can be quite sensitive to changes in the similarity graph so noisy datasets can cause problems.

Choosing the number k of clusters is a general problem for all clustering algorithms, and just like k-means and GMM, it requires the number of clusters to be specified.

2.4. Principal Component Analysis

Principal component analysis (PCA) is a linear dimensionality reduction technique that can be used to extract information from a high-dimensional space by projecting it onto a lower-dimensional subspace. It tries to preserve the essential parts that have more variation of the data and eliminate the non-essential parts with less variation. It does this through a statistical procedure that uses an orthogonal transformation to convert a set of observations of possibly correlated variables (entities each of which takes on several numerical values) into a set of linearly uncorrelated variable values called principal components. In short, what the algorithm does is [5]:

- Standardize the input data (or normalize the variables).
- Get the eigenvectors and eigenvalues of the covariance matrix.
- Sort eigenvalues from high to low and choose d eigenvectors that correspond to d higher eigenvalues (where d is the dimensionality of the new features subspace).
- Construct the projection matrix W with the d eigenvectors selected.
- Transform the original X standardized database via W to obtain the new d-dimensional characteristics.

Thanks to the PCA you can get:

- A measure of how each variable is associated with the others (covariance matrix)
- The direction in which our data is scattered (eigenvectors)
- The relative importance of these different directions (eigenvalues).

In summary, PCA dimensionality reduction causes the least important attribute information to be removed, leaving only the data components with the highest variance, that is, the resulting data retains the maximum data variance. For this reason, although PCA is used to reduce the dimensionality in the data, may also be useful as a visualization tool, for filtering noise and for feature extraction.

2.5. Determine the Number of Clusters and Evaluate Clustering Performance: Silhouette Coefficient

Silhouette Coefficient or silhouette score is a cluster validity measure for evaluating clustering performance. To calculate the Silhouette score for each observation/data

point, the following distances need to be found out for each observation belonging to all the clusters.
- Mean distance between the observation and all other data points in the same cluster. This distance can also be called a mean intra-cluster distance. This mean distance is denoted by $a(i)$
- Mean distance between the observation and all other data points of the next nearest cluster. This distance can also be called a mean nearest-cluster distance. The mean distance is denoted by $b(i)$

Silhouette score, $s(i)$, for each sample is calculated using the following formula:

$$s(i) = \frac{b(i) - a(i)}{\max\{a(i) - b(i)\}} \quad (9)$$

Silhouette coefficient values ranges from -1 to 1. Silhouette coefficients near $+1$ indicate that clusters are well apart from each other and clearly distinguished. A value of 0 indicates that the clusters are indifferent, or we can say that the distance between clusters is not significant and negative values indicate that clustering configuration may have too many or too few clusters. Since silhouette coefficients are used to study the separation distance between the resulting clusters it is possible to use it to select the number of clusters for clustering techniques.

3. Materials & Methods

3.1. Comparative Study of Mexican Universities

The Comparative Study of Mexican Universities [17] is a research project developed by the General Directorate of Institutional Evaluation of the National Autonomous University of Mexico (UNAM) that systematizes, analyzes, and disseminates statistical series, compiled in official sources and recognized databases, which allow to contrast the development of Mexican universities in their substantive functions: higher learning, research, and dissemination of culture.

The CSMU is not a hierarchical classification (ranking) of Mexican higher education institutions but rather it is presented as an alternative to the existing rankings; because its objective is not to rate the universities or build regulations under certain assumptions about the quality or prestige of the institutions and their programs, in contrast, it seeks to provide items of information from public access sources, objective data that covers both the characteristics of institutions such as the substantive functions of university activities.

In this sense, the CSMU favors the presentation of raw data without the use of groupings or weightings, because this type of practice causes the results to always end up being questioned. These characteristics of the CSMU allow users to be responsible for establishing the comparisons and relationships that may exist among the different existing information items, or building indicators based on their own needs and analysis perspectives. Likewise, users are responsible for adapting their interpretations to the different characteristics that Mexican universities have among them [18].

The CSMU data in this study include 60 Mexican universities (45 public and 15 private) but the UNAM from 2009 to 2017. These universities concentrate more than 50 percent of Mexico's higher education enrollment. The database provides information on the following items:
- Teachers, tuition, and academic programs.
- Production of patents by Mexican institutions. It includes data on patents applied for and granted, according to the records of the Mexican Institute of Industrial Protection (MIIP).
- Participation of institutions in documents, articles and citations indexed in international bibliographic databases: ISI, Web of Knowledge, SciVerse, Scopus, etc.

- Participation of institutions in documents and articles indexed in the regional databases (Latin American citations in social sciences and humanities) and Periódica (index of Latin American journals in science).
- Academics of the institutions in the National System of Researchers (SNI) of the National Council of Science and Technology (CONACYT).
- Research journals indexed by Latindex (Latin American Index of Serial Scientific Publications) and the CONACYT Index.
- Academic bodies recognized in the National Program for the Improvement of Teachers (PROMEP), currently known as the Program for Teacher Professional Development (PRODEP) of the Ministry of Public Education (SEP).
- Postgraduate programs recognized in CONACYT's National Register of Quality Postgraduate Programs (PNPC).
- Higher education programs evaluated by the Inter-Institutional Committees for the Evaluation of Higher Education (CIEES) and programs accredited by agencies recognized by the Council for the Accreditation of Higher Education (COPAES).

The results of the study for each of these nine items are published on a dynamic web page with systematized information, which can be consulted through the Data Explorer of the Comparative Study of Mexican Universities.

3.2. Application Instance: 60 ExECUM Universities

The ExECUM database was split into two independent databases taking into account the factors of higher learning and research. The information contained into the higher learning independent database is as follows:

Teachers instructing

- Contract: Full time, 3/4-time, 1/2-time, hourly hired.
- Academic degree: Higher Technical University, bachelor's degree, specialty, master's degree, doctorate.

Number of graduated students.

- Level: Bachelor's degree, specialty, master's degree, doctorate.

Academic programs offered.

- Level: Bachelor's degree, specialty, master's degree, doctorate.

On the other hand, the information contained into the research independent database is described below:

SNI researchers

- Researchers: Candidate, level I, level II, level III

PROMEP academic bodies

- Consolidated, in consolidation, in formation.

ISI

- Articles: Institutional production, analysis by author, collaborators, citations.
- Documents: Institutional production, analysis by author, collaborators, citations.

SCOPUS

- Articles: Institutional production, analysis by author, collaborators, citations.
- Documents: Institutional production, analysis by author, collaborators, citations.

Patents

- Pending or granted journals.
- Latindex or CONACYT index

PNPC postgraduates

- Doctorate: International competence, consolidated, developing, newly created
- Master's Degree: International competence, consolidated, developing, newly created.

- Specialty: International competence, consolidated, developing, newly created

3.3. Proposed Matrix Model

Among many activities that take place within a university (management, dissemination of culture, sports activities, among others), the most important areas were the training of undergraduate and graduate students, as well as research. Only these two last items were taken into consideration for the proposed model. The available data referring to higher learning were used, such as: number of full-time or part-time teachers, maximum degree of studies, number of enrolled students, number of graduated students, and academic programs offered. While in research part, the number of research articles that are in different international indexes (JCR, ISI, Scopus, Latindex, Zentralblat Math, among others) can be considered, as well as the number of patents generated or citations in international journals.

Universities can be classified using the clustering strategies previously described and historical data. From available data it is possible to assign an order from highest to lowest; for example, considering the distance of the centroids with respect to the origin, the centroids closest to the origin imply a lower performance (fewer graduate students or fewer research articles generated).

Considering the dimensions already described, a matrix can be structured where the classification according to higher learning can be shown in the vertical axis and research in the horizontal axis, see Figure 1.

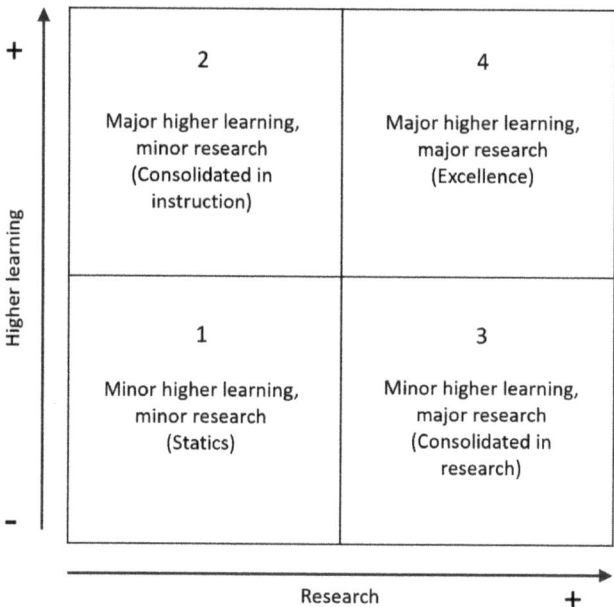

Figure 1. Graphic illustration of the matrix model proposed.

This model is divided into four classification quadrants: the first quadrant will contain static institutions, that is, with minor higher learning and minor research. The second quadrant will have consolidated institutions in higher learning; that is, those institutions with minor research and major higher learning. The third one will house consolidated research institutions, that is, with major research and minor higher learning. Finally, in the fourth quadrant will be the excellence institutions, this means that those universities on this site have the best results in both higher learning and research.

As mentioned above, and in order to locate the institutions, the original database was divided in two parts: part 1 corresponding to higher learning and part 2 corresponding to research. Each part was solved separately using the aforementioned clustering algorithms. In this way, the cases in Table 1 will be had for each clustering technique:

Table 1. Decision table used to locate an institution.

Does It Belong to the Outstanding Group in Higher Learning? (Part 1)	Does It Belong to the Outstanding Group in Research? (Part 2)	Quadrant Where It Will Be Located
No	No	1
Yes	No	2
No	Yes	3
Yes	Yes	4

Using the matrix in Figure 1, arrows will be used to show existing the institutions transitions among the quadrants, indicating at the top of each one the year in which they occurred; on the other hand, the highlighted institutions will be those that remained in the same group throughout the study.

Regarding evaluation, two different types of results were obtained: those that include the UNAM and those that do not include it, its presence represents an imbalance for the instances since this institution is quite far from the others in terms of size and, hence, in their higher learning and research capacity, whereby the distances between this institution and the others are shortened.

To demonstrate the above, first PCA analysis was applied on the databases from higher studies and research, and then, it was analysed how many dimensions are necessary to maintain the largest possible variance of both databases. The results of the PCA analysis based on higher studies database from the years 2009 to 2017 are shown in Figure 2:

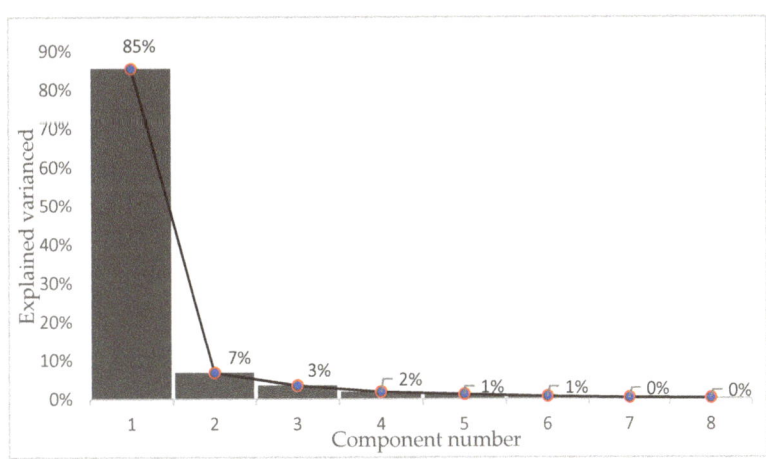

Figure 2. PCA analysis per component from 2009 to 2017 applied to higher learning database.

As it can be seen on Figure 2, only one component represents 85% of variance of the higher studies database. Similarly, an analysis using PCA considering research database is shown in Figure 3:

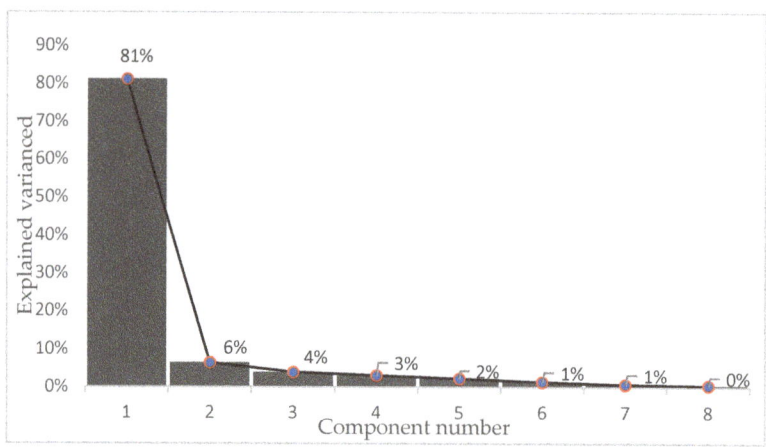

Figure 3. PCA analysis per component from 2009 to 2017 applied to research database.

In Figure 3 is shown that only one component represents around 81% of data variance of the research database. After PCA analysis and considering results in Figures 2 and 3, the graph that can be seen in Figure 4 was created; it might seem remarkable that the sum of the variation in this graph exceeds 100%. This is because, as commented in previous paragraphs, the CSMU database was separated into two databases corresponding to higher learning and research and then, reducing all the higher learning database dimensions to a single principal component that is projected on the ordered axis and reducing all the research database dimensions to a principal component which is projected on the abscissa axis; the graph was created maintaining a total data variance of 85% and 81% for every database respectively, for this reason the sum of both axes exceeds 100%.

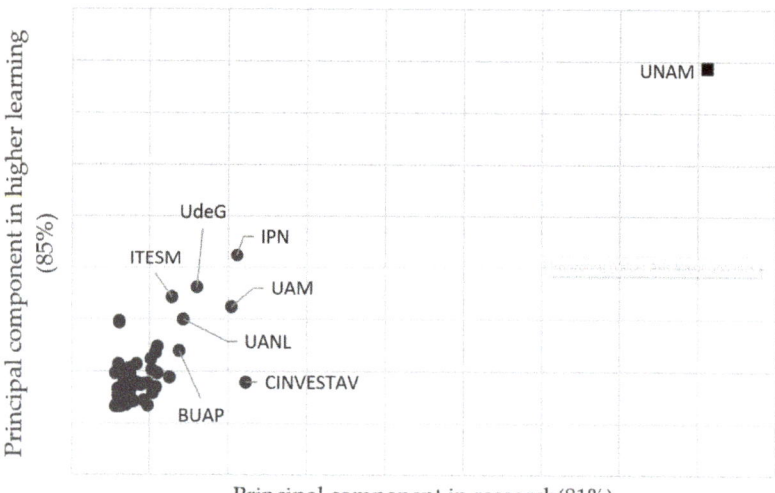

Figure 4. Average PCA from 2009 to 2017 applied to higher learning and research with UNAM.

As it can be seen in Figure 4, UNAM is far away from other institutions and it causes that all of them are seen into a single group; however, by eliminating UNAM, as it can be seen in the Figure 5, a separation among the institutions becomes clear. At first glance the

IPN, UAM and CINVESTAV appear to be the best institutions in research, whereas IPN and UdeG are the best universities in higher learning. It should be mentioned that the previous graphs are only representative of the total data in a certain percentage of the total available information, because, in the case of the component under research, PCA maintains a total data variance of 81% for research database, while in higher learning component, a total data variance of 85% for higher learning database is maintained.

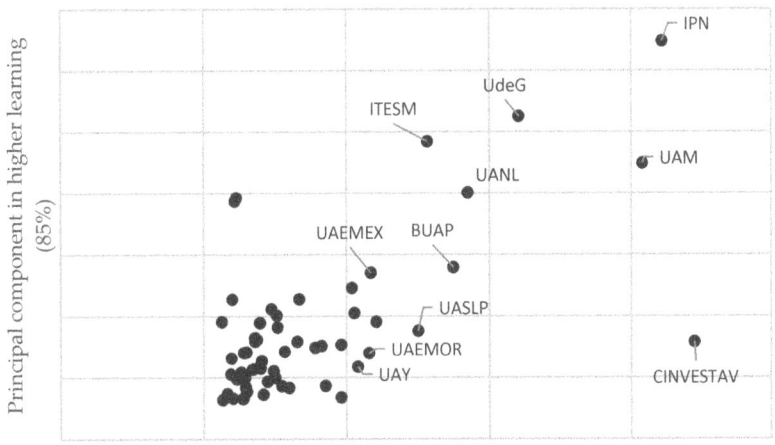

Figure 5. Average PCA from 2009 to 2017 applied to higher learning and research without UNAM.

After setting aside UNAM from the instance, the number of groups to cluster were determined using the silhouette coefficient method. It was applied over research and higher learning instances to determine the number of clusters. The comparisons for the three clustering techniques are on Figures 6 and 7.

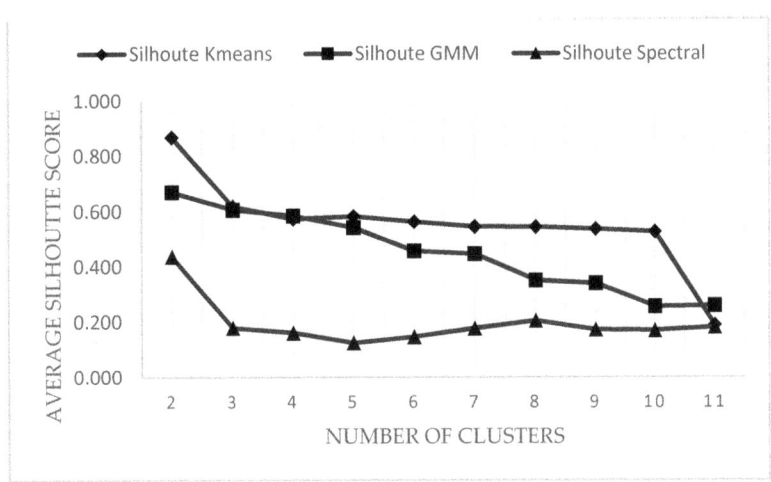

Figure 6. Silhouette coefficient comparisons for three clustering techniques solving research database.

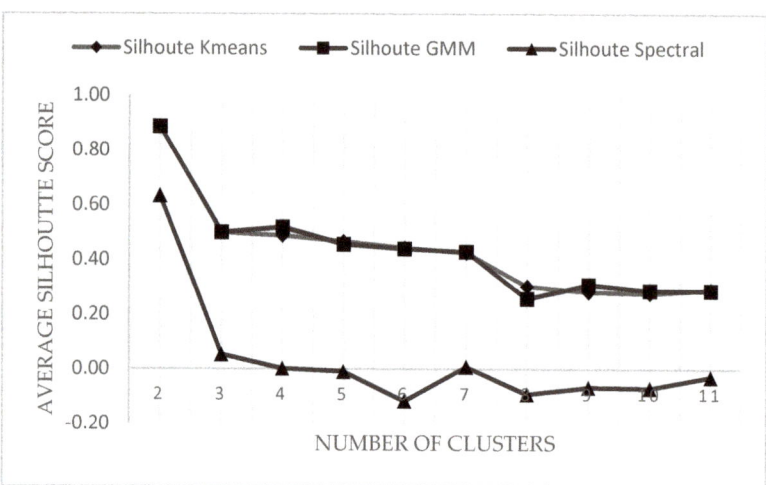

Figure 7. Silhouette coefficient comparisons for three clustering techniques solving higher learning database.

The results given by the silhouette coefficient method show in Figures 6 and 7 that best number to classify the instances is 2. Moreover, this analysis is helpful for evaluating clustering performance where the highest values mean a better performance.

4. Results and Analysis

4.1. k-means Results

This section may be divided by subheadings. It should provide a concise and precise description of the experimental results, their interpretation, as well as the experimental conclusions that can be drawn.

The result of this is that each cluster is associated not with a hard-edged sphere, but with a smooth Gaussian model. Although GMM is categorized as a clustering algorithm, it is technically a generative probabilistic model describing the data distribution; due to this property, there are two important limitations with GMM: the first one is about its computation complexity because it is necessary to calculate the distributions, and whereby the algorithm can fail if the dimensionality of the problem is too high; the second limitation is that in many instances, the number of groups is unknown and it may be necessary to experiment with a number of different groups in order to find the most suitable.

Table 2 shows the classification with *k*-means. Items not included belong to quadrant 1, which includes institutions that are not good in either of the two areas. It can be seen at first glance that there are institutions that remain there throughout time such as The National Polytechnic Institute (IPN), the Metropolitan Autonomous University (UAM) and the University of Guadalajara (UdeG); Likewise, in 2011 the Autonomous University of Nuevo León (UANL) entered this zone and, like the previous institutions, remained there until the last year analyzed. In addition to the UANL, there are two more institutions that follow the same trend, the Monterrey Institute of Technology and Higher Education (ITESM), which has been in this position since 2014, while the Meritorious Autonomous University of Puebla (BUAP) does so from 2015. During the last year of analysis, the Autonomous University of Mexico State (UAMEX) was added to the list. Other institutions can be seen that also vary their position during the analyzed period; these transitions can be observed in Figure 8, where the effort of the institutions to maintain or improve their status is evident throughout the study.

Table 2. Results from 2009 to 2017 applying k-means.

	\multicolumn{9}{c}{k-Means Results Summary}
	2009	2010	2011	2012	2013	2014	2015	2016	2017
Major higher learning, major research	IPN, UAM, UdeG	IPN, UAM, UdeG	IPN, UAM, UANL, UdeG	IPN, UAM, UANL, UdeG	IPN, UAM, UANL, UdeG	IPN, ITESM, UAM, UANL, UdeG	BUAP, IPN, ITESM, UAM, UANL, UdeG	BUAP, IPN, ITESM, UAM, UANL, UdeG	BUAP, IPN, ITESM, UAEMex, UAM, UANL, UdeG
Major higher learning, minor research	ITESM, UANL, UPN, UVM	ITESM, UANL, UPN, UVM	ITESM, UPN, UVM	BUAP, ITESM, LASALLE, UABC, UAEMex, UAS, UP, UPN, UV, UVM	ITESM, UABC, UAEMex, UAS, UPN, UV, UVM	UAEMex, UPN, UVM	UABC, UAEMex, UPN, UV, UVM	LASALLE, UABC, UAS, UP, UPN, UTM, UV, UVM	LASALLE, UABC, UAS, UPN, UTM, UV, UVM
Minor higher learning, major research	CINVESTAV	CINVESTAV	BUAP, CINVESTAV, UASLP	CINVESTAV	CINVESTAV, UASLP	BUAP, CINVESTAV, UASLP	CINVESTAV	CINVESTAV	UASLP, IBERO, CINVESTAV

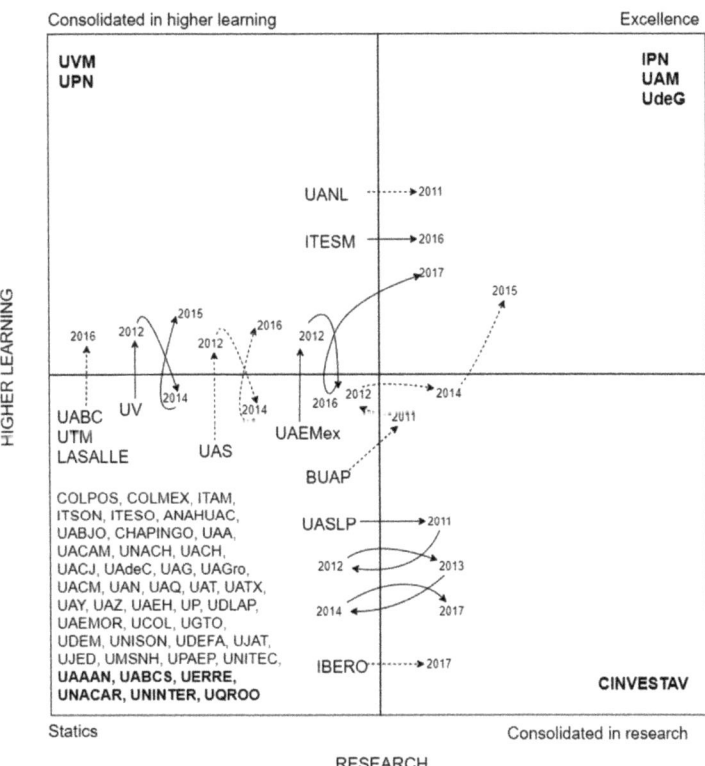

Figure 8. Visual representation of the results from 2009 to 2017 applying k-means.

4.1.1. Invariable Universities in k-Means Analysis

- Excellence

  The invariable universities consolidated in excellence are those that, throughout the analyzed period, that is, from 2009 to 2017, excel in both higher learning and research.

These are: The National Polytechnic Institute (IPN), the Metropolitan Autonomous University (UAM) and the University of Guadalajara (UdeG).
- Consolidated only in higher learning.

  The invariable universities consolidated in higher learning are: The University of the Mexican Valley (UVM) and the National Pedagogical University (UPN).
- Consolidated in research.

  The only institution considered invariably consolidated in research is the Center for Research and Advanced of the IPN (CINVESTAV).
- Static.

  According to k-means classification, approximately 70 percent of educational institutions fall into this category (41 of them). Their names are shown in the lower left of Figure 6.

4.1.2. Universities in Transition in k-Means Analysis

- Universities that have improved in higher learning:

  In 2016 La Salle University (LASALLE), the Autonomous University of Baja California (UABC) and the Technological University of Mexico (UNITEC) became part of the consolidated universities in higher learning.
  The Autonomous University of Sinaloa (UAS) shows a tendency to consolidate higher learning because it begins hovering between static universities at the beginning of the study, and although for 2012 it is consolidated into higher learning, in 2014 it returns to the place where it started; finally, in 2016 it was consolidated again in higher learning and remained on the site until the last year analyzed.
- Universities that have improved in research:

  During the first years of analysis, though the Autonomous University of San Luis Potosí (UASLP) is a static institution, in 2011 it was consolidated in research, the following year it was once again part of the static ones; and this behavior is repeated in the following two years until it remains from 2014 to 2016 as a static institution and during the last year analyzed it is consolidated again in research. This behavior of constant transitions makes evident their interest in consolidating in research.
  The Iberoamerican University (IBERO) remains among the static institutions for almost the entire period analyzed and it is not until the last year that this institution is consolidated in research.
- Universities that became of excellence:

  There are institutions such as the Autonomous University of Nuevo León (UANL) and the Monterrey Institute of Technology and Higher Education (ITESM) which begin the study as consolidated in higher learning, the study shows their interest in being of excellence, achieving their goal for the year 2011 and 2016, respectively.
  The Meritorious Autonomous University of Puebla (BUAP) went from being a static university to one with a better research quality in 2011; However, by 2012 it was once again part of the static institutions, where it remained until 2013. During 2014 its research quality improved again and for the following year it was able to be part of the excellence institutions, maintaining that position until the last year of analysis.
  The Autonomous University of Mexico State (UAMEX) is another institution that begins located among the static universities, but in 2012 it became consolidated in higher learning and only became static again during 2016, because for the following year it passes to be part of the excellence group.

4.2. GMM Results

Table 3 shows the classification with GMM and the elements not included belong to the static institutions, that is, minor in higher learning and research. It is evident that there are institutions that are maintained throughout the time such as: National Polytechnic Institute

(IPN), Autonomous Metropolitan University (UAM) in the case of institutions of excellence; the UPN and UVM for the consolidated institutions in higher learning and the CINVESTAV as the only consolidated institution in research. Likewise, other institutions that are in transition can be observed, as shown in Figure 9, where the efforts of the institutions to maintain or improve their status is evident.

Table 3. Results from 2009 to 2017 applying GMM.

	GMM Results Summary								
	2009	2010	2011	2012	2013	2014	2015	2016	2017
Major higher learning, major research	IPN, UAM, UdeG	IPN, UAM, UANL, UdeG	IPN, UAM, UANL, UdeG	IPN, UAM, UANL, UdeG	IPN, UAM	IPN, UAM	IPN, UAM	IPN, UAM, BUAP, ITESM, UAEMex, UANL, UdeG	IPN, UAM, BUAP, ITESM, UAEMex, UANL, UdeG
Major higher learning, minor research	ITESM, UANL, UPN, UVM	ITESM, UPN, UVM	ITESM, UPN, UVM	ITESM, UPN, UVM	BUAP, ITESM, UANL, UPN, UAEMex, UVM, UdeG	BUAP, ITESM, LASALLE, UABC, UAEMex, UANL, UAS, UPN, UV, UVM, UdeG	BUAP, ITESM, LASALLE, UABC, UAEMex, UANL, UAS, UPN, UV, UVM, UdeG	LASALLE, UABC, UAS, UNITEC, UP, UPN, UV, UVM	
Minor higher learning, major research	CINVESTAV	CINVESTAV, UAEMOR, UGTO, UV	CINVESTAV, UAEMOR, UGTO, UV	CINVESTAV, UAEMOR, UGTO, UV	CINVESTAV	CINVESTAV	CINVESTAV	CINVESTAV	CINVESTAV, IBERO, UASLP

It should be mentioned that the results of this classifier algorithm are quite similar to those of the k-means algorithm; however, perhaps the biggest difference is that the GMM appears to be more sensitive to increases and decreases in the databases.

4.2.1. Invariable Universities in GMM Analysis

- Excellence

  Two of the invariable universities consolidated in excellence by the GMM method coincide with those obtained by k-means.
  For this method, there are two invariable universities consolidated in excellence, these are: IPN and UAM.

- Consolidated in higher learning

  The results of the invariable universities consolidated in higher learning coincide with the results obtained by k-means since they only have UVM and NUP.

- Consolidated in research

  As with the k-means, the only institution regarded as invariable consolidated research is the CINVESTAV.

- Static

  Using GMM, the list of institutions is rather similar to the one provided by k-means; the only changes are UAMOR, UGTO, UP and UNITEC. The first two leaves the first quadrant in 2010 and return in 2013; while the last two leave in 2016 to return in the following year.

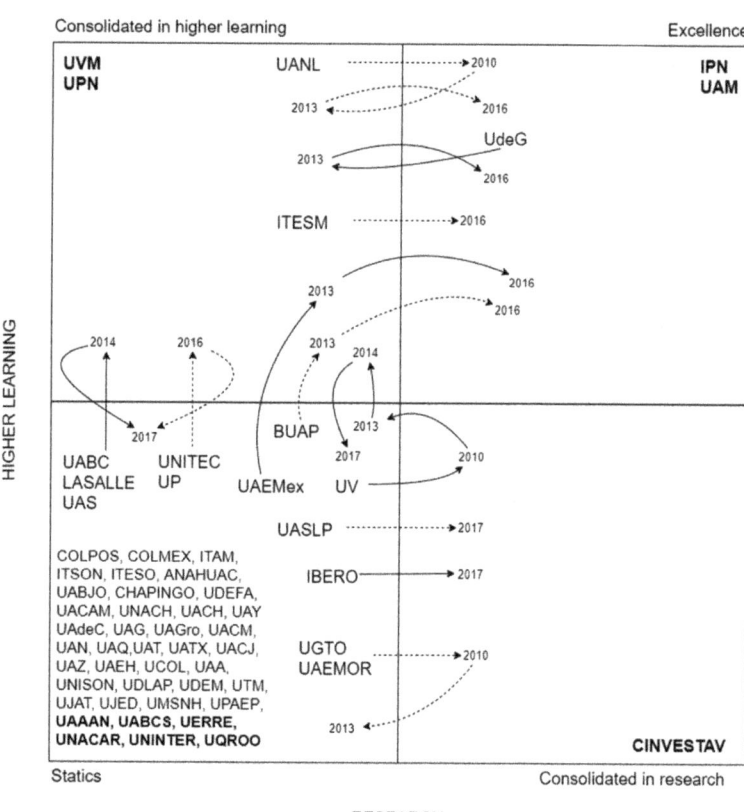

Figure 9. Visual representation of the results from 2009 to 2017 applying GMM.

4.2.2. Universities in Transition in GMM Analysis

- Universities that have improved in research.

    In this analysis, the IBERO and the UASLP begin being static and during 2017 they were consolidated in research.

- Universities that became of excellence:

    Even though UdeG begins as part of the universities of excellence, it does not remain unchanged in its position, because during the period from 2013 to 2015, it is located with the consolidated institutions in higher learning, and it is until 2016 when it finally returns to be part of the excellence institutions.

    The ITESM begins as a consolidated university in higher learning, but its interest focuses on being part of the group of excellence; thus, by 2016 it achieves its goal and becomes a fourth quadrant institution.

    The UANL is a university with a behavior that shows its interest in being part of the g excellence group because it begins being consolidated in higher learning, for 2010 it becomes of excellence and although for the following year until 2015 it returns to the group where it started, in 2016 it is once again part of the group of excellence.

    The BUAP and UAMEX are institutions that, despite starting out as static, focused first on consolidating themselves in higher learning, a group to which they belonged from 2013 to 2015, and then gave the highest to the universities of the fourth quadrant in 2016.

- Universities that became static:

The UABC, LASALLE and the UAS are institutions that begin as part of the static group, but there is an interest in consolidating themselves in higher learning, and although they achieve their goal in the period from 2013 to 2016; finally, in the last year of analysis, these institutions became static again.

The UNITEC and UP are in a similar case, with the only difference that the period in which they are consolidated into higher learning corresponds only to 2016 and return to the static group in 2017.

The University of Guanajuato (UGTO) and the Autonomous University of Morelos State (UAMOR) are institutions that are initially part of the static group and are consolidated in research, in the period that corresponds from 2010 to 2012; after this period, they return to be static institutions until the last year analyzed.

The Veracruzana University (UV) is characterized by its constant transitions; in 2010 it went from being a static university to a consolidated one in research, in 2013 it returned to its starting point; the following year it consolidated its higher learning position, a place where it remained until 2016 and during 2017 it became static again.

4.3. Spectral Clustering Results

Regarding to the results obtained by the *k*-means and GMM algorithms, these have a certain relationship, since it turns out that both are consistent with each other; however, the results of the spectral grouping (see Table 4 and Figure 10) are complementary to the two techniques already mentioned and analyzed, this is because the spectral grouping emphasizes the changes in the less favored universities; where it is shown that some universities in the first quadrant want to improve, either in higher learning or in research, although their efforts are more modest.

4.3.1. Universities in Transition in Spectral Clustering Analysis

Regarding the place where the institutions with good results in research and not so favorable results in higher learning reside, there are the College of Postgraduates (COLPOS) and Autonomous Technological Institute of Mexico (ITAM), which appear from 2009 to 2015 and in 2017; the College of México (COLMEX) that appeared from 2009 to 2012 and from 2014 to 2015; the Autonomous University of Campeche (UACAM) that is presented from 2009 to 2011 and in 2015; the UACM from 2009 to 2011 and from 2014 to 2015; the University of the Americas Puebla (UDLAP) from 2009 to 2012 and in 2014 and Chapingo Autonomous University (CHAPINGO) from 2010 to 2011, from 2014 to 2015 and in 2017. Likewise, the University of Colima (UCOL) only appears twice, in 2015 and 2017 and at the Autonomous University of Yucatán (UAY) only entered in 2015.

A particularity of this area occurs in 2016, when the group with minor higher learning and major research had no members.

For the sector with the most outstanding achievements in higher learning rather than in research, the institutions UPN, UVM, LASALLE, Autonomous University Benito Juarez of Oaxaca (UABJO), Autonomous University of Guadalajara (UAG), Autonomous University of Coahuila (UAdeC), Autonomous University of Chiapas (UANCH) and UNITEC get this place every year, Anahuac University (ANAHUAC) does from 2009 to 2013 and from 2015 to 2016; the University of the Mexican Army and Air force (UDEFA) and the Juarez Autonomous University of Tabasco (UJAT) shown in this place in the years 2009 to 2014 and 2016; the Technological Institute of Sonora (ITSON) from 2009 to 2014; the Autonomous University of Chihuahua (UACH) in the years 2009 to 2013; the Juarez University of Durango State (UJED) from 2009 to 2014 and from 2016 to 2017. The Popular Autonomous University of Puebla State (UPAEP) in the corresponding years from 2010 to 2017; the Autonomous University of Nayarit (UAN) and the UP in the years 2010 to 2014 and 2016 to 2017; the University of Monterrey (UDEM) in the years 2013 to 2017 and the Autonomous University of Tlaxcala (UATX) in the years 2013 to 2014 and 2016 to 2017. There are some other institutions which were not considered because they were not repeated more than three times in the entire period from 2009 to 2017.

Table 4. Results from 2009 to 2017 applying Spectral clustering.

	Spectral Clustering Results Summary								
	2009	2010	2011	2012	2013	2014	2015	2016	2017
Major higher learning, major research	ANAHUAC, ITSON, LASALLE, UAA, UABJO, UACH, UAG, UAdeC, UDEFA, UJAT, UJED, UNACH, UNITEC, UP, UPN, UVM	ANAHUAC, ITSON, LASALLE, UABJO, UACH, UAG, UAN, UAdeC, UDEFA, UJAT, UJED, UNACH, UNITEC, UP, UPAEP, UPN, UVM	ANAHUAC, ITSON, LASALLE, UABJO, UACH, UAG, UAN, UAdeC, UDEFA, UJAT, UJED, UNACH, UNITEC, UP, UPAEP, UPN, UVM	ANAHUAC, ITSON, LASALLE, UAA, UABJO, UACH, UAG, UAGro, UAN, UATx, UAZ, UAdeC, UDEFA, UDEM, UJAT, UJED, UNACH, UNITEC, UP, UPAEP, UPN, UVM	ANAHUAC, IBERO, ITSON, LASALLE, UAA, UABJO, UACH, UAEH, UAG, UAGro, UAN, UAS, UAT, UATx, UAZ, UAdeC, UCOL, UDEFA, UDEM, UJAT, UJED, UNACH, UNITEC, UP, UPAEP, UPN, UVM	ITSON, LASALLE, UAA, UABJO, UAG, UAGro, UAN, UAT, UATx, UAdeC, UDEFA, UDEM, UJAT, UJED, UNACH, UNITEC, UP, UPAEP, UPN, UVM	ANAHUAC, LASALLE, UABJO, UAG, UAdeC, UDEM, UNACH, UNITEC, UPAEP, UPN, UVM	ANAHUAC, LASALLE, UABJO, UACH, UACJ, UAG, UAN, UAT, UATx, UAY, UAZ, UAdeC, UDEFA, UDEM, UJAT, UJED, UNACH, UNITEC, UP, UPAEP, UPN, UVM	ITESO, LASALLE, UAA, UABJO, UACJ, UAG, UAN, UATx, UAdeC, UDEM, UJED, UNACH, UNITEC, UP, UPAEP, UPN, UVM

Table 4. *Cont.*

	Spectral Clustering Results Summary							
Minor higher learning, major research	COLMEX, COLPOS, ITAM, UACAM, UACM, UDLAP	CHAPINGO, COLMEX, COLPOS, ITAM, UACAM, UACM, UDLAP	CHAPINGO, COLMEX, COLPOS, ITAM, UACAM, UACM, UDLAP	COLMEX, COLPOS, ITAM, UDLAP	COLPOS, ITAM	CHAPINGO, COLMEX, COLPOS, ITAM, UACM, UDLAP	CHAPINGO, COLMEX, COLPOS, ITAM, UACAM, UAY, UCOL	CHAPINGO, COLPOS, UTM, ITAM, UCOL
Minor higher learning, minor research	CHAPINGO, ITESO, UAAAN, UABCS, UAN, UATx, UDEM, UERRE, UNACAR, UNINTER, UQROO	ITESO, UAAAN, UABCS, UATx, UTM, UDEM, UERRE, UNACAR, UNINTER, UQROO	ITESO, UAAAN, UABCS, UATx, UTM, UDEM, UERRE, UNACAR, UNINTER, UQROO	CHAPINGO, COLMEX, ITESO, UAAAN, UABCS, UTM, UACAM, UACM, UDLAP, UERRE, UNACAR, UNINTER, UQROO	ITESO, UAAAN, UABCS, UTM, UACAM, UERRE, UNACAR, UNINTER, UQROO	CHAPINGO, COLMEX, COLPOS, ITAM, ITESO, ITSON, UAA, UAAAN, UABCS, UTM, UACAM, UAGro, UAN, UATx, UDEFA, UDLAP, UERRE, UJED, UNACAR, UNINTER, UQROO	CHAPINGO, COLMEX, COLPOS, ITAM, ITESO, ITSON, UTM, UAA, UAGro, UAAAN, UABCS, UACAM, UCOL, UDLAP, UERRE, UNACAR, UNINTER, UQROO	COLMEX, ITSON, UAAAN, UABCS, UACAM, UACM, UDEFA, UDLAP, UERRE, UNACAR, UNINTER, UQROO

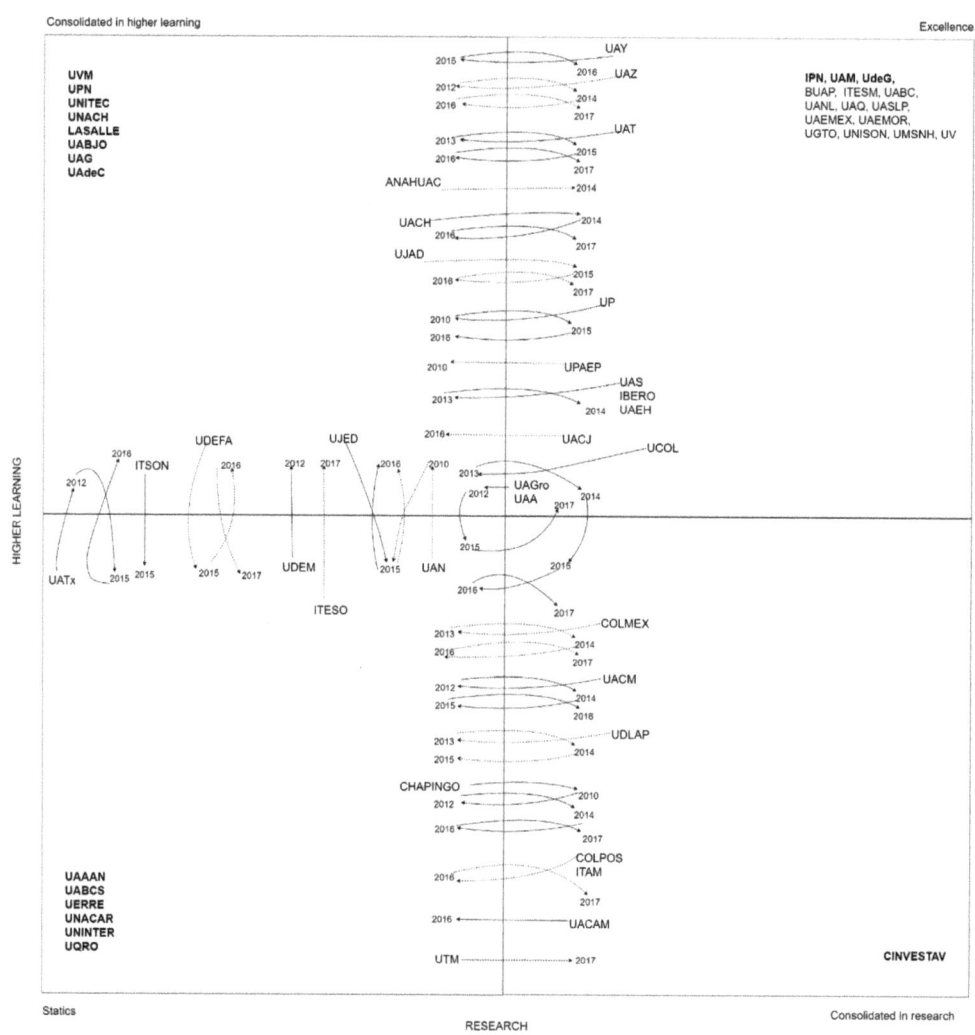

Figure 10. Visual representation of the results from 2009 to 2017 applying Spectral clustering.

4.3.2. Static Universities in Spectral Clustering Analysis

Finally, the results obtained by the spectral classifier with respect to disadvantaged institutions both in research and higher learning aspects. show that the Autonomous Agrarian University Antonio Narro (UAAAN), Autonomous University of Baja California Sur (UABCS), Regiomontana University (UERRE), Autonomous University del Carmen (UNACAR), UNINTER and the University of Querétaro (UQROO) have this site all years; Western Institute of Technology and Higher Studies (ITESO) does so in the period from 2009 to 2016; CHAPINGO in the years 2009, 2012, 2013 and 2016; UATX and UDEM in the period from 2009 to 2011, but reappeared in 2015 and in 2013. The UACAM from 2012 to 2017 and the UDLAP from 2013 and during the period from 2015 to 2017. Although there are other institutions that occupy this place, these were not mentioned because their presence was not repeated for at least four years.

5. Conclusions

The analysis of the data with k-means shows congruent results year after year, this is demonstrated with the institutions of excellence that remain unchanged, such as: National Polytechnic Institute (IPN), the Metropolitan Autonomous University (UAM) and the University of Guadalajara (UdeG). The group of excellence was joined by the Autonomous University of Nuevo León (UANL) in 2011, the Monterrey Institute of Technology and Higher Education (ITESM) in 2014, the Meritorious Autonomous University of Puebla (BUAP) in 2015 and the Autonomous University of Mexico State (UAMEX) in 2017.

The k-means results also indicate that the IPN Center for Research and Advanced Studies (CINVESTAV) remains unchanged within the group of consolidated research institutions; BUAP enters this group, although intermittently in 2011 and 2014; as well as the Autonomous University of San Luis Potosí (UASLP) in the years 2011, 2013, 2014 and 2017. Also, for the last year analyzed, the Iberoamerican University (IBERO) enters the group.

Regarding to the group of institutions consolidated in higher learning, the invariable members are: The National Pedagogical University (UPN) and the University of the Mexican Valley (UVM). The Autonomous University of Baja California (UABC), La Salle University (LASALLE), the Technological University of Mexico (UNITEC) and the Autonomous University of Sinaloa (UAS) were added to it in 2016.

As with k-means, the results with GMM also show to be consistent from year to year. for example, in the institutions of excellence that remain unchanged there are two: the IPN and the UAM. This group is joined by the UdeG that occupies this position from 2009 to 2012 and from 2016 to 2017; the UANL which appears in the years 2010 to 2012 and in 2016 and 2017; as well as the ITESM, the BUAP and the UAMEX who enter during the year 2016.

Regarding the group of consolidated research institutions, CINVESTAV is an invariable member; IBERO and UASLP, which entered in the last year of analysis; while the Autonomous University of Morelos State (UAMOR), the University of Guanajuato (UGTO) and the UV only occupy this place from 2010 to 2012 and do not occupy it again in subsequent years.

In contrast, the consolidated institutions in higher learning that remain unchanged are: the UPN and the UVM and although there are also other institutions in the group, they have occupied this place only in certain periods; Such is the case of the Autonomous University of Sinaloa (UAS) in conjunction with LASALLE and the UABC, which are integrated in the period from 2014 to 2016; BUAP and UAMEX only in 2013 and UNITEC with UP, who only joined in 2016.

Just as k-means and Gauss highlight which are the best institutions, the spectral grouping highlights which are the institutions with the lowest levels of higher learning and research. That is why, despite showing grouped results in a different way, this technique also shows congruent results in the same way as k-means and GMM.

For spectral grouping, in the institutions with good higher learning results, but few research results, there are La Salle University (LASALLE), the Autonomous University Benito Juarez of Oaxaca (UABJO), and the Autonomous University of Guadalajara (UAG), the Autonomous University of Coahuila (UAdeC), the Autonomous University of Chiapas (UANCH), the Technological University of Mexico (UNITEC), the UP and the UVM as permanent members throughout the analyzed period; other institutions such as the Technological Institute of Sonora (ITSON), the Autonomous University of Aguascalientes (UAA), the Autonomous University of Chihuahua (UACH), the University of the Mexican Army and Air force (UDEFA), the Juarez Autonomous University of Tabasco (UJAT), Juarez University of Durango State (UJED), Autonomous University of Nayarit (UAN), Popular Autonomous University of Puebla State (UPAEP), Autonomous University of Tlaxcala (UATX), University of Monterrey (UDEM) they have a consistent presence, although not continuous within the group, in addition to these there are others that due to their small number of appearances within the group were not mentioned.

In the case of institutions with less higher learning and more research, there is none that appears uninterruptedly within the group, because in 2016, the group was empty; However, the College of Postgraduates (COLPOS) and the Autonomous Technological Institute of Mexico (ITAM), are constant members of this group except for the mentioned year; other institutions that also appear frequently within this group are the College of México (COLMEX), the Chapingo University (CHAPINGO), the Autonomous University of Mexico City (UACM) and the University of the Americas Puebla (UDLAP).

Finally, in the least prominent group in higher learning and research, there are the Autonomous Agrarian University Antonio Narro (UAAAN), the Autonomous University of Baja California Sur (UABCS), the Regiomontana University (UERRE), the Autonomous University del Carmen (UNACAR), the International University (UNINTER) and the University of Quintana Roo (UQROO) as permanent members, on the other hand, CHAPINGO, Western Institute of Technology and Higher Studies (ITESO), the UATX, although they are not permanent participants, they appear with some frequency; that is, they have more than four appearances in the group.

The results found by k-means and GMM show the constant effort of the institutions to consolidate themselves in higher learning, research or being of excellence; Despite the fact that k-means and GMM are techniques with different approaches, both highlight the best institutions, this is the reason why the results obtained between both are consistent; for example, in the case of institutions of excellence, the IPN and the UAM are constant in both analyzes, as is the CINVESTAV for consolidated research institutions. As for the consolidated institutions in higher learning and static, the universities that remain unchanged are exactly the same in both techniques, these are: UVM and UPN for the first group and CINVESTAV for the second.

The results provided by the spectral grouping emphasize the institutions and their constant effort to improve or maintain their status, whether in aspects of higher learning, research or both subject to the resources they possess.

In contrast to all the results shown, it must be said that the National Autonomous University of Mexico (UNAM) is in all years the best institution in both higher learning and research.

Author Contributions: Conceptualization, S.G.d.-l.-C.; Formal analysis, E.A.R.; Funding acquisition, R.A.M.; Investigation, D.E.U.; Project administration, M.Á.G.; Software, D.E.U.; Supervision, P.L.; Validation, D.E.U. and S.G.d.-l.-C.; Writing—original draft, D.E.U.; Writing—review & editing, P.L., M.Á.G. and E.A.R. All authors have read and agreed to the published version of the manuscript.

Funding: This research received no external funding.

Institutional Review Board Statement: Not applicable.

Informed Consent Statement: Not applicable.

Data Availability Statement: Estudio Comparativo de las Universidades Mexicanas—Explorador de datos (ExECUM). Is available at http://www.execum.unam.mx/ (accessed on 5 January 2020). ExECUM 2009–2017. Is available at https://github.com/crownirv/execum (accessed on 2 July 2021).

Conflicts of Interest: The authors declare no conflict of interest.

Abbreviations

ANAHUAC	Anahuac University (Sistema Universidad Anahuac)
BUAP	Meritorious Autonomous University of Puebla (Benemérita Universidad Autónoma de Puebla)
CINVESTAV	IPN Center for Research and Advanced Studies (Centro De Investigación y de Estudios Avanzados del IPN)
COLMEX	The School of Mexico (El Colegio De México)
COLPOS	Postgraduate College (Colegio de Posgraduados)
CHAPINGO	Chapingo Autonomous University (Universidad Autónoma Chapingo)
IBERO	Iberoamerican University System (Sistema Universidad Iberoamericana)
IPN	National Polytechnic Institute (Instituto Politécnico Nacional)
ITAM	Autonomous Technological Institute of Mexico (Instituto Tecnológico Autónomo De México)
ITESM	Monterrey Institute of Technology and Higher Education (Sistema Instituto Tecnológico y de Estudios Superiores de Monterrey)
ITESO	Estudios Superiores de Occidente
ITSON	Technological Institute of Sonora (Instituto Tecnológico de Sonora)
LASALLE	La Salle University (Sistema Universidad La Salle, AC)
UAA	Autonomous University of Aguascalientes (Universidad Autónoma de Aguascalientes)
UAAAN	Autonomous Agrarian University Antonio Narro (Universidad Autónoma Agraria Antonio Narro)
UABC	Autonomous University of Baja California (Universidad Autónoma de Baja California)
UABCS	Autonomous University of Baja California Sur (Universidad Autónoma de Baja California Sur)
UABJO	Autonomous University Benito Juarez of Oaxaca (Universidad Autónoma Benito Juárez de Oaxaca)
UACAM	Autonomous University of Campeche (Universidad Autónoma de Campeche)
UACJ	Autonomous University of Juarez City (Universidad Autónoma de Ciudad Juárez)
UACM	Autonomous University of Mexico City (Universidad Autónoma de la Ciudad de México)
UACH	Autonomous University of Chihuahua (Universidad Autónoma de Chihuahua)
UAdeC	Autonomous University of Coahuila (Universidad Autónoma de Coahuila)
UAEH	Autonomous University of Hidalgo State (Universidad Autónoma del Estado de Hidalgo)
UAG	Autonomous University of Guadalajara (Universidad Autónoma de Guadalajara)
UAGRO	Autonomous University of Guerrero (Universidad Autónoma de Guerrero)
UAM	Metropolitan Autonomous University (Universidad Autónoma Metropolitana)
UAEMEX	Autonomous University of Mexico State (Universidad Autónoma del Estado de México)
UAEMOR	Autonomous University of Morelos State (Universidad Autónoma del Estado de Morelos)
UAN	Autonomous University of Nayarit (Universidad Autónoma de Nayarit)
UANL	Autonomous University of Nuevo Leon (Universidad Autónoma de Nuevo León)
UAQ	Autonomous University of Queretaro (Universidad Autónoma de Querétaro)
UAS	Autonomous University of Sinaloa (Universidad Autónoma de Sinaloa)
UASLP	Autonomous University of San Luis Potosi (Universidad Autónoma de San Luis Potosí)
UAT	Autonomous University of Tamaulipas (Universidad Autónoma de Tamaulipas)
UATX	Autonomous University of Tlaxcala (Universidad Autónoma de Tlaxcala)
UAY	Autonomous University of Yucatan (Universidad Autónoma de Yucatán)
UAZ	Autonomous University of Zacatecas (Universidad Autónoma de Zacatecas)
UCOL	University of Colima (Universidad de Colima)
UDEFA	University of The Mexican Army and Air force (Universidad del Ejército y Fuerza Aérea Mexicana)
UdeG	University of Guadalajara (Universidad de Guadalajara)

UDEM	University of Monterrey (Universidad de Monterrey)
UDLAP	University of The Americas Puebla (Universidad de Las Américas Puebla, AC)
UERRE	Regiomontana University (Universidad Regiomontana, AC)
UGTO	University of Guanajuato (Universidad de Guanajuato)
UIC	Intercontinental University (Universidad Intercontinental)
UJAT	Juarez Autonomous University of Tabasco (Universidad Juárez Autónoma de Tabasco)
UJED	Juarez University of Durango State (Universidad Juárez del Estado de Durango)
UMSNH	Michoacana University of San Nicolas from Hidalgo (Universidad Michoacana de San Nicolás de Hidalgo)
UN	Naval University (Universidad Naval)
UNACAR	Autonomous University Del Carmen (Universidad Autónoma del Carmen)
UNACH	Autonomous University of Chiapas (Universidad Autónoma de Chiapas)
UNAM	National Autonomous University of Mexico (Universidad Nacional Autónoma de México)
UNISON	University of Sonora (Universidad de Sonora)
UNITEC	Technological University of Mexico (Universidad Tecnológica de México)
UP	Panamerican University (Universidad Panamericana)
UPAEP	Popular Autonomous University of Puebla State (Universidad Popular Autónoma del Estado de Puebla)
UPN	National Pedagogical University (Universidad Pedagógica Nacional)
UQROO	University of Quintana Roo (Universidad de Quintana Roo)
UTM	Technological University of La Mixteca (Universidad Tecnológica de la Mixteca)
UV	Veracruz University (Universidad Veracruzana)
UVM	University of The Mexican Valley (Sistema Universidad del Valle de México)

References

1. Palma, E. Percepción y Valoración de la Calidad Educativa de Alumnos y Padres en 14 Centros Escolares de la Región Metropolitana de Santiago de Chile. *REICE Rev. Iberoam. Sobre Calid. Efic. Cambio Educ.* **2016**, *6*, 1. Available online: http://www.redalyc.org/articulo.oa?id=55160106 (accessed on 5 January 2020).
2. Jiménez Galán, M.; Hernández Jaime, M.; Ortega Pacheco, M. ¿Forman los programas de formación docente? *Rev. Investig. Educ.* **2014**, *19*, 1–27. [CrossRef]
3. Espinosa, E.M.; Gutiérrez, F.C.; Muñoz, V.M.R. Estudiantes Frente al Espejo: Percepciones de la Calidad Educativa en Programas de Licenciatura y Posgrado en México. Universidad de Guadalajara. 2015. Available online: https://www.sinectica.iteso.mx/index.php?cur=38&art=38_07 (accessed on 7 January 2020).
4. Montes, E.; Mora, R.A.; Obregón, B.; de-los-Cobos, S.G.; Rincón, E.A.; Lara, P.; Gutiérrez, M.Á. Mexican University Ranking Based on Maximal Clique. In *Educational Networking*; Peña-Ayala, A., Ed.; Springer International Publishing: Berlin/Heidelberg, Germany, 2020; pp. 327–395. [CrossRef]
5. Van der Plas, J. *Python Data Science Handbook: Essential Tools for Working with Data*; O'Reilly Media, Inc.: Sebastopol, CA, USA, 2016; ISBN 9781491912058.
6. Ripley, B.D. *Pattern Recognition and Neural Networks*; Cambridge University Press: New York, NY, USA, 2007. [CrossRef]
7. MacQueen, J.B. Some Methods for Classification and Analysis of Multivariate Observations. In Proceedings of the 5th Berkeley Symposium on Mathematical Statistics and Probability, Berkeley, CA, USA, 27 December 1965–7 January 1966; University of California Press: Berkeley, CA, USA, 1967; Volume 1, pp. 281–297. Available online: https://projecteuclid.org/euclid.bsmsp/1200512992 (accessed on 7 January 2020).
8. Drineas, P.; Frieze, A.M.; Kannan, R.; Vempala, S.; Vinay, V. Clustering in Large Graphs and Matrices. In Proceedings of the SODA'99: Proceedings of the Tenth Annual ACM-SIAM Symposium on Discrete Algorithms, Baltimor, MA, USA, 17–19 January 1999. Available online: https://dl.acm.org/doi/10.5555/314500.314576 (accessed on 7 January 2020).
9. Jain, A.K.; Dubes, R.C. *Algorithms for Clustering Data*; Prentice-Hall, Inc.: Upper Saddle River, NJ, USA, 1988. Available online: https://dl.acm.org/doi/book/10.5555/46712 (accessed on 7 January 2020).
10. Mao, J.; Jain, A.K. A self-organizing network for hyperellipsoidal clustering (HEC). *IEEE Trans. Neural Netw.* **1996**, *7*, 16–29. [CrossRef] [PubMed]
11. Bilmes, J.A. *A Gentle Tutorial of the EM Algorithm and Its Application to Parameter Estimation for Gaussian Mixture and Hidden Markov Models*; International Computer Science Institute: Berkeley, CA, USA, 1998.

12. Bishop, P. *Pattern Recognition and Machine Learning*; Springer: Berlin/Heidelberg, Germany, 2006; p. 430. Available online: https://search.ebscohost.com/login.aspx?direct=true&db=cat07429a&AN=ulpgc.547268&lang=es&site=eds-live&scope=site (accessed on 6 January 2020).
13. Von Luxburg, U. A tutorial on spectral clustering. *Stat. Comput.* **2007**, *17*, 395–416. [CrossRef]
14. Jin, R.; Kang, F.; Ding, C.H. A Probabilistic Approach for Optimizing Spectral Clustering. In *Advances in Neural Information Processing Systems*; The MIT Press: Cambridge, MA, USA, 2006; pp. 571–578. Available online: http://papers.nips.cc/paper/2952-a-probabilistic-approach-for-optimizing-spectral-clustering.pdf (accessed on 5 January 2020).
15. Ng, A.; Jordan, M.; Weiss, Y. On spectral clustering: Analysis and an algorithm. *Adv. Neural Inf. Process. Syst.* **2001**, *14*, 849–856.
16. Yin, W.; Zhu, E.; Zhu, X.; Yin, J. Landmark-Based Spectral Clustering with Local Similarity Representation. In *Theoretical Computer Science*; Du, D., Li, L., Zhu, E., He, K., Eds.; Springer: Singapore, 2017; Volume 768, pp. 198–207. [CrossRef]
17. Estudio Comparativo de las Universidades Mexicanas—Explorador de datos (ExECUM). 2017. Available online: http://www.execum.unam.mx/ (accessed on 7 January 2020).
18. Márquez, A. Estudio comparativo de universidades mexicanas (ECUM): Otra mirada a la realidad universitaria. *Rev. Iberoam. Educ. Super.* **2010**, *I*, 148–156. [CrossRef]

Article

Corruption Shock in Mexico: fsQCA Analysis of Entrepreneurial Intention in University Students

Fernando Castelló-Sirvent [1] and Pablo Pinazo-Dallenbach [2,*]

[1] Departamento de Organización de Empresas, Universitat Politècnica de València, Camí de Vera, s/n, 46022 Valencia, Spain; fercassi@upv.es

[2] Área Empresa, Valencian International University, C/Pintor Sorolla, 21, 46002 Valencia, Spain

* Correspondence: pablo.pinazo@campusviu.es

Abstract: Entrepreneurship is the basis of the production network, and thus a key to territorial development. In this line, entrepreneurial intention has been pointed out as an indicator of latent entrepreneurship. In this article, the entrepreneurial intention of university students is studied from a configurational approach, allowing the study of the combined effect of corruption perception, corruption normalization, gender, university career area, and family entrepreneurial background to explain high levels of entrepreneurial intention. The model was tested with the fsQCA methodology according to two samples of students grouped according to their household income (medium and high level: $N = 180$; low level: $N = 200$). Stress tests were run to confirm the robustness of the results. This study highlights the negative impact produced by corruption among university students' entrepreneurial intention. Furthermore, the importance of family entrepreneurial background for specific archetypes like female, STEM, and low household income students is pointed out, as well as the importance of implementing education programs for entrepreneurship in higher education, and more specifically in STEM areas. Policies focused on facilitating the access to financial resources for female students and low household income students, and specific programs to foster female entrepreneurship, are also recommended.

Keywords: corruption perception; corruption normalization; gender; entrepreneurial intention; STEM; family entrepreneurial background; fsQCA; household income

Citation: Castelló-Sirvent, F.; Pinazo-Dallenbach, P. Corruption Shock in Mexico: fsQCA Analysis of Entrepreneurial Intention in University Students. *Mathematics* **2021**, *9*, 1702. https://doi.org/10.3390/math9141702

Academic Editors: Laura Gonzalez-Vila Puchades and Jorge de Andres Sanchez

Received: 30 June 2021
Accepted: 16 July 2021
Published: 20 July 2021

Publisher's Note: MDPI stays neutral with regard to jurisdictional claims in published maps and institutional affiliations.

Copyright: © 2021 by the authors. Licensee MDPI, Basel, Switzerland. This article is an open access article distributed under the terms and conditions of the Creative Commons Attribution (CC BY) license (https://creativecommons.org/licenses/by/4.0/).

1. Introduction

Studies on territorial development point to the production network as one of the fundamental pillars to generate well-being and prosperity [1,2]. Thus, the study of entrepreneurial activity takes on special relevance, since it is the variable on which the productive network is based [3]. However, the microeconomic characteristics of each country prevent a single solution from being offered for the promotion of entrepreneurship from public policies [4], which underlines the need to adapt these policies to the different casuistry that surrounds entrepreneurial activity in each territory [5]. In this context, many authors point out the importance of studying entrepreneurial intention to predict entrepreneurship, as it is considered an indicator of latent entrepreneurship [6,7]. This is because it represents "a self-acknowledged conviction by a person that they intend to set up a new business venture and consciously plan to do so at some point in the future" [8] (p. 676).

The phenomenon of entrepreneurship has been studied by considering the internal determinants, as well as the effect of the context in which it develops, since the environment surrounding the entrepreneur exerts an important influence [9,10]. Furthermore, the external and internal factors that affect entrepreneurship are interrelated [11]. In the same way, the entrepreneurial intention is determined by the characteristics and abilities of each individual, their life experiences, and their context [12]. Mexico has high levels of corruption in its institutions, as reflected in the transparency index prepared by Transparency International [13], in which it ranks 124th out of a total of 180 countries analyzed. A deficient or inadequate regulation favors the appearance of corruption [14,15], and this is the case in Mexico concerning the ease of doing business, as highlighted by the World Bank in *Doing Business 2020* [16]. Corruption affects companies both directly and indirectly, and therefore also entrepreneurial activity [17]. However, in certain specific contexts in which there is a bad business climate, some studies have found a positive relationship between corruption and entrepreneurship [18], which justifies delving into this anomaly in the stage preceding the entrepreneurial event. In the same way, in Mexico, there is still a gender gap in terms of entrepreneurship, as reflected in the Global Entrepreneurship Monitor [19], which highlighted that in 2019, the country had a female/male ratio of Total early-stage Entrepreneurial Activity—TEA of 0.91. This gap is accentuated in the businesses related to STEM—the acronym for "Science, Technology, Engineering and Mathematics" [20,21]—which, however, are the ones that contribute the most to the development of the territories due to their great capacity to generate innovation [22,23]. Finally, the literature points out the interrelationship between individuals' family entrepreneurial background and their probability to start an entrepreneurial project [24].

Studies focused on the entrepreneurial phenomenon have traditionally used samples from higher-education students [25], since these individuals, due to their age and educational level, are the most likely to carry out an entrepreneurial project [26]. This also occurs in the specific case of the study of entrepreneurial intention [27], in which studies carried out with samples of university students are very common [28,29].

Nowadays, the importance of offering person-oriented research to explain the entrepreneurial phenomenon is highlighted [30], and more specifically to clarify the role of internal variables in the configuration of entrepreneurial intention [25,31]. Along these lines, authors such as Gartner [32] suggest that research in the area of entrepreneurship should focus on showing profiles in which the internal variables of individuals are combined and interact with variables of the social context surrounding entrepreneurs. Thus, this study addresses, from a cognitive point of view, the combined effect of several internal variables, some influenced by the environment, which may explain high levels of entrepreneurial intention among Mexican students. This article provides a novelty in the literature by addressing a field of study with a very limited amount of empirical evidence, contributing to the construction of the theory through an adequate methodology for the development of multilevel theory. We chose fsQCA as the methodology to carry out this analysis since, unlike symmetry-based methodologies such as regressions, it offers the possibility of studying the combined effect of different individual characteristics, allowing equifinality and causal asymmetry, and thus providing information that other statistical–inferential methods do not contemplate [33]. This research addresses, from a configurational approach, the effect of two internal variables related to corruption (corruption perception and corruption normalization), two socio-demographic variables (gender and the family entrepreneurial background), and the degree area (STEM or not STEM), on the entrepreneurial intention among students of the Technological University of Zacatecas (Mexico). The analysis was carried out on two subsamples differentiated by the level of household income (High Household Income—HHI, and Low Household Income—LHI), to obtain more specific information and make recommendations for policymakers.

This paper is structured as follows. First, it highlights the importance of studying entrepreneurial intention in the context of economic development, exposing the paradigm in which the research is conducted. In addition, as part of the literature review, a deep insight is offered on corruption perception, corruption normalization, gender, STEM studies, and family entrepreneurial background as factors that can determine an individual's entrepreneurial intention. In the same way, the propositions related to the study are presented. In the following sections, the applied methodology is explained, the research results are presented, and the main findings are discussed. Finally, the conclusions, limitations of the study, and future lines of research are presented.

2. Theoretical Background, Conditions, and Propositions

2.1. Economic Development, Entrepreneurship, and Entrepreneurial Intention (EI)

Regions, and specifically countries, boost their economic growth thanks to entrepreneurship, which favors the economic dynamics of the territories [34]. Furthermore, favoring the development of entrepreneurial activity has positive effects on economic development, since, as pointed out by Levie and Autio [35], entrepreneurship stimulates production factors and their efficiency, and favors the appearance of innovations [36]. The capacity of the entrepreneurial phenomenon to encourage innovations and create jobs is of special relevance for the economic development of the territories [37], since it generates prosperity and well-being for citizens [35,38]. In this way, entrepreneurship, as the basis of the production network, is a key element for territorial development [2] and justifies the study of its dynamics.

Moriano [39] stated that entrepreneurship is a sequential process that goes beyond the start-up phase. Specifically, it incorporates a pre-launch phase, a launch phase, and a post-launch phase. Stages prior to the act of entrepreneurship are difficult to observe, as they remain in the field of intentions [10,39]. In this sense, the intention to start an entrepreneurial project represents the first stage of the entrepreneurial process [40] and is of crucial relevance, since it represents the prior decision that an individual must make before becoming an entrepreneur [41,42]. Thus, understanding entrepreneurial intention is key to understand the entrepreneurial process [41]. This fact explains why EI is considered as a predictor of entrepreneurial behavior [8,43,44] and an indicator of latent entrepreneurship [6,7]. In this study, EI is considered as the conscious will of a person to start a business in the future [8,31].

2.2. Corruption Perception (COPER) and Corruption Normalization (CONOR)

The existence of corruption in public institutions has negative consequences on the well-being of citizens, increases inequality, and harms entrepreneurship [45]. Specifically, the institutional environment directly affects entrepreneurial activity [17]. On one hand, to the extent that there are institutions that favor the emergence of opportunities and generate economic growth, this will have a positive effect on business activity, greatly facilitating the success of entrepreneurs [46,47]. On the other hand, when corruption exists within institutions, entrepreneurial activity is seriously affected [48]. This is because corruption lowers institutions' quality to the extent that it reduces the correct redistribution of resources, increases capital cost, and slows the production-network development [49]. Furthermore, the existence of corruption generates uncertainty around investment projects, which increases transaction-related costs and increases inefficiency of market and business activity [50,51]. Consequently, the absence of corruption control reduces the performance of companies and their ability to generate innovations [52,53].

The evidence obtained shows that there is an inverse relationship between corruption and entrepreneurial activity; however, there are differences at the cognitive level between individuals from the same territory that affect their COPER differently. This fact causes differences between the perception of different individuals, and also between the levels of real corruption and what is perceived by people [54–56]. Since this study is person-oriented, it focuses on the subjective element of corruption; that is, COPER and its possible inhibiting effects on EI. In this way, this research proposes the existence of a causal relationship between the COPER of university students and their EI level.

However, some authors found a positive relationship between corruption and entrepreneurship [18], and explained that when there is a bad business climate, characterized by inefficient or inadequate regulation of business activity, individuals perceive corruption as an opportunity to accelerate the bureaucratic processes that exist around entrepreneurial activity. In this way, Dreher and Gassebner [18] indicated that in these countries, entrepreneurs perceive the possibility of using bribes with public officials as a very interesting shortcut that reduces the problems and times of these procedures. This "grease the wheels" argument can be explained by the normalization of corruption that appears in countries where people coexist with corruption for a long time [57] (Guerber et al., 2016). This normalization of corruption occurs in societies through three interrelated effects: the institutionalization, rationalization, and socialization of corruption [57,58]. The first refers to the fact that a corrupt act becomes routine within institutional structures and processes, the second refers to the development of selfish behaviors that justify and even give value to corrupt acts, and the third indicates the phenomenon by which newcomers are led to think that corruption is permissible or even desirable in society [58]. In this context, it could be argued that when COPER coexists with corruption normalization, this has a positive effect on EI, since corruption is presented as an opportunity and not as a threat. Thus, this study proposes the existence of a causal relationship between the normalization of corruption in university students and their level of EI.

2.3. Gender (GEN)

Regarding EI, the differences between men and women are largely produced by the influence of the environment that surrounds them, which is accentuated in the case of more traditional cultures [59]. This phenomenon explains that, worldwide, women are less likely to start a business [60,61], and that in the case of Latin America, this probability is lower in comparison with those of other more-developed regions [62–64].

Women in Latin America have fewer incentives to undertake a business project than men [64,65]: they are more affected by the corruption of institutions [66,67], find difficulties in accessing services related to business development and training systems [68–70], and have problems joining professional networks in which they can access essential information to start and manage a business, as well as find easier access to financial and technological resources and marketing channels [69,70]. Furthermore, the inequality present in the region penalizes women even when they are trained, motivated, and have the necessary knowledge to be able to start a business [65]. One of the main obstacles they must face is related to obtaining financial resources [71], which, in the end, forces them to focus on their ability to save before starting an entrepreneurial project [72].

This set of barriers that women in Latin America must face negatively conditions their perception of entrepreneurship and reduces, in most cases, their motivation to start a business to the cases in which they are driven by necessity [73], pointing to self-employment as the main motivation of women who undertake a business venture [69]. Thus, considering the influence of the environment surrounding women in Latin America that interposes all these barriers on their path to entrepreneurship, this study proposes the existence of a causal relationship between the gender of university students and their EI.

2.4. STEM Studies

The areas of knowledge related to the fields of Science, Technology, Engineering and Mathematics are known as STEM. This type of knowledge area is directly related to the generation of innovations, competitiveness improvements, and economic and social growth, which ends up having a positive impact on well-being [74,75]. Thus, several studies have linked the creation of new technology-based firms, focused on STEM areas, with economic growth and development [4,76], since innovative entrepreneurship contributes significantly to value creation and the improvement of economic dynamics. This is due, among other things, to its job creation capacity [77]. This explains the growing interest in this type of entrepreneurship, not only within academia, but also within national and supranational organizations such as the Organization for Economic Co-Operation and Development—OECD or the United Nations, which have developed some programs focused on attracting and retaining people within STEM areas [21].

However, there are not many studies that have been developed relating STEM areas and EI [78]. In some cases, they select samples of students from the area of business vs. non-business [79], others use samples from the field of engineering [28], and on very few occasions, both groups are compared [80,81]. In the same way, few studies indicate the degree subject showing the dichotomy of STEM/no STEM as an element to consider when measuring the level of EI [78]. Given the interest in the STEM collective and its potential in territorial development, this study proposes, in an exploratory way, the existence of a causal relationship between the degree subject and the level of EI. In the same way, the combination of this condition with some others in the analysis will show specific profiles that characterize potential entrepreneurs and, in this way, offer interesting information for policymakers. In this sense, the relationship between gender and STEM should be highlighted, since, although women are underrepresented in this area in terms of entrepreneurship, "women entrepreneurs have a 5% greater likelihood of innovativeness than men" [82] (p. 9), which reflects the potential of this group.

2.5. Family Entrepreneurial Background (FEB)

The literature offers evidence that suggests that the probability that an individual feels an interest in entrepreneurship increases if they come from a family in which other members have undertaken entrepreneurial projects [83,84]. Furthermore, various studies find that entrepreneurs often come from entrepreneurial families [85,86], a fact by which the option of self-employment becomes more attractive [87,88] and increases the EI [89]. In this way, it can be suggested that the family environment can function as an antecedent for entrepreneurship [24].

The influence of the family, and more specifically in the case of family entrepreneurs, has been studied from a sociological perspective in which entrepreneurs make their social capital and social networks available to potential entrepreneurs [90,91]. However, this is not the only way in which the FEB exerts an influence on individuals, since having entrepreneurial relatives also serves as a learning model for individuals, improving their attitudes and behaviors related to entrepreneurial activity [92]. In this way, the values and norms of close family members can determine the EI of individuals [93]. Thus, parents act as role models for their children [92]. This means that, for example, these children have had more experiences related to proactivity, risk-taking, and innovation [93]. In this context, Marques et al. [93] highlighted that the learning processes that take place in the family environment favor and reinforce the appearance of strong attitudes and intentions related to entrepreneurship, since children who have grown up in business environments have more learning experiences related to entrepreneurship. Thus, living in a family with a business background makes individuals progressively enter the world of entrepreneurship [94] and offers the option of doing things differently, becoming a motivational factor for the child [95]. Mexican culture is very collectivist, which implies that individuals interact regularly with their extended family members with whom they maintain strong ties [96,97]. As a consequence, to study the influence of the FEB, in this research the extended family model was used [98], including grandparents and uncles in the family unit.

Considering the evidence presented, this study proposes the existence of a causal relationship between the university students' FEB and their level of EI.

The theoretical framework allows the formulation of the following causal model and propositions:

$$EI = f(COPER, CONOR, GEN, STEM, FEB)$$

Proposition 1. *None of the five causal conditions (COPER, CONOR, GEN, STEM, FEB) is necessary to merit a prediction of high levels of EI among university students.*

Proposition 2. *The five causal conditions form multiple configurations that are sufficient to predict a high level of EI among university students.*

3. Materials and Methods

A questionnaire with 23 questions (Appendix A) was provided, including in the first part items of academic (degree name) and socio-demographic (gender, age, household income and family entrepreneurial background) types.

Following Liñan and Chen [26], the questionnaire included items for the analysis of the EI of university students according to a type 1–3 Likert scale, where 1 meant "totally disagree" and 3 "totally agree". According to the Transparency International Corruption Perception Index [99,100], students were asked about their perception of corruption and their degree of normalization while assessing environmental corruption. These questions were evaluated by means of a Likert scale of type 1–5, where 1 meant "totally disagree" and 5 "totally agree". In our study, we used 1–3 Likert scales to obtain the polarized information that was requested in the EI variable and 1–5 Likert scales for COPER and CONOR variables (Appendix B).

The students' responses to question 1 (degree name) were tabulated to obtain the variable type of university career (STEM or not STEM), generating a dichotomous variable (0, 1). Likewise, the students' responses regarding question 4 (household income) were used to divide the sample ($N = 380$) into two subsamples (medium and high level: $N = 180$; low level: $N = 200$), applying a threshold of MXN 11,600.

The variables EI, COPER, and CONOR were constructed using the eigenvalues resulting from two exploratory factor analyses (EFAs) performed with the statistical software IBM SPSS (v24). The EI EFA was carried out on questions 6–10 (Appendix B), and the EFA on corruption perception and normalization of environmental corruption (COPER, CONOR) was carried out according to questions 11–23 (Appendix B). The first EFA gave rise to a factor (7–10 questions), the eigenvalues of which were taken to configure the EI variable. The second EFA resulted in two factors (questions 12–15 and questions 20–23, respectively), whose two eigenvalues were taken to configure the variables COPER and CONOR.

Table 1 reports on the values of the Bartlett sphericity test and the sample adequacy measure for the factors obtained after the analysis of EI and the perception and normalization of environmental corruption (COPER, CONOR). Appendix B reports the methodological detail of the EFAs carried out in this study.

Table 1. Exploratory factor analysis (EFA) results.

	Entrepreneurial Intention	Corruption	
Factors	1	2	
Variables	EI	COPER	CONOR
Items	4	4	4
Cronbach's alpha	0.867	0.889	
KMO (Kaiser–Meyer–Olkin)	0.829	0.805	
Barlett test (sig.)	0.000	0.000	

Source: authors' elaboration.

The correlation analysis between the variables EI and COPER corroborated a positive relationship ($r = 0.170$; $p < 0.001$). Table 2 reports on the descriptive statistics for the variables differentiated by degree type (STEM vs. not STEM) and by gender.

Table 2. Descriptive statistics.

	STEM	GEN	IE	COPER	CONOR
Mean	Non-STEM	Male	0.224	0.203	−0.0182
		Female	−0.0633	−0.0758	−0.0530
	STEM	Male	0.0248	0.108	0.0590
		Female	−0.156	-0.273	0.0248
Standard deviation	Non-STEM	Male	0.878	0.882	1.02
		Female	0.952	0.915	0.930
	STEM	Male	0.926	0.970	0.894
		Female	0.939	1.00	1.01

Source: authors' elaboration.

The proposed model, designed to explain the result (EI), included five variables: two continuous variables (COPER and CONOR) and three categorical variables (GEN, STEM,

FEB). To test the model (EI = f (COPER, CONOR, GEN, STEM, FEB)), we carried out a qualitative comparative analysis of fuzzy sets (fsQCA).

fsQCA is a methodology designed for the systematic analysis of cases that allows researchers to find causal patterns that determine the result or outcome [101,102]. This methodology was originally designed for the analysis of small or medium-sized samples [103,104]. However, fsQCA does not offer any mathematical limitation for its application in large samples, guaranteeing valid results for this type of analysis [105,106]. This methodology is based on Boolean logic and allows the identification of need relationships and sufficiency relationships between a set of independent variables (conditions or attributes) and the dependent variable (outcome). A condition is necessary when it must be present for the outcome studied to occur. The absence of necessary conditions implies the existence of multiple combinations of conditions that can give rise to the outcome studied. The factual analysis of the available cases makes it possible to identify the pathways followed by the profiles of students who manifest a high level of EI.

In addition, fsQCA allows the identification of counterfactual evidence, described as combinations of attributes that could occur and have not been observed within the available sample [102]. Therefore, fsQCA is an adequate methodology for the development and profiling of a theoretical scheme [107], as well as the evolution of multilevel theory [108]. In addition to facilitating the validation of hypotheses, it allows the generation of new knowledge based on the analysis of the different causal relationships of the observed phenomenon. This insight is especially useful in case studies in which the research field falls between diverse and complex theoretical frameworks. When the research areas find contradictory evidence on many occasions or, as in our study, when it comes to novel and little explored relationships. In all cases, the research design must guarantee methodological control [109] and facilitate the understanding of the criteria applied throughout the process, especially in the calibration phase of the variables.

fsQCA is a variant of the original QCA methodology, and is applied in fuzzy sets. These types of sets retain most of the essential mathematical properties of sharp sets [110]. Unlike other methodologies, fsQCA does not compare individual variables, but rather analyzes complete combinations of simultaneous conditions, and allows researchers to overcome the limitations of inferential statistical techniques [33]. Furthermore, this methodology perfectly captures the idea of causal asymmetry [105], because a certain attribute that occurs in a specific directionality does not necessarily offer the same result as its opposite directionality.

Following Ragin [102], it is known that fsQCA allows the deconstruction of a single symmetric analysis in two different asymmetric analyzes of set theory, one focused on sufficiency and the other on necessity. In addition, this methodology allows the display of the advantages of equifinality in the analysis [111], since the same phenomenon (outcome) can be explained through different combinations of attributes grouped in different causal configurations (pathways or recipes). fsQCA has been used in multiple areas of the social sciences (e.g., [112]), highlighted especially in those of management and entrepreneurship [30].

Our fsQCA analysis was performed with the fs/QCA 3.0 software, and the methodological procedure followed four sequential steps [102,105,113]:

1. Calibration: in this step, the variables (dependent and independent) must be calibrated using a logarithmic function so that they can be analyzed based on set theory. The calibration process involves recalculating all continuous variables so that they are integrated into the continuum 0–1, establishing thresholds that allow determining the membership of a value to one of the following three sets: (1) fully inside; (0.5) maximum ambiguity; (0) fully outside. Following Misangyi and Acharya [114], this study used the continuous variable calibration method with percentiles, with the percentiles proposed by Climent-Serrano et al. [115]: 90%, 50%, and 10%, to delimit the thresholds (1), (0.5), and (0) of the variables EI, COPER, and CONOR, respectively.

2. Construction of the "truth table": in this step, a data matrix must be constructed (the "truth table"), which includes 2k rows, where k is the number of causal conditions proposed by the model.
3. Initial reduction of the number of rows: in this step, two sequential criteria are applied. First, from a factual approach, pathways that do not collect cases are eliminated. Second, from a reliability approach, pathways that do not achieve a consistency threshold equal to or greater than 0.75 are eliminated [102].
4. Final reduction of the number of rows: In this step, a final reduction of the number of rows of the "truth table" is carried out by applying the Quine–McCluskey algorithm (Ragin, 2008).

Tables 3 and 4 show the calibration thresholds and descriptive statistics of the variables used in the model for the high and low household subsamples.

Table 3. Calibration and descriptive statistics (HHI).

Variables	Calibration			Descriptive Statistics		
	Fully Inside	Maximum Ambiguity	Fully Outside	Max	Min	Median
EI	0.8086	0.6025	−1.2717	0.8086	−3.3520	0.6025
COPER	1.1718	0.3103	−0.9450	1.2184	−2.1499	0.3103
CONOR	1.3265	−0.1676	−1.2256	1.8893	−1.2407	−0.1676
GEN	1	-	0	-	-	-
STEM	1	-	0	-	-	-
FEB	1	-	0	-	-	-

Source: authors' elaboration.

Table 4. Calibration and descriptive statistics (LHI).

Variables	Calibration			Descriptive Statistics		
	Fully Inside	Maximum Ambiguity	Fully Outside	Max	Min	Median
EI	0.8086	0.2180	−1.2717	0.8086	−3.3520	0.2180
COPER	1.0899	−0.1961	−1.4167	1.1667	−2.2011	−0.1961
CONOR	1.4191	0.1200	−1.2083	1.9712	−1.2421	0.1200
GEN	1	-	0	-	-	-
STEM	1	-	0	-	-	-
FEB	1	-	0	-	-	-

Source: authors' elaboration.

4. Results

4.1. Analysis of Necessary Conditions

This study analyzed the EI of Mexican male and female university students, taking as a reference the concurrence, COPER, and normalization, based on gender, university career, and FEB. The analysis was carried out in two subsamples of students, grouped by the household income of their family units. The analysis of necessary conditions (Table 5) reported information on causal conditions that are necessary for the investigated outcome to occur. Ragin [102] accepted that a condition is necessary if it exceeds the threshold of 0.9.

Table 5. Analysis of necessary conditions.

Condition Tested	HHI		LHI	
	Consistency	Coverage	Consistency	Coverage
fs_COPER	0.6241	0.7140	0.6254	0.6194
~fs_COPER	0.5088	0.5253	0.5289	0.5656
fs_CONOR	0.4356	0.4744	0.5340	0.5929
~fs_CONOR	0.6942	0.7509	0.6286	0.6020
fs_GEN	0.3549	0.5630	0.6196	0.5057
~fs_GEN	0.6451	0.5321	0.3805	0.5287
fs_STEM	0.2958	0.4692	0.4398	0.4916
~fs_STEM	0.7042	0.5809	0.5602	0.5335
fs_FEB	0.6192	0.6081	0.4152	0.6280
~fs_FEB	0.3808	0.4620	0.5848	0.4556

Source: authors' elaboration.

The analysis of necessary conditions reports that the presence (or absence) of any condition is not mandatory for a university student (male or female) to increase their EI, for both subsamples (high and low family income). This finding implies the existence of multiple combinations of factors that can lead to high levels of EI. To get closer to a more extensive knowledge beyond the observed reality, we must delve into the analysis of cases by conducting the sufficiency analysis. This fact confirms both Propositions 1 and 2.

4.2. Analysis of Sufficient Conditions

The consistency of the solutions (complex, intermediate, and parsimonious) was satisfactory. Table 6 reports the results of the analysis of sufficiency for the outcome (EI) for two subsamples (high and low household income). The relevant information was synthesized to understand the solutions of the model (consistency and coverage) and the characteristics that helped us to understand the EI of the different profiles of university students. Black circles indicate the presence of a condition, and white circles indicate its absence. The criterion proposed by Fiss [105] was followed to graphically represent the presence (or absence) of conditions in each pathway, reporting the "core conditions" with large circles and the peripheral conditions with small ones. When a condition is not marked (blank spaces), it indicates "don't care".

Appendices C and D offer more detailed information.

The solutions to the models (HHI and LHI) had an adequate level of consistency (HHI: 0.8301; LHI: 0.7991), higher than the minimum threshold required by Ragin [102] (2008). Both subsamples took cut-off thresholds for the selection of cases in the "truth table" that were higher than 0.75, as established by the best practices in fsQCA [102].

The coverage levels varied depending on the database used. The HHI subsample had a coverage of 0.6992, and the LHI subsample had a coverage of 0.2926. In both cases, the minimum threshold of 0.75 was exceeded [102]. The differences in coverage between the two subsamples must be understood in the Mexican social context, in which opportunity-driven entrepreneurship is scarce among people with less income, developing a search for resources that in many cases leads to necessity-driven entrepreneurship.

Table 6. Analysis of sufficiency for the outcome (EI).

	Intermediate Solutions for the Proposed Model [1]										
	HHI						LHI				
Conditions/Pathways	H1	H2	H3	H4	H5	H6	L1	L2	L3	L4	L5
COPER	●		○	○	●	●		○	●		○
CONOR	○	○	○	●	●	○	○			●	○
GEN	○	○	●	○	●	●	●	●	○	●	○
STEM		○	○	●	○	●	○	○	○	●	●
FEB	○	●	○	●	●	●	●	●	●	●	●
Raw coverage	0.1971	0.2289	0.0492	0.0448	0.1029	0.0462	0.1295	0.0873	0.0609	0.0503	0.0243
Unique coverage	0.1971	0.2289	0.0492	0.0448	0.1029	0.0462	0.0698	0.0276	0.0609	0.0503	0.0243
Consistency	0.8409	0.8197	0.8509	0.8341	0.7640	1	0.7870	0.7512	0.9163	0.9103	0.7699
Intermediate solution											
Coverage	0.6692						0.2926				
Consistency	0.8301						0.7991				
Cutoff											
Frequency	1						1				
Consistency	0.7564						0.7649				
Directional expectations [2]	(-, -, -, -, -)						(-, -, -, -, -)				

[1] Black circles indicate the presence of a condition, and white circles indicate its absence. According to Fiss [105], large circles indicate "core conditions", small ones indicate peripheral conditions, and blank spaces indicate "don't care". [2] Given the exploratory nature of this study, the methodology was not conditioned by directionalities established a priori. Source: authors' elaboration.

The sufficiency for the outcome (EI) analysis in HHI reported six different profiles of university students with high EI. Men with HHI were concentrated in three different profiles (H1, H2, and H4). Pathway H1 explained the behavior of one in five students (unique coverage: 0.1971) and included an archetype of male students with a family background in business creation. They were students with COPER and without corruption normalization. The male students explained by the H2 pathway were not STEM students, and their profile was not permissive regarding environmental corruption either. However, this archetype of entrepreneurial students did not have a family background in the creation of companies, and their perception of environmental corruption was not decisive to explain their EI. One in five college students with HHI fit this archetype (unique coverage: 0.2289).

The H4 solution contributed to improving the understanding of the profile of male students pursuing STEM degrees, with a family background and who were characterized by not having COPER, but having corruption normalization.

The female students in STEM disciplines (H6) had a family background, the entrepreneurial culture was affected by their high COPER, and they were not flexible regarding the corrupt environment, because they did not normalize corruption, showing integrity in the values of social behavior. Solutions H3 and H5 (unique Coverage: 0.0492 and 0.1029, respectively) explained the behavior of female students pursuing non-STEM degrees with a family background and high perception and normalization of corruption (H5) and, on the contrary, without a family background and without perception or normalization of corruption (H3).

The analysis of conditions carried out under the approach of core and peripheral conditions [105] reported important findings. The family background was a core condition for generating EI among STEM degree students, both men and women. In the case of men, it was also a core condition not to declare high levels of COPER (H4). In the case of women, conversely, the core sine qua non condition being the presence of COPER and, simultaneously, the absence of CONOR (H6). Women who clearly perceived the level of environmental corruption (core condition) developed high levels of EI when their perception of corruption was internalized and normalized (H5). Other relevant core conditions were the absence of normalization of corruption in the case of men (H1, H2, H3), which occurred particularly in non-STEM grade students (H2, H3), and the core condition that these students did not have an entrepreneurial background in their family (H3) also was relevant.

The analysis of the subsample of students with LHI revealed the core condition for the presence of a family background for all the profiles identified. In this sense, the archetype of male students in STEM degree programs developed high levels of EI, relying on an archetype that does not perceive environmental corruption, precisely because it normalizes it (L5). Non-STEM female students did not normalize corruption, and in fact, their COPER was not significant in inhibiting EI (L4).

The L1 and L2 solutions for the LHI subsample helped define and understand the profiles of female students in non-STEM degree programs who do not perceive environmental corruption (L2) or do not normalize such corruption (L1). Finally, there was a highly relevant profile (consistency: 0.9163), whose causal configuration of attributes grouped male students in non-STEM degree programs that did have a high rate of COPER, which reflected the feeling of many young people in Mexico who do not support corruption, are aware of it, and try to keep on with their lives while assuming that corruption is a part of their context.

4.3. Reliability and Robustness Fit

A sensitivity test was performed following Skaaning [116] and Schneider and Wagemann [113]. The cut-off points for calibration were modified as suggested by Fiss [105] and Stevens [117]. The percentile that defined the fully inside point was reduced by 10%, and the point that defined the fully outside point was increased by 10%, establishing the following points for fully inside (80%), maximum ambiguity (50%), and fully outside (20%). This new sensitivity analysis involved applying a stress test to the model to validate its robustness fit.

A model has a robust and acceptable fit only if its consistency level remains within the range (+5%, −5%) after performing the stress test, and only if, simultaneously, the consistency of the three solutions (complex, parsimonious, and intermediate) exceeds the threshold of 0.75 and the minimum coverage of the three solutions is 0.25, according to the criteria established by Ragin [102].

Once the new calibration of the variables was carried out, the model was retested and the evidence found by the new analysis (Table 7) validated the robustness of the proposed model for analyzing the EI of university students in corruption environments.

Table 7. Stress test of the calibration process and methodological robustness of the proposed model For the high family income subsample.

	Solution Indicators		
Solutions/Test	Pathways	Consistency	Coverage
Complex			
Model (90%; 50%; 10%)	6	0.8301	0.6692
Stress Test (80%; 50%; 20%)	6	0.8252	0.3640
Robustness Fit (+5%; −5%)		−0.59%	
Parsimonious			
Model (90%; 50%; 10%)	6	0.8268	0.7358
Stress Test (80%; 50%; 20%)	6	0.8163	0.3661
Robustness Fit (+5%; −5%)		−1.27%	
Intermediate			
Model (90%; 50%; 10%)	6	0.8301	0.6692
Stress Test (80%; 50%; 20%)	6	0.8252	0.3640
Robustness Fit (+5%; −5%)		−0.59%	

Source: authors' elaboration.

The stress test result was positive and guaranteed the robustness of the fit of the model in the subsample of students with high family income. The three solutions generated in fsQCA (complex, parsimonious, and intermediate) had consistency values within the range (0.8140, 0.8716), and a maximum consistency deviation of −1.27%. For more detailed information, see Appendices E and F.

The stress test also reported that the subsample of students with low family income (Table 8) guaranteed the robustness adjustment of the model for the three fsQCA solutions (complex, parsimonious, and intermediate), with consistency values within the range (0.7591, 0.8391), and a maximum consistency deviation of +3.99%. For more detailed information, see Appendix F.

Table 8. Stress test of the calibration process and methodological robustness of the proposed model For the low family income subsample.

Solutions/Test	Pathways	Consistency	Coverage
Complex			
Model (90%; 50%; 10%)	5	0.7991	0.2926
Stress Test (80%; 50%; 20%)	4	0.8310	0.2549
Robustness Fit (+5%; 5%)		+3.99%	
Parsimonious			
Model (90%; 50%; 10%)	5	0.7991	0.2926
Stress Test (80%; 50%; 20%)	4	0.8310	0.2549
Robustness Fit (+5%; −5%)		+3.99%	
Intermediate			
Model (90%; 50%; 10%)	5	0.7991	0.2926
Stress Test (80%; 50%; 20%)	4	0.8310	0.2549
Robustness Fit (+5%; −5%)		+3.99%	

Above the Solutions/Test/Pathways/Consistency/Coverage header, the columns Pathways, Consistency, and Coverage are grouped under **Solutions Indicators**.

Source: authors' elaboration.

After applying the stress test, the solutions to the model for both subsamples remained stable, describing student profiles without significant changes (Appendices E and F). The evidence supported that the solutions of the model remained stable for both subsamples, describing the profiles of the students without significant changes (Appendices E and F). Furthermore, the Castelló-Sirvent [118] Robustness Coefficient was calculated to evaluate the model. In addition to the standards set by Ragin [101,102] (consistency ≥ 0.75; coverage ≥ 0.25), and according to Castelló-Sirvent [118], a model is robust only if it reports an adequate RC-value (RC ≥ 0.95) after the stress test carried out on the cut-off points (Appendix G).

The robustness of a model tested in fsQCA can be established by modifying one or more of the cutoff points (fully inside, maximum ambiguity, fully outside), and can be carried out by modifying the calibration, either with percentiles or manual. The stress test aims to identify the robustness of the model by analyzing the average variation of the consistency. If the stress test has been performed by recalibrating the variables, a threshold of $\pm 10\%$ or $\pm 15\%$ can be used. According to the RC-value, if it was recalibrated $\pm 15\%$, the robustness of the model can be very strong ($0.9900 \leq RC \leq 1$) or strong ($0.9500 \leq RC \leq 0.9899$). If it was recalibrated $\pm 10\%$, the robustness of the model can be strong ($0.9900 \leq RC \leq 1$), moderate ($0.9500 \leq RC \leq 0.9899$), or weak ($0.9000 \leq RC \leq 0.9499$). Standardized symbols are used together with the RC-value (*** very strong; ** strong; * moderate) to indicate the robustness of the model tested with fsQCA [118] (Appendix G). In our study, the RC-value supported moderate robustness (RC = 0.9632 *).

5. Discussion

The analysis of these archetypes allowed us to understand the barriers to entrepreneurship in its immediately preceding stage [43,44] and guide those responsible for formulating policies in the design of public policies for regional economic development.

Our study offers a useful guide for political leaders in Mexico and other countries with similar circumstances of environmental corruption. In general, there are few profiles that present high levels of EI and high levels of perception of corruption. This reinforces the thesis that inversely relates corruption and entrepreneurship [17,45,48]. However, the "grease the wheel" argument [18]) could be present in a specific profile, which should be urgently addressed by policymakers, since this reflects that normalization occurs in a certain sector of the population, and this may generate friction with efforts to reduce institutional corruption. Nevertheless, this case should not overshadow the fact that in general, the causal configurations obtained showed that the normalization of corruption is not generalized among university students, which shows the greater awareness of young people regarding this problem and their determination to end it. Thus, the fight against corruption and the improvement of accountability mechanisms must be included in the basic development objectives of the economic model. Institutions must provide a suitable framework for private action, protecting innovation, ideas and private property, and improving the effectiveness of the bureaucratic processes that surround the entrepreneurial process [16]. In the same way, public institutions (state and federal) must develop actions aimed at taking advantage of intangible capital, inhibiting emigration rates for highly educated individuals and reversing the "brain drain" [119], as corruption has been related directly to higher propensity to emigrate within students with high levels of EI [120].

It is important to note that university students with a high level of EI, coming from families with low family income, have comparatively fewer possibilities of developing business projects driven by opportunity, and are more oriented to entrepreneurship driven by necessity [26]. In fact, students with low family income who reported having an entrepreneurial background in their family were more likely to come from a family in which parents and/or other family members started a business out of necessity, not opportunity, which explains the LHI. According to [121], this impacts on EI and, consequently, on regional economic development [6,7]. The study of this subsample showed that these individuals have higher EI when the have a FEB. In fact, all the profiles obtained showed the presence of this condition. This is explained by the importance of household income to start a business venture. Students with LHI will only be willing to undertake a business project if their relatives have enough expertise in the field of entrepreneurship to guarantee higher levels of success.

Another interesting finding is related to the profiles associated with STEM degrees. The results showed that STEM students have minor levels of EI compared to non-STEM students, in line with previous research [78,79]. This could be explained by the lack of entrepreneurial competencies within STEM degrees' curricula. It is interesting to point out that this empirical evidence showed the existence of a FEB in profiles with high levels of EI among STEM students, highlighting a similar phenomenon shown by the LHI students. On the contrary, in the case of non-STEM students, high levels of EI occurred without the need of having a FEB. This reflects that learning processes related to entrepreneurship are necessary to increase EI, either coming from relatives or formal education [93].

Furthermore, the design of the policy should allow the promotion of entrepreneurial projects among STEM students. The available evidence suggests that these types of entrepreneurs have the capacity to generate much greater added value with their projects, positively impacting the income level and long-term economic development [74,75]. Previous research highlighted the positive impact of entrepreneurial education over higher-education students, which proves the efficacy of this kind of training, and reinforces the proposal of introducing this knowledge within the curricula of STEM degrees [79,122].

In addition, the design of the policy must prioritize female entrepreneurship, especially in disadvantaged contexts, assuming the fact that women must face higher barriers than men when it comes to entrepreneurship [19].

In the same way, raising awareness about the role of women in business creation, successful examples of local entrepreneurs (STEM and non-STEM) should be formed as a measure of normalization of female entrepreneurship, making these examples of women entrepreneurs visible, and making it possible to make them visible within local and regional communities. In this line, it should be noted that women are more likely to innovate than men [82]; hence, taking advantage of their entrepreneurial potential in STEM areas can have a significant impact on the economies of the region [21].

Finally, as access to financial resources is one of the principal obstacles for women [71,72], policymakers should focus on promoting bank and alternative financing mechanisms, business angel networks, seed capital, and venture capital. In fact, these measures should focus on female and LHI students, as this can help them to develop their entrepreneurial venture. In Mexico's case, these kinds of programs have been implemented with good results [123,124].

6. Conclusions

Since entrepreneurship is the basis that nurtures the growth of the production network, the interest in the dynamics that surround it keep increasing. Several studies reported entrepreneurial intention as an indicator of latent entrepreneurship. This is because, prior to undertaking a business project, individuals present the intention to do so. In this study, the entrepreneurial intention of university students was studied from a configurational approach thanks to the use of the fsQCA. This allowed the study of the combined effect of several internal variables, some affected by the environment, to explain high levels of entrepreneurial intention.

This study reaffirmed the negative effects that corruption has on entrepreneurship and highlights the importance of implementing policies aimed at eradicating it from institutions. In the same way, it indicates how, for a specific profile of students, corruption can be seen as an opportunity to accelerate processes related to entrepreneurship. However, the results also reflect that, in general, Mexican students do not tolerate corruption. In the same way, the analysis of the results showed the importance of implementing education programs for entrepreneurship in higher education, and more specifically in STEM areas. Finally, the need to implement programs to facilitate access to financial resources for both LHI students and female students was highlighted. In the latter case, it is necessary to set up specific programs dedicated to promoting female entrepreneurship.

The results and recommendations offered can serve as inputs for policymakers in countries with similar characteristics to Mexico, with the countries of the Latin American and Caribbean regions being good examples. These recommendations can favor the dynamics of the productive network and have positive effects on regional development. The design of public policies must focus on corruption control and focus especially on the university context when working against corruption normalization. The evidence found suggested a deep impact on the generation of business opportunities. In addition, the promotion of female entrepreneurship must be highlighted among the priorities of the policymakers. Both male and female students must have access to specific resources to avoid the perverse dynamics of brain drain. Similarly, university students from lower-income households must find adequate support in the federal and state public administrations for the development of early vocations that modernize the production network and promote innovation and long-term economic growth in the region. This article provides important information to guide the achievement of the Sustainable Development Goals (SDGs) and the implementation of the 2030 Agenda for Sustainable Development.

This research had some limitations. On the one hand, the sample studied came from a single university, which offers the possibility for future studies to delve into these results by studying samples obtained from other universities. On the other hand, the limitations of the methodology used did not allow us to combine too many variables. Future works should address these results by combining other methodologies that allow delving into these and other variables.

Author Contributions: Conceptualization, P.P.-D.; methodology, F.C.-S. and P.P.-D.; software, F.C.-S.; validation, F.C.-S.; formal analysis, P.P.-D. and F.C.-S.; investigation, P.P.-D. and F.C.-S.; resources, P.P.-D. and F.C.-S.; data curation, P.P.-D. and F.C.-S.; writing—original draft preparation, P.P.-D. and

F.C.-S.; writing—review and editing, P.P.-D. and F.C.-S. All authors have read and agreed to the published version of the manuscript.

Funding: This research received no external funding.

Data Availability Statement: The data sets generated and/or analyzed during the current study are available from the corresponding author upon reasonable request.

Acknowledgments: This research is a result of the post-doctoral activities in the Academic Unit in Development Studies of the Universidad Autónoma de Zacatecas (UAZ) under the tutelage of Carlos Mallorquín Suzarte. Special thanks to the professors of the Universidad Tecnológica del estado de Zacatecas (UTZAC) who collaborated with this research: Guillermo González Ibarra—head of the Quality Management System department; Tirso Noel Pacheco Delgado—director of the Business Development department and Human Capital Administration; Alicia del Rocio Rosales Zapata—director of the Information and Communication Systems and Technologies department; Manlio Favio Velazco García—director of the Mechatronics, Renewable Energies, and Sustainable Agriculture department; and Roberto Rodríguez Muñoz—Director of the Department of Maintenance, Industrial Processes, and Mining.

Conflicts of Interest: The authors declare no conflict of interest.

Appendix A. Questionnaire

University career

Question 1: Please write down the name of the degree you are studying.

Socio-demographic questions

Question 2: Sex (male or female)

Question 3: Age

Household income question

Question 4: Please indicate the monthly amount corresponding to your household income.

Options	Check
More than 85,000 pesos per month	
Between 35,000 and 84,999 pesos per month	
Between 11,600 and 34,999 pesos per month	
Between 6800 and 11,599 pesos per month	
Between 2700 and 6799 pesos per month	
Between 0 and 2699 pesos per month	

Family entrepreneurial background

Question 5: Do you have a close family member who is or has been an entrepreneur (parents, brothers, grandparents, uncles)?

Options	Check
Yes	
No	

Entrepreneurial intention

Please indicate your degree of intention regarding the following statements, with 1 being 'totally disagree' and 3 'totally agree'.

Questions	1	2	3
Question 6: I have serious doubts about whether I will ever start a business			
Question 7: It is very likely that I will start a business in the future			
Question 8: I am willing to do whatever it takes to be an entrepreneur			
Question 9: I am determined to start a business in the future			
Question 10: My professional goal is to be an entrepreneur			

Corruption perception

Please indicate your degree of agreement with the different statements, with 1 being 'totally disagree' and 5 'totally agree'.

Questions	1	2	3	4	5
Question 11: There are clear accountability procedures and mechanisms that apply to the allocation and use of public funds.					
Question 12: It is common for politicians/civil servants to appropriate public funds for personal or partisan purposes					
Question 13: There are special funds for which there is no accountability					
Question 14: There is widespread abuse of public resources					
Question 15: There is a professional career in the public sector or there are a large number of civil servants who are directly appointed by the government					
Question 16: There is an independent body that audits the administration of public finances					
Question 17: There is an independent judiciary power with the competences to judge public ministers/civil servants who commit abuses.					
Question 18: It is traditionally resorted to paying bribes for awarding contracts or obtain favors.					

Corruption normalization

Please indicate if you consider the following actions to be justified or unjustified, with 1 being 'never justified' and 5 'always justified'.

Questions	1	2	3	4	5
Question 19: Claim state benefits to which you are not entitled such as scholarships, grants, benefits, or subsidies					
Question 20: Avoid paying the ticket in some public transport					
Question 21: Avoid paying the bill in a restaurant or cafeteria					
Question 22: Cheat on paying taxes					
Question 23: That someone accept a bribe in the performance of their duties.					

Appendix B. Exploratory Factor Analysis (EFA) for the Construction of the Variables Entrepreneurial Intention (EI), Corruption Perception (COPER), and Normalization of Corruption (CONOR)

Entrepreneurial Intention (EI)

For years, there has been a classic debate about the number of options that Likert scales should incorporate [125]. However, there is consensus on the adequacy of 1–3 Likert scales [126], since it is of great importance that any scale includes a central point [127]. In our study, we used a 1–3 Likert scale to obtain the polarized information requested. The reliability analysis allowed us to validate the internal consistency of the scale proposed by Liñán and Chen [25]. The first step was to recode question 85, the sense of which was the reverse. However, Cronbach's alpha improved substantially when eliminating the inverse variable of this question.

Table A9. Variable statistics.

	Statistics Total-Item			
	Mean of Scale If Item Is Removed	Scale Variance If Item Is Removed	Corrected Item—Total Correlation	Cronbach's Alpha If Item Is Removed
Inverse variable of Question 6	10.930818	3.204	0.031	0.892
Question 7	10.364780	2.853	0.647	0.509
Question 8	10.314465	2.976	0.628	0.527
Question 9	10.301887	2.997	0.651	0.524
Question 10	10.389937	2.910	0.574	0.536

Cronbach's Alpha calculated after eliminating the inverse variable of question 6, an adequate scale reliability, with a value greater than 0.8.

Table A10. Reliability statistics.

Cronbach's Alpha	Items
0.867	4

An exploratory factor analysis was carried out with the following remaining variables, and since the variables did not present normality, following Liñan and Chen [25], a main axis factorization was applied as an extraction method. The resulting factor analysis presented an adequate goodness of fit, the level of significance of Bartlett's sphericity test is less than 0.05, and the Kaiser–Meyer–Olkin index (KMO) exceeded the minimum value of 0.50 and was very close to 1, so its significance was optimal.

Table A11. KMO and Bartlett's test.

Kaiser–Meyer–Olkin measure of sampling adequacy		0.829
Bartlett's test of sphericity	Chi-squared	709.035
	df	6
	Sig.	0.000

There was only one factor with an eigenvalue greater than 1. As a consequence, a single factor was extracted.

Table A12. Total variance explained.

Factor	Initial Eigenvalues			Sums of the Squared Saturations of the Extraction		
	Total	Variance	Accumulated	Total	Variance	Accumulated
1	2.863	71.580	71.580	2.488	62.198	62.198
2	0.419	10.478	82.058			
3	0.401	10.033	92.091			
4	0.316	7.909	100.000			

Extraction method: factorization of main axes.

The factorial loads are shown in Table A13.

Table A13. Factor matrix.

Factor Matrix	
Questions	Factor
	1
7	0.809
8	0.767
9	0.826
10	0.750

Extraction method: factorization of the principal axis. One factor was extracted, and six iterations were required.

Next, the factor scores of each student for the entrepreneurial intention factor (EI) were calculated using regression. The coefficients of the factor scores were as follows:

Table A14. Coefficient matrix for calculating factor scores.

Questions	Factor
	1
7	0.300
8	0.243
9	0.333
10	0.223

Extraction method: factorization of the principal axis. Rotation method: varimax normalization with Kaiser. Factor score method: regression.

Corruption Perception (COPER) and Corruption Normalization (CONOR)

The main axis factorization [25] was applied as an extraction method because the available data did not show normality. In addition, following Kaiser [128], a varimax rotation was applied, with the aim of ensuring that the factors had a few high saturations in the variables and many that were almost zero. Thus, there were factors with high correlations and with a small number of variables and with null correlations, thus leaving the variance of the factors redistributed.

The first factorial analysis presented a correct goodness of fit: The significance level of the Bartlett sphericity test was less than 0.05, and the Kaiser–Meyer–Olkin index (KMO) reached a minimum value greater than 0.50 and was very close to 1, therefore its significance is very high.

Table A15. KMO and Bartlett's test.

Kaiser–Meyer–Olkin measure of sampling adequacy		0.791
Bartlett's test of sphericity	Chi-squared	2778.113
	df	78
	Sig.	0.000

The correspondence of factors with corruption perception (Factor 1, COPER) and another that corresponded to corruption normalization (Factor 2, CONOR) was noted.

Table A16. Rotated factor matrix.

	Factor			
	1	2	3	4
Question 11	0.275	0.206	0.259	0.511
Question 12	0.787	0.006	0.061	0.047
Question 13	0.837	0.030	0.089	0.105
Question 14	0.889	0.013	0.090	−0.003
Question 15	0.726	0.040	0.162	0.131
Question 16	0.233	0.130	0.839	0.100
Question 17	0.100	0.120	0.790	0.107
Question 18	0.575	0.050	0.501	−0.417
Question 19	0.006	0.350	0.310	−0.168
Question 20	0.054	0.830	0.065	0.151
Question 21	−0.045	0.819	0.204	−0.013
Question 22	0.000	0.815	0.181	−0.061
Question 23	0.132	0.803	−0.135	0.346

Extraction method: factorization of the principal axis. Rotation method: varimax normalization with Kaiser. The rotation converged in five iterations.

The empirical evidence showed two additional factors that were not used in this study. Consequently, the items that did not load on the two indicated factors were ignored, and the exploratory factor analysis was carried out again for questions 12–15 and 20–23.

In the second analysis, the goodness-of-fit tests were also adequate:

Table A17. KMO and Bartlett's test.

Kaiser–Meyer–Olkin measure of sampling adequacy		0.805
Bartlett's test of sphericity	Chi-squared	1774.709
	df	28
	Sig.	0.000

Two factors with eigenvalues greater than 1 were reported:

Table A18. Total variance explained.

Factor	Initial Eigenvalues			Sums of the Squared Saturations of the Extraction			Sum of the Squared Saturations of the Rotation		
	Total	Variance	Accumulated	Total	Variance	Accumulated	Total	Variance	Accumulated
1	3.263	40.789	40.789	2.941	36.759	36.759	2.708	33.849	33.849
2	2.774	34.669	75.458	2.455	30.693	67.452	2.688	33.603	67.452
3	0.443	5.542	81.000						
4	0.404	5.046	86.047						
5	0.364	4.546	90.593						
6	0.294	3.677	94.270						
7	0.267	3.335	97.605						
8	0.192	2.395	100.000						

Extraction method: factorization of main axes.

The factor load analysis of the rotated factor matrix showed a correspondence with the factors of corruption perception (Factor 1, COPER) and corruption normalization (Factor 2, CONOR).

Table A19. Rotated factor matrix.

Questions	Factor	
	1	2
12	0.797	0.015
13	0.861	0.051
14	0.876	0.020
15	0.736	0.062
20	0.068	0.843
21	−0.040	0.836
22	−0.002	0.811
23	0.129	0.783

Extraction method: factorization of the principal axis. Rotation method: varimax normalization with Kaiser. The rotation converged in three iterations.

As indicated, the factor scores of each individual were calculated for the COPER factor and for the CONOR factor by means of a regression. The coefficients for the calculation are shown in Table A20.

Table A20. Coefficient matrix for calculating factor scores.

	Factor 1	Factor 2
12	0.208	−0.014
13	0.336	−0.001
14	0.377	−0.022
15	0.153	0.002
20	−0.001	0.316
21	−0.028	0.290
22	−0.026	0.261
23	0.019	0.222

Extraction method: factorization of the principal axis. Rotation method: varimax normalization with Kaiser. Factor score method: regression.

The reliability analysis allows the checking of the internal consistency of both factors. For this, Cronbach's alpha was used. As we can see, the scale was reliable, and it was not necessary to eliminate any variable in both cases. The reliability of the COPER scale was correct, since Cronbach's alpha was greater than 0.8. Cronbach's alpha did not improve when removing an item.

Table A21. Reliability statistics.

Cronbach's Alpha	Items
0.889	4

Table A22. Statistics total-item.

Questions	Mean of Scale If Item Is Removed	Scale Variance If Item Is Removed	Corrected Item—Total Correlation	Cronbach's Alpha If Item Is Removed
12	10.836842	11.862	0.741	0.865
13	10.823684	11.945	0.796	0.842
14	10.734211	12.148	0.804	0.840
15	10.836842	12.997	0.691	0.881

The reliability of the COPER scale was correct, since Cronbach's alpha was greater than 0.8. Cronbach's alpha did not improve when removing an item.

Table A23. Reliability statistics.

Cronbach's Alpha	Items
0.889	4

Table A24. Statistics total-item.

	Statistics Total-Item			
Questions	Mean of Scale If Item Is Removed	Scale Variance If Item Is Removed	Corrected Item-Total Correlation	Cronbach's Alpha If Item Is Removed
20	7.492105	14.446	0.782	0.849
21	7.702632	14.020	0.763	0.855
22	7.742105	14.139	0.755	0.858
23	7.600000	13.713	0.731	0.869

Appendix C. Model Proposed, Truth Table Analysis, Quine–McCluskey Algorithm, and HHI Subsample

--- COMPLEX SOLUTION ---

frequency cutoff: 1

consistency cutoff: 0.756379

	raw coverage	unique coverage	consistency
~GEN*fs_COPER*~fs_CONOR*~FEB	0.19714	0.19714	0.840891
~GEN*~fs_CONOR*FEB*~STEM	0.228865	0.228865	0.819662
GEN*~fs_ COPER *~fs_CONOR*~FEB*~STEM	0.0492215	0.0492214	0.850855
~GEN*~fs_COPER*fs_CONOR*FEB*STEM	0.0448499	0.0448498	0.834135
GEN*fs_COPER*fs_CONOR*FEB*~STEM	0.102893	0.102893	0.764039
GEN*fs_COPER*~fs_CONOR*FEB*STEM	0.0461908	0.0461907	1

solution coverage: 0.66916

solution consistency: 0.830081

--- PARSIMONIOUS SOLUTION ---

frequency cutoff: 1

consistency cutoff: 0.756379

	raw coverage	unique coverage	consistency
~fs_CONOR*~FEB*~STEM	0.165402	0.0498559	0.862856
~GEN*~fs_COPER*FEB	0.211691	0.0678778	0.909417
fs_COPER*~fs_CONOR*STEM	0.170757	0.0438911	0.827329
GEN*fs_COPER*fs_CONOR	0.118344	0.102893	0.761817
~GEN*~fs_CONOR*~STEM	0.335593	0	0.831482
~GEN*fs_COPER*~fs_CONOR	0.344406	0	0.797524

solution coverage: 0.735795

solution consistency: 0.826768

--- INTERMEDIATE SOLUTION ---

frequency cutoff: 1

consistency cutoff: 0.756379

	raw coverage	unique coverage	consistency
~GEN*fs_COPER*~fs_CONOR*~FEB	0.19714	0.19714	0.840891
~GEN*~fs_CONOR*FEB*~STEM	0.228865	0.228865	0.819662
GEN*~fs_COPER*~fs_CONOR*~FEB*~STEM	0.0492215	0.0492214	0.850855
~GEN*~fs_COPER*fs_CONOR*FEB*STEM	0.0448499	0.0448498	0.834135
GEN*fs_COPER*fs_CONOR*FEB*~STEM	0.102893	0.102893	0.764039
GEN*fs_COPER*~fs_CONOR*FEB*STEM	0.0461908	0.0461907	1

solution coverage: 0.66916

solution consistency: 0.830081

Appendix D. Model Proposed. Truth Table Analysis. Quine-McCluskey Algorithm. LHI Subsample

--- COMPLEX SOLUTION ---

frequency cutoff: 1

consistency cutoff: 0.764857

	raw coverage	unique coverage	consistency
fs_GEN*~fs_CONOR*fs_FEB*~fs_STEM	0.129524	0.0697712	0.787042
fs_GEN*~fs_COPER*fs_FEB*~fs_STEM	0.0873433	0.02759	0.751201
~fs_GEN*fs_COPER*fs_FEB*~fs_STEM	0.0609087	0.0609087	0.916262
Fs_GEN*fs_CONOR*fs_FEB*fs_STEM	0.05029	0.0502899	0.910388
~fs_GEN*~fs_COPER*~fs_CONOR*fs_FEB*fs_STEM	0.0242887	0.0242887	0.769856

solution coverage: 0.292602

solution consistency: 0.799093

--- PARSIMONIOUS SOLUTION ---

frequency cutoff: 1

consistency cutoff: 0.764857

	raw coverage	unique coverage	consistency
fs_GEN*~fs_CONOR*fs_FEB*~fs_STEM	0.129524	0.0697712	0.787042
fs_GEN*~fs_COPER*fs_FEB*~fs_STEM	0.0873433	0.02759	0.751201
~fs_GEN*fs_COPER*fs_FEB*~fs_STEM	0.0609087	0.0609087	0.916262
fs_GEN*fs_CONOR*fs_FEB*fs_STEM	0.05029	0.0502899	0.910388
~fs_GEN*~fs_COPER*~fs_CONOR*fs_FEB*fs_STEM	0.0242887	0.0242887	0.769856

solution coverage: 0.292602

solution consistency: 0.799093

--- INTERMEDIATE SOLUTION ---

frequency cutoff: 1

consistency cutoff: 0.764857

	raw coverage	unique coverage	consistency
fs_GEN*~fs_CONOR*fs_FEB*~fs_STEM	0.129524	0.0697712	0.787042
fs_GEN*~fs_COPER*fs_FEB*~fs_STEM	0.0873433	0.02759	0.751201
~fs_GEN*fs_COPER*fs_FEB*~fs_STEM	0.0609087	0.0609087	0.916262
fs_GEN*fs_CONOR*fs_FEB*fs_STEM	0.05029	0.0502899	0.910388
~fs_GEN*~fs_COPER*~fs_CONOR*fs_FEB*fs_STEM	0.0242887	0.0242887	0.769856

solution coverage: 0.292602

solution consistency: 0.799093

Appendix E. Stress Test of the Calibration Process. Truth Table Analysis. Quine-McCluskey Algorithm. HHI Subsample

--- COMPLEX SOLUTION ---

frequency cutoff: 1

consistency cutoff: 0.779939

	raw coverage	unique coverage	consistency
~fs_GEN*~fs_COPER*~fs_CONOR*~fs_STEM	0.0763878	0.0158165	0.852571
~fs_GEN*~fs_COPER*fs_FEB*~fs_STEM	0.118284	0.06063	0.860816
fs_GEN*fs_COPER*fs_FEB*fs_STEM	0.0300117	0.0300117	0.915681
fs_GEN*fs_COPER*~fs_CONOR*~fs_FEB*~fs_STEM	0.0415437	0.0415435	0.779939
~fs_GEN*fs_COPER*fs_CONOR*~fs_FEB*~fs_STEM	0.0220568	0.0191393	0.791179
~fs_GEN*fs_COPER*~fs_CONOR*~fs_FEB*fs_STEM	0.136314	0.136314	0.788508

solution coverage: 0.364027

solution consistency: 0.825238

--- PARSIMONIOUS SOLUTION ---

frequency cutoff: 1

consistency cutoff: 0.779939

	raw coverage	unique coverage	consistency
~fs_GEN*~fs_COPER*~fs_CONOR*~fs_STEM	0.0763878	0.0158165	0.852571
~fs_GEN*~fs_COPER*fs_FEB*~fs_STEM	0.118284	0.06063	0.860816
fs_GEN*fs_COPER*~fs_CONOR*~fs_FEB	0.0435986	0.0415435	0.719418
fs_COPER*~fs_CONOR*~fs_FEB*fs_STEM	0.138369	0.136314	0.767807
fs_GEN*fs_COPER*fs_FEB*fs_STEM	0.0300117	0.0300116	0.915681
~fs_GEN*fs_COPER*fs_CONOR*~fs_FEB*~fs_STEM	0.0220568	0.0191393	0.791179

solution coverage: 0.366082

solution consistency: 0.816318

--- INTERMEDIATE SOLUTION ---

frequency cutoff: 1

consistency cutoff: 0.779939

	raw coverage	unique coverage	consistency
~fs_GEN*~fs_COPER*~fs_CONOR*~fs_STEM	0.0763878	0.0158165	0.852571
~fs_GEN*~fs_COPER*fs_FEB*~fs_STEM	0.118284	0.06063	0.860816
fs_GEN*fs_COPER*fs_FEB*fs_STEM	0.0300117	0.0300117	0.915681
fs_GEN*fs_COPER*~fs_CONOR*~fs_FEB*fs_STEM	0.0415437	0.0415435	0.779939
~fs_GEN*fs_COPER*fs_CONOR*~fs_FEB*~fs_STEM	0.0220568	0.0191393	0.791179
~fs_GEN*fs_COPER*~fs_CONOR*~fs_FEB*fs_STEM	0.136314	0.136314	0.788508

solution coverage: 0.364027

solution consistency: 0.825238

Appendix F. Stress Test of the Calibration Process. Truth Table Analysis. Quine-McCluskey Algorithm. LHI Subsample

--- COMPLEX SOLUTION ---

frequency cutoff: 1

consistency cutoff: 0.788089

	raw coverage	unique coverage	consistency
fs_GEN*~fs_CONOR*fs_FEB*~fs_STEM	0.119856	0.119856	0.772482
~fs_GEN*fs_COPER*fs_FEB*~fs_STEM	0.0609165	0.0609164	0.931511
fs_GEN*fs_CONOR*fs_FEB*fs_STEM	0.0508182	0.0508181	0.895114
~fs_GEN*~fs_COPER*~fs_CONOR*fs_FEB*fs_STEM	0.0233059	0.0233059	0.792562

solution coverage: 0.254896

solution consistency: 0.83101

--- PARSIMONIOUS SOLUTION ---

frequency cutoff: 1

consistency cutoff: 0.788089

	raw coverage	unique coverage	consistency
fs_GEN*~fs_CONOR*fs_FEB*~fs_STEM	0.119856	0.119856	0.772482
~fs_GEN*fs_COPER*fs_FEB*~fs_STEM	0.0609165	0.0609164	0.931511
fs_GEN*fs_CONOR*fs_FEB*fs_STEM	0.0508182	0.0508181	0.895114
~fs_GEN*~fs_COPER*~fs_CONOR*fs_FEB*fs_STEM	0.0233059	0.0233059	0.792562

solution coverage: 0.254896

solution consistency: 0.83101

--- INTERMEDIATE SOLUTION ---

frequency cutoff: 1

consistency cutoff: 0.788089

	raw coverage	unique coverage	consistency
fs_GEN*~fs_CONOR*fs_FEB*~fs_STEM	0.119856	0.119856	0.772482
~fs_GEN*fs_COPER*fs_FEB*~fs_STEM	0.0609165	0.0609164	0.931511
fs_GEN*fs_CONOR*fs_FEB*fs_STEM	0.0508182	0.0508181	0.895114
~fs_GEN*~fs_COPER*~fs_CONOR*fs_FEB*fs_STEM	0.0233059	0.0233059	0.792562

solution coverage: 0.254896

solution consistency: 0.83101

Appendix G. Components and Calculation According to the Castelló-Sirvent (2021) Robustness Coefficient

Table A25. Components and calculation.

RC Components
RC: Robustness coefficient
CG: Consistency gap
AC: Average consistency
MC_i: Model consistency
STC_i: Stress-Test consistency
N: Total number of outcomes and subsamples
RC Calculation
$RC = 1 - (|CG|/AC)$
$AC = \frac{\sum_{i=1}^{n} MC_i}{N}$
$CG = \frac{\sum_{i=1}^{n} (MC_i - STC_i)}{N}$

Note: RC-value includes intermediate solution consistency scores for both subsamples (HHI and LHI) used to test the model in fsQCA. Source: adapted from Castelló-Sirvent [118].

Table A26. Robustness analysis of the model in fsQCA using RC-value.

Recalibration				RC-Value	Robustness	Symbol [*]
Percentile Variation	Fully Inside	Maximum Ambiguity	Fully Outside			
±0.15	−0.15	0	+0.15	$0.9900 \leq RC \leq 1$	Very Strong	***
				$0.9500 \leq RC \leq 0.9899$	Strong	**
±0.10	−0.10	0	+0.10	$0.9900 \leq RC \leq 1$	Strong	**
				$0.9500 \leq RC \leq 0.9899$	Moderate	*
				$0.9000 \leq RC \leq 0.9499$	Weak	

[*] Weak robustness is pointed out without symbol. Source: adapted from Castelló-Sirvent [118].

References

1. Warf, B.; Storper, M. *The Regional World: Territorial Development in a Global Economy*; Guilford Press: New York, NY, USA.
2. Alburquerque, F.; Pérez, S. El desarrollo territorial: Enfoque, contenido y políticas. *Rev. Iberoam. Gob. Local RIGL.* **2013**, *4*, 1–24.
3. Minniti, M. The role of government policy on entrepreneurial activity: Productive, unproductive, or destructive? *Entrep. Theory Pract.* **2008**, *32*, 779–790. [CrossRef]
4. Acs, Z.J.; Szerb, L. Entrepreneurship, economic growth and public policy. *Small Bus. Econ.* **2007**, *28*, 109–122. [CrossRef]
5. Boettke, P.J.; Coyne, C.J. Context matters: Institutions and entrepreneurship. *Found. Trends Entrep.* **2007**, *5*, 135–209. [CrossRef]
6. Blanchflower, D.G.; Oswald, A.; Stutzer, A. Latent entrepreneurship across nations. *Eur. Econ. Rev.* **2001**, *45*, 680–691. [CrossRef]

7. Verheul, I.; Thurik, R.; Grilo, I. Determinants of Self-Employment Preference and Realization of Women and Men in Europe and the United States. Available online: https://core.ac.uk/download/pdf/7074434.pdf (accessed on 22 April 2021).
8. Thompson, E.R. Individual entrepreneurial intent: Construct clarification and development of an internationally reliable metric. *Entrep. Theory Pract.* **2009**, *33*, 669–694. [CrossRef]
9. Aldrich, H.E. Entrepreneurial strategies in new organizational populations. In *Entrepreneurship: The Social Science View*; Swedberg, R., Ed.; Oxford University Press: Oxford, UK, 2000; pp. 211–288.
10. Brush, C.G.; Duhaime, I.M.; Gartner, W.B.; Stewart, A.; Katz, J.A.; Hitt, M.A.; Alvarez, S.A.; Meyer, G.D.; Venkataraman, S. Doctoral education in the field of entrepreneurship. *J. Manag.* **2003**, *29*, 309–331. [CrossRef]
11. Fritsch, M.; Storey, D.J. Entrepreneurship in a regional context: Historical roots, recent developments and future challenges. *Reg. Stud.* **2014**, *48*, 939–954. [CrossRef]
12. Liñán, F.; Moriano, J.A.; Jaén, I. Individualism and entrepreneurship: Does the pattern depend on the social context? *Int. Small Bus. J. Res. Entrep.* **2016**, *34*, 760–776. [CrossRef]
13. Transparency Internacional. Corruption Perceptions Index. 2020. Available online: https://www.transparency.org/en/cpi/2020/index/mex (accessed on 22 April 2021).
14. Broadman, H.G.; Recanatini, F. Seeds of corruption—Do market institutions matter? *MOCT MOST Econ. Policy Transit. Econ.* **2001**, *11*, 359–392. [CrossRef]
15. Rose-Ackerman, S. *International Handbook on the Economics of Corruption*; Edward Elgar Publishing: Cheltenham, UK, 2007.
16. World Bank. Doing Business. 2020. Available online: https://openknowledge.worldbank.org/bitstream/handle/10986/32436/9781464814402.pdf (accessed on 22 April 2021).
17. Autio, E.; Fu, K. Economic and political institutions and entry into formal and informal entrepreneurship. *Asia Pac. J. Manag.* **2015**, *32*, 67–94. [CrossRef]
18. Dreher, A.; Gassebner, M. Greasing the wheels? The impact of regulations and corruption on firm entry. *Public Choice* **2013**, *155*, 413–432. [CrossRef]
19. GEM. Entrepreneurial Behaviour and Attitudes. 2020. Available online: https://www.gemconsortium.org/economy-profiles/mexico-2 (accessed on 22 April 2021).
20. Smeding, A. Women in science, technology, engineering, and mathematics (STEM): An investigation of their implicit gender stereotypes and stereotypes' connectedness to math performance. *Sex Roles* **2012**, *67*, 617–629. [CrossRef]
21. Poggesi, S.; Mari, M.; De Vita, L.; Foss, L. Women entrepreneurship in STEM fields: Literature review and future research avenues. *Int. Entrep. Manag. J.* **2019**, *16*, 17–41. [CrossRef]
22. Settles, I.H.; Cortina, L.M.; Malley, J.; Stewart, A.J. The climate for women in academic science: The good, the bad, and the changeable. *Psychol. Women Q.* **2006**, *30*, 47–58. [CrossRef]
23. Díaz-García, C.; González-Moreno, A.; Martínez, F.J.S. Gender diversity within R&D teams: Its impact on radicalness of innovation. *Innovation* **2013**, *15*, 149–160. [CrossRef]
24. Aldrich, H.E.; Cliff, J.E. The pervasive effects of family on entrepreneurship: Toward a family embeddedness perspective. *J. Bus. Ventur.* **2003**, *18*, 573–596. [CrossRef]
25. Liñán, F.; Chen, Y.-W. Development and cross-cultural application of a specific instrument to measure entrepreneurial intentions. *Entrep. Theory Pract.* **2009**, *33*, 593–617. [CrossRef]
26. Reynolds, P.D.; Camp, S.M.; Bygrave, W.D.; Autio, E.; Hay, M. Global Entrepreneurship Monitor Gem 2001 Summary Report. 2002. Available online: https://www.gemconsortium.org/report/gem-2001-global-report (accessed on 22 April 2021).
27. De Clercq, D.; Honig, B.; Martin, B. The roles of learning orientation and passion for work in the formation of entrepreneurial intention. *Int. Small Bus. J. Res. Entrep.* **2012**, *31*, 652–676. [CrossRef]
28. Luthje, C.; Franke, N. The making of an entrepreneur: Testing a model of entrepreneurial intent among engineering students at MIT. *R D Manag.* **2003**, *33*, 135–147. [CrossRef]
29. Franke, N.; Lüthje, C. Entrepreneurial intentions of business students—A benchmarking study. *Int. J. Innov. Technol. Manag.* **2004**, *1*, 269–288. [CrossRef]
30. Kraus, S.; Ribeiro-Soriano, D.; Schüssler, M. Fuzzy-set qualitative comparative analysis (fsQCA) in entrepreneurship and innovation research—The rise of a method. *Int. Entrep. Manag. J.* **2018**, *14*, 15–33. [CrossRef]
31. Fayolle, A.; Liñán, F.; Moriano, J.A. Beyond entrepreneurial intentions: Values and motivations in entrepreneurship. *Int. Entrep. Manag. J.* **2014**, *10*, 679–689. [CrossRef]
32. Gartner, W.B. A new path to the waterfall: A narrative on a use of entrepreneurial narrative. *Entrep. Organ.* **2016**, *28*, 326–339. [CrossRef]
33. Woodside, A.G. Moving beyond multiple regression analysis to algorithms: Calling for adoption of a paradigm shift from symmetric to asymmetric thinking in data analysis and crafting theory. *J. Bus. Res.* **2013**, *66*, 463–472. [CrossRef]
34. Acs, Z.J.; Armington, C. Employment growth and entrepreneurial activity in cities. *Reg. Stud.* **2004**, *38*, 911–927. [CrossRef]
35. Levie, J.; Autio, E. A theoretical grounding and test of the GEM model. *Small Bus. Econ.* **2008**, *31*, 235–263. [CrossRef]
36. López-Claros, A.; Altinger, L.; Blanke, J.; Drzeniek, M.; Mía, I. Assessing latin american competitiveness: Challenges and opportunities. In *The Latin America Competitiveness Review 2006*; López-Claros, A., Ed.; World Economic Forum: Cologny, Switzerland, 2006; pp. 3–36.

37. Audretsch, D.B. What's new about the new economy? Sources of growth in the managed and entrepreneurial economies. *Ind. Corp. Chang.* **2001**, *10*, 267–315. [CrossRef]
38. Audretsch, D.B. *The Entrepreneurial Society*; Oxford University Press: Oxford, UK, 2007.
39. Moriano, J.A.; Trejo, E.; Palací, F.J. El perfil psicosocial del emprendedor: Un estudio desde la perspectiva de los valores. *Int. J. Soc. Psychol.* **2001**, *16*, 229–242. [CrossRef]
40. Kessler, A.; Frank, H. Nascent entrepreneurship in a longitudinal perspective. *Int. Small Bus. J. Res. Entrep.* **2009**, *27*, 720–742. [CrossRef]
41. Bird, B. Implementing entrepreneurial ideas: The case for intention. *Acad. Manag. Rev.* **1988**, *13*, 442–453. [CrossRef]
42. Liñán, F.; Nabi, G.; Krueger, N. British and Spanish entrepreneurial intentions: A comparative study. *Rev. Econ. Mund.* **2013**, *33*, 73–103.
43. Kautonen, T.; Van Gelderen, M.; Fink, M. Robustness of the theory of planned behavior in predicting entrepreneurial intentions and actions. *Entrep. Theory Pract.* **2015**, *39*, 655–674. [CrossRef]
44. Krueger, N.F. Entrepreneurial intentions are dead: Long live entrepreneurial intentions. In *Revisiting the Entrepreneurial Mind*; Brännback, M., Carsrud, A., Eds.; Springer: Berlin, Germany, 2017; pp. 13–14.
45. Chowdhury, F.; Desai, S.; Audretsch, D.B. Corruption, entrepreneurship, and social welfare. In *Corruption, Entrepreneurship, and Social Welfare*; Chowdhury, S., Desai, S., Audretsch, D.B., Eds.; Springer: Berlin, Germany, 2018; pp. 67–94.
46. Saxenian, A. *Regional Advantage: Culture and Advantage in Silicon Valley and Route 128*; Harvard University Press: Cambridge, MA, USA, 1996.
47. Kenney, M. Technology, entrepreneurship and path dependence: Industrial clustering in Silicon Valley and Route 128. *Ind. Corp. Chang.* **1999**, *8*, 67–103. [CrossRef]
48. Dutta, N.; Sobel, R. Does corruption ever help entrepreneurship? *Small Bus. Econ.* **2016**, *47*, 179–199. [CrossRef]
49. Estrin, S.; Korosteleva, J.; Mickiewicz, T. Which institutions encourage entrepreneurial growth aspirations? *J. Bus. Ventur.* **2013**, *28*, 564–580. [CrossRef]
50. Teece, D.J. The market for know-how and the efficient international transfer of technology. *Ann. Am. Acad. Political Soc. Sci.* **1981**, *458*, 81–96. [CrossRef]
51. Luhmann, N. Familiarity, confidence, and trust: Problems and alternatives. In *Trust: Making and Breaking Cooperative Relations*; Gametta, D., Ed.; Basil Blackwell: Oxford, UK, 1988; pp. 94–108.
52. Rose-Ackerman, S. Trust, honesty and corruption: Reflection on the state-building process. *Eur. J. Sociol.* **2001**, *42*, 526–570. [CrossRef]
53. Block, J.; Sandner, P.; Spiegel, F. How do risk attitudes differ within the group of entrepreneurs? The role of motivation and procedural utility. *J. Small Bus. Manag.* **2013**, *53*, 183–206. [CrossRef]
54. Olken, B.A. Corruption perceptions vs. corruption reality. *J. Public Econ.* **2009**, *93*, 950–964. [CrossRef]
55. Donchev, D.; Ujhelyi, G. What do corruption indices measure? *Econ. Politics* **2014**, *26*, 309–331. [CrossRef]
56. Gutmann, J.; Padovano, F.; Voigt, S. Perception vs. experience: Explaining differences in corruption measures using microdata. *Eur. J. Political Econ.* **2020**, *65*, 101925. [CrossRef]
57. Guerber, A.; Rajagoplan, A.; Anand, V. The influence of national culture on the rationalization of corruption. In *Crime and Corruption in Organizations: Why It Occurs and What to Do about It*; Burke, R.J., Tomlinson, E.C., Cooper, C.L., Eds.; Routledge: London, UK, 2016; pp. 143–162.
58. Ashforth, B.E.; Anand, V. The normalization of corruption in organizations. *Res. Organ. Behav.* **2003**, *25*, 1–52. [CrossRef]
59. Setti, Z. Entrepreneurial intentions among youth in MENA countries: Effects of gender, education, occupation and income. *Int. J. Entrep. Small Bus.* **2017**, *30*, 308. [CrossRef]
60. Autio, E. GEM Report on High-Expectation Entrepreneurship. 2005. Available online: http://negocios.udd.cl/gemchile/files/2010/12/HEE-2005.pdf (accessed on 22 April 2021).
61. Schøtt, T.; Kew, P.; Cheraghi, M. Future Potential: A GEM Perspective on Youth Entrepreneurship, Global Entrepreneurship Monitor. Youth Economic Opportunities. 2015. Available online: https://youtheconomicopportunities.org/resource/2744/future-potential-gem-perspective-youth-entrepreneurship (accessed on 22 April 2021).
62. Weeks, J.R.; Seiler, D. Women's Entrepreneurship in Latin America: An Exploration of Current Knowledge. Inter-American Development Bank. 2001. Available online: http://publications.iadb.org/handle/11319/5065 (accessed on 22 April 2021).
63. Acs, Z.J.; Desai, S.; Klapper, L.F. What does "entrepreneurship" data really show? *Small Bus. Econ.* **2008**, *31*, 265–281. [CrossRef]
64. Allen, E.; Elam, A.; Langowitz, N.; Dean, M. The Global Entrepreneurship Monitor (GEM) 2007 Report on Women and Entrepreneurship. Center for Women's Leadership, Babson College. 2008. Available online: http://sites.telfer.uottawa.ca/womensenterprise/files/2014/06/GEM-2003_Eng.pdf (accessed on 22 April 2021).
65. Amorós, J.E.; Pizarro, O. Women entrepreneurship context in Latin America: An exploratory study in Chile. In *The Perspective of Women's Entrepreneurship in the Age of Globalization*; Markovic, M.R., Ed.; Information Age Publishing: Charlotte, NC, USA, 2007; pp. 107–126.
66. Boehm, F.; Sierra, E. The Gendered Impact of Corruption: Who Suffers More? Men or Women? 2015. Available online: http://hdl.handle.net/11250/2475280 (accessed on 22 April 2021).
67. Goel, R.K.; Nelson, M.A. Corrupt encounters of the fairer sex: Female entrepreneurs and their corruption perceptions/experience. *J. Technol. Transf.* **2021**, *46*, 1–22. [CrossRef]

68. Bruhn, M. Female-owned firms in Latin America : Characteristics, performance, and obstacles to growth. *Policy Res. Work. Paper Ser.* **2009**, *11*, 5122. [CrossRef]
69. World Bank. Mujeres Empresarias: Barreras Y Oportunidades en el Sector Privado Formal en América Latina Y el Caribe. 2010. Available online: https://dds.cepal.org/redesoc/publication?id=2177 (accessed on 22 April 2021).
70. CEPAL. Mujeres Emprendedoras en América Latina Y el Caribe: Realidades, Obstáculos Y Desafíos. 2010. Available online: http://repositorio.cepal.org/handle/11362/5818 (accessed on 22 April 2021).
71. DE Vita, L.; Mari, M.; Poggesi, S. Women entrepreneurs in and from developing countries: Evidences from the literature. *Eur. Manag. J.* **2014**, *32*, 451–460. [CrossRef]
72. Smith-Hunter, A.; Leone, J. Evidence on the characteristics of women entrepreneurs in Brazil: An empirical analysis. *Int. J. Manag. Mark. Res.* **2010**, *3*, 85–102.
73. Terjesen, S.; Amorós, J.E. Female entrepreneurship in Latin America and the Caribbean: Characteristics, drivers and relationship to economic development. *Eur. J. Dev. Res.* **2010**, *22*, 313–330. [CrossRef]
74. Halabisky, D. Policy Brief on Women's Entrepreneurship. 2018. Available online: https://dspace.ceid.org.tr/xmlui/handle/1/900 (accessed on 15 June 2021).
75. OECD. Transformative Technologies and Jobs of the Future. 2018. Available online: https://www.oecd.org/innovation/transformative-technologies-and-jobs-of-the-future.pdf (accessed on 15 June 2021).
76. Carree, M.A.; Thurik, A.R. The impact of entrepreneurship on economic growth. In *Handbook of Entrepreneurship Research: An Interdisciplinary Survey and Introduction*, 2nd ed.; Acs, Z.J., Audretsch, D.B., Eds.; Springer: Berlin, Germany, 2010; pp. 557–594.
77. Decker, R.; Haltiwanger, J.; Jarmin, R.; Miranda, J. The role of entrepreneurship in US job creation and economic dynamism. *J. Econ. Perspect.* **2014**, *28*, 3–24. [CrossRef]
78. López-Delgado, P.; Iglesias-Sánchez, P.P.; Jambrino-Maldonado, C. Gender and university degree: A new analysis of entrepreneurial intention. *Educ. Train.* **2019**, *61*, 797–814. [CrossRef]
79. Sušanj, Z.; Jakopec, A.; Miljković Krečar, I. Verifying the model of predicting entrepreneurial intention among students of business and non-business orientation. *Manag. J. Contemp. Manag. Issues* **2015**, *20*, 49–69.
80. Solesvik, M.Z. Entrepreneurial motivations and intentions: Investigating the role of education major. *Educ. Train.* **2013**, *55*, 253–271. [CrossRef]
81. Maresch, D.; Harms, R.; Kailer, N.; Wimmer-Wurm, B. The impact of entrepreneurship education on the entrepreneurial intention of students in science and engineering versus business studies university programs. *Technol. Forecast. Soc. Chang.* **2016**, *104*, 172–179. [CrossRef]
82. Kelley, D.J.; Baumer, B.S.; Brush, C.; Greene, P.G.; Mahdavi, M.; Majbouri, M.; Cole, M.; Dean, M.; Heavlow, R. Global Entrepreneurship Monitor. Women's Entrepreneurship 2016/2017 Report. 2017. Available online: http://www.fundacionmicrofinanzasbbva.org/revistaprogreso/wp-content/uploads/2017/12/gem-womens-2016-2017-report-v11df-1504758645.pdf (accessed on 15 June 2021).
83. Matthews, C.H.; Moser, S.B. A longitudinal investigation of the impact of family background and gender on interest in small firm ownership. *J. Small Bus. Manag.* **1996**, *34*, 29–43.
84. Morrison, A. Entrepreneurship: What triggers it? *Int. J. Entrep. Behav. Res.* **2000**, *6*, 59–71. [CrossRef]
85. Krueger, N. The impact of prior entrepreneurial exposure on perceptions of new venture feasibility and desirability. *Entrep. Theory Pract.* **1993**, *18*, 5–21. [CrossRef]
86. Crant, J.M. The proactive personality scale as a predictor of entrepreneurial intentions. *J. Small Bus. Manag.* **1996**, *34*, 42–49.
87. Drennan, J.; Kennedy, J.; Renfrow, P. Impact of childhood experiences on the development of entrepreneurial intentions. *Int. J. Entrep. Innov.* **2005**, *6*, 231–238. [CrossRef]
88. Fayolle, A.; Gailly, B.; Lassas-Clerc, N. Assessing the impact of entrepreneurship education programmes: A new methodology. *J. Eur. Ind. Train.* **2006**, *30*, 701–720. [CrossRef]
89. Escolar-Llamazares, M.C.; Luis-Rico, I.; Torre-Cruz, T.; Herrero, Á.; Jiménez, A.; Palmero-Cámara, C.; Jiménez-Eguizábal, A. The socio-educational, psychological and family-related antecedents of entrepreneurial intentions among spanish youth. *Sustainability* **2019**, *11*, 1252. [CrossRef]
90. Hurley, A.E. Incorporating feminist theories into sociological theories of entrepreneurship. *Women Manag. Rev.* **1999**, *14*, 54–62. [CrossRef]
91. Arregle, J.-L.; Batjargal, B.; Hitt, M.A.; Webb, J.W.; Miller, T.; Tsui, A.S. Family ties in entrepreneurs' social networks and new venture growth. *Entrep. Theory Pract.* **2015**, *39*, 313–344. [CrossRef]
92. Dyer, J.W.G.; Handler, W. Entrepreneurship and family business: Exploring the connections. *Entrep. Theory Pract.* **1994**, *19*, 71–83. [CrossRef]
93. Marques, C.; Santos, G.; Galvão, A.; Mascarenhas, C.; Justino, E. Entrepreneurship education, gender and family background as antecedents on the entrepreneurial orientation of university students. *Int. J. Innov. Sci.* **2018**, *10*, 58–70. [CrossRef]
94. Moriano, J.A.; Palací, F.J.; Morales, J.F. El perfil psicosocial del emprendedor universitario. *J. Work Organ. Psychol.* **2006**, *22*, 75–99.
95. Kirkwood, J. Igniting the entrepreneurial spirit: Is the role parents play gendered? *Int. J. Entrep. Behav. Res.* **2007**, *13*, 39–59. [CrossRef]

96. Davila, A.; Hartman, A.M. Tradition and modern aspects of Mexican corporate culture. In *Mexican Business Culture: Essays on Tradition, Ethics, Entrepreneurship and Commerce and the State*; Coria-Sánchez, C.M., Hyatt, J.T., Eds.; McFarland & Company, Inc.: Ashe County, NC, USA, 2016; pp. 26–37.
97. Pittino, D.; Chirico, F.; Baù, M.; Villasana, M.; Naranjo-Priego, E.E.; Barron, E. Starting a family business as a career option: The role of the family household in Mexico. *J. Fam. Bus. Strat.* **2020**, *11*, 100338. [CrossRef]
98. Kuratko, D.F. Family business succession in Korean and US firms. *J. Small Bus. Manag.* **1993**, *31*, 132–137.
99. Transparency International. Índice de Percepción de la Corrupción 2017. 2017. Available online: https://transparencia.org.es/wp-content/uploads/2018/02/fuentes_datos_ipc-2017.pdf (accessed on 15 June 2021).
100. Transparency International. Corruption Perception Index 2018 Executive Summary. 2019. Available online: https://files.transparency.org/content/download/2383/14554/file/2018_CPI_Executive_Summary.pdf (accessed on 15 June 2021).
101. Ragin, C.C. *The Comparative Method: Moving beyond Qualitative and Quantitative Methods*; University of California press: Berkeley, CA, USA, 1987.
102. Ragin, C.C. *Redesigning Social Inquiry: Fuzzy Sets and Beyond*; University of Chicago Press: Chicago, IL, USA, 2008.
103. Cezar, R.F. Compliance in "exceptional" trade disputes: A set-theoretical approach. *Rev. Brasil. Política Int.* **2020**, *63*. [CrossRef]
104. McLevey, J. Think tanks, funding, and the politics of policy knowledge in Canada. *Can. Rev. Sociol. Can. de Sociol.* **2014**, *51*, 54–75. [CrossRef]
105. Fiss, P.C. Building better causal theories: A fuzzy set approach to typologies in organization research. *Acad. Manag. J.* **2011**, *54*, 393–420. [CrossRef]
106. Woodside, A.G. Proposing a new logic for data analysis in marketing and consumer behavior: Case study research of large-N survey data for estimating algorithms that accurately profile X (extremely high-use) consumers. *J. Glob. Sch. Mark. Sci.* **2012**, *22*, 277–289. [CrossRef]
107. Redding, K.; Viterna, J.S. Political demands, political opportunities: Explaining the differential success of left-libertarian parties. *Soc. Forces* **1999**, *78*, 491. [CrossRef]
108. Lacey, R.; Fiss, P.C. Comparative organizational analysis across multiple levels: A set-theoretic approach. In *Research in the Sociology of Organizations (Studying Differences between Organizations: Comparative Approaches to Organizational Research)*; King, B., Felin, T., Whetten, D., Eds.; Emerald Group Publishing Limited: Bingley, UK, 2009; pp. 91–116.
109. Rihoux, B. Diseños de investigación en QCA. In *Rihoux, Análisis Cualitativo Comparado (QCA)*; Medina, I., Castillo-Ortiz, P.J., Alamos-Concha, P., Rihoux, B., Eds.; Centro de Investigaciones Sociológicas: Madrid, Spain, 2017; pp. 53–66.
110. Ragin, C.C.; Pennings, P. Fuzzy sets and social research. *Sociol. Methods Res.* **2005**, *33*, 423–430. [CrossRef]
111. Fiss, P.C. A set-theoretic approach to organizational configurations. *Acad. Manag. Rev.* **2007**, *32*, 1180–1198. [CrossRef]
112. Arias-Oliva, M.; de Andrés-Sánchez, J.; Pelegrín-Borondo, J. Fuzzy set qualitative comparative analysis of factors influencing the use of cryptocurrencies in Spanish households. *Mathematics* **2021**, *9*, 324. [CrossRef]
113. Schneider, C.Q.; Wagemann, C. *Set-Theoretic Methods for the Social Sciences. A Guide to Qualitative Comparative Analysis*; Cambridge University Press: Cambridge, UK, 2012.
114. Misangyi, V.F.; Acharya, A.G. Substitutes or complements? A configurational examination of corporate governance mechanisms. *Acad. Manag. J.* **2014**, *57*, 1681–1705. [CrossRef]
115. Climent-Serrano, S.; Bustos-Contell, E.; Labatut-Serer, G.; Rey-Martí, A. Low-cost trends in audit fees and their impact on service quality. *J. Bus. Res.* **2018**, *89*, 345–350. [CrossRef]
116. Skaaning, S.-E. Assessing the robustness of crisp-set and fuzzy-set QCA results. *Sociol. Methods Res.* **2011**, *40*, 391–408. [CrossRef]
117. Stevens, A. Configurations of corruption: A cross-national qualitative comparative analysis of levels of perceived corruption. *Int. J. Comp. Sociol.* **2016**, *57*, 183–206. [CrossRef]
118. Castelló-Sirvent, F. Environmental and socio-technical transitions in IBEX 35 companies: fsQCA analysis of the media representation of innovation and sustainability. *Int. J. Innov. Sustain. Dev.* **2021**, in press.
119. Lowell, L.; Findlay, A.; Stewart, E. Brain Strain. Optimising Highly Skilled Migration from Developing Countries. 2004. Available online: https://www.ippr.org/files/images/media/files/publication/2011/05/brainstrain_1365.pdf (accessed on 22 April 2021).
120. Pinazo-Dallenbach, P.; Castelló-Sirvent, F. The effect of insecurity and corruption on opportunity-driven entrepreneurship in Mexico: An fsQCA analysis. *Acad. Rev. Latinoam. Adm.* **2020**, *34*, 105–121. [CrossRef]
121. Raijman, R. Determinants of entrepreneurial intentions: Mexican immigrants in Chicago. *J. Socio Econ.* **2001**, *30*, 393–411. [CrossRef]
122. Mei, H.; Lee, C.-H.; Xiang, Y. Entrepreneurship education and students' entrepreneurial intention in higher education. *Educ. Sci.* **2020**, *10*, 257. [CrossRef]
123. González, J.D.J.; Medina, F.E.V.; García, M.L.S. Factores de éxito en el financiamiento para pymes a través del crowdfunding en México. *Rev. Mex. Econ. Y Finanz.* **2020**, *16*, 1–23. [CrossRef]
124. Guiñez-Cabrera, N.; Mansilla-Obando, K.; Jeldes-Delgado, F. La transparencia publicitaria en los influencers de las redes sociales. *Retos* **2020**, *10*, 265–281. [CrossRef]
125. Matell, M.S.; Jacoby, J. Is there an optimal number of alternatives for Likert-scale items? Effects of testing time and scale properties. *J. Appl. Psychol.* **1972**, *56*, 506–509. [CrossRef]
126. Jacoby, J.; Matell, M.S. Three-point likert scales are good enough. *J. Mark. Res.* **1972**, *8*, 495–500. [CrossRef]

127. Chyung, S.Y.; Roberts, K.; Swanson, I.; Hankinson, A. Evidence-based survey design: The use of a midpoint on the likert scale. *Perform. Improv.* **2017**, *56*, 15–23. [CrossRef]
128. Kaiser, H.F. The varimax criterion for analytic rotation in factor analysis. *Psychometrika* **1958**, *23*, 187–200. [CrossRef]

Article

Pythagorean Membership Grade Aggregation Operators: Application in Financial knowledge

Fabio Blanco-Mesa [1,*], Ernesto León-Castro [2,3] and Jorge Romero-Muñoz [1]

[1] Facultad de Ciencias Económicas y Administrativas, Escuela de Administración de Empresas, Universidad Pedagógica y Tecnológica de Colombia, Av. Central del Norte, 39-115, Tunja 150001, Colombia; Jorge.romero@uptc.edu.co

[2] Facultad de Ciencias Administrativas y Sociales, Universidad Autonoma de Baja California, Boulevard Zertuche y Boulevard de los Lagos S/N Fracc, Valle Dorado, Ensenada 22890, Baja California, Mexico; ernesto.leon.castro@uabc.edu.mx

[3] Facultad de Ciencias Económicas y Administrativas, Escuela de Administración de Empresas, Universidad Católica de la Santísima Concepción, Alonso de Ribera 2850, Concepción, Chile

* Correspondence: fabio.blanco01@uptc.edu.co

Abstract: This paper presents the Pythagorean membership grade induced ordered weighted moving average (PMGIOWMA) operator with some particular cases and theorems. The main advantage of this new operator is that can include the knowledge, expectation, and aptitude of the decision maker into the Pythagorean membership function by using a weighting vector and induced variables. An application in financial knowledge based on a survey conducted in 13 provinces in Boyacá, Colombia, is presented.

Keywords: pythagorean membership; OWA operator; financial knowledge; decision-making

Citation: Blanco-Mesa, F.; León-Castro, E.; Romero-Muñoz, J. Pythagorean Membership Grade Aggregation Operators: Application in Financial knowledge. *Mathematics* 2021, 9, 2136. https://doi.org/10.3390/math9172136

Academic Editor: Antonio Francisco Roldán López de Hierro

Received: 24 July 2021
Accepted: 31 August 2021
Published: 2 September 2021

Publisher's Note: MDPI stays neutral with regard to jurisdictional claims in published maps and institutional affiliations.

Copyright: © 2021 by the authors. Licensee MDPI, Basel, Switzerland. This article is an open access article distributed under the terms and conditions of the Creative Commons Attribution (CC BY) license (https://creativecommons.org/licenses/by/4.0/).

1. Introduction

Many different proposals and approaches have been developed in decision-making under uncertainty. Among these approaches, those related to the theory of aggregation functions have been highlighted. [1]. These functions provide compensatory properties, where the low values of some inputs are compensated for by the high values of the others [1]. In this sense, an average result can be obtained that is a representative value of the inputs [1,2]. One such function that has been extensively studied is the ordered weighted average (OWA) operator, which associates weights not with a particular input, but rather with its value [1,3]. Based on this function, proposals have been developed that allow different types of data to be aggregated. For example, some operators focus on probability [4], distance measures [5–7], linguistic [8,9] and induced variables [10], prioritized items [11], Bonferroni means [12], Choquet integrals [13], moving averages [14], Pythagorean operators [15], etc.

Here, we focused on progress in the induced variables, the Pythagorean operator and moving averages. The authors of [10] introduced the Induced OWA (IOWA) operator that uses induced values in the reordering process instead of using the values of the arguments. However, the authors of [15] introduced Pythagorean membership grades in combination with an OWA operator as a nonstandard Pythagorean fuzzy subset whose membership grades are pairs (a, b) that satisfy the requirement $a^2 + b^2 \leq 1$. However, [14] introduced the moving average, which is a classical formulation in statistics but can be used in a wide range of problems and can be combined with the OWA operator to generate new possibilities for data analysis by becoming the ordered weighted moving average (OWMA) operator [16,17]. Based on these methods, proposals have been developed along different lines. For example, along the IOWA operator line, many proposals have taken a variety of approaches, which have used linguistic variables [18,19], fuzzy preference [20], intuitionistic fuzzy sets [21–23], distance measures [24,25], and heavy and prioritized operators [16,26,27], with others applying means such as Bonferroni means [28], VIKOR [29], Choquet [30], etc. For the OWMA operator, [31] generalized moving averages, distance measures and OWA; [16] proposed induced heavy moving averages, which can be applied in forecasting approaches [17,32–34]. Finally, the Pythagorean membership grade operator has focused on Pythagorean fuzzy sets, which have been extensively studied. Relevant studies have focused on multicriteria decision-making [35–37], and several applications have been developed to solve problems in finance [38,39] and business [40]. In this sense, we observed the potential of these methods and found a gap that allowed us to propose a new extension that can combine these operators into one.

The main aim of the present study was to present the Pythagorean membership grade induced ordered weighted moving average (PMGIOWMA) operator, with some cases and theorems. To achieve this, an aggregation operator [2] is proposed as a new extension of the ordered weighted average (OWA) operator [3], with Pythagorean membership grades [15] proposed on the basis of including the induced variables [10] and moving averages [32]. The objective of this new operator, called the Pythagorean membership grade induced ordered weighted moving average (PMGIOWMA) operator, is used to combine the reordering process of the OWA operator, based on induced values, in a set of arguments that needs to be analyzed as the moving average of a series, to analyze the Pythagorean membership grade. The most important theorems and formulations of the PMGIOWMA operator have been developed. Moreover, cases applying the Pythagorean membership grade induced ordered weighted average (PMGIOWA) operator and the Pythagorean membership grade ordered weighted moving average (PMGOWMA) operator are presented. Its mathematical application was focused on an analysis of financial knowledge based on a survey of 1914 individuals from 13 different provinces in the department of Boyacá, Colombia, with different educational levels. Specifically, the survey explored if their perceptions of savings and credits were related to their membership grade. One of the main advantages of this new formulation is that the data can be analyzed in a more complex way than with the usual average, moving average or OWA operator by itself. One of the disadvantages of the new formulation is that it can be too complex to apply, and more information is needed by the decision-maker. Therefore, if the problem is considered to be simple in terms of the elements behind the analysis, the use of an average may be sufficient to solve it.

The remainder of the study is organized as follows. In Section 2, the main formulations and definitions of the OWA operator, some of its extensions and Pythagorean membership grades are presented. Section 3 presents the formulations of the PMGIOWMA operator and its cases; the main theorems are also shown. Section 4 presents an application in Boyacá, Colombia. Finally, Section 5 summarizes the main conclusions of the paper.

2. Preliminaries

This section presents the basic concepts that have been used throughout the paper including OWA, IOWA, OWMA operators and Pythagorean membership grades, their formulations and their main characteristics.

2.1. The OWA Operator and Its Extensions

The main advantage of the OWA operator developed by [3] is that the process of reordering the weights is based on the values of the attributes. Thus, it is possible to obtain the maximum and minimum results. The formulation is as given below.

Definition 1. *An OWA operator of dimension n is a mapping* $OWA : R^n \to R$ *that has an associated weighting vector* W *of n dimensions with* $w_j \in [0,1]$ *and* $\sum_{j=1}^{n} w_j = 1$, *such that:*

$$OWA(a_1, a_2, \ldots, a_n) = \sum_{j=1}^{n} w_j b_j, \tag{1}$$

where b_j is the jth largest of a_i.

Instead of basing its reordering process on the values of the attributes, the induced OWA (IOWA) operator [10] uses an induced vector. This makes it possible to discriminate some arguments by using weights based on the information and knowledge of the decision-maker. The formulation is as follows:

Definition 2. *An IOWA operator of n dimensions is an application $IOWA : R^n \times R^n \to R$ that has an associated weighting vector W of n dimensions, where the sum of the weights is 1 and $w_j \in [0,1]$, where an induced set of ordering variables is included (u_i) such that the formula is:*

$$IOWA(\langle u_1, a_1 \rangle, \langle u_2, a_2 \rangle, \ldots, \langle u_n, a_n \rangle) = \sum_{j=1}^{n} w_j b_j, \qquad (2)$$

where b_j is the a_i value of the OWA pair $\langle u_i, a_i \rangle$ with the jth largest u_i, u_i is the order inducing variable and a_i is the argument variable.

The moving average is a method used to forecast the future for different variables by using historical data, which is why it is a common technique in economics and statistics [41]. The moving average can be defined as follows [42]:

Definition 3. *Given $\{a_i\}_{i=1}^{N}$, the moving average of n dimensions is defined as the sequence $\{s_i\}_{i=1}^{N-n+1}$, which is obtained by taking the arithmetic mean of the sequence of n terms, such that:*

$$s_i = \frac{1}{n} \sum_{j=i}^{i+n-1} a_j. \qquad (3)$$

It is important to note that, in every case, $n < N$.

Moreover, following the idea of moving averages and the OWA operator, the ordered weighted moving average (OWMA) and the induced ordered weighted moving average can be proposed [31].

Definition 4. *An OWMA operator of m dimensions is a mapping $OWMA : R^m \to R$ that has an associated weighting vector W of m dimensions with $w_j \in [0, 1]$ and $\sum_{j=1+t}^{m+t} w_j = 1$, such that:*

$$OWMA(a_{1+t}, a_{2+t}, \ldots, a_{n+t}) = \sum_{j=1+t}^{m+t} w_j b_j, \qquad (4)$$

where b_j is the jth largest argument of a_i, m is the total number of arguments considered in the whole sample and t indicates the movement of the average from the initial analysis.

Definition 5. *An IOWMA operator of m dimensions is a mapping $IOWMA : R^M \times R^M \to R$ that has an associated weighting vector W of m dimensions with $w_j \in [0, 1]$ and $\sum_{j=1+t}^{m+t} w_j = 1$, such that:*

$$IOWMA(\langle u_{1+t}, a_{1+t} \rangle, \langle u_{2+t}, a_{2+t} \rangle, \ldots, \langle u_{n+t}, a_{n+t} \rangle) = \sum_{j=1+t}^{m+t} w_j b_j, \qquad (5)$$

where b_j is the a_i value of the IOWMA pair u_i, a_i is the jth largest u_i, u_i is the order-inducing variable, a_i is the argument variable, m is the total number of arguments considered in the whole sample and t indicates the movement of the average from the initial analysis.

2.2. Pythagorean Membership Grades Used in Multiple Criteria Decision-Making

Definition 6. *MCDM considers a finite collection X of alternatives and a set of q criteria that we desire to satisfy. These criteria are referred to as c_j for $j = 1$ to q. Each criterion c_j is associated with an importance weight w_j such that $w_j \in [0,1]$ and $\sum w_j = 1$. Likewise, $c_j(x)$ indicates the degree of satisfaction of criterion c_j by alternative x.*

$$C(x) = \left(\sum_{j=1}^{q} w_j c_j(x) \right), \tag{6}$$

Let us now consider the situation in which the values of $c_j(x)$ are Pythagorean membership grades [15]. Here, each $c_j(x) = (a(x), b(x))$, where $a(x)$ and $b(x) \in [0,1]$ and $a_j(x)^2 + b_j(x)^2 \leq 1$.

$$C(x) = \left(\sum_{j=1}^{q} w_j a_j(x), \sum_{j=1}^{q} w_j b_j(x) \right), \tag{7}$$

where $a_j(x)$ and $b_j(x)$ indicate the degree of satisfaction of criterion c_j by alternative x.

Additionally, this formulation is completed via the following function, which is based on fuzzy rules [15].

$$F(r, \theta) = \frac{1}{2} + r \left(\frac{1}{2} - \frac{2\theta}{\pi} \right). \tag{8}$$

3. Pythagorean Membership Grade Aggregation Operators

This section presents new operators that combine the IOWA and OWMA operators with Pythagorean membership grades.

3.1. Extensions of the Pythagorean OWA

Extensions of the Pythagorean OWA are new propositions that combine the characteristics of IOWA and OWMA operators and Pythagorean membership grades. These new formulations are important because, when imprecise and ambiguous information is present in the problem, this needs to be analyzed, Pythagorean and OWA operators have proven to be useful [43–46]. Due to this, expanding the formulations by using more complex situations, such as those that can be analyzed with the IOWA and OWMA operators, presents a good opportunity to generate new results by considering a new reordering process based on induced values or problems that use time series. The new formulations are as follows. Since we present a new formulation, the notation $B_j(x) = (B(x), B'(x))$ to $B_j(x)^2 + B'_j(x)^2 \leq 1$ corresponds to the notation $a_j(x)^2 + b_j(x)^2 \leq 1$ to distinguish the new contribution.

Proposition 1. *A Pythagorean membership grade induced OWA operator (PMGIOWA) is an extension of the OWA operator. Thus, an PMGIOWA operator is a map $R^n \to R$ that is associated with a weight vector w, with $w_j \in [0,1]$ and $\sum_{j=1}^{n} w_j = 1$. Additionally, each $B_j(x) = (B(x), B'(x))$, where $B(x)$ and $B'(x) \in [0,1]$ and $B_j(x)^2 + B'_j(x)^2 \leq 1$, such that:*

$$PMGIOWA(x) = \left(\sum_{j=1}^{q} w_j B_j(x), \sum_{j=1}^{q} w_j B'_j(x) \right), \tag{9}$$

where $B_j(x)$ and $B'_j(x)$ indicate the degree of satisfaction of criterion $B_j(x)$ by alternative x. Thus, $B_j(x)$ is the a_i value of the OWA pair $\langle u_i, a_i \rangle$ with the jth largest u_i, u_i is the order-inducing variable and a_i is the argument variable.

Proposition 2. *A Pythagorean membership grade OWMA operator (PMGOWMA) is a mapping $PMGOWMA : R^M \times R^M \to R$ that has an associated weighting vector W of m dimensions with $w_j \in [0,1]$ and $\sum_{j=1+t}^{m+t} w_j = 1$. Additionally, each $B_j(x) = (B(x), B'(x))$, where $B(x)$ and $B'(x) \in [0,1]$ and $B_j(x)^2 + B'_j(x)^2 \leq 1$, such that:*

$$PMGOWMA(x) = \left(\sum_{j=1+t}^{m+t} w_j B_j(x), \sum_{j=1+t}^{m+t} w_j B'_j(x) \right), \tag{10}$$

where $B_j(x)$ and $B'_j(x)$ indicate the degree of satisfaction of criterion $B_j(x)$ by alternative x. Thus, $B_j(x)$ is the jth largest argument of a_i, m is the total number of arguments considered in the whole sample and t indicates the movement of the average from the initial analysis.

Proposition 3. *A Pythagorean membership grade induced OWMA operator (PMGIOWMA) is a map $PMGIOWMA : R^M \times R^M \to R$ that that has an associated weighting vector W of m dimensions with $w_j \in [0,1]$ and $\sum_{j=1+t}^{m+t} w_j = 1$. Additionally, each $B_j(x) = (B(x), B'(x))$, where $B(x)$ and $B'(x) \in [0,1]$ and $B_j(x)^2 + B'_j(x)^2 \leq 1$, such that:*

$$PMGIOWMA(x) = \left(\sum_{j=1+t}^{m+t} w_j B_j(x), \sum_{j=1+t}^{m+t} w_j B'_j(x) \right), \tag{11}$$

where $B_j(x)$ and $B'_j(x)$ indicate the degree of satisfaction of criterion $B_j(x)$ by alternative x. Thus, $B_j(x)$ is the a_i value of the OWA pair $\langle u_i, a_i \rangle$ with the jth largest u_i, u_i is the order-inducing variable and a_i is the argument variable. Likewise, m is the total number of arguments considered in the whole sample and t indicates the movement of the average from the initial analysis.

The Pythagorean membership grade has the property that the sum of the squares must be less than 1 [15]. We now prove that the new operators meet that condition.

Theorem 1. *If for $i = 1, 2, \ldots, q$, we have $B_i, B'_j \in [0,1]$ and $w_i \in [0,1]$ with $(u_i, u_i)^2 + (u'_i, a'_i)^2 \leq 1$ and $\sum_l w_l = 1$, then $\left(\sum_l w_l(u_l, a_l) \right)^2 + \left(\sum_l w_l(u'_l, a'_l) \right)^2 \leq 1$.*

Proof.

$(\sum_i w_i(u_i, a_i))^2 + (\sum_i w_i(u'_i, a'_i))^2 = \left(\sum_i w_i^2(u_i, a_i)^2 \right)^2 + \sum_{1 \neq j, i < j} 2 w_i(u_i, a_i) w_j(u_j, a_j) + \left(\sum_i w_i^2(u'_i, a'_i)^2 \right) + \sum_{1 \neq j, i < j} 2 w_i(u'_i, a'_i) w_j(u'_j, a'_j).$

$(\sum_i w_i(u_i, a_i))^2 + (\sum_i w_i(u'_i, a'_i))^2 = \sum_i w_i^2 \left((u_i, a_i)^2 + (u'_i, a'_i)^2 \right) + \sum_{1 \neq j, i < j} 2 w_i w_j \left((u_i, a_i)(u_j, a_j) + (u'_i, a'_i)(u'_j, a'_j) \right).$

$(\sum_i w_i(u_i, a_i))^2 + (\sum_i w_i(u'_i, a'_i))^2 \leq \sum_i w_i^2 + \sum_{1 \neq j, i < j} 2 w_i w_j \left((u_i, a_i)(u_j, a_j) + (u'_i, a'_i)(u'_j, a'_j) \right).$

□

Theorem 2. *For a moving average, if for $i = 1, 2, \ldots, q$, we have $B_i, B'_i \in [0,1]$ and $w_i \in [0,1]$ with $B_{i+t}^2 + B'^2_{i+t} \leq 1$ and $\sum_{j=1+t}^{m+t} w_j = 1$, then $(\sum_i w_i B_{i+t})^2 + (\sum_i w_i B'_{i+t})^2 \leq 1$.*

Proof.

$$(\sum_i w_i B_{i+t})^2 + (\sum_i w_i B'_{i+t})^2 = (\sum_i w_i^2 B_{i+t}^2) + \sum_{1\neq j, i<j} 2w_i B_{i+t} w_j B_{j+t} + (\sum_i w_i^2 B'^2_{j+t}) + \sum_{1\neq j, i<j} 2w_i B'_{i+t} w_j B'_{j+t}$$

$$(\sum_i w_i B_{i+t})^2 + (\sum_i w_i B'_{i+t})^2 = \sum_i w_i^2 (B_{i+t}^2 + B'^2_{i+t}) + \sum_{1\neq j, i<j} 2w_i w_j (B_{i+t} B_{j+t} + B'_{i+t} B'_{j+t})$$

$$(\sum_i w_i B_{i+t})^2 + (\sum_i w_i B'_{i+t})^2 \leq \sum_i w_i^2 + \sum_{1\neq j, i<j} 2w_i w_j (B_{i+t} B_{j+t} + B'_{i+t} B'_{j+t})$$

□

Theorem 3. *For an induced moving average, if for* $i = 1, 2, \ldots, q$*, we have* $B_i, B'_i \in [0,1]$ *and* $w_i \in [0,1]$ *with* $(u_{i+t}, a_{i+t})^2 + (u'_{i+t}, a'_{i+t})^2 \leq 1$ *and* $\sum_{j=1+t}^{m+t} w_j = 1$*, then*

$$(\sum_i w_i (u_{i+t}, a_{i+t}))^2 + (\sum_i w_i (u'_{i+t}, a'_{i+t}))^2 \leq 1.$$

Proof.

$$(\sum_i w_i(u_{i+t}, a_{i+t}))^2 + (\sum_i w_i(u'_{i+t}, a'_{i+t}))^2 = \left(\sum_i w_i^2 (u_{i+t}, a_{i+t})^2\right)^2 +$$
$$\sum_{1\neq j, i<j} 2w_i(u_{i+t}, a_{i+t}) w_j(u_{j+t}, a_{j+t}) + \left(\sum_i w_i^2 (u'_{i+t}, a'_{i+t})^2\right) +$$
$$\sum_{1\neq j, i<j} 2w_i(u'_{i+t}, a'_{i+t}) w_j(u'_{j+t}, a'_{j+t})$$

$$(\sum_i w_i(u_{i+t}, a_{i+t}))^2 + (\sum_i w_i(u'_{i+t}, a'_{i+t}))^2 = \sum_i w_i^2 \left((u_{i+t}, a_{i+t})^2 + (u'_{i+t}, a'_{i+t})^2\right) +$$
$$\sum_{1\neq j, i<j} 2w_i w_j ((u_{i+t}, a_{i+t})(u_{j+t}, a_{j+t}) + (u'_{j+t}, a'_{j+t})(u'_{j+t}, a'_{j+t}))$$

$$(\sum_i w_i(u_{i+t}, a_{i+t}))^2 + (\sum_i w_i(u'_{i+t}, a'_{i+t}))^2 \leq \sum_i w_i^2 +$$
$$\sum_{1\neq j, i<j} 2w_i w_j ((u_{i+t}, a_{i+t})(u_{j+t}, a_{j+t}) + (u'_{i+t}, a'_{i+t})(u'_{j+t}, a'_{j+t}))$$

□

3.2. Numerical Example

In this example, we have three criteria (C_1, C_2, and C_3) with the importance weights $w_1 = 0.25$, $w_2 = 0.45$, and $w_3 = 0.30$. We must choose between two alternatives, x and y. The two alternatives satisfy the criteria as follows, where the induced variables μ are $C_1(\mu) = (5,6)$, $C_2(\mu) = (8,9)$ and $C_3(\mu) = (2,3)$.

x $C_1(x) = (0.7, 0.2)$, $C_2(x) = (0.2, 0.9)$, $C_3(x) = (0.3, 0.5)$

y $C_1(y) = (0.7, 0.4)$, $C_2(y) = (0.5, 0.8)$, $C_3(y) = (0.3, 0.1)$

By using Equations (7)–(9), we calculate the following:

$$PMGIOWA(x) = \left(\sum_{j=1}^{q} w_j B_j(x), \sum_{j=1}^{q} w_j B'_j(x)\right) = (0.38, \ 0.66).$$

$$PMGIOWA(y) = \left(\sum_{j=1}^{q} w_j B_j(y), \sum_{j=1}^{q} w_j B'_j(y)\right) = (0.51, \ 0.51).$$

Now, $r(x)^2$ and $r(y)^2$ can be obtained as: $r(x)^2 = (0.38)^2 + (0.66)^2 = 0.57$ and $r(y)^2 = (0.51)^2 + (0.51)^2 = 0.515$. Here, $r(x) = 0.75$ and $r(y) = 0.72$. Additionally, if trigonometric values are used, it is possible to apply the function $F(r(x), \theta(x))$. In this case, the results are as follows: $\cos(\theta(x)) = \frac{0.38}{0.75} = 0.5236 \ (rad)$ and $\cos(\theta(y)) = \frac{0.51}{0.72} = 0.7679 \ (rad)$; $F(r(x), \theta(x)) = \frac{1}{2} + r\left(\frac{1}{2} - \frac{2\theta}{\pi}\right) = \frac{1}{2} + 0.75 * \left(\frac{1}{2} - \frac{2(0.5236)}{\pi}\right) = 0.6258$ and $F(r(y), \theta(y)) = \frac{1}{2} + r\left(\frac{1}{2} - \frac{2\theta}{\pi}\right) = \frac{1}{2} + 0.72 * \left(\frac{1}{2} - \frac{2(0.7679)}{\pi}\right) = 0.5080$. By following the same process, we can obtain the results of Equations (8) and (9). In PMGOWMA, the moving average is calculated as: $B = (((0.7 \times 0.45) + (0.3 \times 0.3)/2) + (0.2 \times 0.25)/2)$. The same process can be used to obtain B and B' for each PMGOWMA and PMGIOWMA. In all cases, the best option is x (see Table 1).

Table 1. Consolidated results.

	PMGIOWA		PMGOWMA		PMGIOWMA	
	x	y	x	y	x	y
B	0.38	0.51	0.13	0.15	0.11	0.15
B'	0.66	0.51	0.16	0.13	0.18	0.12
r	0.75	0.72	0.21	0.20	0.21	0.19
r^2	0.57	0.52	0.04	0.04	0.05	0.04
$\cos(\theta(x))$	0.5236	0.7679	0.6632	0.6981	0.5585	0.6807
$F(r,\theta)$	0.6258	0.5080	0.5161	0.5113	0.5310	0.5127

As can be seen in Table 1, even when different aggregation operators were used, the results were the same, but an interesting finding is that the results were not the same for all the formulations, and in some the difference between x and y was smaller, but in others it was larger. Therefore, analyzing the information with the use of more data is important for better understanding of the phenomenon under study.

4. Case Study

The financial environment nowadays has become more precarious. The new generation has problems related to mortgages, credit availability, borrowing options, pensions, savings and investing options [47]. In addition to these problems, many citizens do not understand the financial concepts that are crucial for consumers' financial decision-making [48]. The main idea of financial education is to provide individuals with the knowledge, aptitude and skill base necessary to become questioning and informed consumers of financial services and to manage their finances effectively [49]. Among the benefits of having good financial knowledge is that it is directly correlated with self-beneficial financial behavior [50]. Due to this, it is important to analyze the impact of financial education so that it is possible to answer the question as to whether financial knowledge allows people with more financial education to achieve better decisions than those with less [51]. Among the aspects that should be better understood are savings and credit. On the one hand, saving is an action in which a reserve of capital from income is kept for future use, which depends on the level of income, generating a positive or negative rate. On the other hand, credit is a means of financing that can be used to start a new business or project at a personal or business level, although it can also be used when acquiring goods that do not generate income. To provide a better understanding of these concepts, they will be briefly explained below.

4.1. Savings

The importance of the concept of savings in financial education derives from the economy, from both macroeconomic and microeconomic approaches. From the macroeconomic view, savings is the surplus of income over what is spent on consumption [52]. Likewise, good financial intermediation between savers and investors increases the economic development of a nation [53]. In this sense, [54] indicated that, if a nation has high savings rates, this is correlated with economic growth, pointing out that there is a cycle between savings and prosperity. Hence, macroeconomic principles have led several authors to explore diverse concepts that contribute to improving the development of a nation. From a microeconomic view, an individual demonstrates a saving behavior if he/she balances current consumption against expected consumption. This indicates that a well-informed individual will not consume all his/her income and will save for difficult times [55]. Likewise, some approaches have indicated that an environment of uncertainty is relevant to having such savings in the future [56]. Moreover, other studies have incorporated liquidity restrictions derived from imperfections in credit markets [57]. Additionally, savings behavior is driven by externalities produced by inadequate financial intermediation and by imperfections in the credit and insurance markets [58].

4.2. Credit

The concept of credit is related to macroeconomic factors, such as inflation and financial market volatility, which have produced small and inefficient financial systems in Latin America. The authors of [59] stated that the consequences of the macroeconomic environment of a country are reflected in its financial intermediaries (banks, financial corporations, insurance companies, among others) and their characteristics. As indicated by [60], the function of financial intermediaries is to stimulate and collect savings in an economy, and to give credit to those who require capital and thus develop their economic activities.

Another aspect to consider is credit rationing. For [61], this rationing occurs when the borrower is denied credit, even if he/she is willing to meet the provisions of the loan contract. The authors of [62] also suggested that credit rationing occurs in two ways: first, when one or all members of a group are rationed; second, when some members of the group are rationed. Finally, [63] mentioned that credit restrictions can be classified in analyses of the credit market environment into those on the credit offering side and those on credit demanders. In conclusion, credit rationing has several factors that look for imperfections, such as high interest rates, credit ceilings, usury laws, the level of financial intermediation, the number of borrowers and credit risk, among others.

4.3. Process of Analyzing Financial Knowledge Levels

This section presents the main steps taken to analyze financial knowledge levels and the use of the new method of criteria satisfaction to obtain the best alternative. The decision-making problem is associated with savings and credit perceptions through a combination of the features, to determine which of them satisfies personal criteria. For this purpose, this study used a dataset of the 13 provinces of the department of Boyacá, Colombia, regarding financial knowledge. For the selection of the database sample, the provinces were classified into three groups in proportion to the population, and the best offers of financial services in the cities and municipalities were established as the selection criteria. A 95% confidence level and a 5% error were selected for a sample of 1914 people. Following is the step-by-step decision-making process.

Step 1. Analyze and determine the significant characteristics for savings (Sav) and credit (Cred) perceptions, which comprise six actions for Sav and five actions for Cred, which are considered to be characteristics. Each characteristic of the sector is considered as a property (see Table 2).

Table 2. Savings and credit actions.

Acrom	Savings	Acrom	Credit
MB	Money box	FFC	Family or Friend Credit
SF	Savings funds	GG	Gota to gota
SA	Savings account	CC	Credit card
Tr	Trust	ML	Moneylender
SC	Savings chains	BC	Bank credit
FS	Family Savings	PS	Pawnshop
CDT's	CDT's		

Step 2. In this step, the educational level is established as a determining condition for saving or taking credit. Three groups were established for this analysis: Group 1: none and primary; Group 2: high school and technical; Group 3: undergraduate and postgraduate. These classifications are based on the Colombian educational system (see Table 3).

Table 3. Education level groups.

	Savings	Credit
Group 1	None and Primary	
Group 2	High school and Technical	
Group 3	Undergraduate and Postgraduate	

Step 3. Using the database, construct the matrices containing the satisfactory conditions for savings or credit, considering the classifications described in Steps 1 and 2. Here, each value comprised the criterion, knowledge, and the subjectivity of the decision-maker (see Tables 4 and 5).

Table 4. Saving values ratios.

	Group 1		Group 2		Group 3	
MB	0.6000	0.4505	0.4769	0.4271	0.3723	0.2057
SF	0.0000	0.0571	0.0437	0.0507	0.0693	0.0914
SA	0.2000	0.2703	0.2497	0.2875	0.2879	0.3714
Tr	0.0000	0.0120	0.0087	0.0063	0.0108	0.0514
SC	0.0857	0.0781	0.1024	0.1099	0.0887	0.0800
FS	0.0857	0.1021	0.0936	0.0803	0.0996	0.0800
CDT's	0.0286	0.0300	0.0250	0.0381	0.0714	0.1200

Table 5. Credit values ratios.

	Group 1		Group 2		Group 3	
FFC	0.3784	0.3495	0.4861	0.4324	0.4627	0.2986
GG	0.2432	0.0311	0.0362	0.0158	0.0051	0.0000
CC	0.1081	0.1557	0.1560	0.1306	0.1748	0.2500
ML	0.0811	0.0969	0.0738	0.0811	0.0308	0.0486
BC	0.1892	0.3287	0.2368	0.3266	0.3111	0.3889
PS	0.0000	0.0381	0.0111	0.0135	0.0154	0.0139

Step 4. Weight vectors are determined by using the sources of information most often used by people when deciding to save or take out credit. These vectors are represented by w. Likewise, the induced values are established, using the actions most often used to make savings or take credit as a reference. These values are represented by μ. (see Tables 6 and 7).

Table 6. Weighting vectors and induced values for savings.

	IFI	AFM	AFR	OWE	CON	INT	MTR
	W_1	W_2	W_3	W_4	W_5	W_6	W_7
WS	0.2723	0.2213	0.1368	0.1153	0.1021	0.0899	0.0623
	MB	SF	SA	Tr	SC	FS	CDT's
μ	1	6	2	7	4	3	5

WS: weighted savings; IFI: information from financial institutions; AFM: advice from family members; AFR: advice from friends; OWE: own experience; CON: consulting; INT: internet; MTR: marketing, TV and radio.

Table 7. Weighting vectors and induced values for credit.

	IFI	AFM	MTRI	AFR	OWE	CON
	W_1	W_2	W_3	W_4	W_5	W_6
EC	0.2723	0.2213	0.1522	0.1368	0.1153	0.1021
	FFC	GG	CC	ML	BC	PS
μ	1	6	2	7	4	3

WC: weighted credit; IFI: information from financial institutions; AFM: advice from family members; MTRI: marketing, TV, radio and internet; AFR: advice from friends; OWE: own experience; CON: consulting.

5. Results

For analysis of savings and credit perceptions, we used the following: the PMGOWA operator, the PMGIOWA operator, the PMGOWMA operator and the PMGIOWMA operator (see Table 8).

Table 8. Results of the PMGOWA, PMGIOWA, PMGOWMA and PMGIOWMA.

	PMGOWA					
	Group 1		Group 2		Group 3	
	Savings	Credit	Savings	Credit	Savings	Credit
B	0.2090	0.2062	0.1951	0.2026	0.1798	0.1954
B'	0.1927	0.1808	0.1884	0.1912	0.1558	0.1723
r	0.2842	0.2743	0.2712	0.2786	0.2379	0.2605
r2	0.0808	0.0752	0.0736	0.0776	0.0566	0.0679
COS ($\theta(x)$)	0.0962	0.9976	0.9976	0.9976	0.9986	0.9986
F	0.6247	0.4630	0.4634	0.4624	0.4677	0.4646
	PMGIOWA					
	Group 1		Group 2		Group 3	
	Savings	Credit	Savings	Credit	Savings	Credit
B	0.3221	0.2098	0.2172	0.2222	0.2031	0.2280
B'	0.3110	0.2124	0.2132	0.2242	0.1467	0.2183
r	0.4477	0.2986	0.3043	0.3157	0.2505	0.3156
r2	0.2004	0.0891	0.0926	0.0996	0.0628	0.0996
COS ($\theta(x)$)	0.9781	0.9962	0.9962	0.9962	0.9986	0.9962
F	0.4451	0.4599	0.4592	0.4576	0.4660	0.4576
	PMGOWMA					
	Group 1		Group 2		Group 3	
	Savings	Credit	Savings	Credit	Savings	Credit
B	0.0073	0.0128	0.0075	0.0145	0.0087	0.0159
B'	0.0076	0.0177	0.0077	0.0165	0.0097	0.0177
r	0.0105	0.0219	0.0107	0.0220	0.0131	0.0238
r2	0.0001	0.0005	0.0001	0.0005	0.0002	0.0006
COS ($\theta(x)$)	1.0000	1.0000	1.0000	1.0000	1.0000	1.0000
F	0.4986	0.4970	0.4985	0.4970	0.4982	0.4968
	PMGIOWMA					
	Group 1		Group 2		Group 3	
	Savings	Credit	Savings	Savings	Credit	Savings
B	0.0107	0.0142	0.0094	0.0207	0.0120	0.0242
B'	0.0114	0.0266	0.0098	0.0251	0.0153	0.0325
r	0.0157	0.0301	0.0136	0.0325	0.0194	0.0406
r2	0.0002	0.0009	0.0002	0.0011	0.0004	0.0016
COS ($\theta(x)$)	1.0000	1.0000	1.0000	1.0000	1.0000	1.0000
F	0.4979	0.4959	0.4981	0.4956	0.4973	0.4945

With the results obtained, it is possible to graph and observe the change in perceptions about credit and savings. Figure 1 shows the relationship between perceptions for each method. Figure 2 shows the perceptions for each group regarding credit and savings for each method used. Figure 3 shows the perceptions grouped into savings and credit.

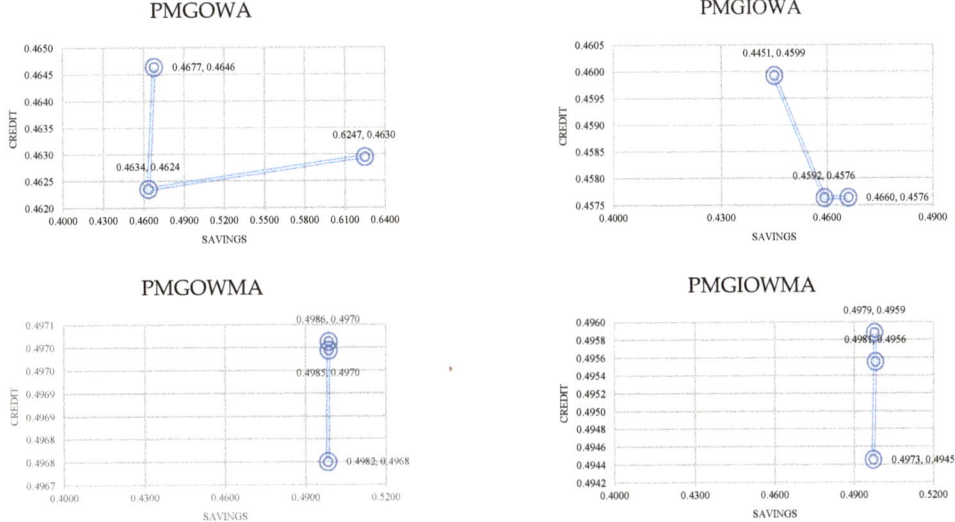

Figure 1. Perception relationships of savings and credit for each method.

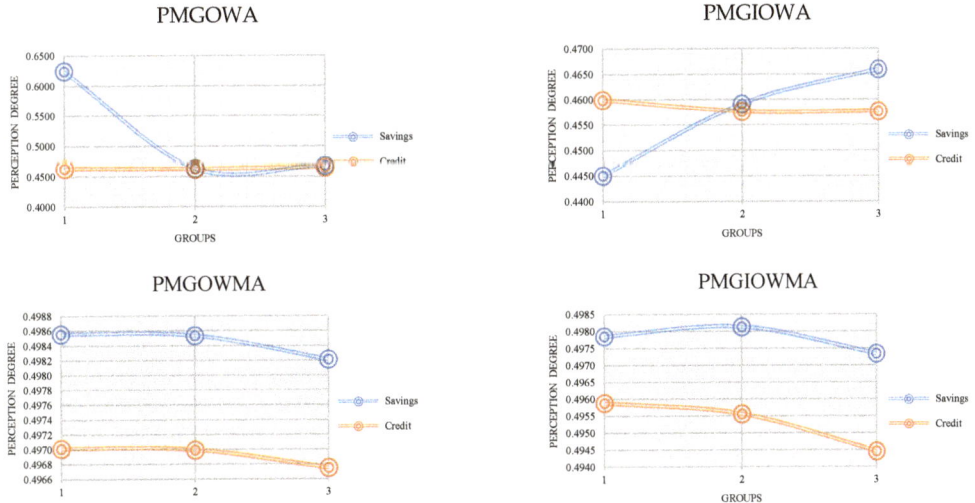

Figure 2. Perceptions of savings and credit by each method.

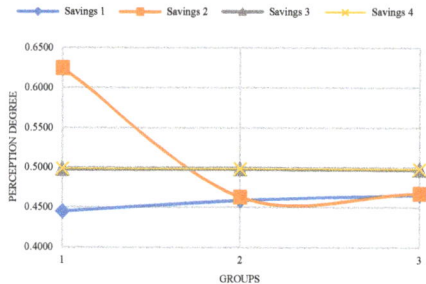 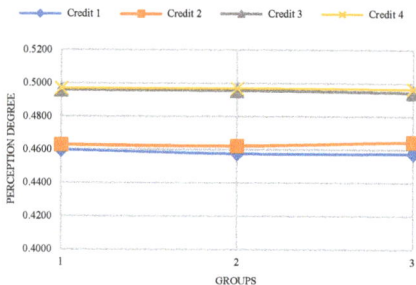

Savings 1: PMGIOWA; Savings 2: PMGOWA; Savings 3: PMGIOWMA; Savings 4: PMGOWMA

Credit 1: PMGIOWA; Credit 2: PMGOWA; Credit 3: PMGIOWMA; Credit 4: PMGOWMA

Figure 3. Perception of savings and credit by group.

Based on the different graphs presented in Figure 1, it is possible to see that, with the PMGOWA operator, the relationship between credit and savings can change drastically from (0.4634, 0.4624) to (0.6247, 0.4630), but if a more complex operator is used, such as the PMGIOWMA operator the changes are not that important and are between (0.4973, 0.4945) and (0.4979, 0.4959). Finally, as a conclusion and considering all the results obtained by the operators, the relationships between the perceptions are 0.4500–0.500 and 0.4500–0.5000; higher results than these are seldom seen.

Based on Figure 2, it is possible to see that the results are very similar between the groups, with exception of savings with the PMGOWA, where perception changed from 0.4634 to 0.6247, with Group 1 giving a higher value to savings.

Figure 3 shows the different results for saving and credit by group. As seen before, the results for saving in Group 1, specifically Savings 1 (PMGOWA operator), had a higher value than the other operators.

Finally, based on all the information provided by Figures 1–3, the main result is that the education of the people who responded to the survey, the importance of savings and credits and the relationship between them were shown to have similar levels. This information can be useful, because sometimes decision-makers believe that this perception can change based on the educational level, but with the information and the operators used in this research, it is possible to affirm that in the case of Boyacá, Colombia, this was not the case.

Since the moving average allows us to calculate averages over time, it is often used in financial problems such as pricing, sales, and others. The main idea of using the moving average in this research was that it can demonstrate the change in perception that it can indicate in a decision problem and that it can show a trend in the collected data, where t is the number of times that an answer can be repeated. Due to this characteristic and in combination with the induced variables and the Pythagorean membership grade, it offers a more complex method that considers the changes in perception in an uncertain system. Hence, the selected case considers that credit and savings can be influenced by people's schooling and their own reasoning, which changes over time. Additionally, it shows that the Cartesian relationship of two variables can be considered as independent variables and used to aggregate variables in the same system, which offers a representative value of the degree of perception and the relationships of perceptions. By analyzing the degree of perception, we seek to highlight the benefit of limited reasoning and the subjectivity of individuals.

6. Conclusions

The objective of this study was to present the Pythagorean membership grade induced ordered weighted moving average (PMGIOWMA) operator. The main characteristic of this new proposition is that it combines a weighting vector with a Pythagorean membership grade to include the knowledge, expertise, and expectations of the decision-maker, as well as a reordering step based on induced values and the application of a series of data based on moving averages.

Additionally, the main definitions and theorems of the PMGIOWMA operator are provided here, as well as some other extensions that are less complex, such as the PMGOWMA or PMGIOWA operators. These two extensions are considered less complex because in the first, the reordering is not based on induced values, and in the second, there is not a moving average. The main idea of presenting fewer complex formulations is based on the problem that needs to be solved. To reduce the uncertainty, it is possible to use a simpler formula. Moreover, sometimes not all the information needed for using the most complex operator can be obtained and the analysis must be carried out with another operator.

The PMGIOWMA operator was used to visualize the perceptions of and relationship between savings and credit based on a survey of 1914 people from 13 provinces of the department of Boyacá, Colombia. Among the main results, it was possible to see that the three different groups had the same perceptions of saving and credits, and the relationship degree was also the same. This idea is important, considering that the groups were based on their educational level, and it was possible to see that, regarding the financial topics of saving and credits, their perceptions were the same. Another interesting result was that the perception of saving in Group 1 (people with no and primary education) was 50% higher with the PMGOWA operator.

Futures research should focus on fuzzy decision-making [64] to propose more extension of the Pythagorean membership grade and OWA operator [65], possibly through the use of distance operators [40], linguistic variables [8,66], logarithmic operators [67], heavy operators [33] or prioritized operators [68].

Author Contributions: Conceptualization, J.R.-M. and F.B.-M.; methodology, F.B.-M. and E.L.-C.; formal analysis, F.B.-M. and E.L.-C.; writing—original draft preparation, J.R.-M. and F.B.-M.; writing—review and editing, F.B.-M. All authors have read and agreed to the published version of the manuscript.

Funding: This research was funded by Universidad Pedagogica y Tecnologica de Colombia (grant number SGI-2589). Author Number 2 acknowledges support from the Chilean Government through FONDECYT initiation grant No. 11190056.

Institutional Review Board Statement: The study was conducted according to the guidelines of the Universidad Pedagogica y Tecnologica de Colombia.

Informed Consent Statement: All subjects gave their informed consent for inclusion before they participated in the study. The study was conducted in accordance with the Declaration of Helsinki.

Data Availability Statement: The database used is from a secondary source provided by the Eugene Fama research group of the Financial Literacy in Boyacá project (2018–2019).

Conflicts of Interest: The authors declare no conflict of interest.

References

1. Beliakov, G.; Bustince Sola, H.; Calvo Sánchez, T. *A Practical Guide to Averaging Functions*, 1st ed.; Kacprzyk, J., Ed.; Springer International Publishing: Cham, Switzerland, 2016; ISBN 9783319247533.
2. Blanco-Mesa, F.; León-Castro, E.; Merigó, J.M. A Bibliometric Analysis of Aggregation Operators. *Appl. Soft Comput. J.* **2019**, *81*, 105488. [CrossRef]
3. Yager, R.R. On Ordered Weighted Averaging Aggregation Operators in Multicriteria Decision-Making. *IEEE Trans. Syst. Man Cybern.* **1988**, *18*, 183–190. [CrossRef]
4. Merigó, J.M. Probabilities in the OWA Operator. *Expert Syst. Appl.* **2012**, *39*, 11456–11467. [CrossRef]
5. Gil-Lafuente, A.M.; Merigó, J.M. The Ordered Weighted Averaging Distance Operator. *Lect. Model. Simul.* **2007**, *8*, 84–95.
6. Xu, Z.; Chen, J. Ordered Weighted Distance Measure. *J. Syst. Sci. Syst. Eng.* **2008**, *17*, 432–445. [CrossRef]
7. Merigó, J.M.; Gil-Lafuente, A.M. On the Use of the OWA Operator in the Euclidean Distance. *Int. J. Comput. Sci. Eng.* **2008**, *2*, 170–176.
8. Herrera, F.; Martinez, L. A 2-Tuple Fuzzy Linguistic Representation Model for Computing with Words. *IEEE Trans. Fuzzy Syst.* **2000**, *8*, 746–752. [CrossRef]

9. Yu, D. A Scientometrics Review on Aggregation Operator Research. *Scientometrics* **2015**, *105*, 115–133. [CrossRef]
10. Yager, R.R.; Filev, D.P. Induced Ordered Weighted Averaging Operators. *IEEE Trans. Syst. Man Cybern. Part B (Cybern.)* **1999**, *29*, 141–150. [CrossRef]
11. Chen, L.; Xu, Z. A Prioritized Aggregation Operator Based on the OWA Operator and Prioritized Measure. *J. Intell. Fuzzy Syst.* **2014**, *27*, 1297–1307. [CrossRef]
12. Yager, R.R. On Generalized Bonferroni Mean Operators for Multi-Criteria Aggregation. *Int. J. Approx. Reason.* **2009**, *50*, 1279–1286. [CrossRef]
13. Modave, F.; Ceberio, M.; Kreinovich, V. *Choquet Integrals and OWA Criteria as a Natural (and Optimal) Next Step after Linear Aggregation: A New General Justification*; Springer: Berlin/Heidelberg, Germany, 2008; pp. 741–753.
14. Yager, R.R. Time Series Smoothing and OWA Aggregation. *IEEE Trans. Fuzzy Syst.* **2008**, *16*, 994–1007. [CrossRef]
15. Yager, R.R. Pythagorean Membership Grades in Multicriteria Decision Making. *IEEE Trans. Fuzzy Syst.* **2014**, *22*, 958–965. [CrossRef]
16. León-Castro, E.; Avilés-Ochoa, E.; Merigó, J.M. Induced Heavy Moving Averages. *Int. J. Intell. Syst.* **2018**, *33*, 1823–1839. [CrossRef]
17. Espinoza-Audelo, L.; Aviles-Ochoa, E.; Leon-Castro, E.; Blanco-Mesa, F. Forecasting Performance of Exchange Rate Models with Heavy Moving Average Operators. *Fuzzy Econ. Rev.* **2019**, *24*, 3–21. [CrossRef]
18. Xu, Z. Induced Uncertain Linguistic OWA Operators Applied to Group Decision Making. *Inf. Fusion* **2006**, *7*, 231–238. [CrossRef]
19. Merigó, J.M.; Casanovas, M.; Palacios-Marqués, D. Linguistic Group Decision Making with Induced Aggregation Operators and Probabilistic Information. *Appl. Soft Comput.* **2014**, *24*, 669–678. [CrossRef]
20. Chiclana, F.; Herrera-Viedma, E.; Herrera, F.; Alonso, S. Some Induced Ordered Weighted Averaging Operators and Their Use for Solving Group Decision-Making Problems Based on Fuzzy Preference Relations. *Eur. J. Oper. Res.* **2007**, *182*, 383–399. [CrossRef]
21. Atanassov, K.T. Intuitionistic fuzzy sets. In *Intuitionistic Fuzzy Sets*; Springer: Berlin/Heidelberg, Germany, 1999; pp. 1–137.
22. Shouzhen, Z.; Qifeng, W.; Merigó, J.M.; Tiejun, P. Induced Intuitionistic Fuzzy Ordered Weighted Averaging: Weighted Average Operator and Its Application to Business Decision-Making. *Comput. Sci. Inf. Syst.* **2014**, *11*, 839–857. [CrossRef]
23. Xu, Y.; Wang, H. The Induced Generalized Aggregation Operators for Intuitionistic Fuzzy Sets and Their Application in Group Decision Making. *Appl. Soft Comput.* **2012**, *12*, 1168–1179. [CrossRef]
24. Merigó, J.M.; Gil-Lafuente, A.M. Decision Making with the Induced Generalized Adequacy Coefficient. *Appl. Comput. Math.* **2011**, *10*, 321–339.
25. Merigó, J.M.; Casanovas, M. Decision-Making with Distance Measures and Induced Aggregation Operators. *Comput. Ind. Eng. Ind. Eng.* **2011**, *60*, 66–76. [CrossRef]
26. Perez-Arellano, L.A.; Leon-Castro, E.; Blanco-Mesa, F.; Fonseca-Cifuentes, G. The Ordered Weighted Government Transparency Average: Colombia Case. *J. Intell. Fuzzy Syst.* **2021**, *40*, 1837–1849. [CrossRef]
27. Merigó, J.M.; Casanovas, M. Induced and Uncertain Heavy OWA Operators. *Comput. Ind. Eng.* **2011**, *60*, 106–116. [CrossRef]
28. Blanco-Mesa, F.; León-Castro, E.; Merigó, J.M.; Xu, Z. Bonferroni Means with Induced Ordered Weighted Average Operators. *Int. J. Intell. Syst.* **2019**, *34*, 3–23. [CrossRef]
29. Liu, H.-C.; Mao, L.-X.; Zhang, Z.-Y.; Li, P. Induced Aggregation Operators in the VIKOR Method and Its Application in Material Selection. *Appl. Math. Model.* **2013**, *37*, 6325–6338. [CrossRef]
30. Tan, C.; Chen, X. Induced Choquet Ordered Averaging Operator and Its Application to Group Decision Making. *Int. J. Intell. Syst.* **2010**, *25*, 59–82. [CrossRef]
31. Merigó, J.M.; Yager, R.R. Generalized Moving Average, Distance Measures and OWA Operators. *Int. J. Uncertain. Fuzziness Knowl. Based Syst.* **2013**, *21*, 533–559. [CrossRef]
32. Olazabal-Lugo, M.; Leon-Castro, E.; Espinoza-Audelo, L.F.; Merigó, J.M.; Gil Lafuente, A.M. Forgotten Effects and Heavy Moving Averages in Exchange Rate Forecasting. *Econ. Comput. Econ. Cybern. Stud. Res.* **2019**, *53*, 79–96. [CrossRef]
33. León-Castro, E.; Avilés-Ochoa, E.; Merigó, J.M.; Gil-Lafuente, A.M. Heavy Moving Averages and Their Application in Econometric Forecasting. *Cybern. Syst.* **2018**, *49*, 26–43. [CrossRef]
34. Fonseca-Cifuentes, G.; Leon-Castro, E.; Blanco-Mesa, F. Predicting the Future Price of a Commodity Using the OWMA Operator: An Approximation of the Interest Rate and Inflation in the Brown Pastusa Potato Price. *J. Intell. Fuzzy Syst.* **2021**, *40*, 1971–1981. [CrossRef]
35. Zhang, X.; Xu, Z. Extension of TOPSIS to Multiple Criteria Decision Making with Pythagorean Fuzzy Sets. *Int. J. Intell. Syst.* **2014**, *29*, 1061–1078. [CrossRef]
36. Peng, X.; Yang, Y. Some Results for Pythagorean Fuzzy Sets. *Int. J. Intell. Syst.* **2015**, *30*, 1133–1160. [CrossRef]
37. Ren, P.; Xu, Z.; Gou, X. Pythagorean Fuzzy TODIM Approach to Multi-Criteria Decision Making. *Appl. Soft Comput. J.* **2016**, *42*, 246–259. [CrossRef]
38. Liu, M.; Zeng, S.; Baležentis, T.; Streimikiene, D. Picture Fuzzy Weighted Distance Measures and Their Application to Investment Selection. *Amfiteatru Econ. J.* **2019**, *21*, 682–695. [CrossRef]
39. Zhang, H.; Liao, H.; Wu, X.; Zavadskas, E.K.; Al-Barakati, A. Internet Financial Investment Product Selection with Pythagorean Fuzzy DNMA Method. *Eng. Econ.* **2020**, *31*, 61–71. [CrossRef]
40. Mariño-Becerra, G.; Blanco-Mesa, F.; León-Castro, E. Pythagorean Membership Grade Distance Aggregation: An Application to New Business Ventures. *J. Intell. Fuzzy Syst.* **2021**, *40*, 1827–1836. [CrossRef]

41. Evans, M.K. *Practical Business Forecasting*; Blackwell Publishers: Oxford, UK, 2002; ISBN 9780631220657.
42. Kenney, J.; Keeping, E. *Moving Averages*; Van Nostrand: Princeton, NJ, USA, 1962.
43. Zhang, X. Multicriteria Pythagorean Fuzzy Decision Analysis: A Hierarchical QUALIFLEX Approach with the Closeness Index-Based Ranking Methods. *Inf. Sci.* **2016**, *330*, 104–124. [CrossRef]
44. Naz, S.; Ashraf, S.; Akram, M. A Novel Approach to Decision-Making with Pythagorean Fuzzy Information. *Mathematics* **2018**, *6*, 95. [CrossRef]
45. Zeng, S.; Chen, J.; Li, X. A Hybrid Method for Pythagorean Fuzzy Multiple-Criteria Decision Making. *Int. J. Inf. Technol. Decis. Mak.* **2016**, *15*, 403–422. [CrossRef]
46. Yager, R.R. Pythagorean Fuzzy Subsets. In Proceedings of the 2013 Joint IFSA World Congress and NAFIPS Annual Meeting (IFSA/NAFIPS), Edmonton, AB, Canada, 24–28 June 2013; IEEE: Piscataway, NJ, USA, 2013; pp. 57–61.
47. Fernandes, D.; Lynch, J.G.; Netemeyer, R.G. Financial Literacy, Financial Education, and Downstream Financial Behaviors. *Manag. Sci.* **2014**, *60*, 1861–1883. [CrossRef]
48. Fox, J.; Bartholomae, S.; Lee, J. Building the Case for Financial Education. *J. Consum. Aff.* **2005**, *39*, 195–214. [CrossRef]
49. Mason, C.L.J.; Wilson, R.M.S. *Conceptualising Financial Literacy*; Business School, Loughborough University: Loughborough, UK, 2000; Volume 7.
50. Mandell, L.; Schmid Klein, L. The Impact of Financial Literacy Education on Subsequent Financial Behavior. *J. Financ. Couns. Plan.* **2009**, *20*, 1–10.
51. Romero-Muñoz, J.; Fonseca-Cifuentes, G.; Blanco-Mesa, F. *Analysis and Evaluation of Financial Education in Boyacá*, 1st ed.; Editorial UPTC: Tunja, Colombia, 2021; ISBN 978-958-660-484-0.
52. Keynes, J.M. *The General Theory of Employment, Interest, and Money*; Stellar Classics: Bloomington, IN, USA, 2016; ISBN 198781780X.
53. Gurley, J.G.; Shaw, E.S. Financial Aspects of Economic Development. *Am. Econ. Rev.* **1955**, *45*, 515–538. [CrossRef]
54. Karlan, D.; Ratan, A.L.; Zinman, J. Savings by and for the Poor: A Research Review and Agenda. *Rev. Income Wealth* **2014**, *60*, 36–78. [CrossRef] [PubMed]
55. Modigliani, F. The Life Cycle Hypothesis of Saving, the Demand for Wealth and the Supply of Capital. *Soc. Res.* **1996**, *33*, 160–217. [CrossRef]
56. Carroll, C.D. Precautionary Saving and the Marginal Propensity to Consume out of Permanent Income. *J. Monet. Econ.* **2009**, *56*, 780–790. [CrossRef]
57. Hayashi, F. Tests for liquidity constraints: A critical survey and some new observations. In *Advances in Econometrics*; Bewley, T.F., Ed.; Cambridge University Press: CT, USA, 2008; pp. 91–120.
58. Liu, L.-Y.; Woo, W.T. Saving Behaviour under Imperfect Financial Markets and the Current Account Consequences. *Econ. J. R. Econ. Soc.* **1994**, *104*, 512–527. [CrossRef]
59. Kiguel, M.A.; Levy Yeyati, E.; Galindo, A.; Panizza, U.; Miller, M.; Rojas-Suárez, L.; Bebczuk, R.N.; López-de-Silanes, F.; Bernal, O.; Auerbach, P.; et al. *Desencadenar El Crédito: Cómo Ampliar y Estabilizar La Banca*; Banco Interamericano de Desarrollo: Washington, DC, USA, 2004.
60. Baldivia Urdininea, J. *Las Microfinanzas: Un Mundo de Pequeños Que Se Agrandan*; Fundación Milenio: La Paz, Bolivia, 2004.
61. Baltensperger, E. The Borrower-Lender Relationship, Competitive Equilibrium, and the Theory of Hedonic Prices. *Am. Econ. Rev.* **1976**, *66*, 401–405.
62. Keeton, W.R. *Equilibrium Credit Rationing*; Garland Publishing: New York, NY, USA; London, UK, 1979.
63. Murcia, A. Determinantes Del Acceso al Crédito de Los Hogares Colombianos. In *Borradores de Economía*; No. 449; Banco de la Republica: Bogotá, Colombia, 2007.
64. Blanco-Mesa, F.; Merigó, J.M.; Gil-Lafuente, A.M. Fuzzy Decision Making: A Bibliometric-Based Review. *J. Intell. Fuzzy Syst.* **2017**, *32*, 2033–2050. [CrossRef]
65. Baez-Palencia, D.; Olazabal-Lugo, M.; Romero-Muñoz, J. Toma de Decisiones Empresariales a Través de La Media Ponderada Ordenada. *Inquiet. Empres.* **2019**, *19*, 11–23.
66. Garg, H. Linguistic Pythagorean Fuzzy Sets and Its Applications in Multiattribute Decision-Making Process. *Int. J. Intell. Syst.* **2018**, *33*, 1234–1263. [CrossRef]
67. Alfaro-García, V.G.; Merigó, J.M.; Gil-Lafuente, A.M.; Kacprzyk, J. Logarithmic Aggregation Operators and Distance Measures. *Int. J. Intell. Syst.* **2018**, *33*, 1488–1506. [CrossRef]
68. Pérez-Arellano, L.A.; León-Castro, E.; Avilés-Ochoa, E.; Merigó, J.M. Prioritized Induced Probabilistic Operator and Its Application in Group Decision Making. *Int. J. Mach. Learn. Cybern.* **2019**, *10*, 451–462. [CrossRef]

Article

Incorporating Fuzzy Logic in Harrod's Economic Growth Model

Joan Carles Ferrer-Comalat [1,*], Salvador Linares-Mustarós [1] and Ricard Rigall-Torrent [2]

[1] Department of Business Administration, University of Girona, C/Universitat de Girona 10, 17071 Girona, Spain; salvador.linares@udg.edu
[2] Department of Economics, University of Girona, C/Universitat de Girona 10, 17071 Girona, Spain; ricard.rigall@udg.edu
* Correspondence: joancarles.ferrer@udg.edu

Abstract: This paper suggests the possibility of incorporating the methodology of fuzzy logic theory into Harrod's economic growth model, a classic model of economic dynamics for studying the growth of a developing economy based on the assumption that an economy with only savings and investment income is in equilibrium when savings are equal to investment. This model was the first precursor to exogenous growth models, which in turn gave rise to endogenous growth models. This article therefore represents a first step towards introducing fuzzy logic into economic growth models. The study concerned considers consumption and savings to depend on income by means of uncertain factors, and investment to depend on the variation of income through the accelerator factor, which we consider uncertain. These conditions are used to determine the equilibrium growth rate of income and investment, as well as the uncertain values for these variables in terms of fuzzy numbers. As a result, the new model is shown to expand the classical model by incorporating uncertainty into its variables.

Keywords: fuzzy logic; fuzzy arithmetic; extension principle; economic models; Harrod's growth

Citation: Ferrer-Comalat, J.C.; Linares-Mustarós, S.; Rigall-Torrent, R. Incorporating Fuzzy Logic in Harrod's Economic Growth Model. *Mathematics* **2021**, *9*, 2194. https://doi.org/10.3390/math9182194

Academic Editors: Jorge de Andres Sanchez and Laura González-Vila Puchades

Received: 29 June 2021
Accepted: 6 September 2021
Published: 8 September 2021

Publisher's Note: MDPI stays neutral with regard to jurisdictional claims in published maps and institutional affiliations.

Copyright: © 2021 by the authors. Licensee MDPI, Basel, Switzerland. This article is an open access article distributed under the terms and conditions of the Creative Commons Attribution (CC BY) license (https://creativecommons.org/licenses/by/4.0/).

1. Introduction

As is well known, macroeconomics studies how economic systems work from an aggregate point of view as a result of the interactions that take place between different economic agents. Given this consideration, we can state that this field of economic science has the following two aims:

(a) To investigate and clarify situations that have taken place and study what may have caused them. These prior experiences should allow us to devise mathematical models that can help explain economic reality;

(b) To put forth predictions, usually in the short and medium run, related to how certain economic variables evolve, including national income, investment, consumption, among others. Consequently, mathematical models created from past experiences must be empirically tested in order to show their suitability for representing the reality studied and thus evaluating that the predictions with the models are reliable.

All economic models rely on a theoretical basis that tries to simplify the complexity of economic systems by stating a set of relationships between variables linked by parameters that are a priori unknown (but in certain cases can be estimated econometrically). The dependence of the model on several parameters implies that when it is necessary to determine the adequacy of the model to reality, the exact values of most of the parameters embedded in the model are not known. This involves taking uncertain quantities in the models as defined, and thus, when the model is applied, a result close to the empirical reality is already expected, but in practice it can often be wrong. Solow [1] already warns of this fact with the statements "all theory depends on assumptions which are not quite true", and "the art of successful theorizing is to make the inevitable simplifying assumptions in such a way that the final results are not very sensitive". The aim of this article is to present a complementary and novel point of view, introducing a fuzzy logic model in order to obtain results based on an infinity of possible inputs. This new modelling based on fuzzy logic [2] allows that empirical reality can be one of the expected results of the model and therefore the actual result is not expected to be outside the range of possible expected values. To sum up, the idea is not to check how much we have deviated from the actual solution with the model, but to confirm that this actual solution is one of the possible solutions derived in the model.

In order to limit the set of possible solutions yielding the best forecast, it is common to resort to certain constructs of the theory of fuzzy sets, known as fuzzy numbers, and which can be associated with numerical values with a degree of possibility [3]. The idea of considering the future in a pessimistic, optimistic and highly plausible scenario is the basis of these types of numbers that collect all values with some degree of possibility greater than zero. It is the ease of use of these mathematical structures [4–7] that allows uncertainty to be incorporated efficiently into any economic model regarding the behavior of economic, social and financial scenarios. This is especially necessary when sudden changes in the values of the variables are expected due to events not reflected in the historical series.

The incorporation of fuzzy logic into economic models has been addressed in a multitude of papers [8–10]. Following this line of research, in recent years the fuzzy logic perspective has been proposed for various growth models [11,12]. The present work, included in this current line of research, proposes the creation of a dynamic model of income behavior based on the classic Harrod model. This model is seminal in the economic growth literature. It is well known that this model has several shortcomings [13,14]. For example, the model assumes that productive capacity is proportional to capital stock, an assumption that has been shown to be false. In fact, the model was overtaken first by the so-called "exogenous growth" models of Solow [1] and Swan [15] and then by the "endogenous growth" models [16]. Essentially, exogenous growth models explain long-term growth using a variable which exogenous to the model: technology. Endogenous growth models try to explain the way in which the evolution of technology allows long-term growth. This article recovers Harrod's original approach in order to show how the contribution of fuzzy logic to a very simple model can change the predictions that stem from the model and bring them closer to reality. This paper must be understood, then, as a first step in a process that will have to continue with the application of fuzzy logic to exogenous growth models, first, and endogenous, later.

In order to meet our primary objective, the main body of the work has been divided into three different blocks. Section 2 presents a brief summary of the fuzzy logic tools used by the new model, specifically the method used to solve equations with fuzzy parameters. Section 3 details the initial Harrod model for determining income through the use of fuzzy parameters. The results are simplified in Section 4 to make them more operational when we analyze the particular case in which uncertain parameters are expressed through triangular fuzzy numbers (TFN). Section 5 provides a numerical example with the proposed new model. Finally, the paper ends with the conclusions and references.

2. Preliminaries

In classical set theory, a set is defined as any well-defined collection of objects. This theory highlights the fact that there can never be any doubt at all as to whether any object belongs to the set or not. If the object belongs to it, it is said to belong to the set in degree 1. If the object does not belong to it, it is said to belong to the set in degree 0, there is no other option. With the birth of fuzzy set theory [2], the membership constraint is relaxed, accepting that objects can partially belong to the set. Thus, for example, a bitten apple may belong to the set of apples in grade 0.5 [17,18]. The sets created with this approach are called fuzzy sets and make up the basic elements of Zadeh's theory. In order to establish the nomenclature that will be used in the present paper, we proceed to define the different concepts related to fuzzy sets.

Definition 1. *Fuzzy subset, membership function and support.*

We consider E an ordinary set that we take as referential set. A fuzzy subset $\tilde{A}$ of E is a set of ordered pairs:

$$\tilde{A} = \{(x, \mu_{\tilde{A}}(x))/x \in E\}$$

where $\mu_{\tilde{A}}$ is a function: $\mu_{\tilde{A}} : E \to [0, 1]$.

Function $\mu_{\tilde{A}}$ is called membership function of fuzzy subset $\tilde{A}$, and given an element $x \in E$, the value $\mu_{\tilde{A}}(x)$ is called membership degree, compatibility degree or truth degree of element x in fuzzy subset $\tilde{A}$.

We call Support of $\tilde{A}$, and we indicate by $Supp(\tilde{A})$ at the ordinary set:

$$Supp(\tilde{A}) = \{x \in E/\mu_{\tilde{A}}(x) > 0\}$$

Definition 2. *Fuzzy number*

A fuzzy number $\tilde{A}$ is defined as a fuzzy subset of the referential R of the set of real numbers such that the membership function fulfils the following conditions:

1a. A minimum of value x exists so that $\mu_{\tilde{A}}(x) = 1$ (normality)
1b. $\mu_{\tilde{A}}(x) \geq min(\mu_{\tilde{A}}(x_1), \mu_{\tilde{A}}(x_2))\ \forall\ x_1, x_2 \in R$ and $\forall\ x \in [x_1, x_2]$ (convexity)
1c. $\mu_{\tilde{A}}(x)$ is a piecewise continuous function in all the points of $\mathbb{R}$, this is, it's a continuous function except perhaps at a finite number of points in its domain.

The fuzzy number is in accordance with a prediction made by a human being. For example, convexity guarantees that as we approach a value x the closer x_0 with $\mu_{\tilde{A}}(x_0) = 1$, then its possibility value of this x is higher. It is important that readers are aware of this restrictive assumption.

Further, note that any real number "m" (which may be referred to as a crisp number) can be considered a particular case of fuzzy number whose membership function is $\mu(x) = \begin{cases} 1\ if\ x = m \\ 0\ if\ x \neq m \end{cases}$.

It is important to know that, by virtue of the representation theorem established by Zadeh [19], every fuzzy number $\tilde{A}$ can be expressed via its α-cuts, and for this reason we can express in a simplified way:

$$\tilde{A} = \{A_\alpha\ ,\ 0 \leq \alpha \leq 1\} \tag{1}$$

where $A_\alpha = \{x \in R/\mu_{\tilde{A}}(x) \geq \alpha\} \subseteq R$.

By virtue of convexity, A_α is reduced to a closed interval of R, which shall be represented thus:

$$A_\alpha = [\underline{A}(\alpha),\ \overline{A}(\alpha)] \tag{2}$$

where $\underline{A}(\alpha) = min\{x/\mu_{\tilde{A}}(x) \geq \alpha\}$ and $\overline{A}(\alpha) = max\{x/\mu_{\tilde{A}}(x) \geq \alpha\}$.

Note that the representation theorem is justified in its formal form trough the following equality:

$$\tilde{A} = \bigcup_{\alpha \in [0,1]} \alpha \cdot A_\alpha$$

understanding the union through the maximum operator, because the membership function verifies:

$$\mu_{\tilde{A}}(x) = \max_{\alpha \in [0,1]} \{\alpha \cdot \mu_{A_\alpha}(x)\}$$

We must also take into account that $\underline{A}(\alpha)$ is an increasing function with regard to α, and that $\overline{A}(\alpha)$ is a decreasing function with regard to α due to convexity.

However, just as algebraic operations can be conducted using real numbers, here our interest lies in determining how common operations on real numbers can incorporate the use of fuzzy numbers.

This can be resolved if we apply the extension principle of operations or composition laws, initially introduced by Zadeh [19], and subsequently modified by other authors [20–23], which considers a general method for extending the usual operations of arithmetic to the case in which uncertain amounts are represented through fuzzy subsets or fuzzy numbers.

In the case that we have a unary operation $f(a)$ or a binary operation $f(a,b)$ between one or two quantities respectively, which are the cases we will study in our model, we will have that the principle of extension of operations to uncertain magnitudes expressed by fuzzy subsets will represent the degree of possibility of each possible solution from the following membership functions:

1. If $\widetilde{A}$ is a fuzzy subset of E, the extension of a unary operation from a set E to another set F:

$$f : E \to F$$

will be given by the fuzzy subset $\widetilde{C} = f(\widetilde{A})$ with the membership function:

$$\mu_{\widetilde{C}}(z) = \begin{cases} \bigvee_{x \in f^{-1}(z)} \mu_{\widetilde{A}}(x) & if \quad f^{-1}(z) \neq \emptyset \\ 0 & if \quad f^{-1}(z) = \emptyset \end{cases} \quad (3)$$

where $\vee$ represents the supremum operator.

2. For the most common case, whereby a binary operation or internal composition law between the elements of E is defined by:

$$f : E \times E \to E$$
$$(x,y) \to f(x,y) = x * y$$

if we consider $\widetilde{A}$ and $\widetilde{B}$ as fuzzy subsets of E, then $\widetilde{C} = f(\widetilde{A}, \widetilde{B}) = \widetilde{A}(*)\widetilde{B}$ is expressed by:

$$\mu_{\widetilde{C}}(z) = \begin{cases} \bigvee_{\{(x,y)/x*y=z\}} [\mu_{\widetilde{A}}(x) \wedge \mu_{\widetilde{B}}(y)] & if \; f^{-1}(z) \neq \emptyset \\ 0 & if \; f^{-1}(z) = \emptyset \end{cases} \quad (4)$$

where $\vee$ represents the supremum and $\wedge$ represents the infimum, which in this case coincides with the minimum due to there being a finite number of values.

The support of $\widetilde{C}$ contains the possible results of the operation obtained from the elements of the respective supports of $\widetilde{A}$ and $\widetilde{B}$, and using the max-min convolution we determine the degree of membership of each possible result of the operation. In fact, when $\widetilde{A}$ and $\widetilde{B}$ are fuzzy numbers, and, therefore, fuzzy convex subsets of R, the degree of membership $\mu_{\widetilde{C}}(z)$ of a possible solution is the maximum for the values of membership $\mu_{\widetilde{A}x\widetilde{B}}(x,y)$ among all elements (x,y), which following the operation give us the result z; that is to say that $f(x,y) = z$.

More symbolically, if it is being understood that if $f^{-1}(z) = \emptyset$ then $\mu_{\widetilde{C}}(z) = 0$ we simply write:

$$\mu_{\widetilde{C}}(z) = \bigvee_{z = x*y} [\mu_{\widetilde{A}}(x) \wedge \mu_{\widetilde{B}}(y)] \quad (5)$$

When we write $\widetilde{C} = \widetilde{A}(*)\widetilde{B}$, observe that the extension to the fuzzy subsets $\widetilde{A}$ and $\widetilde{B}$ from the binary operation $*$ defined in E are denoted by means of the $(*)$ symbol.

In particular, if $\widetilde{A}$ and $\widetilde{B}$ are fuzzy numbers with continuous membership functions $\mu_{\widetilde{A}}$ and $\mu_{\widetilde{B}}$, respectively, and $*$ is a binary operation in R, then the membership function of the extension $\widetilde{A}(*)\widetilde{B}$ is given by:

$$\mu_{\widetilde{A}(*)\widetilde{B}}(z) = \bigvee_{z=x*y} [\mu_{\widetilde{A}}(x) \wedge \mu_{\widetilde{B}}(y)] \quad (6)$$

If the function $f(x,y) = x * y$ which defines the binary operation is continuous, the supremum $\vee$ coincides with the maximum and $\wedge$ is the minimum. Adapting the extension principle to the Equation (6) is therefore also referred to as a max-min convolution operation.

Remark 1. *Compatibility of the extension principle with the α-cuts.*

In the case that $\tilde{A}$ and $\tilde{B}$ are fuzzy numbers, and the binary operation f is a continuous function, Nyugen [24] states that as the supremum of Equation (4) is achieved by an (x,y), then the supremum coincides with the maximum, resulting in the following property being fulfilled:

$$C_\alpha = \left[f\left(\tilde{A}, \tilde{B}\right) \right]_\alpha = f(A_\alpha, B_\alpha) \tag{7}$$

where A_α and B_α are the respective α-cuts of $\tilde{A}$ and $\tilde{B}$, and $f(A_\alpha, B_\alpha)$ indicates, in this case, the image set.

Remark 2 (Dubois and Prade [22]). *If $\tilde{A}$ and $\tilde{B}$ are fuzzy numbers with continuous membership functions $\mu_{\tilde{A}}$ and $\mu_{\tilde{B}}$, respectively, and $*$ is an increasing monotonous (or decreasing monotonous) binary operation in R, then $\tilde{A}(*)\tilde{B}$ is a fuzzy number with a continuous membership function.*

Remark 3 (Buckley [25]). *If $*$ is a binary operation in R defined by $x * y = f(x,y)$, where f is a continuous function, and $\tilde{A}$ and $\tilde{B}$ are fuzzy numbers whose membership functions are continuous, then, if we consider $\tilde{C} = \tilde{A}(*)\tilde{B}$, we have:*

$$C_\alpha = \{ z = x * y / x \in A_\alpha, y \in B_\alpha \} \tag{8}$$

Remark 4 (Moore [26]). *If $*$ is a binary operation in R defined by $x * y = f(x,y)$, where f is a continuous rational function in its domain and each variable appears only once as a maximum and is elevated to the first power, and $\tilde{A}$ and $\tilde{B}$ are fuzzy numbers whose membership functions are continuous, then, if we consider $\tilde{C} = \tilde{A}(*)\tilde{B}$, the α-cuts of the result are:*

$$C_\alpha = A_\alpha(*)B_\alpha \tag{9}$$

where $A_\alpha(*)B_\alpha$ in this case represents the corresponding operation induced by the function f through the elementary operations of the confidence intervals.

Definition 3. *Fuzzy equation.*

A fuzzy equation is an equation in which coefficients or variables are expressed through fuzzy numbers. Without loss of generality, we will write a fuzzy equation:

$$F\left(\tilde{A}, \tilde{X}\right) = \tilde{B} \tag{10}$$

where $\tilde{A}$ and $\tilde{B}$ are fuzzy numbers, which are usually called fuzzy parameters and $\tilde{X}$ is the unknown, which we call fuzzy variable.

We call the crisp equation associated with the fuzzy Equation (10) the equation:

$$F(a, x) = b \tag{11}$$

where we consider that parameters a and b of Equation (2) represent uncertain values that can accept several different values expressed through respective distributions of possibility taken from the fuzzy numbers $\tilde{A}$ and $\tilde{B}$.

Buckley and Qu [27] propose a way of interpreting the fuzzy Equation (10) which is fully consistent with the theory of the possibility.

Let us consider Equation (10): $F\left(\tilde{A}, \tilde{X}\right) = \tilde{B}$.

Buckley and Qu's idea is to interpret Equation (10) as a family of crisp equations:

$$F(a, x) = b \quad a \in Supp(\tilde{A}) \quad \text{and} \quad b \in Supp(\tilde{B}) \tag{12}$$

where we assume that a accepts all possible values given by the fuzzy number $\tilde{A} = \{(a, \mu_{\tilde{A}}(a)/a \in R)\}$ and b all possible values given by the fuzzy number $\tilde{B} = \{(b, \mu_{\tilde{B}}(b)/b \in R)\}$. We then need to find all possible values of x, each with their own degrees of possibility, that satisfy any of Equation (12).

To this end, if it is assumed that F verifies the hypotheses of the theorem of the implicit function, then from $F(a, x) = b$ we can determine x, so that:

$$x = f(a, b) \tag{13}$$

Thus, we obtain a new fuzzy equation:

$$\tilde{X} = f(\tilde{A}, \tilde{B}) \tag{14}$$

whose solution $\tilde{X}$ expresses the solution of Equation (10) in the sense of Buckley and Qu.

Thus, we can understand $f(\tilde{A}, \tilde{B})$ as a binary operation between two quantities of uncertain values, resulting in another uncertain quantity represented by $\tilde{X}$. We must therefore study the possibility of this magnitude taking on a specific value x, considering that several combinations of possible values for a and b will exist so that $f(a, b) = x$. Each of the possible values of x fulfills some of the equations of the family (12).

Let us now observe how the equation is resolved in practice.

With the hypothesis of continuity of function f, and due of the compatibility of the principle of extension with the α-cuts, the α-cuts X_α of $\tilde{X}$ are given by:

$$X_\alpha = \{x/x = f(a,b), a \in A_\alpha, b \in B_\alpha\} \tag{15}$$

When considering f continuous and $\tilde{A}$ and $\tilde{B}$ fuzzy numbers, then the domain of definition of f, that is $A_\alpha \times B_\alpha$, is a compact set in R^2, and therefore, by virtue of Weierstrass extreme value theorem ensures the existence of maximum and minimum of f, resulting in:

$$X_\alpha = [\underline{X}(\alpha), \overline{X}(\alpha)]$$

with:

$$\underline{X}(\alpha) = \min\{x/x = f(a,b), a \in A_\alpha, b \in B_\alpha\} \overline{X}(\alpha) = \max\{x/x = f(a,b), a \in A_\alpha, b \in B_\alpha\} \tag{16}$$

Therefore, in this case X_α is a closed interval, and by extension convex. This means that $\tilde{X}$ is convex. In addition, like $\tilde{A}$ and $\tilde{B}$, $\tilde{X}$ is obviously normal. Indeed, since $\tilde{A}$ and $\tilde{B}$ are normal, there exist a^* and b^* with $\mu_{\tilde{A}}(a^*) = 1$ and $\mu_{\tilde{B}}(b^*) = 1$. Therefore, value $x^* = f(a^*, b^*)$ meet $\mu_{\tilde{X}}(x^*) = 1$. Therefore, the solution $\tilde{X}$ is a fuzzy number. Furthermore, the membership function of $\tilde{X}$ will be expressed by applying the extension principle:

$$\mu_{\tilde{X}}(x) = \bigvee_{\{x/x=f(a,b)\}} (\mu_{\tilde{A}}(a) \wedge \mu_{\tilde{B}}(b)) \tag{17}$$

Keep in mind that when applying the compatibility of the first extension with α-cuts, that the uncertain quantities represented by $\tilde{A}$ and $\tilde{B}$ have no interaction, that is, the value assumed by one of these does not affect that assumed by the other. That being said, in order to compute X_α it is not generally possible to do so by directly applying the arithmetic of intervals and directly substituting A_α and B_α in the expression of the function. That is, if we have the binary operation $f(a,b) = a * b$, and directly compute the α-cuts by applying the arithmetic of intervals:

$$V_\alpha = A_\alpha(*)B_\alpha \qquad (18)$$

it is not generally verified that $X_\alpha = V_\alpha$. However, if f is monotonous with respect to the inclusion of intervals, then the following condition is fulfilled $X_\alpha \subseteq V_\alpha$. Therefore, if the arithmetic of the intervals is directly applied to calculate the α-cuts of $\widetilde{X}$, then wider intervals containing the solution will be generally obtained. If calculating X_α is complicated due to the behavior of the function f, then V_α can be considered as an approximation of the true result.

All of that being said, two common cases can be highlighted in which α-cuts X_α can be calculated easily:

(1) If $f(a,b) = a * b$ represents a rational function in which each variable appears at most once and is raised to the first power, then we are dealing with the hypotheses from Moore's theorem [26], and $X_\alpha = V_\alpha$. Thus, in this case, X_α can be determined by directly calculating $A_\alpha(*)B_\alpha$ using the arithmetic of confidence intervals;

(2) If $f(a,b)$ is a monotonous function with regard to each of the variables, the lower and upper limits $\underline{X}(\alpha)$ and $\overline{X}(\alpha)$ of (16) will obviously be reached at some of the ends of the α-cuts A_α and B_α, as Table 1 shows.

Table 1. Extremes of the α-cuts depending of the monotony of f.

Monotonicity of f with Respect to a	Monotonicity of f with Respect to b	$\underline{X}(\alpha)$	$\overline{X}(\alpha)$
increasing	increasing	$f(\underline{A}(\alpha), \underline{B}(\alpha))$	$f(\overline{A}(\alpha), \overline{B}(\alpha))$
increasing	decreasing	$f(\underline{A}(\alpha), \overline{B}(\alpha))$	$f(\overline{A}(\alpha), \underline{B}(\alpha))$
decreasing	increasing	$f(\overline{A}(\alpha), \underline{B}(\alpha))$	$f(\underline{A}(\alpha), \overline{B}(\alpha))$
decreasing	decreasing	$f(\overline{A}(\alpha), \overline{B}(\alpha))$	$f(\underline{A}(\alpha), \underline{B}(\alpha))$

We will use Buckley and Qu's resolution method in our study into the behavior of Harrod's income growth model in a context of uncertainty, since it agrees with the way in which we interpret the values obtained by income in a fuzzy environment.

3. Study and Solution of the Harrod's Growth Model in Conditions of Uncertainty

Three years after publication of "The General Theory of Employment, Interest and Money" by famed English economist John Maynard Keynes in 1936 [28], another English economist, Henry Roy Harrod studied, following some of Keynes' ideas, the conditions for the harmonious growth of the economy and the factors of instability that may affect it [29].

With this purpose, Harrod determined, assuming a simplified model, the guaranteed rate of growth that keeps over time the balance in the circular flow of income. Years later, along with the parallel work of Domar [30,31] it gave rise to what is known as Harrod–Domar's model of economic growth. As discussed above, this model was overtaken first by the so-called exogenous growth models and then by endogenous growth models, which have always criticized that part of assumptions made by Harrod–Domar do not conform to reality. In these circumstances, in the present article we study how the contribution of fuzzy logic can change the predictions that derive from the model and adjust them more to reality.

The version of the model that we discuss in this article considers that investment constitute a stimulus to aggregate demand (multiplier principle) and, at the same time, cause an increase in productive capacity (accelerator principle). We consider that investment is the result of the growth of productive capacity. For this reason we examine the conditions under which the stimulus of investment towards aggregate demand is exactly offset by the increase in productive capacity that investment entails.

Harrod's index of guaranteed growth indicates a growth path for the economic system according to which the two objectives of investment (stimulating aggregate demand and contributing to the expansion of productive capacity) are in equilibrium, so that greater demand justifies greater productive capacity, as well as greater productive capacity allows meeting greater demand.

It is therefore a question of examining the conditions under which this equilibrium is established. The initial model proposed by Harrod considers a simplified system, without public sector and without relations with other countries. In this case, logically, the equilibrium condition between aggregate demand and productive capacity occurs when saving equals investment.

With respect to saving, as with consumption, the Keynesian condition that establishes income dependence is considered. On the other hand, with respect to investment, due to the complexity of the factors that may influence them, the accelerator principle is used, considering only the induced component that refers to the reaction of investment to changes in the level of income, and thus reflecting the fact that investment constitute an increase in productive capacity. The problem is trying to find the growth rate (according to Harrod's terminology) of income and investment to be in equilibrium.

If we consider a period t, the macromagnitudes National Income (Y_t), Consumption (C_t), Saving (S_t) and Investment (I_t), we have that the growth model of Harrod will be formed by three conditions expressed through three equations:

(a) A Keynesian-type saving and a consumption equation:

$$S_t = a \cdot Y_t \quad C_t = (1-a) \cdot Y_t \quad 0 < a < 1 \quad t = 0, 1, 2, \ldots \tag{19}$$

(b) An investment equation based on the accelerator principle, i.e., the investment is proportional to the rate of change in national income over time:

$$I_t = b \cdot (Y_t - Y_{t-1}) \quad b > 1 \quad t = 1, 2, 3, \ldots \tag{20}$$

where b is a constant (the accelerator) that represents the average ratio between the increase in capital and the increase in production, and is therefore generally considered to have a value greater than unity.

(c) The equilibrium condition based on equality between savings and investment, since:

$$Y_t = C_t + S_t \quad AD_t = C_t + I_t \quad (AD = \text{aggregate demand}) \tag{21}$$

So that equilibrium is given by:

$$S_t = I_t \tag{22}$$

And we get that with the equilibrium condition:

$$a \cdot Y_t = b \cdot (Y_t - Y_{t-1}) \tag{23}$$

that is:

$$(b-a) \cdot Y_t = b \cdot Y_{t-1} \tag{24}$$

Dividing by $b-a$, which is a value strictly greater than zero, gives the relation:

$$Y_t = \left(\frac{b}{b-a}\right) \cdot Y_{t-1} \tag{25}$$

It is an equation in differences of first order that presents immediate solution applying the recurrence, and leads us to determine the expression for income in period t given the initial value Y_0, which results from:

$$Y_t = \left(\frac{b}{b-a}\right)^t \cdot Y_0 = \left(1 + \frac{a}{b-a}\right)^t \cdot Y_0 \tag{26}$$

Thus, the stability of the time trajectory depends on $b/(b-a)$. Since b represents the capital/production ratio, which, as stated previously, is usually greater than 1, and since a represents the marginal propensity to saving, which is greater than zero and smaller than 1, the base $b/(b-a)$ will be greater than 0 and, generally, greater than 1. Therefore, the trajectory of income Y_t is explosive, but not oscillating. Thus, according to the relationships given by the model, income grows indefinitely.

We observe that the guaranteed growth rate is the relative percentage of growth between two consecutive periods, which is obviously:

$$G = \frac{a}{b-a} \qquad (27)$$

Finally, the value of the investment in period t is determined from the equality:

$$S_t - S_{t-1} = a \cdot Y_t - a \cdot Y_{t-1} = a \cdot (Y_t - Y_{t-1}) = \frac{a}{b} \cdot I_t \qquad t = 1, 2, 3, \ldots \qquad (28)$$

Given the equilibrium condition ($S_t = I_t$) we have:

$$I_t - I_{t-1} = \frac{a}{b} \cdot I_t \;\Rightarrow\; \left(1 - \frac{a}{b}\right) \cdot I_t = I_{t-1} \;\Rightarrow\; I_t = \frac{b}{b-a} \cdot I_{t-1} \qquad (29)$$

from which the following relationship is deduced:

$$I_t = \left(\frac{b}{b-a}\right)^t \cdot I_0 = \left(1 + \frac{a}{b-a}\right)^t \cdot I_0 \qquad t = 1, 2, 3, \ldots \qquad (30)$$

which indicates the type of investment growth required to sustain equilibrium according to the model assumptions, which implies full employment. We observe that income and investment growth take the same form, so they must grow at the same required rate to sustain the equilibrium situation.

In this model, we see that if we know the true values of the marginal propensity to save, the accelerator, and the value of income at the initial time, we can determine the subsequent values of income and investment as well as the value of the growth rate to maintain equilibrium.

As we have argued for the models of previous works [11,12], the use of fuzzy numbers to express the marginal propensity to save and the accelerator, makes subsequent calculations difficult, but it allows the possibility of studying the behavior of the model when we operate with uncertain values, thus obtaining more information and giving a wider range of applications of the model. By operating with fuzzy numbers, we will determine the values of income, investment and growth rate as fuzzy numbers under conditions of uncertainty defined through their α-cuts, as well as their function of theoretical membership from the approach and solution of the equation given by the model under conditions of fuzziness.

When the values of a and b are expected to be of a similar amount in all periods, but uncertain, they can be considered as fuzzy numbers $\tilde{a}$ and $\tilde{b}$ and equal for each period; or, in other words, having the same membership function for each value of t. They can then be expressed generically as fuzzy subsets and as α-cuts, thus:

$$\begin{aligned}\tilde{a} &= \{(x, \mu_{\tilde{a}}(x))\} = \left\{ a_\alpha = [\underline{a}(\alpha), \overline{a}(\alpha)] \quad 0 \leq \alpha \leq 1 \right\} \\ \tilde{b} &= \{(x, \mu_{\tilde{b}}(x))\} = \left\{ b_\alpha = [\underline{b}(\alpha), \overline{b}(\alpha)] \quad 0 \leq \alpha \leq 1 \right\}\end{aligned} \qquad (31)$$

where $\mu_{\tilde{a}}$ and $\mu_{\tilde{b}}$ represent their respective membership functions, and a_α and b_α the respective α-cuts at the level α. According to the assumptions of the model, the following conditions are established:

$$0 < \underline{a}(\alpha) \leq \overline{a}(\alpha) < 1 \quad \text{and} \quad \overline{b}(\alpha) \geq \underline{b}(\alpha) > 1 \qquad (32)$$

The value of income is determined from the fuzzy equation, thus:

$$\tilde{Y}_t = \left(\frac{\tilde{b}}{\tilde{b} - \tilde{a}}\right)^t \cdot Y_0 \qquad (33)$$

To solve it we first determine the fuzzy number:

$$\tilde{M} = \frac{\tilde{b}}{\tilde{b} - \tilde{a}} \qquad (34)$$

Given the crisp equality: $M = b/(b-a)$, we can consider M as the result of the binary operation $f(a,b) = a * b = b/(b-a)$. From the principle of extension, we know that if we have a binary operation $f(a,b)$ between two quantities, we can extend the operation to the case where the quantities are uncertain and apply the principle of extension to obtain: $f(\tilde{a}, \tilde{b}) = \tilde{M}$. In this case, if we consider $\mu_{\tilde{a}}$ and $\mu_{\tilde{b}}$ to be continuous membership functions, then, by applying the extension principle to the binary operation $f(a,b)$, the membership function of the fuzzy number $\tilde{M}$ will be given by:

$$\mu_{\tilde{M}}(z) = \bigvee_{\{z/z=f(x,y)\}} (\mu_{\tilde{a}}(x) \wedge \mu_{\tilde{b}}(y)) \qquad (35)$$

Since f is a continuous function, by virtue of Buckley's theorem [25], we have that the α-cuts of $\tilde{M}$ are:

$$M_\alpha = [\underline{M}(\alpha), \overline{M}(\alpha)] = \{z = f(x,y) / \ x \in a_\alpha, \ y \in b_\alpha\} \qquad (36)$$

where logically:

$$\underline{M}(\alpha) = min\{z = f(x,y) / \ x \in a_\alpha, \ y \in b_\alpha\}$$

$$\overline{M}(\alpha) = max\{z = f(x,y) / \ x \in a_\alpha, \ y \in b_\alpha\}$$

Since in our case $f(x,y)$ is a continuous function whose domain of definition is $a_\alpha \times b_\alpha$, which is a compact set in R^2, the existence of $\underline{M}(\alpha)$ and $\overline{M}(\alpha)$ is assured and, thus, M_α is a closed interval and consequently $\tilde{M}$ is convex. The condition of normality of $\tilde{M}$ is immediately deduced from the fact that $\tilde{a}$ and $\tilde{b}$ are normal, meaning $\tilde{M}$ is a fuzzy number.

However, since we are not dealing with the hypotheses from Moore's theorem [26], if the arithmetic of the intervals is directly applied to calculate the α-cuts M_α by calculating $f(a_\alpha, b_\alpha)$, an unwanted result can be obtained in the sense that the range is wider than the true M_α.

Indeed, if we calculate the α-cuts of $\tilde{M}$ from the fuzzy equation:

$$\tilde{M} = \frac{\tilde{b}}{\tilde{b} - \tilde{a}} \qquad (37)$$

We have that since the function $f(a,b) = b/(b-a)$ is continuous in its domain and increasing with respect to $\left(\frac{\partial f}{\partial a} > 0\right)$ and decreasing with respect to $\left(\frac{\partial f}{\partial b} < 0\right)$, when applying the results given by Buckley and Qu [27] we get:

$$\underline{M}(\alpha) = min\{z = f(x,y) / \ x \in a_\alpha, \ y \in b_\alpha\} = f\left(\underline{a}(\alpha), \overline{b}(\alpha)\right) = \frac{\overline{b}(\alpha)}{\overline{b}(\alpha) - \underline{a}(\alpha)}$$

$$\overline{M}(\alpha) = max\{z = f(x,y) / \ x \in a_\alpha, \ y \in b_\alpha\} = f\left(\overline{a}(\alpha), \underline{b}(\alpha)\right) = \frac{\underline{b}(\alpha)}{\underline{b}(\alpha) - \overline{a}(\alpha)} \qquad (38)$$

Instead, with the application of the arithmetic of the intervals, we would obtain:

$$M^*_\alpha = b_\alpha(:)(b_\alpha(-)a_\alpha) = \left[\underline{b}(\alpha), \overline{b}(\alpha)\right] (:) \left(\left[\underline{b}(\alpha), \overline{b}(\alpha)\right] (-)[\underline{a}(\alpha), \overline{a}(\alpha)]\right)$$
$$= \left[\underline{b}(\alpha), \overline{b}(\alpha)\right] (:) \left[\underline{b}(\alpha) - \overline{a}(\alpha), \overline{b}(\alpha) - \underline{a}(\alpha)\right] = \left[\frac{\underline{b}(\alpha)}{\overline{b}(\alpha) - \underline{a}(\alpha)}, \frac{\overline{b}(\alpha)}{\underline{b}(\alpha) - \overline{a}(\alpha)}\right] \quad (39)$$

From this expression we get what Moore [26] the interval extension of α-cuts of $\widetilde{M}$, since it holds:

$$M_\alpha \subseteq M^*_\alpha \quad (40)$$

If we start instead from the crisp equality:

$$N = 1 + \frac{a}{b-a} \quad (41)$$

Then, we set the fuzzy equation:

$$\widetilde{N} = 1 + \frac{\widetilde{a}}{\widetilde{b} - \widetilde{a}} \quad (42)$$

it turns out that in this case the α-cuts of $\widetilde{N}$ coincide with the extension by intervals obtained by directly applying the arithmetic.

Indeed, for the calculation of the α-cuts of $\widetilde{N}$, we consider the function $g(a, b) = 1 + \frac{a}{b-a}$, easily checking that this function is increasing in a and decreasing in b, and by application of monotony in Table 1:

$$N_\alpha = \left[\underline{N}(\alpha), \overline{N}(\alpha)\right]$$

where:

$$\underline{N}(\alpha) = \min\{z = g(x,y) / x \in a_\alpha, y \in b_\alpha\} = 1 + \frac{\underline{a}(\alpha)}{\overline{b}(\alpha) - \underline{a}(\alpha)}$$
$$\overline{N}(\alpha) = \max\{z = g(x,y) / x \in a_\alpha, y \in b_\alpha\} = 1 + \frac{\overline{a}(\alpha)}{\underline{b}(\alpha) - \overline{a}(\alpha)} \quad (43)$$

On the other hand, with the direct application of the arithmetic of the intervals, we obtain in this case:

$$N^*_\alpha = 1(+)[a_\alpha(:)(b_\alpha(-)a_\alpha)] = [1,1](+)\left[[\underline{a}(\alpha), \overline{a}(\alpha)](:)\left(\left[\underline{b}(\alpha), \overline{b}(\alpha)\right](-)[\underline{a}(\alpha), \overline{a}(\alpha)]\right)\right]$$
$$= [1,1](+)\left[[\underline{a}(\alpha), \overline{a}(\alpha)](:)\left[\underline{b}(\alpha) - \overline{a}(\alpha), \overline{b}(\alpha) - \underline{a}(\alpha)\right]\right] \quad (44)$$
$$= [1,1](+)\left[\frac{\underline{a}(\alpha)}{\overline{b}(\alpha) - \underline{a}(\alpha)}, \frac{\overline{a}(\alpha)}{\underline{b}(\alpha) - \overline{a}(\alpha)}\right] = \left[1 + \frac{\underline{a}(\alpha)}{\overline{b}(\alpha) - \underline{a}(\alpha)}, 1 + \frac{\overline{a}(\alpha)}{\underline{b}(\alpha) - \overline{a}(\alpha)}\right]$$

which, as we see, matches the α-cut N_α computed in (43).

We see, therefore, with this particular application, the importance of applying the method of solving fuzzy equations proposed by Buckley and Qu [27], since, otherwise, if we start from Equation (37) and calculate the α-cuts of $\widetilde{M}$ by directly applying the arithmetic of the confidence intervals, we could not ensure that all values of the α-cut are greater than 1, since $\underline{b}(\alpha)$ does not need to be smaller than $\overline{b}(\alpha) - \underline{a}(\alpha)$. On the other hand, calculating the α-cuts from Equation (38), it is ensured that the lower end of each α-cut, that is $\overline{b}(\alpha) / \left(\overline{b}(\alpha) - \underline{a}(\alpha)\right)$ always has a value greater than 1, and thus, it turns out that the base of the exponential function which gives the income growth, despite being an uncertain value, is greater than 1 and gives rise to an increasing trajectory for income.

In addition, another very important question is that, because the hypotheses of Moore's theorem [26] are not fulfilled, from the equality $M = N$ the fuzzy equality $\widetilde{M} = \widetilde{N}$ would not be deduced if we applied the arithmetic of the intervals in the calculation of α-cuts. However, if we determine the α-cuts from Equations (38) and (43), respectively, we can write that $\tilde{M} = \tilde{N}$, because for each level $\alpha \in [0,1]$ we have $M_\alpha = N_\alpha$. Indeed:

$$N_\alpha = [\underline{N}(\alpha), \overline{N}(\alpha)] = \left[1 + \frac{\underline{a}(\alpha)}{\underline{b}(\alpha) - \underline{a}(\alpha)}, 1 + \frac{\overline{a}(\alpha)}{\overline{b}(\alpha) - \overline{a}(\alpha)}\right] \\ = \left[\frac{\underline{b}(\alpha)}{\underline{b}(\alpha) - \underline{a}(\alpha)}, \frac{\overline{b}(\alpha)}{\overline{b}(\alpha) - \overline{a}(\alpha)}\right] = M_\alpha \quad (45)$$

To determine the membership function of the fuzzy number $\tilde{M}$, we apply the extension principle, and obtain the expression:

$$\mu_{\tilde{M}}(z) = \bigvee_{\{z/z = y/(y-x)\}} \left(\mu_{\tilde{a}}(x) \wedge \mu_{\tilde{b}}(y)\right) \quad (46)$$

If we use the partial order relation of the confidence intervals defined by:

$$[a_1, b_1] \leq [a_2, b_2] \Leftrightarrow a_1 \leq b_1 \text{ and } a_2 \leq b_2$$

note that for each level α we have $M_\alpha > [1,1]$. Therefore, because the base exponential function greater than 1 is always increasing; it will turn out that of trajectory given by the equality:

$$\tilde{Y}_t = \tilde{M}^t \cdot Y_0 \qquad t = 0, 1, 2, \ldots \quad (47)$$

we deduce the income membership function in period t:

$$\mu_{\tilde{Y}_t}(x) = \mu_{\tilde{M}^t}\left(\frac{x}{Y_0}\right) = \mu_{\tilde{M}}\left[\left(\frac{x}{Y_0}\right)^{\frac{1}{t}}\right] \quad (48)$$

As well as its α-cuts:

$$(Y_t)_\alpha = [\underline{Y_t}(\alpha), \overline{Y_t}(\alpha)] = \left[(\underline{M}(\alpha))^t \cdot Y_0, (\overline{M}(\alpha))^t \cdot Y_0\right] = \left[\frac{\left(\overline{b}(\alpha)\right)^t \cdot Y_0}{\left(\overline{b}(\alpha) - \underline{a}(\alpha)\right)^t}, \frac{(\underline{b}(\alpha))^t \cdot Y_0}{(\underline{b}(\alpha) - \overline{a}(\alpha))^t}\right] \quad (49)$$

On the other hand, the income growth factor is given by the fuzzy number:

$$\tilde{G} = \frac{\tilde{a}}{\tilde{b} - \tilde{a}} \quad (50)$$

which has membership function:

$$\mu_{\tilde{G}}(x) = \mu_{\tilde{M}}(x - 1) \quad (51)$$

where its α-cuts are:

$$G_\alpha = [\underline{G}(\alpha), \overline{G}(\alpha)] = \left[\frac{\underline{a}(\alpha)}{\overline{b}(\alpha) - \underline{a}(\alpha)}, \frac{\overline{a}(\alpha)}{\underline{b}(\alpha) - \overline{a}(\alpha)}\right] \quad (52)$$

Finally, in an analogous way to income, we determine the membership function of investment from the equation:

$$\tilde{I}_t = \tilde{M}^t \cdot \tilde{I}_0 \qquad t = 1, 2, 3, \ldots \quad (53)$$

considering that $\tilde{I}_0 = \tilde{a} \cdot Y_0$ and, therefore:

$$\mu_{\tilde{I}_0}(x) = \mu_{\tilde{a}}\left(\frac{x}{Y_0}\right) \quad (54)$$

Further, applying the principle of extension, we obtain the expression for the membership function:

$$\mu_{\widetilde{I}_t}(z) = \bigvee_{\{z/z=x\cdot y\}} \left(\mu_{\widetilde{M}^t}(x) \wedge \mu_{\widetilde{I}_0}(y) \right) = \bigvee_{\{z/z=x\cdot y\}} \left(\mu_{\widetilde{M}}\left(x^{\frac{1}{t}}\right) \wedge \mu_{\widetilde{a}}\left(\frac{y}{Y_0}\right) \right) \qquad (55)$$

The corresponding α-cuts for investment in period t are given by:

$$(I_t)_\alpha = [\underline{I_t}(\alpha), \overline{I_t}(\alpha)] = \left[(\underline{M}(\alpha))^t \cdot \underline{a}(\alpha) \cdot Y_0, \; (\overline{M}(\alpha))^t \cdot \overline{a}(\alpha) \cdot Y_0 \right]$$
$$= \left[\frac{(\overline{b}(\alpha))^t \cdot \underline{a}(\alpha) \cdot Y_0}{(\overline{b}(\alpha) - \underline{a}(\alpha))^t}, \; \frac{(\underline{b}(\alpha))^t \cdot \overline{a}(\alpha) \cdot Y_0}{(\underline{b}(\alpha) - \overline{a}(\alpha))^t} \right] \qquad (56)$$

4. Analysis of the Particular Case in Which the Parameters Are Expressed through Triangular Fuzzy Numbers (TFN)

We now study the particular case for which $\widetilde{a}$ and $\widetilde{b}$ are triangular fuzzy numbers (TFN). As it is well known considering $\widetilde{a}$ and $\widetilde{b}$ as TFN is the result of assuming in the process of estimation that for the marginal propensity to save in a given period we can indicate two values a_1 and a_3 which correspond to the minimum and maximum estimations, respectively, and a value a_2 that we believe the most likely. Likewise, we interpret parameter $\widetilde{b}$ as a TFN. Thus, we consider:

$$\widetilde{a} = (a_1, a_2, a_3) \quad \text{and} \quad \widetilde{b} = (b_1, b_2, b_3) \qquad (57)$$

With membership functions $\mu_{\widetilde{a}}$ and $\mu_{\widetilde{b}}$ which are obviously know, as well as their expressions through the α-cuts a_α and b_α.

From $\widetilde{a}$ and $\widetilde{b}$ we can determine the fuzzy number $\widetilde{M}$ that we later use for determining income. Following the general methodology set out in the previous section, we obtain the α-cuts of the fuzzy number $\widetilde{M}$ which, logically, will no longer be triangular due of the quotient that defines it:

$$M_\alpha = [\underline{M}(\alpha), \overline{M}(\alpha)] = \left[\frac{b_3 - (b_3 - b_2)\cdot\alpha}{b_3 - a_1 - (b_3 - b_2 + a_2 - a_1)\cdot\alpha}, \; \frac{b_1 + (b_2 - b_1)\cdot\alpha}{b_1 - a_3 + (b_2 - b_1 + a_3 - a_2)\cdot\alpha} \right] \qquad (58)$$

From this expression we get the membership function:

$$\mu_{\widetilde{M}}(x) = \begin{cases} 0 & \text{if} & x < \frac{b_3}{b_3 - a_1} \\ \frac{(b_3 - a_1)\cdot x - b_3}{(b_3 - b_2 + a_2 - a_1)\cdot x - (b_3 - b_2)} & \text{if} & \frac{b_3}{b_3 - a_1} \leq x \leq \frac{b_2}{b_2 - a_2} \\ \frac{(b_1 - a_3)\cdot x - b_1}{(b_2 - b_1) - (b_2 - b_1 + a_3 - a_2)\cdot x} & \text{if} & \frac{b_2}{b_2 - a_2} \leq x \leq \frac{b_1}{b_1 - a_3} \\ 0 & \text{if} & x > \frac{b_1}{b_1 - a_3} \end{cases} \qquad (59)$$

From which we find the membership function of income in period t by using (48):

$$\mu_{\widetilde{Y}_t}(x) = \mu_{\widetilde{M}}\left[\left(\frac{x}{Y_0}\right)^{\frac{1}{t}} \right]$$

So that the expression for $\widetilde{Y}_t$ by means of α-cuts is:

$$(Y_t)_\alpha = \left[\left(\frac{b_3 - (b_3 - b_2)\cdot\alpha}{b_3 - a_1 - (b_3 - b_2 + a_2 - a_1)\cdot\alpha} \right)^t \cdot Y_0, \; \left(\frac{b_1 + (b_2 - b_1)\cdot\alpha}{b_1 - a_3 + (b_2 - b_1 + a_3 - a_2)\cdot\alpha} \right)^t \cdot Y_0 \right] \qquad (60)$$

Following again the general methodology for this particular case, we determine the α-cuts for the growth factor:

$$G_\alpha = \left[\frac{a_1 + (a_2 - a_1)\cdot\alpha}{(b_3 - a_1) - (b_3 - b_2 + a_2 - a_1)\cdot\alpha}, \frac{a_3 - (a_3 - a_2)\cdot\alpha}{(b_1 - a_3) + (b_2 - b_1 + a_3 - a_2)\cdot\alpha} \right] \qquad (61)$$

from which we compute the membership function:

$$\mu_{\widetilde{G}}(x) = \begin{cases} 0 & \text{if} & x < \frac{a_1}{b_3 - a_1} \\ \frac{(b_3 - a_1)\cdot x - a_1}{(b_3 - b_2 + a_2 - a_1)\cdot x + (a_2 - a_1)} & \text{if} & \frac{a_1}{b_3 - a_1} \leq x \leq \frac{a_2}{b_2 - a_2} \\ \frac{a_3 - (b_1 - a_3)\cdot x}{(b_2 - b_1 + a_3 - a_2)\cdot x + (a_3 - a_2)} & \text{if} & \frac{a_2}{b_2 - a_2} \leq x \leq \frac{a_3}{b_1 - a_3} \\ 0 & \text{if} & x > \frac{a_3}{b_1 - a_3} \end{cases} \qquad (62)$$

as can be clearly seen from the membership function $\widetilde{G}$, it does not maintain the triangular fuzzy number structure.

Finally, if we replace the values in this particular case in the general Equation (56), we would get the expression for investment in each period through its α-cuts. For the function of membership of investment we refer to the general case as it is not a simple operational expression.

5. Example of Application

As an illustrative example of the model that we have just developed, we complement the explanation with a specific numerical case, considering $\widetilde{a}$ and $\widetilde{b}$ as triangular fuzzy numbers with the following values:

$$\widetilde{a} = (0.15,\ 0.17,\ 0.20) \qquad \widetilde{b} = (3.5,\ 4,\ 5)$$

Additionally, considering the initial value for income $Y_0 = 100$, we will determine the fuzzy values of the guaranteed growth index $\widetilde{G}$, as well as income $\widetilde{Y}_t$ and investment $\widetilde{I}_t$ for period $t = 4$.

In this case, the membership functions of $\widetilde{a}$ (marginal propension to save) and $\widetilde{b}$ (accelerator) are formed by linear sections, as seen in Figures 1 and 2, respectively.

The analytical expression for the membership function of the fuzzy number $\widetilde{a}$ is:

$$\mu_{\widetilde{a}}(x) = \begin{cases} 0 & \text{if} & x < 0.15 \\ \frac{x - 0.15}{0.02} & \text{if} & 0.15 \leq x \leq 0.17 \\ \frac{0.20 - x}{0.03} & \text{if} & 0.17 \leq x \leq 0.20 \\ 0 & \text{if} & x > 0.20 \end{cases}$$

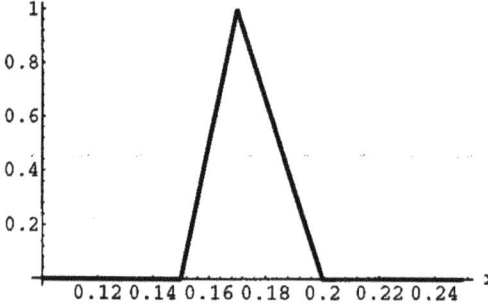

Figure 1. Membership function of marginal propension to save (mps).

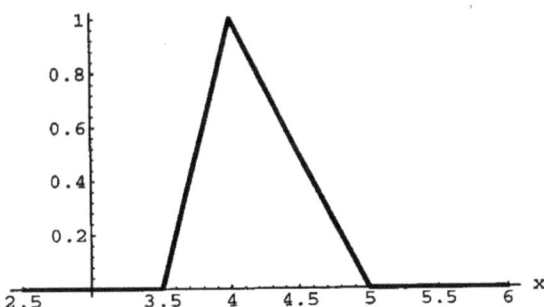

Figure 2. Membership function for the accelerator.

The expression of its α-cuts is:

$$a_\alpha = [0.15 + 0.02 \cdot \alpha,\ 0.2 - 0.03 \cdot \alpha] \qquad 0 \leq \alpha \leq 1$$

On the other hand, for the fuzzy number $\tilde{b}$ we have:

$$\mu_{\tilde{b}}(x) = \begin{cases} 0 & if & x < 3.5 \\ \frac{x-3.5}{0.5} & if & 3.5 \leq x \leq 4 \\ 5 - x & if & 4 \leq x \leq 5 \\ 0 & if & x > 5 \end{cases}$$

For the expression of its α-cuts it is:

$$b_\alpha = [3.5 + 0.5 \cdot \alpha,\ 5 - \alpha]$$

From the fuzzy equation:

$$\tilde{M} = \frac{\tilde{b}}{\tilde{b} - \tilde{a}}$$

Given the membership functions for $\mu_{\tilde{a}}$ and $\mu_{\tilde{b}}$, we get the membership function of the fuzzy number $\tilde{M}$ which, as shown in Figure 3, does not correspond to a triangular fuzzy number, because, as we see, this membership function is made up of nonlinear sections. Its expression is given by:

$$\mu_{\tilde{M}}(x) = \begin{cases} 0 & if & x < 1.030 \\ \frac{4.85x - 5}{1.02x - 1} & if & 1.030 \leq x \leq 1.044 \\ \frac{3.5 - 3.3x}{0.53x - 0.5} & if & 1.044 \leq x \leq 1.060 \\ 0 & if & x > 1.060 \end{cases}$$

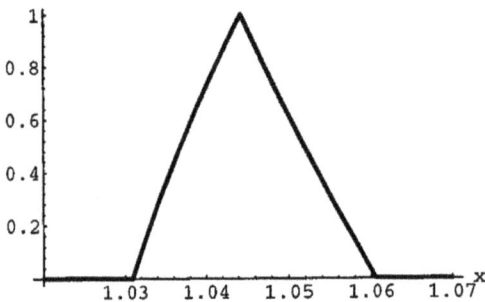

Figure 3. Membership function for $\widetilde{M}$.

From the fuzzy number $\widetilde{M}$ we determine the guaranteed growth rate, expressed by means of the fuzzy number $\widetilde{G}$, with α-cuts:

$$G_\alpha = \left[\frac{0.15 + 0.02 \cdot \alpha}{4.85 - 1.02 \cdot \alpha}, \frac{0.2 - 0.03 \cdot \alpha}{3.3 + 0.53 \cdot \alpha}\right] \quad 0 \leq \alpha \leq 1$$

The membership function $\mu_{\widetilde{G}}$ takes a similar form to $\mu_{\widetilde{M}}$ since as we have seen in Equation (51), they are closely related. Its expression is:

$$\mu_{\widetilde{G}}(x) = \begin{cases} 0 & if & x < 0.030 \\ \frac{4.85x - 0.15}{1.02x + 0.02} & if & 0.030 \leq x \leq 0.044 \\ \frac{0.2 - 3.3x}{0.53x + 0.03} & if & 0.044 \leq x \leq 0.060 \\ 0 & if & x > 0.060 \end{cases}$$

As shown in Figure 4, it is not a triangular fuzzy number because it is made up of rational sections.

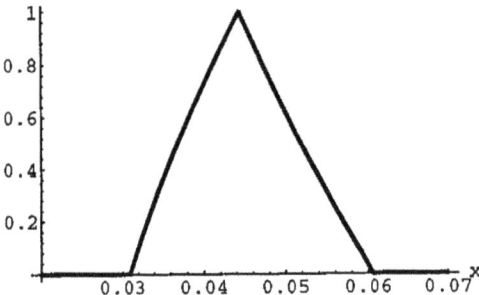

Figure 4. Membership function for the growth index.

Finally, we determine the fuzzy expression for income for $t = 4$, whose α-cuts are:

$$Y_{4\alpha} = \left[\left(\frac{5 - \alpha}{4.85 - 1.02 \cdot \alpha}\right)^4 \cdot 100, \left(\frac{3.5 + 0.5 \cdot \alpha}{3.3 + 0.53 \cdot \alpha}\right)^4 \cdot 100\right]$$

as well as its membership function from Equation (48):

$$\mu_{\widetilde{Y_4}}(x) = \begin{cases} 0 & if & x < 112.55 \\ \frac{1.5337 \cdot \sqrt[4]{x} - 5}{0.3225 \cdot \sqrt[4]{x} - 1} & if & 112.55 \leq x \leq 118.79 \\ \frac{3.5 - 1.0435 \cdot \sqrt[4]{x}}{0.1676 \cdot \sqrt[4]{x} - 0.5} & if & 118.79 \leq x \leq 126.24 \\ 0 & if & x > 126.24 \end{cases}$$

As in all non-triangular membership functions in the example, the coefficients are approximate, but with the degree of accuracy tolerable by the uncertainty situation of the problem.

Finally, to determine the fuzzy expression for investment in period $t = 4$, we take the uncertain value of the initial investment given by the triangular fuzzy number:

$$\widetilde{I_0} = (15, 17, 20) \qquad I_{0\alpha} = [15 + 2 \cdot \alpha,\ 20 - 3 \cdot \alpha]$$

with which, by application of Equation (56), we determine the α-cuts for investment in period $t = 4$ which are given by:

$$I_{4\alpha} = \left[\left(\frac{5 - \alpha}{4.85 - 1.02 \cdot \alpha} \right)^4 \cdot (15 + 2 \cdot \alpha),\ \left(\frac{3.5 + 0.5 \cdot \alpha}{3.3 + 0.53 \cdot \alpha} \right)^4 \cdot (20 - 3 \cdot \alpha) \right]$$

If we consider the α-cut at level $\alpha = 0$, we get the support for investment in the considered period:

$$I_{4,0} = \left[\underline{I_4}(0),\ \overline{I_4}(0) \right] = [16.94,\ 25.30]$$

from which we get the minimum (16.94) and maximum (25.30) estimates for investment in that period.

6. Conclusions

In the paradigm of deterministic models inspired as models of historical construction, these constitute a reflection of the previous behavior at the time of its construction without any possibility to consider sudden changes in the values of the variables due to events not reflected in the historical series. This method of using historical data to make estimates from probability distributions often meets with contradictions in economic reality, which undergoes continual change given the common appearance of new factors not reflected in past behaviors. Therefore, it is convenient, as shown in the present paper, to be able to incorporate in the models future unknown values which can be evaluated by experts with a certain degree of possibility, so that, in this way, models can take into account the continuously changing nature of the economic, social and financial environment. This paper has shown that fuzzy modelling allows introducing inaccuracy by incorporating the variability of model parameters in a reasonable manner. This allows us to obtain a model that includes more information and can therefore produce better future predictions. In this sense, the proposed model allows us to ascertain the impact on the prediction process of income behavior being jointly studied for all possible values of the uncertain parameters considered, each with their own degree of possibility.

Summing up, this paper shows how, by starting with some distributions of possibility for the uncertain magnitudes that are taken as inputs, we can determine the corresponding distribution for any equilibrium values that potentially represent income. Assigning possibility distributions to variables that are not precisely known allows us to provide a proper interpretation of economic reality, thus obtaining forecasts more suited to the real world, since more information is taken into account when applying the model.

On the other hand, from the example based on the classic Harrod's model of economic growth in the context of uncertainty, we highlight the following points:

1. Under the established fuzzy conditions, that is, considering the marginal propensity to save $\tilde{a}$ and the accelerator $\tilde{b}$ as fuzzy numbers, the results obtained with the application of the model have an interpretation within the context of uncertainty in which they are presented, and, in this sense, they give a more generality to the application of the model, which implies, in this context of uncertainty, a closer approach to reality than the still image that results from the classical application. This study in a fuzzy context allows us to incorporate within the model what would be a broad analysis of sensitivity. The study is a logical generalization because if the parameters that we have considered as fuzzy numbers are reduced to crisp numbers, the results obtained coincide with those resulting from the classical application;

2. When analyzing the behavior of the model under the considered hypotheses of uncertainty, we observe that, as in the context of certainty, income and investment must grow at the same rate to be in equilibrium under the conditions imposed by the model. This rate is given by the growth index, which in this study is expressed as a fuzzy number with the corresponding membership function being expressed from the membership functions $\mu_{\tilde{a}}$ and $\mu_{\tilde{b}}$;

3. The direct application of the arithmetic of the intervals is not always suitable for the calculation of the fuzzy growth index, since, depending on the form of the expression of this index, the expression of the intervals that determine the α-cuts with the direct application of arithmetic has much more entropy than the calculation of the α-cuts obtained by using the interpretation proposed by Buckley and Qu, which is appropriate in the application of the model presented;

4. If we use $\tilde{G} = 1 + \tilde{a}/\left(\tilde{b} - \tilde{a}\right)$ as the expression for the growth index, we have found that the applications of the rules of fuzzy arithmetic lead to a correct and interpretable result of the α-cuts of the growth rate. Therefore, by using this expression the usual arithmetic can be applied;

5. As is to be expected, the time factor increases the uncertainty in determining future income and investment values, which is translated into the increase in entropy that characterizes the resulting fuzzy numbers as we move forward in time;

6. The initial value of income Y_0, which we consider as a known and therefore crisp, is a value that influences the results only as a scale factor, as in the classical model;

7. From the knowledge of the membership functions $\mu_{\tilde{a}}$ and $\mu_{\tilde{b}}$, it is possible to determine the membership function for income in any period, and this function adopts a simple and fully operational expression in the case that $\tilde{a}$ and $\tilde{b}$ are triangular fuzzy numbers. On the other hand, for investment values the expression obtained for the membership function is not operational and it is essential in practice to calculate the possible values and their degree of possibility through the expression of their α-cuts;

8. The purpose of the application presented in this paper is how models based on fuzzy numbers work, as well as to analyze how the model behaves when the parameters linked to the variables are fuzzy numbers and need to be operated according to the arithmetic of uncertainty. In further studies it will be necessary to go beyond the simple Harrod model of economic growth and study and analyze models of exogenous and endogenous growth with the aim of building new models based on uncertainty, more suited to the complex economic reality. It should be noted that government expenditure and taxes are not considered in our representation of Harrod's model. Introducing these variables would prove a fairly straightforward task and would not affect the results produced by the model. An approach that involves exogenous and endogenous growth models would be more interesting, since it would allow us to take into account the role of, for example, technology, education, property rights or human capital, relevant variables missing from the Harrod approach;

9. It should also be noted that our representation of Harrod's model considers a closed economy and does not take into account the financial aspect of the economy. Nowadays, money flows around the world as countries trade with one another. Indeed, finance was at the heart of the economic crisis before the world was hit by that of the Covid pandemic.

Therefore, an interesting approach would be to consider fuzzy logic in a model aimed at the open economy (with imports, exports and financial flows). This would represent a valuable addition to the Mundell–Fleming setting, which takes into account deterministic values for exchange rates, interest rates and output;

10. Beyond the long-term considerations of economic growth, our approach is also useful for analyzing the economy in the short term. By incorporating fuzzy logic, it is possible to study economic fluctuations related to the business cycle. Currently, dynamic stochastic general equilibrium models introduce fluctuations way means of random shocks into the economy. Fuzzy numbers could be used to represent these shocks so that the model takes into account the uncertainty pervading the real world;

11. Finally, considering macroeconomic variables as fuzzy numbers may also represent a way of improving the current measurements used by traditional economic models. It is well known that real world measures of national income, consumption, saving, investment, exports or imports are subject to uncertainty. Indeed, national accounts data are subject to frequent revisions as new information emerges. Presenting macroeconomic variables as fuzzy numbers would make the uncertainty inherent in such magnitudes apparent and result in policymakers being able to make more informed decisions.

Consequently, the modeling presented in this article expands the possibilities offered by the current Harrod's growth model [29–31], and continues along the path first followed with the proposal to use fuzzy numbers [11,12].

Author Contributions: This paper is the result of the joint work by all the authors. J.C.F.-C.; (conceptualization, methodology, conclusions); S.L.-M. (literature review, introduction, validation); R.R.-T. (literature review, validation, conclusions). All authors have read and agreed to the published version of the manuscript.

Funding: This research was funded by the Spanish Ministry of Science, Innovation and Universities and FEDER, grant number RTI2018-095518-B-C21.

Data Availability Statement: The data and methods used in the research have been presented in sufficient detail to the work so that other researchers can replicate the work. Numerical data has not been copied from another source.

Conflicts of Interest: The authors declare no conflict of interest.

References

1. Solow, R.M. A contribution to the theory of economic growth. *Q. J. Econ.* **1956**, *70*, 65–94. [CrossRef]
2. Zadeh, L.A. Fuzzy Sets. *Inf. Control* **1965**, *8*, 338–353. [CrossRef]
3. Zadeh, L.A. Fuzzy Sets as a basis for a theory of possibility. *Fuzzy Sets Syst.* **1978**, *1*, 3–28. [CrossRef]
4. Ferrer-Comalat, J.C.; Linares-Mustarós, S.; Corominas-Coll, D. A formalization of the theory of expertons. Theoretical foundations, properties and development of software for its calculation. *Fuzzy Econ. Rev.* **2016**, *21*, 23–39. [CrossRef]
5. Ferrer-Comalat, J.C.; Linares-Mustarós, S.; Corominas-Coll, D. A generalization of the theory of expertons. *Int. J. Uncertain. Fuzziness Knowl. Based Syst.* **2018**, *26* (Suppl. 1), 121–139. [CrossRef]
6. Linares-Mustarós, S.; Ferrer-Comalat, J.C.; Corominas-Coll, D.; Merigó, J.M. The ordered weighted average in the theory of expertons. *Int. J. Intell. Syst.* **2019**, *34*, 345–365. [CrossRef]
7. Linares-Mustarós, S.; Ferrer-Comalat, J.C.; Corominas-Coll, D.; Merigó, J.M. The weighted average multiexperton. *Inf. Sci.* **2021**, *557*, 355–372. [CrossRef]
8. Billot, A. *Economic Theory of Fuzzy Equilibria*; Springer: Berlin, Germany, 1992.
9. Mansur, Y. *Fuzzy Sets and Economics*; Edward Elgar Publishing: Cheltenham, UK, 1995.
10. Ponsard, C. Fuzzy mathematical models in economics. *Fuzzy Sets Syst.* **1988**, *28*, 273–283. [CrossRef]
11. Ferrer-Comalat, J.C.; Linares-Mustarós, S.; Corominas-Coll, D. Fuzzy logic in economic models. *J. Intell. Fuzzy Syst.* **2020**, *38*, 5333–5342. [CrossRef]
12. Ferrer-Comalat, J.C.; Linares-Mustarós, S.; Corominas-Coll, D. A Fuzzy Economic Dynamic Model. *Mathematics* **2021**, *9*, 826. [CrossRef]
13. Easterly, W. *Ghost of the Financing Gap: How the Harrod-Domar Model Still Haunts Development Economics*; World Bank Development Research Group: Washington, DC, USA, 1997.
14. Solow, R.M. Perspectives on Growth Theory. *J. Econ. Perspect.* **1994**, *8*, 45–54. [CrossRef]
15. Swan, T.W. Economic growth and capital accumulation. *Econ. Rec.* **1956**, *32*, 334–361. [CrossRef]
16. Romer, P.M. The Origins of Endogenous Growth. *J. Econ. Perspect.* **1994**, *8*, 3–22. [CrossRef]

17. Kosko, B. *Fuzzy Thinking: The New Science of Fuzzy Logic*; Hyperion: New York, NY, USA, 1993.
18. Linares Mustarós, S.; Viladevall Valldeperas, Q.; Llacay Pintat, T.; Ferrer-Comalat, J.C. An introduction to the foundations of fuzzy logic through art. *CUAd C* **2018**, *20*, 133–156.
19. Zadeh, L.A. Similarity relations and fuzzy orderings. *Inf. Sci.* **1971**, *3*, 177–200. [CrossRef]
20. Zadeh, L.A.; Fu, K.S.; Tanaka, K.; Shimura, M. *Fuzzy Sets and Their Applications to Cognitive and Decision Processes*; Academic Press: New York, NY, USA, 1975.
21. Jain, R. Tolerance analysis using fuzzy sets. *Int. J. Syst. Sci.* **1976**, *7*, 1393–1401. [CrossRef]
22. Dubois, D.; Prade, H. *Fuzzy Sets and Systems: Theory and Applications*; Academic Press: New York, NY, USA, 1980.
23. Zimmerman, H.J. *Fuzzy Set Theory and Its Applications*; Kluwer Academic Publishers: Boston, FL, USA, 2000.
24. Nguyen, H.T. A note on the extensions principle for fuzzy sets. *J. Math. Anal. Appl.* **1978**, *64*, 369–380. [CrossRef]
25. Buckley, J.J. On using α-cuts to evaluate fuzzy equations. *Fuzzy Sets Syst.* **1990**, *38*, 309–312. [CrossRef]
26. Moore, R. *Methods and Applications of Interval Analysis*; SIAM: Philadelphia, PA, USA, 1979.
27. Buckley, J.J.; Qu, Y. Solving fuzzy equations: A new solution concept. *Fuzzy Sets Syst.* **1991**, *39*, 291–301. [CrossRef]
28. Keynes, J.M. *The General Theory of Employment, Interest, and Money*; Springer: Berlin, Germany, 2018.
29. Harrod, R.F. An essay in dynamic theory. *Econ. J.* **1939**, *49*, 14–33. [CrossRef]
30. Domar, E.D. Capital expansion, rate of growth, and employment. *Econom. J. Econom. Soc.* **1946**, *14*, 137–147. [CrossRef]
31. Domar, E.D. Expansion and employment. *Am. Econ. Rev.* **1947**, *37*, 34–55.

A Multidimensional Fuzzy Quality Function Deployment Design for Brand Experience Assessment of Convenience Stores

Tsuen-Ho Hsu [1] and Ling-Zhong Lin [2,*]

[1] Department of Marketing and Distribution Management, National Kaohsiung University of Science and Technology, Kaohsiung 824005, Taiwan; thhsu@nkust.edu.tw
[2] Department of Marketing Management, Kaohsiung Campus, Shih Chien University, Kaohsiung 84550, Taiwan
* Correspondence: lingzhong@g2.usc.edu.tw

Abstract: In the past, few studies have explored brand attachment and how to deliver the content of a strategic management program through brand experience. The purpose of this study is to construct an integrated model of consumer brand attachment and brand experience. It also applies the fuzzy quality function deployment (FQFD) method to develop a value competitive strategy model of convenience stores. The respondents were split across two stages. In the first stage, 265 consumers were surveyed on the importance of brand attachment; in the second stage, 38 experts and scholars were invited to evaluate the content of a strategic management program. The results showed that for three convenience store brands, the levels of brand attachment of the young and older customer groups each had their own advantages and must be strengthened. Taking 7-Eleven as a practical implications example, we can learn from the analysis that young customers believe "brand prominence" is the most important factor, followed by "brand passion", "brand affection", and "brand–self connection". As for older customers, the order from most to least important was "brand–self connection", "brand affection", "brand prominence", and "brand passion". The originality value is that this study extends previous research findings in measuring and building brand attachment frameworks and importance assessments while providing a sound demonstration of enhancement strategy solutions.

Keywords: enhancement strategy; brand attachment; convenience stores; fuzzy quality function deployment

1. Introduction

Brand experience and brand attachment are two important factors in the relationship between consumers and brands. Many studies have already explained the importance of brand experience, indicating that brand experience affects consumer loyalty and satisfaction [1–3]. Kang et al. [4] also maintained that, in addition to increasing the value of products and services, the consumer experience created by enterprises will become more substantial.

However, the idea of measuring consumer brand attachment has not received much attention. Despite increased customer-oriented marketing efforts in the service industry, relatively little attention has been given to the processes between brand attachment and brand experience. Although the general methodology or technology has focused on measuring customer perceptions of service quality and satisfaction [2,5], few studies have analyzed brand attachment for specific guidelines on how to design services of brand experience to meet the quality standards expected by consumers.

Although brand has the characteristic of material, it is also an important marketing tool for creating consumers' experiences [6]. However, consumers cannot build profound and meaningful relationships with enterprises only through brand experience [7]. They seem to lack an emotional attachment to brand [8], and they use their attachment to brand to show their personalities, characteristics [9], and social self-identities [10]. Therefore, we believe that through the special attachment consumers hold for a brand, the brand can arouse certain types of experiences for consumers, who can also gain satisfaction. There are other ways for brands to attract consumers' attention rather than only brand experience. In other words, in the retail or service industry, brand attachment (such as sense, emotion, self-connection, captivation, or social identity) can act as a connection and a conversation between consumers and brand experience.

In the light of this, Das et al. [11], Tafesse [12], and Schmitt [13] all believe that business strategies should be adjusted toward advantageous brand experience delivery. They think that there are already many companies that successfully manage their performance of organization through brand experience. They can also hold exiting consumers with product quality, market share, and profit rate. Therefore, we believe that the message sent by these issues does not simply emphasize that enterprises should use "brand experience" as the basis of competing advantages. Enterprises should further elevate and discover how to create brand experience and make consumers form an authentic value of brand attachment. Hence, the purpose of this study is to probe the meaning of brand experience activities. Apart from delivering brand messages, it is more important to listen to consumers' most genuine voices at the scene. Through the conversation between consumer and brand, enterprises can find the core of attachment that consumers care about the most and create a greater actual benefit of brand experience activities.

The originality of this study was that we used the process of quality function deployment (QFD) for enterprises to truly hear every voice of the brand attachment demand attribute and further acknowledged the brand attachment demand and value treasured by consumers. In other words, through the planning and design of products or services, along with the execution of the management process, enterprises can further understand the consumer's cognitive feel of brand attachment. By using the multiple procedures of QFD, the consumer's genuine demand in their mind can be expressed by the internal relational structure matrix.

Hence, this study tried to integrate the procedure designs of QFD and fuzzy linguistic preference relations (FLPR) into fuzzy quality function deployment (FQFD). By doing so, we try to understand the brand attachment that consumers long for and the value that consumers demand and hence turn them into a major reference for brand experience strategy planning. The expected results and specific highlights of this study included the following: (1) establishment of a hierarchy of brand attachment through a literature review and discussion with experts; (2) construction of a QFD framework of brand attachment and a brand experience management strategy, as well as a correlation analysis between these two components; and (3) execution of a performance evaluation of different convenience stores based on the perceived difference between the desired performance and the current performance of brand experience strategy management.

The paper is organized as follows. A relevant literature review is provided in Section 2. Section 3 presents the construction of the brand attachment FQFD analysis process. The research design and empirical study are presented and discussed in Sections 4 and 5, respectively. Finally, Section 6 provides the conclusion and implications.

2. Literature Review

2.1. Brand Attachment (BA)

The definition of brand attachment is the emotional bond between consumers and brands [14,15], and the concept of brand attachment refers to the strength of the bond connecting the brand with the self, which is very powerful in leading to satisfied, trusting, and committed relationships. The strength of the attachment varies on the basis of differences in the link between consumers and brands, and brand attachment has a significant impact on consumer behavior [16].

Carroll and Ahuvia [17] considered brand attachment as a way for consumers to show their obsession with and loyalty to products, which is a kind of profound, heartfelt emotional attachment behavior. Mishra et al. [18] defined brand attachment as the strength of the bond linking brands with the self. Park et al. [19] believed that brand attachment reflects the emotional bond between people and brands. Moreover, it enriches consumers' memory network and the perception of the brand-self relationship. Bian and Haque [16] maintained that if consumers have a stronger brand attachment, they are more willing to engage in challenging activities that require time, money, energy, and reputation. Combining all these research results, we conclude that the external brand attachment behavior shown by consumers is actually an expression of satisfaction of their internal demands.

Hung and Lu [20] regarded brand attachment as an integration of the structure of emotions, which can be divided into positive and negative emotions. The results of their research demonstrate that the positive emotion of brand attachment is an effective predictive factor of brand repurchase intentions and word-of-mouth behavior. The dimensions of positive emotions are affection, passion, the brand–self connection, and brand prominence. The four dimensions of positive emotion proposed by Hung and Lu [20] are based on the pioneering empirical research on affection and brand attachment conducted by Thomson et al. [21], who elaborated that the bond of brand affection is characterized by deep connection, affection, and consumer passion. In addition, Park et al. [19] regarded the brand–self connection and brand prominence as two key factors of the brand attachment concept. On the basis of the research results above, this study adopts four key factors as brand attachment dimensions: affection, passion, the brand–self connection, and brand prominence.

Regarding the attributes of each dimension, the affection dimension is based on the study of Thomson et al. [21], who evaluated consumers' brand affection using a scale. This factor reflects consumers' warmth toward brands, which can be affectionate, loving, friendly, and peaceful. Among them, the peaceful attribute cannot be controlled by either consumers or businesses; therefore, it is excluded from this study. The passion dimension mainly reflects consumers' ardor, excitement, and positive emotions, which can include passion, delight, and captivation. The brand–self connection dimension is mainly based on the brand function evaluation dimension proposed by Rio et al. [22]. It includes the brand guarantee, self-identity, social identity, and status. Among these, the status function corresponds to the personal desire to gain recognition from others. However, it does not represent the groups recognized by brands; consequently, it is excluded from this study. The brand prominence dimension is based on the brand awareness concept proposed by Aaker [23]. Aaker conceptualized brands and divided brand awareness into six attributes on the basis of brand equity: brand cognition, brand memory, brand opinion, brand dominance, top-of-mind brand awareness, and brand knowledge. Brand cognition, brand memory, and brand knowledge were chosen as the evaluative attributes of brand prominence in the present study to match the questions regarding brand prominence. The content above is organized in Table 1.

Table 1. The dimensions and attributes of brand attachment.

Goal	Dimension	Attribute	Description	Reference
Brand Attachment	Affection	Brand affection	Tending to feel or show affection or tenderness.	Thomson et al. [21]
		Brand love	Having deep emotions for a brand, just like love. Not having it will cause great disappointment.	
		Brand friendliness	Friendly relationship with the brand.	

Table 1. *Cont.*

Goal	Dimension	Attribute	Description	Reference
	Passion	Brand passion	Having strong affection or emotion toward the brand.	Hung and Lu [20], Thomson et al. [21]
		Brand delight	Feeling delighted about the brand.	
		Brand captivation	Without any strong favor or even hope to consume a product at any time, but still having strong desire for a certain product/brand.	
	Brand-self connection	Brand guarantee	Based on the reliability maintained by the brand to identify with the commitment and guarantee of the brand's product and fulfill personal expectations.	Rio et al. [22]
		Brand self-identity	Consistency between consumers' self-image and brand image.	
		Brand social identity	Integrating the communication tools of the brand, and the consumers hope to become part of the group.	
	Brand prominence	Brand cognition	When a consumer comes into contact with related clues of a familiar brand, he/she thinks of it spontaneously.	Aaker [23], Xu et al. [24]
		Brand memory	The ease and possibility for a consumer to recall the brand in his/her memory.	
		Brand knowledge	How much a consumer knows about the brand.	

Ko et al. [25] pointed out that once a business has begun to add value, the next step is trying to increase its competitive advantages. The best way to achieve this goal is to build consumers' emotional attachment to the business and its products. One way to build attachment is through experiential marketing. The purpose of experiential marketing is to fully understand consumers and think from the standpoint of the consumers to gain feedback from genuine experience, which can be converted into a precious resource to make brands grow.

2.2. Strategic Management Plan for Brand Experience

Brand experience is the sum of the product, shopping, service, and consumption experience gained by consumers when interacting and cooperating with brands with a clear intention to consume products and services [1,2,26]. Regarding consumers, compared with real products and intangible services, experience is the most unforgettable [27]. Bleier et al. [28] considered that there are two kinds of experiences: direct experiences, such as shopping, purchasing, and consuming products, and indirect experiences, such as consumers receiving commercials and marketing communication. Brakus et al. [2] believed that brand experience is formed by the brand-related stimuli (names, slogans, and pictures) received by consumers when searching for, purchasing, and consuming brands. These brand-related stimuli form the main source of consumers' subjective reactions. The foundation of brand experience lies within the brand itself. Many studies have mentioned that brand experience constitutes the best chance for consumers to have a sense of passion for and differentiate between brands and that the interaction between experience and brands is very close [29].

The meaning of a brand involves creating explicit consumer reactions; therefore, deeper relationships with consumers can be developed [27]. Experience can create brand image and offer opportunities for consumers' curiosity to be aroused and for consumers to explore, purchase, and become involved, thus creating a sense of favorability [30]. Jimenez-Barreto et al. [31] also proposed that brand experience can effectively arouse consumers' interest in brands and products and further arouse their desire to purchase. Mclean et al. [32] also conducted a related study on brand experience activities and developed a mobile application customer experience model in which consumers keep circulating and creating valuable things with brands. Molinillo et al. [33] observed the effect of brand experience on loyalty to retailers. They believed that the purpose of brand experience activities is not only to deliver brand messages but also to listen to the real voice of consumers through the activities. Only if a brand finds the core of what consumers care the most about through conversation can brand activities create more substantial benefits.

Schmitt [34] developed strategic experiential modules (SEMs), which divided experience into five forms: sensing, feeling, thinking, acting, and relating. With regard to sensing, feeling, and thinking, consumers' expectations of brands go far beyond the brand-related function and practicality. Maehle et al. [35] tried to satisfy consumer demands, considered that experience can achieve full customer satisfaction, and divided experience into five dimensions: definition, notification, imagination, immersion, and interest arousal. At the same time, Schmitt [34] pointed out the difference between products and experience and confirmed that the concept of brand experience genuinely depicts the spirit of a brand. He further explained that consumers' expectations of brands, which integrate consumers' feelings, sensations, and wisdom, go far beyond the brand-related function and practicality.

Many scholars continue to investigate experience issues, and they all propose their own views. On the basis of past research on consumers and marketing, Lam et al. [36] offered another perspective: when consumers search for products, go shopping, receive services, and consume products, experience automatically happens at the same time as such behaviors. They divided experience into sensing, feeling, cognition, and behavior. Nysveen et al. [37] believed that in addition to the four dimensions of sensing, feeling, cognition, and behavior, relating is essential and constitutes a fifth dimension. The measurement base of overall brand experience in this study adopted the brand experience factor dimensions proposed by Schmitt [9], which are recognized by major scholars and include five dimensions: sensing, feeling, thinking, acting, and relating. The five dimensions of brand experience proposed by Lam et al. [36] and Nysveen et al. [37] were also adopted, and the mobile application brand experience research conducted by Kim and Yu [38] was also included. Excluding the sensing dimension, which has little effect, this study adopted four dimensions as the structure of brand experience and strategic management planning, i.e., feeling, thinking, acting, and relating, as shown in Table 2.

Table 2. Strategic management plan of brand experience.

Goal	Dimension	Attribute	Strategic Management Plan	Reference
Brand Experience	Feeling experience	Emotionality	Pictures that can stimulate sense and arouse consumers' emotions	Beig and Khan [39], Brakus et al. [2], Gavilanes et al. [40], Kim and Yu [38], Morrison and Crane [41], Nysveen et al. [37]
		Affectivity	Small games that are appealing and capable of stirring consumers' emotions	
		Participation	Functions of commenting, liking, and sharing	
	Cognition experience	Knowledge	Informative articles and messages	Beig and Khan [39], Brakus et al. [2], Malär et al. [15]
		Utility	Consumers actively search for special offers	
		Curiosity	Special videos or commercials	
	Acting experience	Function experience	Functions supplied by apps (ex: mobile wallet)	Bleier et al. [28], Brakus et al. [2], Dwivedi et al. [30], Hwang et al. [42], McLean et al. [32],
		Passive involvement in activities	Messages of special offers	
		Active involvement in activities	Point gathering activities	
	Relating experience	Community belongingness	Consumers interact in the brand community and gain belongingness toward it	Jouzaryan et al. [43], Kang et al. [4], Kim and Yu [37], Mishra et al. [18], Nysveen et al. [36], Schmitt and Zarantonello [44], Yasin et al. [45]
		Friend sharing	Consumers actively share messages with friends	
		Brand identity	Makes consumers identify with the brand and think they are part of it	

2.3. Fuzzy Quality Function Deployment

To truly understand consumers' reactions and voices with regard to brand attachment, the QFD procedure makes it possible to understand the brand attachment felt by consumers [46] and, in turn, provide a service experience. The QFD concept originated in Japan in the 1960s and had a widespread impact on Western countries in the 1980s. It is particularly important for the business industry when trying to understand consumers' versatile demands [47]. Through the planning and design of products or services and the execution of management procedures, QFD can help business owners understand differences in the cognition of consumers, especially when business owners are providing new services and products [48]. This multistep procedure can be expressed by an internal relational structure matrix, and the easiest method is to use the HOQ framework [48]. The relationship between WHATs and HOWs and the weight and performance values of every element can be presented in an HOQ matrix. Figure 1 shows the six parts of the HOQ matrix: (1) customer requirements (WHATs), (2) technical requirements (HOWs), (3) the relationship matrix (correlation of WHATs and HOWs), (4) the customer needs assessment, (5) competence analysis, and (6) prioritized technical requirements. However, when the demands of consumers become complex and versatile, the message provided by the HOQ will subsequently become fuzzy. For example, the key demand attribute may be overestimated, or it may not be possible to precisely show its importance. Therefore, through the explanation of fuzzy sense, the process design can be more flexible.

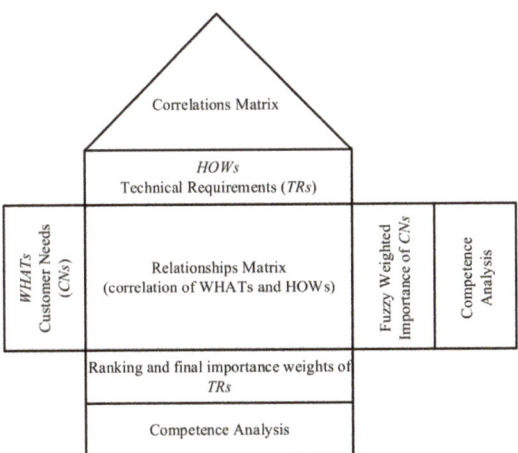

Figure 1. The framework of HOQ.

Thus, this study attempts to use the QFD procedure to analyze the brand attachment attributes felt by consumers. Subsequently, it makes the analyzed data the basis for designing a brand experience strategy. Applying QFD involves collecting semantic data, which include the subjective judgment or emotional uncertainty of consumers or experts. Therefore, regarding the issue of brand attachment, consumers will be inconsistent in the judgment standard of favorability because of different cognition of their internal demands. If an evaluation is performed using precise numerical values, it is possible that the study team will be unable to know consumers' reactions and genuine feelings; thus, the evaluation may also be unreliable. In fact, using semantic data to show emotional characteristics, such as liking or attachment, is fundamentally fuzzy and uncertain. However, when the values and demands of consumers become complex and versatile, the message provided by the HOQ will subsequently become fuzzy [49]. To process these data properly, this study adopts fuzzy set theory to cope with the problem posed by the fact that traditional statistical evaluation cannot precisely measure the emotions of attachment, attitudes, and ideas.

Regarding FQFD research, Hsu and Tang [49] believed that expressing consumers' demands through natural semantics involves fuzziness and uncertainty. Therefore, using the concept of fuzzy logic can overcome some limitations of the demand side of the HOQ and make consumers' imprecise demands easier to understand. Yuen [50] found some limitations in the traditional QFD method: (1) the traditional QFD method cannot provide a reasonable explanation of how a consensus is reached in the process of member evaluation; (2) the presentation of traditional QFD information cannot explain the subject's cognitive conflict; and (3) the data used in traditional QFD are assumed to be clear values, which ignores the fuzziness and differences in subjects' attitudes and behaviors. Therefore, Yuen [50] maintained that integrating fuzzy set theory into the HOQ will make complex decision-making data easier to organize and represent in a more flexible way. In addition, Haber et al. [51] used both fuzzy values and clear values to make comparisons and see what uncertainty would cause in the demand attribute of the HoQ. Their results showed that using clear values enhanced the effect caused by uncertainty.

In recent years, studies that have combined fuzzy set theory and QFD include the following. Mehtap and Karsak [52] combined QFD and fuzzy multicriteria decision making (MCDM) to select suppliers. Hsu and Tang [53] combined QFD and the fuzzy analytic hierarchy process (FAHP) and considered the combination of customer relationship interest and the business relationship linking strategy. Ultimately, they developed a model of customer relationship management strategy. Lam and Lai [54] used the fuzzy analytic network process (FANP) and QFD to develop and design new products. Dat et al. [55] combined

FQFD with the technique for order preference by similarity to ideal solution (TOPSIS) to maximize company interest and compare market segmentation, adding characteristics of different markets to companies' existing advantages. Lu et al. [56] combined the FAHP, QFD, and the brand knowledge concept proposed by Keller to develop a model of brand revitalization. Kayapınar and Erginel [57] combined a fuzzy multiobjective decision-making model and FQFD to design airport services. Xie et al. [58] proposed a multiobjective decision-making method to choose the technical index in the HOQ; the purpose of this method was to calculate and fix the fundamental importance of the technical index. They also used mobile phone product design, for example, to verify the utility and effectiveness of the method. Zaitsev and Dror [59] applied QFD to examine the relationship between multiple indexes of corporate social responsibility and their results. Lizarelli et al. [60] used fuzzy approaches and QFD to support improvement decisions in an entrepreneurial education service. Chen et al. [61] proposed a new integrated MCDM method for improving QFD, which integrated the hesitant fuzzy linguistic term set under an uncertainty environment. Haiyun et al. [62] proposed the innovation strategies in the energy industry using the QFD-based hybrid interval valued intuitionistic fuzzy decision approach combined with a novel hybrid methodology.

Analyzing the studies above, we learn that the messages that consumers send often contain ambiguity and multiplicity in the sensing dimension. Moreover, the data or attributes oriented toward qualitative data often show subjective uncertainty in evaluations. Hence, it seems that the analytical framework of traditional QFD cannot truly represent the importance of the evaluation of consumer demand attributes. Therefore, applying the method of fuzzy set theory to analyze the semantic data expressed by the subjects and using fuzzy numbers to represent consumers' feelings toward their favorite targets will make the analysis more objective and evidence-based.

3. Brand Attachment FQFD Analysis Process

The objective of developing the FQFD framework in this study is to solve the above-mentioned problems encountered by traditional QFD [50]. One of the main advantages of using consistency fuzzy preference relation analysis [63] in this study is that rather than the traditional method of comparing $n(n-1)/2$ attributes multiple times, only $(n-1)$ evaluations are required. This can reduce the number of importance comparisons between attributes. Another advantage is that the FQFD analysis process allows the multiple semantic data presented by the subjects to be properly measured and avoids the subjective judgment and uncertainty of consumers that may be implied by the semantic communication in the evaluation process of traditional HOQ [55]. The strategic process of strengthening brand experience was constructed by FQFD and based on brand attachment, as shown in Figure 2. Each step is explained below.

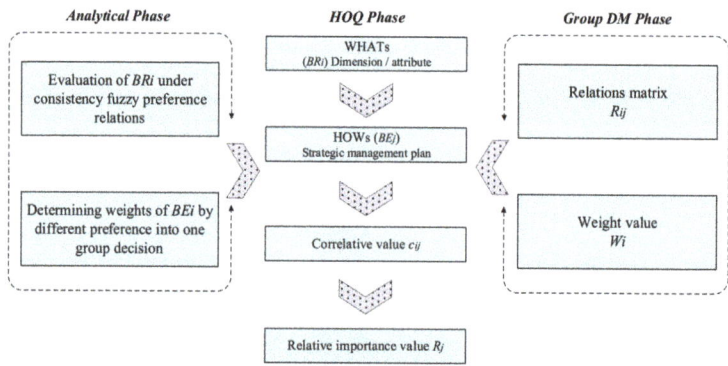

Figure 2. Experience strengthens the strategic importance evaluation process of brand attachment demands.

3.1. Step 1: Building Brand Attachment Demand Attributes (WHATs) and Experience Strategic Management Plans (HOWs)

In the FQFD analysis process, the brand attachment demand attributes were WHATs, and HOWs were experience strategic management plans that could satisfy or make consumers form brand attachment. We used both a literature review and expert interviews to collect and confirm the content of each brand attachment attribute. BR_i represents the brand attachment demand dimension/attribute, and BE_j represents the experience strategic management plan.

3.2. Step 2: Analysis Process Platform of Brand Attachment

In the FQFD analysis platform, this study used MATLAB software with Excel to analyze the data, and the calculation process applied was as follows: (1) the calculation of the relative importance value of WHATs (W_i), (2) the calculation of the value of the relationship between WHATs and HOWs (R_{ij}), and (3) the calculation of the final relative importance value of HOWs (R_j). This study did not include the HOWs value of c_{ij} in the calculation, and therefore, the final integrated calculation focused on the value of R_j.

3.3. Step 3: Assessing the Consistency Fuzzy Preference Relations of Brand Attachment

This study used the concept of fuzzy linguistic variables and consistency fuzzy preference relations, in addition to the nine linguistic scales proposed by Herrera-Viedma et al. [64], to build a consistent fuzzy preference relations matrix. On a practical level, consistent fuzzy preference relations can effectively rectify the inconsistency in the evaluation outcome as a result of increasing the number of constructs examined. This method can effectively solve the inconsistency problem that may occur when interviewed experts complete a questionnaire, and in turn, it increases the overall effectiveness and accuracy. This study used the characteristic of the fuzzy linguistic preference relation addition reciprocal to calculate and used defuzzification [65] to acquire the crisp relative weight value of the brand attachment attributes ($w_1, w_2, \ldots, w_n$).

3.4. Step 4: Developing the Technical Items of Brand Experience Strategic Management Plans

This study used the brand experience dimensions proposed by Schmitt [9] as the basis of developing a whole brand experience strategy, including sensing, feeling, thinking, acting, and relating. This study also combined the five brand experience dimensions proposed by Nysveen et al. [37] and the mobile application brand attachment items developed by Kim and Yu [38] to build the four dimensions and strategic management plan of this study. We also assumed that $T_1, T_2, \ldots, T_j$ are the technical items of the brand experience strategic management plan.

3.5. Step 5: Evaluating the Incidence Matrix of Brand Attachment Attributes and Brand Experience Strategic Management Plans

This study assumed that the number of subjects is k, they evaluate the ith brand attachment attribute and the jth brand experience strategic management plan, and the fuzzy incidence is $\widetilde{R}_{ijk} = (L_{ijk}, M_{ijk}, U_{ijk})$. A triangular fuzzy number is demoted simply as $(L_{ijk}, M_{ijk}, U_{ijk})$. The parameters L_{ijk}, M_{ijk}, and U_{ijk} indicate the smallest possible value, the most promising value, and the largest possible value, respectively. Therefore, the average formula $\overline{\widetilde{R}}_{ijk}$ below can be used to organize the views from K subjects and to convert fuzzy linguistics into crisp numbers through defuzzification. Finally, we can obtain the crisp incidence matrix value, R_{ijk}.

$$\overline{\widetilde{R}}_{ijk} = \left(\frac{\sum_{k=1}^{m} L_{ijm}}{m}, \frac{\sum_{k=1}^{m} M_{ijm}}{m}, \frac{\sum_{k=1}^{m} U_{ijm}}{m} \right), k = 1, 2, m$$

3.6. Step 6: Evaluating the Importance of Brand Experience Strategic Management Plan Items

The target value of brand experience strategic management plan items was calculated from the incidence matrix value, R_{ijk}, multiplied by the relative weight value of the brand attachment attributes ($w_1, w_2, \ldots, w_n$), and we acquired the relative weight value of the brand experience strategic management plan R_j. To evaluate all brand experience strategic management plans on the same basis, this study used a formula to standardize each value. Then, this study obtained the technical importance evaluation value of the target brand experience strategic management plan.

4. Research Design

The subjects of this study were three major chain convenience stores: 7-Eleven, Family Mart, and Hi-Life. We discussed the differences among the three rival enterprises and analyzed the differences in various consumer groups' brand attachments. The subjects of the questionnaire were loyal consumers who used the apps of these brands. After discussions with three experts from 7-Eleven, Family Mart, and Hi-Life, we learned that the best way to distinguish the main consumer groups of convenience stores was to divide these consumers into groups by age. Taking convenience into consideration, the experts suggested dividing these consumers into a young consumer group and an older consumer group. The young consumer group consisted of consumers 30 years of age or younger. Meanwhile, the older consumer group consisted of consumers 31 years of age or older. The analysis was performed on the basis of these definitions.

The main reference of the questionnaire was the concept of brand attachment proposed by Malär et al. [15] and Thomson et al. [21], which was adjusted to fit the purpose of this study. By interviewing chain convenience store owners, we developed a questionnaire about convenience store brand attachment containing four parts. The first part followed the standard and method of classifying brand attachment attributes, the second part tested the importance of brand attachment attributes, the third part analyzed consumers' shopping characteristics, and the last part assessed the subjects' shopping patterns and basic information.

In addition, this FQFD-based research was divided into two steps. The first step consisted of building the hierarchical structure of consumer brand attachment on the basis of a literature review and applying it as the basis of the evaluation of the importance of attachment attributes. We targeted consumers of 7-Eleven, Family Mart, and Hi-Life to complete the questionnaire and obtained the weight sets of all kinds of dimensions and attributes. According to McLean et al. [32], consumers who have frequently used an app for more than 6 months have more awareness of experience with the brand's app. Therefore, if one meets the conditions listed below, he/she was defined as a consumer expert: (1) the subject "agreed" that he/she was a loyal consumer of a brand, (2) the subject had been using the brand's app 11 or more times per month for more than 6 months, and (3) the subject had engaged in consumption using the brand's app. The questionnaire was distributed at stores of the three brands and on the Internet. The number of questionnaires distributed at 7-Eleven stores was 71; 63 were valid, and 8 were invalid. The number of questionnaires distributed at Family Mart stores was 86; 60 were valid, and 26 were invalid. The number of questionnaires distributed at Hi-Life stores was 108; 60 were valid, and 48 were invalid. The organized data are listed in Table 3.

Table 3. Analysis of returned valid copies of questionnaires from consumer experts.

Brand	Distributed Copies	Return Copies	Return Rate	Valid Copies	Invalid Copies	Valid Rate
7-Eleven	71	71	100%	63	8	89%
Family Mart	86	86	100%	60	26	70%
Hi-Life	108	108	100%	60	48	56%

The second step consisted of the literature review; we organized the strategic management plan items used by brands when creating a consumer brand experience, and we used those items as test targets for the questionnaire evaluation tool. This part of the questionnaire was for senior employees and managers who were well acquainted with the operations and app of the brand. Through the questionnaire, we were able to understand the degree of relevance between consumers' brand attachment attributes and brands' experience strategic management plan items, as well as the self-performance evaluation of each experience strategic management plan item. In the current study, 13 questionnaires were completed by personnel from 7-Eleven, 13 were completed by personnel from Family Mart, and 12 were completed by personnel from Hi-Life. The profile of the experts, including their educational level, age group, and education, is shown in Table 4.

Table 4. Expert profile statistics.

Categories		Numbers
Gender	Male	18
	Female	20
Age	29 years old and under	3
	30 to 49 years old	27
	50 years old and above	8
Education	High school or below	0
	College/university	23
	Master's degree	11
	Doctorate	4
Expertise	Store Manager	6
	Store Deputy Manager	6
	Store Consultant	3
	Project Specialist	13
	Store Management Specialist	10

5. Empirical Study

5.1. Building the Hierarchical Structure of Brand Attachment Attributes and Conducting Importance Evaluation

On the basis of a literature review and consultation plus discussion with experts, we developed the "basic criterion of convenience store brand attachment" and deliberated on all kinds of principles for the evaluating factors, including brand affection, brand passion, the brand-self connection, and brand prominence. Therefore, we established the evaluation criteria of convenience store brand attachment that consumers value the most, which is illustrated in Figure 3.

This study analyzed young and older groups of consumers of three brands of convenience stores (7-Eleven, Family Mart, and Hi-Life) by fuzzy linguistic calculation and performed calculations using a fuzzy pairwise comparison matrix questionnaire completed by the subjects. Subsequently, a fuzzy average formula was used to integrate the views of all subject groups (to calculate the average). This study also obtained the fuzzy average of the consumer brand attachment factors of the young and older consumer groups and established an explicit FLPR matrix of consumer brand attachment factors. In the same way, this study continued to use the same operating procedure used to obtain the consumer brand attachment factors (second level) for calculation, and it conducted an orderly numerical value evaluation of attributes under the consumer brand attachment factors (third level): brand affection, brand passion, brand-self connection, and brand prominence. Finally, using the method of defuzzification, this study built consumer preferences for each brand's attachment factors and attributes. Hence, the data on 7-Eleven are presented in Tables 5 and 6. Table 5 indicates the weight value and sequence of brand attachment dimensions, and Table 6 indicates the weight value of consumer brand attachment attributes.

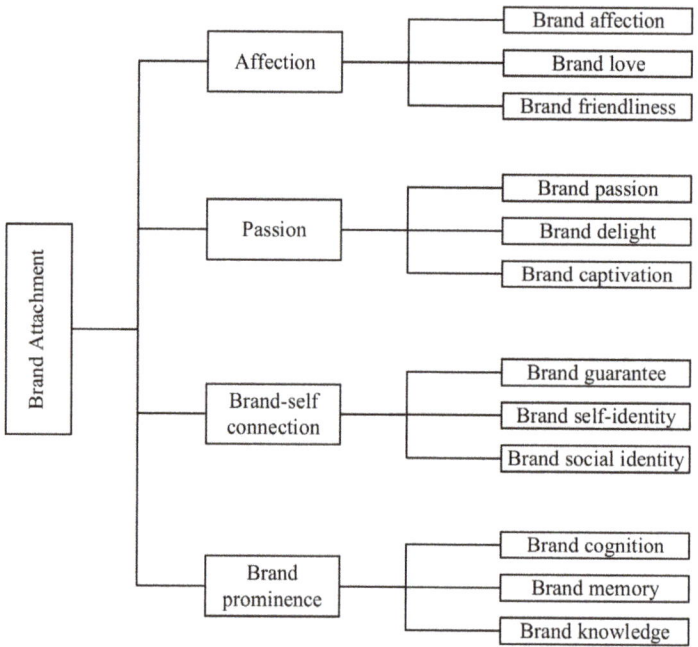

Figure 3. Brand attachment hierarchical structure graph of convenience stores.

Table 5. Weight value of brand attachment dimensions of 7-Eleven's young/older consumers.

Dimensions of Brand Attachment	Fuzzy Weight Values (Young/Older)	Defuzzification of Weight Values (Young/Older)	Rankings (Young/Older)
Affection	(0.245, 0.244, 0.249)/ (0.255, 0.259, 0.271)	0.246/0.261	3/2
Passion	(0.259, 0.253, 0.249)/ (0.234, 0.231, 0.235)	0.254/0.233	2/4
Brand–self connection	(0.244, 0.239, 0.237)/ (0.267, 0.265, 0.265)	0.222/0.266	4/1
Brand prominence	(0.253, 0.265, 0.265)/ (0.243, 0.245, 0.230)	0.261/0.239	1/3

Table 5 shows that brand prominence was the most important for young consumers, followed by brand passion and brand affection. For older consumers, brand-self connection mattered the most, followed by brand affection and brand prominence. Brand–self connection was the factor that involved the greatest difference between the two groups.

Table 6 shows that for young consumers of 7-Eleven, the top three final weightings were brand delight (0.0904), brand friendliness (0.0895), and brand cognition (0.0887). The top three final weightings for older consumers were brand guarantee (0.0910), brand self-identity (0.0886), and brand friendliness (0.0882). The two groups had different preferences for the brand attachment attributes, especially in the attachment degree of the brand assurance function; in this respect, the preference of young consumers was significantly different from that of the older group.

Table 6. Weight value of consumer brand attachment attributes of 7-Eleven's young/older consumers.

Goal	Weight Values of Dimensions (a) (Young/Older)	Ranking of Dimensions (Young/Older)	Weight Values of Attributes (b) (Young/Older)	Final Weight Values (a × b) (Young/Older)	Importance Ranking (Young/Older)
Brand Attachment	Affection (0.246/0.261)	3/2	Brand affection (0.301/0.337)	0.0740/0.0864	11/5
			Brand love (0.336/0.325)	0.0827/0.0848	7/6
			Brand friendliness (0.364/0.338)	0.0895/0.0882	2/3
	Passion (0.254/0.233)	2/4	Brand passion (0.323/0.317)	0.0820/0.0739	8/12
			Brand delight (0.356/0.352)	0.0904/0.0820	1/7
			Brand captivation (0.322/0.331)	0.0818/0.0771	9/11
	Brand–self connection (0.222/0.266)	4/1	Brand guarantee (0.308/0.342)	0.0739/0.0910	12/1
			Brand self-identity (0.332/0.333)	0.0797/0.0886	10/2
			Brand social identity (0.360/0.327)	0.0846/0.0870	6/4
	Brand prominence (0.261/0.239)	1/3	Brand cognition (0.340/0.328)	0.0887/0.0784	3/10
			Brand memory (0.329/0.339)	0.0859/0.0810	5/8
			Brand knowledge (0.331/0.333)	0.0864/0.0796	4/9

5.2. Consumer's Cognitive Difference in Brand Attachment Demand Attributes

Regarding the three main brands of convenience stores, we also evaluated the brand attachment performance of the young and older consumer groups. In order to understand the current and expected performance and its relative position in the market and identify priorities for further improvement, the young and older consumers were asked to rate the relative performance of some similar commodities and services of the three convenience stores in terms of the 12 attributes of brand attachment. The results of the consumer groups' assessments and calculations are shown in Table 7, columns 3 and 4. Taking 7-Eleven as an example, the consumers gave their current and expected performance on "brand passion" of 7-Eleven with a triangular fuzzy rating of (4, 5, 6) and (5, 6, 7), which implies that the consumers considered their performances on "brand passion" as "good" and between "fair" and "good". The fuzzy integrated average formula was used to calculate the performance values of brand attachment and convert them into crisp numbers through defuzzification. We see from the analysis in Table 7 that for young consumers, brand cognition (0.379), brand memory (0.37), and brand friendliness (0.358) had the top 3 performance values. Additionally, brand captivation (1.1), brand passion (1.086), and brand self-identity (1.079) had the top 3 improvement rates. Meanwhile, for older consumers, brand friendliness (0.358), brand social identity (0.351), and brand guarantee (0.348) had the top 3 performance values, and brand knowledge (1.069), brand guarantee (1.061), and brand delight (1.06) had the top 3 improvement rates.

Table 7. Brand attachment performance value and improvement rate of 7-Eleven's young and older consumer groups.

Attributes	Weights (A) (Young/Older)	Current Performance (B) (Young/Older)	Expected Performance (C) (Young/Older)	Performance Value (A × B) (Young/Older)	Improvement Rate (C/B) (Young/Older)
Brand affection	0.074/0.088	4.000/3.941	4.192/4.059	0.296/0.347	1.048/1.03
Brand love	0.0827/0.085	3.846/4.029	4.077/4.059	0.318/0.342	1.06/1.007
Brand friendliness	0.089/0.088	4.000/4.088	4.192/4.000	0.358/0.361	1.008/0.978
Brand passion	0.082/0.074	3.469/3.852	3.769/3.882	0.309/0.287	1.056/1.008
Brand delight	0.090/0.082	3.692/3.882	3.923/4.118	0.334/0.318	1.063/1.082
Brand captivation	0.082/0.077	3.462/3.824	3.808/3.853	0.283/0.295	1.038/1.008
Brand guarantee	0.074/0.091	3.923/3.824	4.077/4.059	0.29/.0348	1.039/1.071
Brand self-identity	0.080/0.088	3.885/3.971	4.192/4.088	0.31/0.349	1.059/1.029
Brand social identity	0.086/0.087	3.462/4.029	3.731/3.941	0.299/0.351	1.025/0.978
Brand cognition	0.089/0.078	4.269/4.059	4.308/4.206	0.379/0.318	1.009/1.086
Brand memory	0.086/0.081	4.308/3.941	4.192/4.088	0.370/0.333	0.973/1.037
Brand knowledge	0.0864/0.0796	3.846/3.853	4.115/4.118	0.332/0.307	1.07/1.079

This study compared the performance of brand attachment with the numerical analysis of improvement rates to determine whether brand attachment achieved the expected performance. The data analysis of improvement rates in Table 7 clearly shows that for the young consumer group, 7-Eleven's current performance almost meets the expected performance. However, for the older consumer group, 7-Eleven's improvement rates in brand captivation, brand delight, brand cognition, and brand knowledge were all greater than 1.07, and all failed to meet the expected performance. 7-Eleven can work on making improvements in these areas. The HOQ in Figure 4 shows the overall performance of the three convenience stores. The young consumers of Family Mart were familiar with 7-Eleven, and the older consumers of Family Mart were slightly lacking in brand cognition, but the current performance in brand knowledge was far above the expected performance. This is a great advantage that should be carefully maintained. The older consumers of Hi-Life had excellent performance in brand love and brand captivation, but there was a severe lack of brand passion, brand cognition, and brand knowledge. These are the areas that need to be improved.

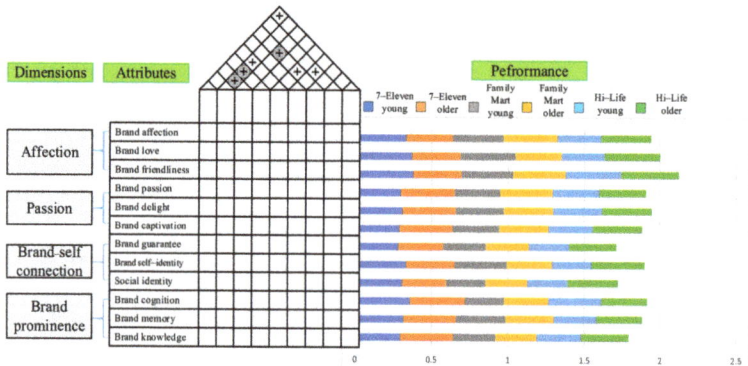

Figure 4. Performance difference analysis of brands of convenience stores.

5.3. Associative Analysis of Brand Attachment and Brand Experience Management Strategy

We listed 12 consumer brand attachment attributes (Table 1) and 12 brand experience strategic management plans (Table 2) and asked the brand manager subjects to evaluate the degree of correlation that they have with each other. We also integrated all subjects' incidence matrices (taking the average) and ultimately obtained the fuzzy incidence matrix of brand attachment attributes and experience strategic management plans. Then, through defuzzification and by multiplying the relevant overall brand attachment attribute weight value, we obtained the correlation value of the attachment attributes and the experience strategic management plans using the brand experience concept. Hence, we present the data on 7-Eleven in Appendix A. The data in the upper left corner are the values after defuzzification, and the data in the bottom right corner are the correlation values of the attachment attributes and experience strategic management plans based on the brand experience concept. Ultimately, we summed the values in the bottom right corner and acquired the initial weight value of the brand experience strategic management plans. Through standardization, we obtained the critical relative weight value of each brand experience strategic management plan.

We compared all analyses of the three main brands of convenience stores mentioned above to understand how consumers connect brand attachment through brand experience and how enterprises maintain their advantage. Enterprises can increase their competitive advantages by comparing the advantages and disadvantages of rival brands, as shown in Figure 5.

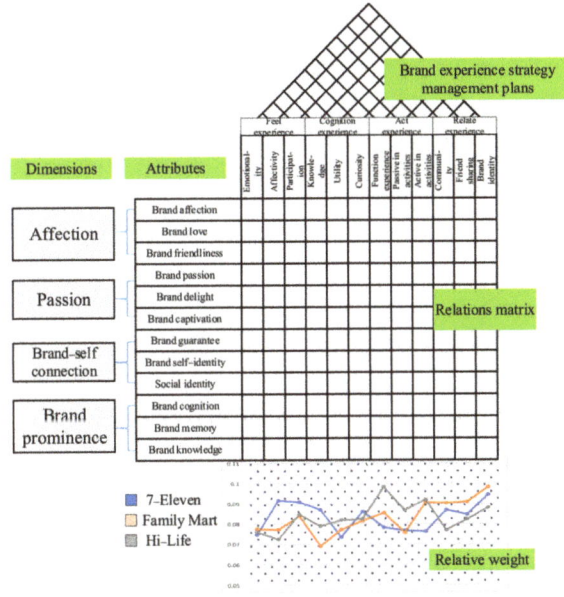

Figure 5. Relative weight value of convenience store brand experience strategy management plans.

Figure 5 shows that 7-Eleven was outstanding in knowledge, affection, and participation. The highest relative weight values of the convenience store brand experience strategy management plans for Family Mart were brand identity, active involvement in activities, community belongingness, and friend sharing. For Hi-Life, the function experience was the most excellent, followed by active involvement in activities and passive involvement in activities.

5.4. Performance Evaluation of the Brand Experience Strategic Management Plan of Each Brand

We used scores of 1 to 5 as evaluation scores to measure the cognitive perception of professional managers of the subject convenience stores to understand the differences among the execution, expected performance, and current performance of the brand experience strategic management plans. The higher the evaluation score was, the more highly the subjects rated the performance. Through the specific value calculation of expected performance and current performance, we observed differences in the subjects' evaluation of overall performance and the improvement rate for all brand experience strategic management plans. Appendix A shows that the best current performance of 7-Eleven was in friend sharing (4.62), the best expected performance was in brand identity (4.62), the best performance was in participation (3.49), and the highest improvement rate was for function experience (1.286). Hence, the performance evaluation of brand experience strategic management plans is illustrated in Figure 6.

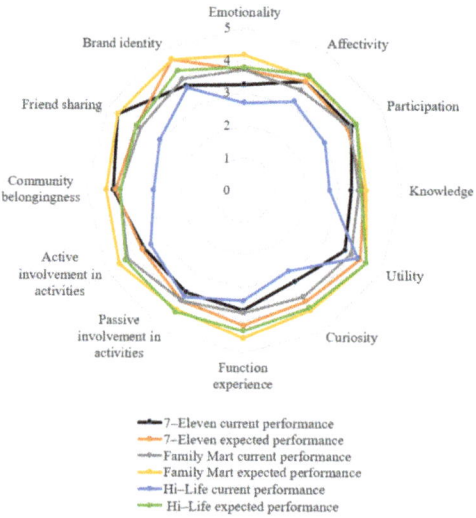

Figure 6. Performance evaluation of brand experience strategic management plans of the main brands of convenience stores.

Figure 6 shows that the best current performance of 7-Eleven was in friend sharing, followed by participation. The current performance of both already surpasses the original expected performance, which was outstanding. However, with regard to brand identity, function experience, and utility, there is still much room for improvement. For Family Mart, the current performance of participation and active involvement in activities was the closest to the expected performance, but with regard to friend sharing and function experience, there is still much room for improvement. Finally, we clearly observed that for Hi-Life, only utility performed well, and all the other brand experience strategic management plans did not meet expectations, which was very different from 7-Eleven and Family Mart.

5.5. Discussion of Results

This study performed a traditional AHP calculation of brand attachment attribute items to observe different attributes' order and differences in importance attachment in different evaluation models. We found that the value of brand attachment attributes showed greater variability and fluctuation amplitude in the calculation process of traditional AHP, especially for older consumers. As for consistency fuzzy preference relation analysis, the fluctuation amplitude of both young and older consumers was relatively smooth, which showed the consistency and stabilization of this method. In addition, in the consistency fuzzy preference relation analysis, both young and older consumers had higher attachment levels toward "brand cognition" and "brand memory"; however, in the traditional AHP, higher attachment levels for young consumers were found for "brand memory" and "brand friendliness", whereas higher levels were found for "brand friendliness" and "brand affection" for older consumers. Hence, this study found that the priority order of brand attachment attributes had a greater fluctuation amplitude in the traditional AHP.

In addition, this study found that brand experience could be driven by brand attachment to create consumer value, and according to the psychological characteristics of consumer attachment to services or products, it could be considered by business organizations as an important component of brand experience strategic management plans. These results are consistent with Shamim and Butt's [66] study and Yu and Yuan's [67] study, which validated differences in the psychological constructs of brand experience strategies on consumer attachment. Following Shamim and Butt's study, this study further extends the results in the service industry. Furthermore, in line with the findings of Huang et al. [68], they suggested that brand attachment is a mutually reinforcing outcome of the brand experience strategies. On this basis, this study integrated the attributes of brand attachment with brand experience management strategies to develop a framework for the analysis process of FQFD.

Finally, the FQFD analysis of this study revealed that brand attachment has a significant influence on the choice of brand experience strategy management plans, both in the data sets analyzed and in different groups (young and older consumers) [69]. In other words, when consumers experience a brand, it evokes positive and multiple psychological attachments and has a favorable impact on the emotional bond between consumers and the brand [70]. This confirms the findings of Japutra et al.'s [71] exploratory study, which found that young consumers in the UK showed attachment to Fatface Clothing because of their enjoyment of the service experience offered by the retailer. The results of this study are also consistent with Dolbec and Chebat's [72] findings that young North American consumers develop more attachment to brands that offer an impressive fashion brand experience. This study thus provides a new perspective on brand experience management strategies. In contrast to previous studies that focused on satisfaction, brand equity, or word-of-mouth as influential factors of brand experience, we considered more emotional bonds, i.e., brand attachment. This is the expression of emotion, passion, and connection that is associated with valuable behavioral consequences, such as improving brand cognition and creating favorable consumer reactions [73].

6. Conclusions and Implications

6.1. Conclusions

This study discussed consumers' degree of brand attachment to different brands of convenience stores and analyzed how business owners use brand experience strategic management plans to meet consumers' cognitive demands and reactions. We used the FLPR method and the analysis procedure of QFD to properly measure multilinguistic data provided by subjects and avoided consumers' subjective judgment and uncertainty hidden in linguistic communication in the traditional HOQ evaluation process. We performed analysis through FQFD to determine the relevance between brand attachment and brand experience strategic management plans and discover useful coping strategies for convenience store brand owners. By using useful brand experience strategies, convenience store brand owners can cause consumers to have stronger brand attachment, which they can convert into a competitiveness advantage to boost their own brands.

The theoretical implications and the linkage with previous studies are as follows. First, this study extended previous research findings on measuring and building brand attachment frameworks and importance assessments [74,75]. Second, this study supported the theoretical concept from quantitative empirical findings that the importance of brand attachment should be integrated into the interaction between consumers' requirements and brand values, which could be embedded in the content of the brand experience [2]. Last but not least, this study also found explicit theoretical evidence for the relevance and importance of brand attachment in the development of brand experience management strategies in the literature [76–78].

The conclusions are listed below:

(1) Understanding the cognitive demands and perceptions associated with consumers' brand attachment: The results showed that young consumers of 7-Eleven had a

stronger perception of brand prominence. Brand prominence is the degree of perception of brand affection and memory in consumers' minds, reflecting the significance of the connection between consumers' brand affection and self-recognition. This finding also showed that 7-Eleven had good results in developing young consumers and causing them to be attached to the brand, making it possible for consumers to maintain long-term loyalty to the brand. Regarding young consumers of Family Mart and Hi-Life, both consumer groups had stronger reactions to the brand-self connection, which means that the level at which these consumers integrate these brands into their self-concept is higher. Compared with 7-Eleven, this could be an advantage for Family Mart and Hi-Life. Combining the analyses of the young and older consumer groups, we noted that although the consumers of the three brands valued the factors of brand attachment differently, most of them still thought highly of brand prominence and the brand–self connection. This finding means that if retailing enterprises can connect with consumers through the brand-self connection and integrate the brand-self connection into their perception, affection, and self-concept, they will flourish because these elements are crucial.

(2) Analyzing the differences in brand attachment between different consumer groups: The results showed that for the young consumer group, 7-Eleven had the best performance in brand cognition and brand memory; for the older consumer group, it had the best performance in brand friendliness and brand social identity. However, the young consumer group believed that the performance in brand attachment was not good enough and that some improvements needed to be made. Meanwhile, the older consumer group thought that brand knowledge and brand guarantee needed to be respected and that they play important roles in linking with consumers. For Family Mart, the young consumer group maintained that the aspect that needed to be improved was brand self-identity, whereas for the older consumer group, brand guarantee was the aspect that needed to be improved. Young consumers of Hi-Life had a strong perception of brand cognition, but brand friendliness and brand passion were lacking. By contrast, the older consumer group had a strong perception of brand love and brand memory, but brand cognition was lacking. Using the results of this study, enterprises can direct marketing strategies to different consumer groups to achieve the best brand attachment.

(3) Comparing the brand experience strategic management plans of different brands: Among all experience strategic plans of the three different brands, 7-Eleven had great advantages in brand identity, function experience, and utility. Although 7-Eleven is the leading convenience store enterprise in Taiwan, there are still strong competitors. In the future, 7-Eleven can enhance the experience strategic plans mentioned above to effectively arouse brand attachment and consumers' shopping desire. Moreover, the participation and friend sharing of 7-Eleven showed strong performance and should be maintained. Family Mart performed well in active involvement in activities, passive involvement in activities, and knowledge, but other aspects, such as brand identity and friend sharing, were behind those of 7-Eleven. The performance in function experience was plain, but it had higher relevance to consumers' attachment attribute, which means that function experience is the experience strategic item that Family Mart should enhance and invest in. Hi-Life's performance in utility was better than that of 7-Eleven and Family Mart, and it is a great advantage and should be maintained. Function experience was also highly relevant to consumers' attachment attributes, as with Family Mart, and Hi-Life should invest in it. Except for utility and function experience, all the other brand experience strategic management plans of Hi-Life were suboptimal. It must dedicate resources and improve its brand experience to compete with 7-Eleven and Family Mart.

6.2. Implications for Practice

In regard to the discussion of the research findings above, this study makes the following recommendations for the three convenience stores to implement brand experience strategy management programs.

(1) Suggestions for 7-Eleven to enhance its brand experience strategy: 7-Eleven's important factors of brand attachment are brand friendliness, brand social identity, and brand delight, and the most relevant strategic management plans are affection, knowledge, and brand identity. Therefore, the suggestions are as follows: (a) design interesting games to stimulate consumers' feeling experience and form friendly relationships with them; (b) provide informative articles and messages through apps to make consumers think, identify, and feel pleasant in the hope that they can understand more about the brand and become part of it.

(2) Suggestions for Family Mart to enhance its brand experience strategy: Family Mart's important factors of brand attachment are brand social identity, brand knowledge, and brand self-identity, and the most relevant strategic management plans are brand identity, friend sharing, and function experience. Therefore, the suggestions are as follows: (a) Promote campaigns to encourage consumers to actively share brand messages (such as sharing messages to qualify for a lucky draw). Through friend sharing to promote the brand, more people can have a deeper understanding of the brand. (b) Provide versatile and convenient apps (such as a take-out-later service, a mobile wallet, and online shopping). Consumers will identify more with the brand when they become used to the convenient app and even make associations with the brand when they see those functions.

(3) Suggestions for Hi-Life to enhance its brand experience strategy: Hi-Life's important factors of brand attachment are brand guarantee, brand cognition, and brand self-identity, and the most relevant strategic management plans are function experience and community belongingness. Therefore, the suggestions are as follows: (a) Provide a convenient app to attract consumers and let them identify with the brand's promise and commitment. Although Hi-Life performs well in function experience, trending and special campaigns are still lacking. We suggest that Hi-Life should discover what kinds of campaigns attract consumers the most and "find excuses" to attract consumers to enter stores and shop. (b) Interact with consumers on brand social media to arouse their belongingness and identification with the brand. Through the connection of the subscription economy and product presales, consumers will have the feeling that they can obtain discounts when shopping and pick them up when they have needs. This could allow them to realize the concept that businesses give priority to consumers' demands through subscription.

6.3. Limitations and Future Research

Brand attachment is a subjective concept of the consumer. In the highly competitive market, the impression consumers hold for brand attachment is constantly changing. In addition, convenience stores are highly dense and have different styles and characteristics according to different neighborhoods. Under the influence of the external environment, the brand positioning of convenience stores and attachment links of consumers can easily be altered, which makes the classification and weight of the original attachment element change accordingly. For future research, researchers can duly retest the brand attachment dimension element and check whether the brand experience strategic management plan is suitable or not to make sure that the brand experience strengthening strategy is effective during the execution process.

In addition, because the subject experts of this study were brand professional managers who were mostly in charge of southern Taiwan, they probably cannot fully understand the market environment of different areas. We can broaden the range of subjects for further studies to add to the completeness of the study. Finally, in the area of brand experience strategic management plans, we lack a rolling discussion with experts and scholars. For further research, researchers can use the Fuzzy Delphi Method to acquire the brand attachment consensus dimension factor identified by experts, screen common attachment attribute factors, and revise the content of the brand experience strategic management plan. By carrying out the steps above, we can make the management plans closer to the real market.

Author Contributions: Conceptualization, T.-H.H. and L.-Z.L.; methodology, L.-Z.L.; software, L.-Z.L.; validation, L.-Z.L.; formal analysis, L.-Z.L.; investigation, T.-H.H.; resources, T.-H.H.; data curation, T.-H.H.; writing—original draft preparation, T.-H.H. and L.-Z.L.; writing—review and editing, T.-H.H. and L.-Z.L.; visualization, L.-Z.L.; supervision, T.-H.H.; project administration, T.-H.H.; funding acquisition, L.-Z.L. All authors have read and agreed to the published version of the manuscript.

Funding: This research received no external funding.

Institutional Review Board Statement: Not applicable.

Informed Consent Statement: Not applicable.

Conflicts of Interest: The authors declare no conflict of interest. The funders had no role in the design of the study; in the collection, analyses, or interpretation of data; in the writing of the manuscript; or in the decision to publish the results.

Appendix A

Table A1. 7-Eleven's incidence matrix of whole consumer group brand attachment and brand experience strategic management plans.

			Strategic Management Plan of Brand Experience													
			Feel Experience			Cognition Experience				Act Experience		Relate Experience			
			Emotionality	Affectivity	Participation	Knowledge	Utility	Curiosity	Function Experience	Passive Involvement in Activities	Active Involvement in Activities	Community Belongingness	Friend Sharing	Brand Identity
Brand Attachment	Affection	Brand affection	4.23	6.00	5.62	4.15	4.38	5.23	4.77	4.15	4.15	5.69	5.54	6.69
			0.34	0.49	0.45	0.34	0.36	0.42	0.39	0.34	0.34	0.46	0.45	0.54
		Brand love	4.54	5.92	5.15	5.69	4.54	5.15	4.92	4.00	4.31	5.69	4.92	5.92
			0.38	0.50	0.43	0.48	0.38	0.43	0.41	0.34	0.36	0.48	0.41	0.50
		Brand friendliness	4.85	6.08	5.62	5.38	4.38	5.54	4.92	4.46	5.08	4.92	5.38	5.69
			0.43	0.54	0.50	0.48	0.39	0.49	0.44	0.40	0.45	0.44	0.48	0.51
	Passion	Brand passion	4.54	4.92	5.77	3.92	4.77	5.08	4.62	3.69	4.77	5.54	5.08	5.85
			0.35	0.38	0.45	0.31	0.37	0.40	0.36	0.29	0.37	0.43	0.40	0.46
		Brand delight	4.54	5.85	5.62	5.85	4.08	5.69	4.77	5.85	4.77	5.69	5.54	6.00
			0.39	0.51	0.49	0.51	0.35	0.49	0.41	0.51	0.41	0.49	0.48	0.52
		Brand captivation	4.62	5.69	5.62	5.85	4.00	5.23	4.77	5.54	5.38	4.77	5.85	4.77
			0.37	0.45	0.45	0.47	0.32	0.42	0.38	0.44	0.43	0.38	0.47	0.38
Brand-self connection	Brand-self connection	Brand guarantee	5.08	5.62	5.92	6.54	5.00	5.08	5.85	5.54	5.08	5.69	5.38	6.15
			0.42	0.46	0.49	0.54	0.41	0.42	0.48	0.46	0.42	0.47	0.44	0.51
		Brand self-identity	4.85	5.85	5.92	6.23	4.69	5.85	4.92	4.77	5.08	6.00	6.00	6.15
			0.40	0.48	0.49	0.51	0.39	0.48	0.40	0.39	0.42	0.49	0.49	0.51
		Brand social identity	4.54	4.92	6.23	6.38	5.85	5.23	4.62	4.77	4.31	5.08	4.46	6.00
			0.38	0.41	0.52	0.54	0.49	0.44	0.39	0.40	0.36	0.43	0.38	0.51
Brand prominence	Brand prominence	Brand cognition	4.69	6.38	5.92	4.85	4.77	5.38	4.77	5.54	4.92	4.77	5.54	6.00
			0.41	0.56	0.52	0.42	0.41	0.47	0.41	0.48	0.43	0.41	0.48	0.52
		Brand memory	4.69	5.54	5.46	5.00	4.38	5.38	4.92	4.54	4.46	5.69	4.31	5.69
			0.39	0.46	0.46	0.42	0.37	0.45	0.41	0.38	0.37	0.48	0.36	0.48
		Brand knowledge	4.69	5.77	5.15	5.15	4.15	5.54	4.77	4.46	4.77	5.54	5.38	5.85
			0.39	0.48	0.43	0.43	0.34	0.46	0.40	0.37	0.40	0.46	0.45	0.49
Evaluations		Current performance	3.23	3.85	3.92	3.38	3.69	3.23	3.69	3.62	3.62	4.15	4.62	3.69
		Expected performance	3.69	3.85	3.77	3.77	4.23	3.92	4.15	3.92	3.69	4.08	3.92	4.62
		Improvement rate	1.143	1	0.961	1.115	1.146	1.063	1.125	1.083	1.019	0.983	0.848	1.252
		Initial weights	4.66	5.72	5.67	5.43	4.59	5.37	4.89	4.79	4.76	5.42	5.28	5.9
		Relative weights	0.0746	0.0916	0.0908	0.0869	0.0734	0.086	0.0782	0.0766	0.0762	0.0868	0.0846	0.0945
		Performance value	0.254	0.315	0.349	0.248	0.342	0.293	0.294	0.322	0.304	0.32	0.333	0.375

References

1. Chattopadhyay, A.; Laborie, J.L. Managing brand experience: The market contact audit. *J. Advert. Res.* **2005**, *45*, 9–16. [CrossRef]
2. Brakus, J.J.; Schmitt, B.H.; Zarantonello, L. Brand experience: What is it? How is it measured? Does it affect loyalty? *J. Mark.* **2009**, *73*, 52–68. [CrossRef]
3. Kamboj, S.; Sarmah, B.; Gupta, S.; Dwivedi, Y. Examining branding co-creation in brand communities on social media: Applying the paradigm of Stimulus-Organism-Response. *Int. J. Inf. Manag.* **2018**, *39*, 169–185. [CrossRef]
4. Kang, J.; Manthiou, A.; Sumarjan, N.; Tang, L. An investigation of brand experience on brand attachment, knowledge, and trust in the lodging industry. *Int. J. Hosp. Manag.* **2017**, *26*, 1–22. [CrossRef]
5. Li, C.Y.; Fang, Y.H. Predicting continuance intention toward mobile branded apps through satisfaction and attachment. *Telemat. Inform.* **2019**, *43*, 201–248. [CrossRef]
6. Holt, D.B. Why do brands cause trouble? A dialectical theory of consumer culture and branding. *J. Consum. Res.* **2002**, *29*, 70–90. [CrossRef]
7. Fournier, S. Consumers and their brands: Developing relationship theory in consumer research. *J. Consum. Res.* **1998**, *24*, 343–373. [CrossRef]
8. Lim, X.J.; Cheah, J.H.; Cham, T.H.; Ting, H.; Memon, M.A. Compulsive buying of branded apparel, its antecedents, and the mediating role of brand attachment. *Asia. Pacific. J. Mark. Logist.* **2020**, *32*, 1539–1563. [CrossRef]
9. Swaminathan, V.; Stilley, K.M.; Ahuluwalia, R. When brand personality matters: The moderating role of attachment styles. *J. Consum. Res.* **2009**, *35*, 985–1002. [CrossRef]
10. Escalas, J.E.; Bettman, J.R. Self-construal, reference groups, and brand meaning. *J. Consum. Res.* **2005**, *32*, 378–389. [CrossRef]
11. Das, G.; Agarwal, J.; Malhotra, N.K.; Varshneya, G. Does brand experience translate into brand commitment? A mediated-moderation model of brand passion and perceived brand ethicality. *J. Bus. Res.* **2019**, *95*, 479–490. [CrossRef]
12. Tafesse, W. Conceptualization of brand experience in an event marketing context. *J. Promot. Manag.* **2016**, *22*, 34–48. [CrossRef]
13. Schmitt, B. The consumer psychology of brands. *J. Consum. Psychol.* **2012**, *22*, 7–17. [CrossRef]
14. Japutra, A.; Ekinci, Y.; Simkin, L. Positive and negative behaviours resulting from brand attachment. *Eur. J. Mark.* **2018**, *52*, 1185–1202. [CrossRef]
15. Malär, L.; Krohmer, H.; Hoyer, W.D.; Nyffenegger, B. Emotional brand attachment and brand personality: The relative importance of the actual and the ideal self. *J. Mark.* **2011**, *75*, 35–52. [CrossRef]
16. Bian, X.; Haque, S. Counterfeit versus original patronage: Do emotional brand attachment, brand involvement, and past experience matter? *J. Brand. Manag.* **2020**, *27*, 438–451. [CrossRef]
17. Carroll, B.A.; Ahuvia, A.C. Some antecedents and outcomes of brand love. *Mark. Lett.* **2006**, *17*, 79–89. [CrossRef]
18. Mishra, A.S.; Roy, S.; Bailey, A.A. Exploring brand personality-celebrity endorser personality congruence in celebrity endorsements in the Indian context. *Psychol. Mark.* **2015**, *32*, 1158–1174. [CrossRef]
19. Park, C.W.; MacInnis, D.J.; Priester, J.; Eisingerich, A.B.; Iacobucci, D. Brand attachment and brand attitude strength: Conceptual and empirical differentiation of two critical brand equity drivers. *J. Mark.* **2010**, *74*, 1–17. [CrossRef]
20. Hung, H.Y.; Lu, H.T. The rosy side and the blue side of emotional brand attachment. *J. Consum. Behav.* **2018**, *17*, 302–312. [CrossRef]
21. Thomson, M.; MacInnis, D.J.; Park, C.W. The ties that bind: Measuring the strength of consumers' emotional attachments to brands. *J. Consum. Psychol.* **2005**, *15*, 77–91. [CrossRef]
22. Rio, A.B.; Vazquez, R.; Iglesias, V. The effects of brand associations on consumer response. *J. Consum. Mark.* **2001**, *18*, 410–425.
23. Aaker, D.A. Measuring brand equity across products and markets. *Calif. Manag. Rev.* **1996**, *38*, 102–120. [CrossRef]
24. Xu, X.; Xue, K.; Wang, L.; Gursoy, D.; Song, Z. Effects of customer-to-customer social interactions in virtual travel communities on brand attachment: The mediating role of social well-being. *Tour. Manag. Perspect.* **2021**, *38*, 100790. [CrossRef]
25. Ko, E.; Phau, I.; Aiello, G. Luxury brand strategies and customer experiences: Contributions to theory and practice. *J. Bus. Res.* **2016**, *69*, 5749–5752. [CrossRef]
26. Kumar, A.; Paul, J. Mass prestige value and competition between American versus Asian laptop brands in an emerging market—Theory and evidence. *Int. Bus. Rev.* **2018**, *27*, 969–981. [CrossRef]
27. Barnes, S.J.; Mattsson, J.; Sorensen, F. Destination brand experience and visitor behavior: Testing a scale in tourism context. *Ann. Tour. Res.* **2014**, *48*, 121–139. [CrossRef]
28. Bleier, A.; Harmeling, C.M.; Palmatier, R.W. Creating effective online customer experiences. *J. Mark.* **2019**, *83*, 98–119. [CrossRef]
29. Chen, H.; Papazafeiropoulou, A.; Chen, T.K.; Duan, Y.; Liu, H.W. Exploring the commercial value of social networks: Enhancing consumers' brand experience through Facebook pages. *J. Ent. Inform. Manag.* **2014**, *27*, 576–598. [CrossRef]
30. Dwivedi, A.; Nayeem, T.; Murshed, F. Brand experience and consumers' willingness-to-pay (WTP) a price premium: Mediating role of brand credibility and perceived uniqueness. *J. Retail. Consum. Serv.* **2018**, *44*, 100–107. [CrossRef]
31. Jimenez-Barreto, J.; Sthapit, E.; Rubio, N.; Campo, S. Exploring the dimensions of online destination brand experience: Spanish and north American tourists' perspectives. *Tour. Manag. Perspect.* **2019**, *31*, 348–360. [CrossRef]
32. McLean, G.; Al-Nabhani, K.; Wilson, A. Developing a mobile applications customer experience model (MACE)-implications for retailers. *J. Bus. Res.* **2018**, *85*, 325–336. [CrossRef]

33. Molinillo, S.; Navarro-García, A.; Anaya-Sánchez, R.; Japutra, A. The impact of affective and cognitive app experiences on loyalty towards retailers. *J. Retail. Consum. Serv.* **2020**, *10*, 19–48. [CrossRef]
34. Schmitt, B. The concept of brand experience. *J. Brand. Manag.* **2009**, *16*, 417–419. [CrossRef]
35. Maehle, N.; Otnes, C.; Supphellen, M. Consumers' perceptions of the dimensions of brand personality. *J. Consum. Behav.* **2009**, *10*, 290–303. [CrossRef]
36. Lam, S.K.; Ahearne, M.; Schillewaert, N. A multinational examination of the symbolic–instrumental framework of consumer–brand identification. *J. Int. Bus. Stud.* **2012**, *43*, 306–331. [CrossRef]
37. Nysveen, H.; Pedersen, P.E.; Skard, S. Brand experiences in service organizations: Exploring the individual effects of brand experience dimensions. *J. Brand. Manag.* **2013**, *20*, 404–423. [CrossRef]
38. Kim, J.; Yu, E. The holistic brand experience of branded mobile applications affects brand loyalty. *Soc. Behav. Pers.* **2016**, *44*, 77–87. [CrossRef]
39. Beig, F.A.; Khan, M.F. Impact of social media marketing on brand experience: A study of select apparel brands on Facebook. *Vision* **2018**, *22*, 264–275. [CrossRef]
40. Gavilanes, J.M.; Flatten, T.C.; Brettel, M. Content strategies for digital consumer engagement in social networks: Why advertising is an antecedent of engagement. *J. Adv.* **2018**, *47*, 4–23. [CrossRef]
41. Morrison, S.; Crane, F.G. Building the service brand by creating and managing an emotional brand experience. *J. Brand. Manag.* **2007**, *14*, 410–421. [CrossRef]
42. Hwang, J.; Choe, J.Y.; Kim, H.M.; Kim, J.J. Human baristas and robot baristas: How does brand experience affect brand satisfaction, brand attitude, brand attachment, and brand loyalty? *Int. J. Hosp. Manag.* **2021**, *99*, 103050. [CrossRef]
43. Jouzaryan, F.; Dehbini, N.; Shekari, A. The impact of brand personality, brand trust, brand love and brand experience on consumer brand loyalty. *Int. J. Life Sci. Res.* **2015**, *5*, 69–76.
44. Schmitt, B.; Zarantonello, L. Consumer experience and experiential marketing: A critical review. *Rev. Mark. Res.* **2013**, *10*, 25–61.
45. Yasin, M.; Liébana-Cabanillas, F.; Porcu, L.; Kayef, R.N. The role of customer online brand experience in customers' intention to forward online company-generated content: The case of the Islamic online banking sector in Palestine. *J. Retail. Consum. Serv.* **2020**, *52*, 10–19. [CrossRef]
46. Park, S.H.; Ham, S.; Lee, M.A. How to improve the promotion of Korean beef barbecue, bulgogi, for international customers: An application of quality function deployment. *Appetite* **2012**, *59*, 324–332. [CrossRef]
47. Bossert, J.L. *Quality Function Deployment: A Practitioner's Approach*; ASQC Quality Press: Milwaukee, WI, USA, 1991.
48. Shaojing, C.; Hong-Bin, Y. A systematic fuzzy QFD model and its application to hotel service design 2016. In Proceedings of the 13th international Conference on Service Systems and Service Management (ICSSSM), Kunming, China, 24–26 June 2016; pp. 1–6.
49. Hsu, T.H.; Tang, J.W. Development of hierarchical structure and analytical model of key factors for mobile app stickiness. *J. Innov. Knowl.* **2020**, *5*, 68–79. [CrossRef]
50. Yuen, K.K.F. A hybrid fuzzy quality function deployment framework using cognitive network process and aggregative grading clustering: An application to cloud software product development. *Neurocomputing* **2014**, *142*, 95–106. [CrossRef]
51. Haber, N.; Fargnoli, M.; Sakao, T. Integrating QFD for product-service systems with the Kano model and fuzzy AHP. *Total. Qual. Manag. Bus. Excell.* **2020**, *31*, 929–954. [CrossRef]
52. Mehtap, D.; Karsak, E.E. A QFD-based fuzzy MCDM approach for supplier selection. *Appl. Math. Model.* **2013**, *37*, 5864–5875.
53. Hsu, T.H.; Tang, J.W. An analytic model for developing strategies of customer relational management. *Manag. Rev.* **2014**, *33*, 1–17.
54. Lam, J.S.L.; Lai, K. Developing environmental sustainability by ANP-QFD approach: The case of shipping operations. *J. Clean. Prod.* **2015**, *105*, 275–284. [CrossRef]
55. Dat, L.Q.; Phuong, T.T.; Kao, H.P.; Chou, S.; Nghia, P.V. A new integrated fuzzy QFD approach for market segments evaluation and selection. *Appl. Math. Model.* **2015**, *39*, 3653–3665. [CrossRef]
56. Lu, C.F.; Lin, L.Z.; Yeh, H.R. A multi-phased FQFD for the design of brand revitalization. *Total. Qual. Manag. Bus. Excell.* **2019**, *30*, 848–871. [CrossRef]
57. Kayapınar, S.; Erginel, N. Designing the airport service with fuzzy QFD based on SERVQUAL integrated with a fuzzy multi-objective decision model. *Total. Qual. Manag. Bus. Excell.* **2019**, *30*, 1429–1448. [CrossRef]
58. Xie, J.; Qin, Q.; Jiang, M. Multiobjective decision-making for technical characteristics selection in a house of quality. *Math. Probl. Eng.* **2020**, *12*, 1–12. [CrossRef]
59. Zaitsev, N.; Dror, S. A corporate social responsibility (CSR) model—A QFD-based approach. *Total. Qual. Manag. Bus. Excell.* **2020**, *31*, 137–148. [CrossRef]
60. Lizarelli, F.L.; Osiro, L.; Ganga, G.M.D.; Mendes, G.H.S.; Paz, G.R. Integration of SERVQUAL, Analytical Kano, and QFD using fuzzy approaches to support improvement decisions in an entrepreneurial education service. *Appl. Soft. Comput.* **2021**, *112*, 107786. [CrossRef]
61. Chen, Y.; Ran, Y.; Huang, G.; Xiao, L.; Zhang, G. A new integrated MCDM approach for improving QFD based on DEMATEL and extended MULTIMOORA under uncertainty environment. *Appl. Soft. Comput.* **2021**, *105*, 107222. [CrossRef]
62. Haiyun, C.; Zhixiong, H.; Yüksel, S.; Dinçer, H. Analysis of the innovation strategies for green supply chain management in the energy industry using the QFD-based hybrid interval valued intuitionistic fuzzy decision approach. *Renew. Sust. Energ. Rev.* **2021**, *143*, 110844. [CrossRef]

63. Wang, T.C.; Chen, Y.H. Applying fuzzy linguistic preference relations to the improvement of consistency of fuzzy AHP. *Inf. Sci.* **2008**, *178*, 3755–3765. [CrossRef]
64. Herrera-Viedma, E.; Herrera, F.; Chiclana, F.; Luque, M. Some issues on consistency of fuzzy preference relations. *Eur. J. Oper. Res.* **2004**, *154*, 98–109. [CrossRef]
65. Opricovic, S.; Tzeng, G.H. Defuzzification within a multicriteria decision model. *Int. J. Uncertain. Fuzz.* **2003**, *11*, 635–652. [CrossRef]
66. Shamim, A.; Butt, M.M. A critical model of brand experience consequences. *Asia Pacific J. Mark. Logist.* **2013**, *25*, 102–117. [CrossRef]
67. Yu, X.; Yuan, C. How consumers' brand experience in social media can improve brand perception and customer equity. *Asia Pacific J. Mark. Logist.* **2019**, *31*, 1233–1251. [CrossRef]
68. Huang, R.; Lee, S.H.; Kim, H.; Evans, L. The impact of brand experiences on brand resonance in multi-channel fashion retailing. *J. Res. Interact. Mark.* **2015**, *9*, 129–147. [CrossRef]
69. Ramirez, R.H.; Merunka, D. Brand experience effects on brand attachment: The role of brand trust, age, and income. *Eur. Bus. Rev.* **2019**, *31*, 610–645. [CrossRef]
70. Borges, A.P.; Vieira, E.; Lopes, J.M. Emotional Intelligence Profile of Tourists and Its Impact on Tourism. *J. Qual. Assur. Hosp. Tour.* **2021**. [CrossRef]
71. Japutra, A.; Ekinci, Y.; Simkin, L. Exploring brand attachment, its determinants and outcomes. *J. Strateg. Mark.* **2014**, *22*, 616–630. [CrossRef]
72. Dolbec, P.Y.; Chebat, J.C. The impact of a flagship vs. a brand store on brand attitude, brand attachment and brand equity. *J. Retail.* **2013**, *89*, 460–466.
73. Nierobisch, T.; Toporowski, W.; Dannewald, T.; Jahn, S. Flagship stores for FMCG national brands: Do they improve brand cognitions and create favorable consumer reactions? *J. Retail. Consum. Serv.* **2017**, *34*, 117–137. [CrossRef]
74. Eggers, F.; O'Dwyer, M.; Kraus, S.; Vallaster, C.; Güldenberg, S. The impact of brand authenticity on brand trust and SME growth: A CEO perspective. *J. World. Bus.* **2013**, *48*, 340–348. [CrossRef]
75. Paulssen, M. Attachment orientations in business-to-business relationships. *Psychol. Mark.* **2009**, *26*, 507–533. [CrossRef]
76. Burmann, C.; Zeplin, S. Building brand commitment: A behavioural approach to internal brand management. *J. Brand. Manag.* **2005**, *12*, 279–300. [CrossRef]
77. Fastoso, F.; González-Jiménez, H. Materialism, cosmopolitanism, and emotional brand attachment: The roles of ideal self-congruity and perceived brand globalness. *J. Bus. Res.* **2020**, *121*, 429–437. [CrossRef]
78. Trudeau, H.S.; Shobeiri, S. The relative impacts of experiential and transformational benefits on consumer-brand relationship. *J. Prod. Brand. Manag.* **2016**, *25*, 586–599. [CrossRef]

Article

Constructing the Audit Risk Assessment by the Audit Team Leader When Planning: Using Fuzzy Theory

Luis Porcuna-Enguix [1,*], Elisabeth Bustos-Contell [2], José Serrano-Madrid [3] and Gregorio Labatut-Serer [2]

[1] Departament of Economics and Social Sciences, CEGEA (Centre for Research in Business Management), Universitat Politècnica de València, 46022 Valencia, Spain
[2] Department of Accounting, Universitat de València, 46010 Valencia, Spain; elisabeth.bustos@uv.es (E.B.-C.); gregorio.labatut@uv.es (G.L.-S.)
[3] Department of Financial Economics and Accounting, Universidad de Murcia, 30003 Murcia, Spain; jose.serrano@bnfix.com
* Correspondence: lporeng@esp.upv.es

Abstract: The aim of this study is to construct the assessment of the expected audit risk by the audit team leader (ATL) during the planification phase of the audit. The ATL plays an important role within the audit, and even more so regarding small and medium-sized (SME) audit firms. The audit risk assessment is critical as relying more (less) on internal controls implemented by the client leads to performing less (more) substantive audit procedures. This is determined by the ATL based on their professional judgement and previous experience. The use of fuzzy theory has powerful potential into the audit arena, as the audit risk assessment (outcome) is critically related to the auditors' judgement and perception. We argue that ATL characteristics are core conditions in determining the audit risk assessment when planning. Using hand-collected and private data from Spanish SME audit firms, we find that a comprehensive set of conditions must be given for perceived high audit risk. The results indicate that female and inexperienced ATLs planning the audit of indebted firms with high proportions of capital assets, less profitability, and with a larger board sizes, as they are expected to have bad internal control. The same conditions are met when expecting errors, as well as shorter audit tenures. Finally, conditions such as the ATL's experience gains importance in expecting irregularities. This paper extends our understanding of the role of ATL characteristics on the audit risk assessment when planning and raising awareness on studying SME audit firm behavior.

Keywords: audit team leader; audit risk assessment; small- and medium-sized audit firms; planification; fuzzy theory

Citation: Porcuna-Enguix, L.; Bustos-Contell, E.; Serrano-Madrid, J.; Labatut-Serer, G. Constructing the Audit Risk Assessment by the Audit Team Leader When Planning: Using Fuzzy Theory. *Mathematics* **2021**, *9*, 3065. https://doi.org/10.3390/math9233065

Academic Editors: Jorge de Andres Sanchez and Laura González-Vila Puchades

Received: 31 October 2021
Accepted: 26 November 2021
Published: 28 November 2021

Publisher's Note: MDPI stays neutral with regard to jurisdictional claims in published maps and institutional affiliations.

Copyright: © 2021 by the authors. Licensee MDPI, Basel, Switzerland. This article is an open access article distributed under the terms and conditions of the Creative Commons Attribution (CC BY) license (https://creativecommons.org/licenses/by/4.0/).

1. Introduction

Auditing is a collective process conducted by a professional accounting team with a range of skills, experience, and emotions [1]. An audit consists of collecting evidence to express and issue an audit opinion about whether the financial statements are prepared, in all material aspects, in accordance with an applicable financial reporting framework (ISA 200). Such opinion is on whether the financial statements are presented fairly, in all material respects, or given a true and fair view in accordance with the framework (ISA 200). The fundamental goal of an audit is to reduce uncertainty and enhance the degree of confidence of intended users in the financial statements. As stablished by the reference number 4 of ISA 315, "the objective of the auditor is to identify and assess the risks of material misstatements, whether due to fraud or error, at the financial statement and assertion levels, (. . .) including the entity's internal control (. . .)". Hence, the audit staff must evaluate and respond to risks caused by uncertainty in the information. Here, three key elements pop up because of their relevance on the auditing process: (i) the audit plan; (ii) the audit staff; and (iii) the audit risk. Of course, these concepts are interdependent.

The auditor shall undertake some preliminary engagement activities at the beginning of the audit engagement, such as performing procedures required by ISA 220, evaluating compliance with ethical and independence requirements, and understanding the engagement. According to ISA 300, the audit plan is more accurate than the overall audit strategy. Planning takes place over the course of the audit, so changes to early planning decisions may happen. Planning activities entail, among others, the auditor's risk assessment procedures, the subsequent nature, timing and extent of further audit procedures and the consideration of those factors that, in the auditor's professional judgement, are significant in directing the engagement team's efforts.

Even though the audit partner is the person who firstly contacts the client, the role of the audit team leader (ATL) is critical, as they must allocate resources efficiently to lower audit risk up to an acceptable level, as indicated by the International Audit Standards (ISA). It is true that the audit partner will sign the audit report with an opinion, but such an opinion is the final outcome after a great audit effort. The audit effort is mainly carried out by those who are not the partners, and ATLs (named by the audit partner) are decisive to opt for and issue an opinion or other in the audit report [2,3]. Sincerely, lower-level audit team members gather most audit evidence, however, upward communication and knowledge sharing are decisive for effective and efficient audits [4,5]. In this aspect, audit team leaders play a critical role in easing such information exchange and in encouraging audit team members [6]. Audit leadership is expected to affect group performance [7]. Leadership style not only influences the levels of satisfaction, motivation, and performance of audit team individuals, but also may improve the audit quality practices. Furthermore, appropriate risk audit assessment of the client from the beginning of the audit by audit team leaders may be momentous for an effective and efficient audit execution. Relying more (less) on internal controls implemented by the client leads to performing less (more) substantive audit procedures, which is determined by audit team leaders based on their professional judgement and previous experience [8].

With respect to audit risk assessment, some clarifications must be made in order to understand the meaning of audit risk. According to the IAASB Glossary of Terms (1), audit risk is defined as the risk that the auditor expresses an inappropriate audit opinion when the financial statements are materially misstated, so it is a function of material misstatement and detection risk. This term is difficult to measure, as it may be only estimated after the audit work is complete. For this reason, we center our attention in the audit risk assessment. The audit risk assessment is the imperative identification, analysis, and management of risks relevant to prepare the financial statements [9], and its procedures take place in the planning period of the audit. During the risk assessment time, the auditor inquires with management, entity staff, performs analytical procedures, observations and inspections, and understands business risks that may result in material misstatements. The audit risk assessment is fundamental to the audit process for several reasons. In accordance with ISAs, audits must follow the risk-based approach, because auditors may not check all transactions and must minimize the chance of giving an inappropriate audit opinion. Addressing insignificant risks in a high level of detail would be inefficient, and no one is prepared to pay for the auditors to do such amounts of work.

Setting aside the inherent risk for the difficulty in its estimation, we focus on the assessment of internal controls and the probability of detecting errors and irregularities by ATLs when planning. These terms are defined in ISA 315, and will be explained in more detail later. The audit risk assessment in the planification phase by the audit team leader (ATL) is highly based on subjective judgements. The ATL must evaluate internal control and the expected existence of errors and irregularities in the financial statements, so they determine if a risk is classified a key risk, material risk, or insignificant risk (it is not black or white, but there might be shades of gray). The application of fuzzy theory is particularly meaningful, as audit risk assessment involves people thoughts, inference, and perception [10], so fuzzy method retains a certain degree of fuzziness in such aspect. Compared to econometric analysis (symmetric approach), fuzzy method (asymmetric approach) does not draw a conclusion based on a single correct answer (reject or not a null hypothesis) but provides different solutions (decisions) based on consistency and frequency thresholds. In addition, fuzzy technique does not either show estimates of each hypothesized relationship through the net effects, but covers the influence of other variables, leading to various combinations of conditions.

The aim of this study is to construct the assessment of the expected audit risk by the audit team leader (ATL) during the planification phase of the audit. To do so, the use of fsQCA has powerful potential into the audit arena, as the audit risk assessment is critically related to auditors' judgement and perception. We argue that ATL characteristics are core conditions in determining the audit risk assessment when planning.

This research has been made with data coming from a Spanish sample. Spain is classified as a code law country [11], unlike Anglo-Saxon countries such as USA or UK (common law) where the accounting and auditing culture is more developed by tradition and where private professional bodies played an important role in the organization of the audit profession from the beginning [12]. Differently, in Spain the main organisms in this sense are public. The Spanish environment is characterized by weak investor protection and low risk of litigation, where the fundamental stakeholders are families, banks, and industrial companies [13,14]. The banking sector dominates over the capital market, as a result, agency problems may arise. The average cost of debt of Spanish SMEs is 22%, compared to the 7% of the Eurozone as a whole [15]. According to Kim et al. [16], audits may contribute to reducing information asymmetries, and thus the cost of financing, so it may generate incentives to financial entities to preserve their dominance. In this context, no market-based institutional incentives exist (i.e., litigation costs, reputation loss, etc.) and auditor decisions may be influenced by managers and other parties. Despite the existence of great efforts to empower audit legislation and construct cohesive professional audit infrastructure aiming at solving these questions, pressures on the audit risk assessment by ATLs when planning can occur. In Spain, more than the 88% of Spanish audit firms are small- and medium-sized, shaping the real Spanish audit market behavior, unlike most of studies exploring the actions of the Big Four or second-tier audit firms. Audits of SME firms are likely to be conducted by a very small audit team. Presumably, complexity of audits should be lower (ISA 300), but coordination and audit team leader implication must be higher, so ATL characteristics (together with the pressure and progression within the audit firm) are supposed to directly affect audit risk assessment and audit team behavior. These reasons come to justify the relevance of the audit risk assessment (internal control, errors, and irregularities) in the Spanish scenario.

We use hand-collected and private data taken from legal applications declared by audit firms to the national public oversight board (ICAC), spanning 2001 to 2015. The sample consists of a balanced panel of 1338 Spanish firm-year observations. The audit risk assessment is set to be the outcome. Consistent with ISA 315, we use three outcomes: (i) the absence of internal control; (ii) the presence of expected errors; and (iii) the presence of expected irregularities. A complete set of contrasted conditions has been used to explain the focus outcomes. The analysis of necessity reveals that there is no exclusive necessary condition, but a comprehensive set of conditions must be given for perceived high audit risk when planning (expected bad internal control and expected errors and irregularities in the financial statements). More specifically, four conditions are the most significant drivers of the outcome, either for their higher consistency values or for the wider difference between their presence and absence: the presence of female ATLs, inexperienced ATLs, old-audited clients, and large board sizes. The sufficient analyses show that more than one possible configuration (i.e., combination of conditions) happens, and that causal asymmetry may occur. Regarding the absence of internal control, the results demonstrate that female and inexperienced ATLs planning the audit of firms with a high proportion of property, plants and equipment, high leverage ratios, less profitability and large board sizes are expected to have bad internal control. For the presence of expected errors, the likelihood of expecting error in the financial statements increases as perceived internal control worsens (same conditions are met), and also when the audit tenure is shorter. Lastly, for the presence of irregularities, conditions such as experience gain importance. Irregularities are intended errors, and experienced ATLs are capable of expecting or forecasting irregularities because of their widespread background and prior accumulated know-how.

The contribution of this work is threefold. First, we focus the research on the audit risk assessment by the audit team leaders (ATLs), who are the pivotal audit members in the planification phase, and thus in the subsequent test of controls and substantive audit procedures. Secondly, the use of fuzzy-set modelling allows us to deal with the fuzziness when assessing audit risk. In addition, primary objective data connected to internal control and the probability to uncover errors or irregularities have been used. As far as we know, this is the first study doing so. One may find the application of fuzzy theory fundamentals to auditing so as to explain the formation of an effective quality audit team [17], to construct and check the effectiveness and efficiency of an artificial intelligence audit tool compared to human auditors [18], to design a more precise audit detection risk assessment system [10], to gain additional insights into the relationship between firm characteristics and audit fees [19], or to propose a client acceptance method in order to improve audit firms' risk management [20], among others. However, no one refers to the ambiguity or vagueness when assessing audit risk in the planification regarding internal control and the expectation to find errors and/or irregularities. Last, but not least, this study is responsive to the call from regulators to consider audit team attributes, in particular, ATL characteristics, as important factors that affect audit quality practices [21], which is directly influenced by the audit risk assessment when planning.

The remainder of the paper is structured as follows. Section 2 outline the basics of fuzzy-set qualitative comparative analysis aiming at showing its potentiality in the auditing arena. In Section 3, a prior study about ATL designation and their audit risk assessment of the client in the planification stage is re-visited. We briefly describe the role of ATLs of SMEs audit firms on the audit risk assessment and also explain the selection of the outcome and conditions. Sample and descriptive statistics state in Section 4. Section 5 shows the empirical fuzzy-set results, and finally Section 6 concludes, with a brief discussion on the treated topic and some limitations are exposed.

2. Fuzzy-Set Qualitative Comparative Analysis (fsQCA)

2.1. Fuzzy Set Basics

The fuzzy set theory seeks to solve fuzzy data in realistic context. It means a set signifying conditions with specific and apparently clear properties but imprecise boundaries. We may mathematically articulate the fuzzy set as follows:

The universal set or universe of discourse is called "U"; a set "A" is any well-defined collection of objects; an object in a set is called "a_i", which represents an element or member of that set. Sets are defined by simple statements and the fact whether a particular object having a certain property belongs to a particular set may be described as A = {a_1, a_2, a_3, ..., a_n}. If the elements "a_i" (i = 1, 2, 3, ..., n) of a set "A" are subset of universal set "U", then set "A" may be mapped for all members x ϵ X by its characteristic function μ_A (x) = 1 if x ϵ X, 0 otherwise (note that the fuzzy subset A on U means that for any X ϵ U, there is a real number) However, fuzzy logic consists of many degrees of membership [0, 1], and all are allowed. In this sense, when the universe of number of A is {0,1}, then the membership function of A becomes $\mu_{\tilde{A}}$ (x): A → [0, 1] to x ϵ U, where the maximum degree of membership is the top heigh of the fuzzy set (equals to 1) and the minimum degree of membership is the bottom heigh of the fuzzy set (equals to 0). Note that Ã is one fuzzy subset on U and $\mu_{\tilde{A}}$ (x) ϵ [0, 1] is one number for each x ϵ U to show membership degree of x for Ã, which is the membership degree of U.

2.2. Key Points to Keep in Mind

Firstly, it is required to define all elements that will constitute the pathway of the fuzzy logic. To that end, Table 1 is presented, following Valaei et al. [22]. We display three columns: (i) symmetric concepts that are mostly used in regression terminology; (ii) asymmetric concepts that are most frequently used in fuzzy set nomenclature; and (iii) a brief explanation of such terms.

Table 1. Symmetric vs. asymmetric concepts.

Symmetric Concepts	Asymmetric Concepts	Explanation
Dependent variable	Outcome	Consequence or variable of interest. The researcher may decide to either compute the presence (full membership) or the absence or negation (full non-membership) of the outcome.
Independent variable	Condition	Variable, element, object, factor, cause, or predictor that is supposed to drive the outcome. Configuration refers to a combination of conditions to obtain the outcome.
Measurement	Calibration	Re-express data to values between 0 and 1 (where 0 denotes full non-membership and 1 full membership).
Net effects	Causal recipes, solutions	Variables that are traditionally considered as control variables may be part of the solutions and be combined to explain the outcome. There are three types of solution sets: ■ *Complex*: represents all the possible combinations of conditions, but its interpretation becomes rather difficult and often impractical. The number of configurations is usually large, whose interpretation becomes difficult and impractical for the researcher. ■ *Parsimonious*: simplified version of the complex solution, based on simplified assumptions. It presents the most relevant conditions (*core conditions*) which cannot be left out from any solution. ■ *Intermediate*: the result when performing counterfactual analysis on the complex and parsimonious solutions. It should be consistent with the theoretical and empirical knowledge, and by default, the presence or absence (negation) of the variables is computed. While *core conditions* appear in both parsimonious and intermediate solutions, *peripheral conditions* only appear in the intermediate solution, which enhances the complexity in favor of increased consistency.
Correlation matrix	Truth table	It computes all possible combinations of conditions that may happen, providing 2^k rows, where k represents the number of outcome predictors and each row every possible configuration
Coefficient of determination (R2)	Coverage	It represents to what extent the fact of belonging to an input group (*recipe*) drives the output (*outcome*). Hence, it indicates the empirical relevance of a solution (the higher the better). In other words, it displays how many cases with the outcome are represented by a particular causal condition.
Correlation	Consistency	It responds to what extent the hypothesis is coherent. That is, the degree to which configurations that share conditions lead to the same outcome. In order to decide which combinations to include in the final solution, we must pick a cut-off value for consistency scores, so rows with high consistency indicate combinations that almost always lead to the outcome.
N/A	Boolean signs	Fuzzy set operations. For instance, given X to be the universe set and $\tilde{A}$ and $\tilde{E}$ to be fuzzy sets with $\mu_{\tilde{A}}(x)$ and $\mu_{\tilde{E}}(x)$ as their respective membership functions: ■ Union: $\mu_{\tilde{A} \cup \tilde{E}}(x) = \max\{\mu_{\tilde{A}}(x), \mu_{\tilde{E}}(x)\}$ ■ Intersection: $\mu_{\tilde{A} \cap \tilde{E}}(x) = \min\{\mu_{\tilde{A}}(x), \mu_{\tilde{E}}(x)\}$ ■ Complement: $\mu_{\tilde{A}}(x) = 1 - \mu_{\tilde{A}}(x)$

Note: symmetric concepts are just tabulated for informational and comparative purpose. The explanation alludes to asymmetric concepts.

Secondly, apart from the previously defined concepts, for appropriate use of the fsQCA, some assumptions or application hypotheses must be taken into consideration [23]:

I. **Model specification** [24]. A theoretical basis is imperative so as to demonstrate the causal impact of each condition on the outcome variable. Additionally, avoid non-contrasted supposition to improve artificially the results.

II. **Construct clarity** [25]. The use of new definitions or metrics is not recommended to avoid an additional undue "forking point". In addition, wherever possible, measure conditions directly and objectively with no subjectivity.

III. **Difficulty of interpreting fsQCA results**. Results (outcomes) are hardly ever driven by a single condition (cause), but a combination of them. Different configurations of causes may trigger the same focus outcome, but it is not possible to extract all possible combinations. In other words, a causal combination is not usually 100% sufficient.

IV. **Parameter sensitivity** [26]. When sample size is small (n < 50), set coverage and consistency thresholds should be higher. It is generally accepted that cut-off consistency values ranging 0.7 to 0.8 are good starting points to look at [27,28], but the higher the cut-off point (higher consistency), the lower the coverage will be (lower empirical importance of a solution). This is particularly challenging regarding large samples.

V. **Unfounded claiming**. There are two main mistakes to avoid: (i) claiming that antecedents variables are necessary conditions when they do not, and (ii) claiming that the absence or negation of a condition means "non-importance". First, running necessity analysis is fundamental to assert that antecedent variables (conditions) are effectively drivers of the outcome; second, the absence or negation of a condition is also important to the outcome.

The third aspect to bear in mind is the order and number of steps that are cardinal so as to run appropriately the fuzzy pathway. The indispensable stages are as follows:

Step 1—Substantial theoretical and empirical knowledge of the matter of interest. The state of art is essential to be controlled. Fuzzy logic is far away from mixing many ingredients (conditions) to cook a recipe (outcome), so "more" does not mean "better". It is true that in qualitative analysis certain subjective bias can be introduced by researchers, but their own knowledge of the field of study also lead to a richer analysis and understanding of the data.

Step 2—Data calibration. Data treatment can be considered the most important step in fuzzy modelling. Unlike traditional approaches, data are transformed from $]-\infty, +\infty[$ to $[0, 1]$, that is, into degrees of membership in the target set, where 0 meaning full non-membership (absence/negation) and 1 full membership (presence). To avoid misunderstandings, researchers have to clearly define the cut-off points that will determine the degree of membership. Differently from crisp sets, whose variables may discretely take values of 0 or 1, conditions are usually calibrated through an n-value fuzzy set, specifying the values of an interval-scale variable that correspond to n-qualitative breakpoints [27]. For instance, a three-value fuzzy set is one of the most used direct log-odd transformation methods, where the 95th percentile (0.95) is the threshold for full membership, 5th percentile (0.05) is the full non-membership threshold, and the cross-over point (0.50) is the threshold for "ambiguous" or "do not care" cases.

Step 3—Necessity conditions. The analysis of necessity for the presence and the absence or negation of the outcome is prerequisite before proceeding with the scrutiny.

Step 4—Truth table. To undertake sufficiency analysis, we must generate the truth table that reveals which combinations of conditions or configurations are sufficient for the focal outcome to occur.

Step 5—Boolean algebra. Once the Truth table is generated, all configurations (2k rows) are not significant, so the number of cases and Boolean logic must be employed. Researcher must remove those low frequency settings and leave out redundant clauses to reduce the initial 2k configurations to simplified combinations. For this purpose, consistency and coverage must be defined.

Thus, having said the above, the theoretical an empirical framework of this study is to apply the fuzzy theory to construct the assessment of the expected audit risk by the audit team leader (ATL) during the planification stage of the audit work of their audited clients.

3. Re-Examining the Data of a Prior ATLs and Audit Risk Assessment Study

The goal of this study is not to provide a fully comprehensive and instructional guide about fuzzy-set Qualitative Comparative Analysis (fsQCA). One may find in the existing literature complete guides of fuzzy-set modelling on management (e.g., [29]), entrepreneurship (e.g., [23,30]), artificial intelligence, expert systems, control engineering, and multi-principle decision making and risk assessment [31–33] for the presence of ambiguity. Undoubtfully, the problem of imprecision also appears in accounting and auditing (e.g., [34]), where concepts such as materiality, fair value, reliability or risk assessment are not essentially two-dimensional but a scale of degrees. It is noticeable that it does not refer to the uncertainty of the occurrence or not occurrence of an event, but the lack of clarity in the assessment and judgements, which directly affects usefulness and predicting power of accounting information. Instead, we pursue to sketch out the method for those who are interested in employing this qualitative approach to their fields of study, especially in the auditing arena.

The word "fuzzy" means "vagueness" or "ambiguity", and fuzziness occurs when the boundary of a piece of information in not clear-cut. Zadeh [35] proposed the fuzzy set theory, which is an extension of classical set theory where components have a degree of membership. Classical set theory establishes evident sharp (crisp or exact) boundaries with apparently no uncertainty associated (i.e., black–white, true–false, good–bad, etc.); contrarily, fuzzy set theory considers ambiguous boundaries (e.g., the gray color exists between black and white; describing human reasoning by true or false is frequently insufficient; audit risk evaluation is not only good or bad). Hence, running fsQCA facilitates to incorporate imprecise values to obtain non-symmetric relationships (necessary and sufficient conditions).

The fsQCA approach is based on complexity theory following the principles of conjunction, equifinality, and causal asymmetry. Conjunction refers to the notion that conditions interacts interdependently reveling combinations of sufficient conditions for the outcome. Equifinality denotes the existence of multiple equally effective combinations of conditions to reach the same high outcome score. Causal asymmetry indicates that conditions driving an outcome in one combination may be simultaneously unrelated or even oppositely related in another configuration linked with the same outcome [36].

The aim of this study is, therefore, to spotlight the potentiality that fsQCA have into the auditing arena, in particular, for the audit risk assessment by audit team leaders (ATLs) during the planification level. Next, we will use this powerful qualitative technique to re-visit data from earlier research that had used symmetric quantitative approaches. We select the investigation carried out by [37], who operated with multiple regression analysis, consistent with what is commonly performed in the risk aversion associated with female gender [38–40], female appointment decisions (e.g., [41]), and gender diversity in audit risk assessment and quality (e.g., [42,43]).

The audit risk assessment is highly based on audit personnel's subjective judgements. Auditing is a process of collecting evidence to express and issue an audit opinion about on whether the financial statements are prepared, in all material aspects, in accordance with an applicable financial reporting framework (ISA 200). Such opinion is on whether the financial statements are presented fairly, in all material respects, or give a true and fair view in accordance with the framework (ISA 200). The general purpose is to reduce uncertainty and enhance the degree of confidence of intended users in the financial statements. To do so, the audit staff must evaluate and respond to risks caused by uncertainty in the information. The application of fuzzy theory is particularly meaningful, as audit risk assessment involves people' thoughts, inferences, and perception [10], so fuzzy method retains a certain degree of fuzziness in such respect.

We selected the Porcuna-Enguix et al. [37] study for four main reasons. First, it represents the first work in analyzing audit team leader (ATL) attributes, while most studies look to audit partners. ATLs have a substantial role when issuing the audit report, since they have direct contact with the audit client and the potential risks compared to the signing audit partners. Second, the first work that explores the effect of the gender attribute of ATLs when appointed by the audit partner to an audit engagement, especially, in the audit risk assessment. Third, the Spanish context illustrates a peculiar but representative setting worldwide, in spite of limited related SME studies. Fourthly, using the same data as that earlier study allows us to both demonstrate whether or not fsQCA may report different results about alternative routes to audit risk assessments, and avoid subjectivity that would be linked to collecting a new sample.

In Porcuna-Enguix et al. [37], the audit risk assessment in the planification phase is measured in three ways as required by ISA 315. The evaluation of internal control (intcontrol) uses a Likert scale ranging discretely from 1 to 5, with 1 being bad internal control and 5 good internal control. The probability of expected errors (error) follows the same Likert-scale as intcontrol, with 1 being unlikely to uncover errors and 5 more likely. The probability of expected irregularity, with 1 being less likely to uncover irregularities and 5 more likely. Therefore, these are expected or perceived "audit risk assessment" in the planification stage, unlike the term "audit risk". The study has several "interest" and "control" variables seeking to determine whether female ATLs are appointed to riskier clients and whether female ATLs are more likely to be conservative or prudent when assessing audit risk in the planification stage of the audit engagement.

Table 2 presents the results from the ordered probit sample selection model (regression equation). The variable of interest "GENDER" was previously estimated (selection equation), but untabulated, and was then incorporated into the regression equation. This symmetric procedure was chosen so as to control for existing endogeneity problems [44]. We found that riskier audited clients were assigned to female ATLs (untabulated results). From the regression equation, we observed that female ATLs are more likely to uncover deficiencies in the internal control of the client and are more likely to find expected errors and irregularities during the planification phase. In addition, further analysis, also untabulated, demonstrates a more indulgent or tolerant attitude of experienced female ATLs towards possible internal control weaknesses and to inform about irregularities, attitudes traditionally attributed to males. Note that auditing is male-conditioned environment, so there might exist an intrinsic motivation to enhance their job opportunities such as promotion or just to not jeopardize their surveillance in the audit firm.

In order to align the fsQCA with that Heckman modelling study, we considered including all the same antecedent variables used by Porcuna-Enguix et al. [37] as conditions in our configural model, with the exception of "SIZE", "CATA" and "COMPULSORY" for two reasons: (i) SIZE (log of total assets) and CATA (current assets to total assets) were not statistically significant in all cases; and (ii) COMPULSORY was removed because the final balanced fuzzy set consists of firms that must be audited, as indicated in Spanish audit law. The three dependent variables of audit risk assessment represent the outcomes.

Table 2. Ordered probit sample selection results for Audit Risk Assessment [37].

	EVAL_IC			PROB_ERROR			PROB_IRREG		
	Coef.		(z-Stat)	Coef.		(z-Stat)	Coef.		(z-Stat)
INTERCEPT	−2.558		(−1.56)	−2.610	**	(−2.18)	−2.695		(−1.63)
GENDER	−0.742	***	(−3.33)	0.960	***	(4.08)	0.513	***	(3.42)
EXP	−0.124	***	(−4.08)	−0.001		(−0.05)	0.088	***	(2.79)
GENDER*EXP	0.098	***	(3.00)	0.049	**	(1.98)	−0.048		(−1.50)
SIZE	0.079		(1.64)	−0.030		(−0.78)	−0.093	**	(−2.19)
PPE	0.142	**	(2.19)	0.187	**	(2.39)	0.300	***	(5.75)
CATA	0.131		(0.69)	−0.029		(−0.14)	−0.457	**	(−2.45)
LEV	−0.814	***	(−2.69)	0.929	***	(2.94)	−0.172		(−0.51)
ROA	−0.042	*	(−1.67)	0.520	***	(3.98)	0.146	***	(5.46)
COMPULSORY	−0.412		(−1.40)	0.428		(1.39)	0.197		(0.92)
YAUDITED	−0.049	***	(−5.68)	0.026	***	(3.29)	0.031	***	(3.86)
NEW	0.040		(0.26)	−0.134		(−0.54)	−0.004		(−0.02)
BOD	−0.383	***	(−6.00)	0.243	***	(6.87)	0.394	***	(5.40)
N	1559			1559			1,559		
Censored	811			811			811		
Uncensored	748			748			748		
White correction	Yes			Yes			Yes		
Year dummies	Yes			Yes			Yes		
Industry dummies	Yes			Yes			Yes		
Log Pseudolikelihood	−1285.18			−1452.04			−1386.91		
	Chi2		p-value	Chi2		p-value	Chi2		p-value
Wald test of indep. eqns.	11.090		0.001	16.630		0.000	11.680		0.001

***, **, and * indicate statistical significance at the 1%, 5%, and 10% level, respectively.

3.1. The Role of the Audit Team Leaders (ATLs) of Some Audit Firms on the Audit Risk Assessment

Audit work is labor-intensive, resulting in high staff costs [45]. Audit firms often adopt aggressive pricing strategies [46], which place pressure on audit engagement. As a result, efforts by audit teams to work with short deadlines and to lower costs may compromise audit quality [47], and undoubtedly this price–time–effort conflict influences audit team members' behavior and thus audit practices. Of course, the audit partner possesses a high degree of autonomy to exert professional judgment [48], and this autonomy includes the possibility to alter audit prices (i.e., maximize audit fees) and planned hours (i.e., minimize audit effort). The audit partner also holds the decision of designating the audit team leaders [ATLs] [49], who are presumed to be the main character of the audit team appointed. When an audit partner agrees with the client to carry out an audit engagement, they must adjudicate which audit member is going to be responsible (audit team leader) for the audit team. This arrangement is not meaningless, as individual characteristics are vital for the planification and audit risk management [50]. Therefore, the determination of the ATL is significantly affected by audit evidence collected when planning the audit. In turn, audit evidence hinges on detection risk, among others.

The role of the ATL is critical as they must allocate resources efficiently to lower audit risk up to an acceptable level, according to the International Audit Standards (ISA). Even though the audit report is signed by the audit partner, the audit effort is mainly carried out by those who are not the partners, and ATLs are decisive to opt for and issue an opinion or other in the audit report [2,3]. The ATLs act as the pivotal character in an audit engagement. The ATLs duties are mainly associated with assigning functions to audit team members, evaluating risk management activities within the client, and ensuring compliance with established internal control procedures by examining records, operating practices, reports, documentation, etc.

Human resources are the most important asset for all audit firms, and a balance between the skills of audit members and the complexity of audit engagements is needed. The optimal point is to avoid under-auditing (not taking unnecessary risks) or over-auditing (keep costs at a competitive level). This time and economic effectiveness must be properly provided by the ATL [8], which depends on their professional judgement and prior accumulated experience. Their leadership must prevail, facilitating upward communication and knowledge sharing, while also promoting good group relationships within the audit team members to perform more effective and efficient audits [5,6].

To allocate resources, the audit team members have to gather information and the ATL must keep the audit risk at an acceptable level. The assessment of the audit risk in the planification phase by the ATL is critical. This really determines the confidence in the internal control of the client and the subsequent substantive audit procedures. For instance, after the initial evaluation of the audit risk when planning, if the ATL concludes that internal controls are more effective than they actually are and that a material misstatement in the financial statement assertions does not exist when in fact it does, it affects the audit effectiveness, leading to issuing an inappropriate audit opinion (under-auditing). Contrarily, if the ATL concludes that internal controls are less effective than they actually are and that material misstatements in the financial statements assertions exists when in fact it does not, it hurts the audit efficiency increasing audit costs (over-auditing). A deficient audit risk assessment boosts sampling risk, that is, the risk that the auditor's conclusion based on a sample may be different from the conclusion if the entire population were subjected to the same audit procedure (ISA 530), inducing to the two already mentioned erroneous conclusions.

ISA 300 establishes information about "considerations specific to smaller entities". The above explanation applies to all audit firms and audit engagement, nevertheless, small- and medium-sized audit firms are singular for their available resources. In an audit engagement, the entire audit may be conducted by a small audit team compared to others. With a smaller team, the role of the ATL is essential to favor coordination, communication, and member motivation. The audit is not so complex and time-consuming, but losing clients because of a deficient audit risk assessment is costly. The market share of small- and medium-sized audit firms is strikingly lower than that of the Big Four and second-tier audit firms. Small- and medium-sized audit firms usually have a lower diversified client portfolio, so if things go wrong, they spread through the market by word of mouth very easily and quickly. According to these discrepancies, we may appreciate two main differences between ATLs in SME audit firms from those in Big Four or second-tier audit firms: (i) pressure and (ii) progression. First, time pressure is considerably more notable for ATLs in SME audit firms. Being efficient is imperative and the opposite might be very costly. Dysfunctional auditing behaviors might be expected because of such pressures [51]. Consequently, evaluating those factors or combination of them affecting audit risk assessment by ATLs in SME audit firms is determinant as they are the most representative figure (in number) in the audit market, shaping market behavior. Secondly, Big Four and second-tier audit firms have many employees at different levels. For instance, n-degrees of entry-level employees (e.g., assistant auditors), n-degrees of senior level (e.g., auditor, senior auditor, lead auditor or audit team leader, etc.), a partner with their own degrees of advancement, and finally director or managing director level (top management positions). In case of SME audit firms, progression is shorter with fewer levels such as assistant, audit team leader, and partner. As seen, promotion in SME audit firms is not as long, slow, or tough as their larger mates. Why is this relevant? Investors perceive a lower quality of accounting information when the audited client's management exerts influence on auditors (e.g., [52]), and auditor–client identification on audit client's acquiescence to client-preferred treatment occurs in non-Big 4 firms [53]. Understanding the factors that might alter or modify audit risk assessment when planning is crucial as it determines the subsequent audit procedures. Moreover, client familiarity is desirable but may threaten auditor objectivity, and this is closely linked with the auditor position within the firm, affecting the audit quality practices.

Having exposed the above, we find contrasting results in the existing literature on this matter. Small- and medium-sized audit firms usually face the handicap of offering a lower quality audit in comparison to larger ones. There is a certain and questionable "bigness syndrome" because size indicates sufficient staff to carry out the audit engagement, which may ensure quality or independence [54,55] or higher prices when clients go public [56]. Other arguments reinforce the appointment of larger audit firms [57]: to reduce agency costs; to reduce information asymmetry; etc. However, less complexity ("the larger the more to lose" or "size matters" hypotheses) does not imply less attention or worse audit risk assessments. In fact, Lawrence et al. [58] show that the differences in proxies for audit quality between Big 4 and non-Big 4 auditors are more likely due to client characteristics. Boone et al. [59] explained that little evidence exists of a difference in audit quality between Big 4 and second-tier audit firms. Comprix and Huang [60] find no evidence that small audit firms are associated with real activity manipulation. Oppositely, Svanström [51] described, using a sample of Swedish small audit firms, a positive association between perceived time pressure (time budget pressure and time deadline pressure) and dysfunctional auditing behavior (e.g., superficial reviews of client documents, premature signing-off or accepting weak client explanations). Even though less audit complexity in small audit firms exists, time pressure is more marked as resources are more limited. Furthermore, Svanström [51] finds that training activities may reduce such dysfunctional audit behavior, nevertheless, small audit firms have limited opportunities to arrange such training activities with invited experts.

Therefore, we argue that ATLs of small- and medium-sized audit firms play a crucial role in assessing the audit risk in the planification phase. Furthermore, the ATL characteristics are core conditions to evaluate and determine management risks.

3.2. Outcome and Conditions

Based on a review of prior research, the aim of this study is to forecast the expected or perceived audit risk assessment by the ATL, especially by female ATLs. This outcome is supposed to depend on the presence or negation/absence of several conditions. Therefore, we expect several configurations to explain the same outcome as audit risk assessment involves skepticism and judgments by auditors. As established by reference number 4 of the ISA 315, "the objective of the auditor is to identify and assess the risks of material misstatements, whether due to fraud or error, at the financial statement and assertion levels, (. . .) including the entity's internal control (. . .)". In this sense, our main focus outcomes will be the assessment of internal control (intcontrol), the probability to expect errors (error) and the likelihood to expect irregularities (irreg) in the planification phase. Conditions, together with outcomes, are briefly described in Table 3, and explained in detail below.

The focus outcomes are three to enrich the analysis. As established in ISA 315, the internal control is the internal "process designed, implemented and maintained by those charged with governance, management and other personnel to provide reasonable assurance about the achievement of an entity's objectives regarding reliability of financial reporting, effectiveness and efficiency of operations, and compliance with applicable laws and regulations. The term "controls" refers to any aspects of one or more of the components of internal control". Material misstatements due to errors or mistakes are supposed to be unintended, with no fraud, such as arithmetic errors, misinterpretations, inadvertences, or incorrect application of accounting principles and standards without intent to cause harm by act or omission. Unlike errors, irregularities relate to intended acts or omissions with fraud, such as record and document manipulation, falsification or alteration, misappropriation, and irregular use of assets, record fictitious operations, or the improper and intentional application of accounting principles and standards.

With respect to the economic and financial features of the audited client, the ratio property, plant and equipment to total assets (ppe), leverage ratio (lev), and return on assets (roa) have been selected. Capital assets are purchased (bought or constructed) to be used in a business and are often the largest accounting item on a balance sheet. Even though the value is usually high, the risk associated is often low or moderate. Overall, these assets are depreciated (accounting expense record) over their useful life and are presented in a balance sheet by the book value (cost minus accumulated depreciation). Though appreciation in market value is not allowed, decreases are booked known as impairments, which may be reversed if those circumstances that originated the impairment disappear. Property, plant and equipment walkthrough consists of looking for ways that this item might be overstated, though understatements may also occur. Concerns about who authorizes the purchase, reconciliations, consistent depreciation methods, adequate records, the existence of controls, segregation of duties, period physical inventories and custody, schedules upon sale, capitalization of repair expenses (not expensed), etc. are questions to look at (primary risks). The higher capital assets, the wider gap for discretion or managerial choice.

Table 3. Description of outcomes and conditions.

Variables	Label	Description
Outcomes		
Expected internal control	intcontrol	Discrete evaluation of internal control ranging from 1 to 5, with 1 being bad internal control and 5 being good internal control
Expected error	error	Discrete probability of expected error ranging from 1 to 5, with 1 being unlikely and 5 being likely
Expected irregularity	irreg	Discrete probability of expected irregularity ranging from 1 to 5, with 1 being unlikely and 5 being likely
Conditions		
Economic and financial features		
Property, Plant and Equipment	ppe	Ratio property, plant and equipment to total assets
Leverage	lev	Leverage ratio
Return on Assets	roa	Return on assets
Audit team leader features		
Gender	gender	It takes value 1 if the audit team leader is female and 0 if male
Experience	exp	Years of experience
Audit client features		
Years audited	yaudited	Years that the client is being audited by the audit firm
New audit	new	It takes value 1 if the client is new and 0 otherwise
Board of Directors	bod	Personnel belonging to board of directors

Information asymmetry and uncertainty about the use of funds (loans) by insiders, such as managers or shareholders, are problems that all creditors (debtholders) face when granting such loans. Insiders may take actions that contribute to their wellness at the expense of creditors (agency conflicts). Basically, once insiders obtain the financing, they may undertake riskier investments or underinvest (e.g., [61]). Riskier decisions sometimes threaten future repayments to creditors and renouncing positive net present value projects because they may benefit creditors. It is true that this opportunistic behavior may be limited through covenants in creditors' debt contracts in terms of accounting numbers. However, again, restrictions may encourage managerial accounting choices such as income-increasing or -decreasing practices [62]. The higher the leverage, the greater the probability for discretion and the more need for reliable monitoring of accounting information.

Related to managerial accounting choices, those more profitable audited firms may have used discretion to achieve such strong economic positions. Profitability does not necessarily mean bad practices, however, high profitability may be the result of confusing accounting practices, and may increase the likelihood of uncovering material misstatements in the financial statements (unreliable accounting numbers) by independent auditors. Proving this behavior, Hardies et al. [42] found evidence that higher return-on-assets values increase the likelihood that an auditor issues a going concern opinion, as an indicator of audit quality.

Regarding audit team leader (ATL) features, we may distinguish two characteristics: gender and experience. The gender diversity on ATLs is not trivial. The ATL may be chosen by the audit firm or the audited client to meet operating and contracting environment [63]. Attitudes towards corporate social responsibility [64], preference to overlap audit quality to cost reduction [65], or riskier positions may precipitate female appointments' decisions [41], because of their more prudent posture when assessing audit risk when planning. Moreover, women take more time to plan in an audit engagement, which directly influences the subsequent work execution [66]. Experience is another ATL attribute that affects audit work and the professional judgment of the ATL to assess audit risk. The experience (competence characteristic) of the audit personnel is listed as one of the main audit quality indicators [67,68]. Experience promotes general and industry expertise and enhances the ability of auditors to identify and assess audit risk. Professional background and experience make auditors stay alert and focused and make their clients feel higher audit quality which leads to audit premium. Experienced auditors are more accurate in interpreting, judging, and assessing information, so the audit risk assessment becomes more reasonable and targeted to carry out further audit procedures [69]. Therefore, the accumulation of this professional competence facilitates interpretation, judgment, problem-solving skills, the allocation of audit resources and the evaluation of audit risk, which is the cornerstone. Experimental studies such as Koch et al. [70], for example, demonstrate that experienced auditors are more likely to issue unqualified audit opinions, as a signal of either client internal control deficiencies and/or the existence of uncorrected material misstatements in the financial statements, due to fraud or error.

Associated to audit client features, we have chosen the number of years that the client is being audited by the audit firm (yaudited), whether the audit work for certain client is new (new) and the total board of directors (bod). The length of the audit–client relationship (audit tenure) is fully related to information and insurance roles and firm risk affects auditor behavior [71]. Nevertheless, the directional effect is still unanswered [72] because time might (not) threaten auditor independence. For instance, on the one hand, some argue that longer audit tenure erodes agency costs (less information asymmetry), which leads to better audit quality [54]. Shorter audit tenure denotes lack of adequate knowledge of the clients and the industry during the early years of the audit engagement [73] and widens the lag between hiring an auditor and the motivation to engage in earnings management by the client [74]. In this respect, Myers et al. [75] found that the use of discretionary accruals diminishes as audit tenure increases. In addition, changing auditors is costly and may influence audit quality because less effort may trigger less audit adjustments as a consequence of a weaker audit risk assessment. Meanwhile, others proclaim that longer audit tenures jeopardize objectivity and skepticism, which also influences audit quality. According to these different points of view, the Spanish audit law 22/2015 (which incorporated the Directive 2014/56/UE into the national law) tried to find the midpoint. Thereby, overall, the duration of an audit engagement may be no shorter than three years and no longer than nine years counting from the date on which the first year to be audited starts (Article 22.1, Audit law 22/2015). The explanation being new clients are in line with the precedent reasoning. Finally, despite considering small- and medium-sized firms, our sample consists of compulsory audit engagement, so differences between ownership and control should presumably be assumed. This is the so-called board size effect: communication and coordination problems rise as group size increases [76], which broadens management–board conflicts (agency problems).

4. Sample and Descriptive Statistics

The data to run the fuzzy analysis came from 24 small- and medium-sized audit firms, with the information taken from their legal application declared to the national public oversight board, that is, to the Spanish Accounting and Auditing Institute (ICAC) spanning 2001 to 2015. These audit firms were randomly chosen regarding their size and areas of activity. Finally, the required information about the assessment of audit risk was sent by 13 audit firms, including their evaluation of internal control, error, and irregular probabilities of the client, and were assessed by the audit team leader during the planification phase.

Our hand-collected data values are original and no estimation process which would imply a high subjectivity charge has been employed in our sample. Our sample consists of a balanced panel of 1338 firm-year observations.

Untabulated distribution by economic sector shows that service is the most audited industry, followed by construction, manufacturing, agriculture and supplies. In addition, an untabulated correlation matrix does not suggest any multicollinearity problem associated with quantitative variables.

Table 4 displays the descriptive statistics of the original quantitative data and of the calibration cut-off points. As shown, we have used three qualitative breakpoints at 95%, 50% and 5%. It is important to say the dummy variables such as gender and new have not been calibrated for logical reasons. On the one side, the gender attribute of an audit team leader (ATL) will take value 1 if female and 0 if male, so "do not care" values do not exist; on the other hand, variable new will take 1 if first-time audit by the ATL for a specific client and 0 otherwise, so "ambiguity" values do not exist either.

Table 4. Balanced sample descriptive statistics and calibration parameters for fsQCA.

Variable	N	Mean	s.d.	Min.	Median	Max.	Full In (95%)	Max'm Ambig. (50%)	Full Out (5%)
intcontrol	1338	3.556	1.362	1.000	3.000	5.000	5.000	3.000	1.000
error	1338	2.093	1.091	1.000	2.000	5.000	4.000	2.000	1.000
irreg	1338	2.050	1.110	1.000	2.000	5.000	4.000	2.000	1.000
gender	1338	0.646	0.478	0.000	1.000	1.000	1.000	1.000	0.000
exp	1338	9.609	4.802	0.000	9.609	25.000	19.000	9.609	3.000
ppe	1338	0.269	0.450	0.000	0.200	9.707	0.681	0.200	0.002
lev	1338	0.129	0.143	0.000	0.080	0.850	0.410	0.080	0.000
roa	1338	0.151	1.860	−1.880	0.020	58.550	0.150	0.020	−0.080
yaudited	1338	7.611	5.970	1.000	6.000	25.000	20.000	6.000	1.000
new	1338	0.122	0.327	0.000	0.000	1.000	1.000	0.000	0.000
bod	1338	3.256	2.496	1.000	2.000	12.000	12.000	2.000	1.000

5. Empirical Results

5.1. Necessary Analysis of Conditions for Audit Risk Assessment Outcomes

Table 5 indicates the necessity conditions. As introduced in previous sections, the necessity analysis comes to explain whether necessary conditions exist. The consistency score has to be higher than 0.9 for a condition to be necessary [77]. To simplify the necessary analysis, we focus the research on the expectation of detecting audit risk in the planification phase. The objective of the auditor is to "identify and assess the risks of material misstatements, whether due to fraud or error, at the financial statement and assertion levels, (...) including the entity's internal control (...)" (ISA 315, reference [4]). In this sense, the fact of uncovering risks prevails, that is, the existence of expected significant risks, if true, is more relevant than expecting insignificant risks. Therefore, we are interested in those necessary conditions to expect higher audit risk levels, that is, to expect bad internal control (~fs_intcontrol—absence of internal control), to be likely to expect errors (fs_error—presence of expected errors), and to be likely to expect irregularities (fs_irreg—presence of expected irregularities).

Table 5. Analysis of necessity for the absence of internal control and for the presence of expected errors and irregularities.

Conditions Tested	Outcome: ~fs_intcontrol Consistency	Coverage	Outcome: fs_error Consistency	Coverage	Outcome: fs_irreg Consistency	Coverage
gender	**0.662917**	0.381944	**0.668419**	0.465185	**0.672636**	0.450081
~gender	0.337084	0.354009	0.331583	0.420634	0.327368	0.399284
fs_exp	0.567980	0.468703	0.580491	0.578624	0.534463	0.512216
~fs_exp	**0.757912**	0.513483	**0.705986**	0.577750	**0.683201**	0.537560
fs_ppe	0.675169	0.542541	0.638614	0.619863	0.642138	0.599266
~fs_ppe	0.649335	0.449872	0.626441	0.524248	0.584106	0.469984
fs_lev	0.651747	0.548364	0.600049	0.609837	0.566567	0.553621
~fs_lev	0.656667	0.437981	0.661832	0.533205	0.660421	0.511566
fs_roa	0.728020	0.525049	0.702243	0.611758	0.658915	0.551895
~fs_roa	0.684792	0.526255	0.644054	0.597855	0.609810	0.544256
fs_yaudited	0.638448	0.516251	0.597671	0.583759	0.595263	0.559004
~fs_yaudited	0.696602	0.480041	0.673107	0.560291	0.629113	0.503492
new	0.118160	0.360859	0.131964	0.486810	0.121565	0.431166
~new	**0.881840**	0.373599	**0.868036**	0.444212	**0.878437**	0.432213
fs_bod	**0.860242**	0.608017	**0.789237**	0.673813	**0.759116**	0.623124
~fs_bod	0.700299	0.550116	0.639280	0.606594	0.568642	0.518776

Note: The symbol "~" indicates logical NOT which means the absence/negation of a condition or the absence/negation of the outcome. All variables are defined in Table 2. Those conditions that have been calibrated are preceded by the "fs_" prefix, so dummy conditions such as gender or new do not have it.

As shown in Table 5, there is no exclusive necessary condition, so audit risk assessment does not happen because of any one single condition. Therefore, a comprehensive set of conditions must be given for expected high audit risk (i.e., bad internal control, and expected errors and irregularities in the financial statements). More specifically, four conditions are the most significant drivers of the outcome, either for their higher consistency values or for the wider difference between their presence and absence to explain the expected audit risk in the planification phase. At a glance, we firstly observe that female ATLs (gender) are more demanding when evaluating audit risks; secondly, less experience (~fs_ppe) propitiates conservative attitudes reinforcing the detection of risks; thirdly, displaying the highest consistency value (87.61% in average), the attribute of being an old client (~new) clearly indicates a stricter audit risk assessment. According to our data, old clients (~new) are audited by inexperienced ATLs (less than 3 years) in the short term (less than 1 year), so their knowledge about their client is still scarce and they are more rigorous at work. Finally, in fourth place, a crowded board of directors (fs_bod) leads to a more meticulous audit risk assessment in the planification stage. Setting aside board composition, board size matters. Traditional literature emphasizes the board size effect: communication and coordination problems increase as group size boosts, thereby leading to potential agency problems [76,78,79]. In our sample, the board size exceeds 12 people (full membership) and, because all sample firms have to be audited compulsorily, conflicts of interests may easily arise stressing agency problems between management and control. This circumstance may encourage accounting discretion by managers and thus earning management practices (e.g., [79]) and also may decrease firm value [76,78]. In this scenario, some procedures of the internal control may fail, be adulterated, or even not exist, and/or the financial statements as a whole are not likely to be free from material misstatements, whether due to fraud or error.

5.2. Sufficiency Analyses for Audit Risk Assessment Outcomes

In this section, we will explore the causal relationships obtained in the different equally effective configurations. It is important to say that, in all three focus outcomes, there is more than one plausible combination, and that causal asymmetry may occur. The combinations leading to the absence of internal control and to the presence of expected errors and irregularities are presented in Tables 6–8, respectively. In order to improve the presentation and readability of the findings, we transformed the solutions from fsQCA output into tables that are easy to interpret. Following Fiss [80], the presence of a condition is indicated with a black circle (●), absence/negation with a crossed-out circle (⊗), and the "do not care" condition with a blank space. The distinction between core and peripheral is made by using large and small circles, respectively. Of course, overall consistency and coverage values will be also presented.

Table 6. Configurations sufficient for the absence of internal control (Outcome: ~fs_intcontrol).

		Solution (Frequency Cutoff: 1; Consistency Cutoff: 0.801661)								
Configuration		1	2	3	4	5	6	7	8	9
Economic and financial. features	fs_ppe				●	●	●	⊗		●
	fs_lev	●	●	●			●		●	⊗
	fs_roa		●		⊗	⊗	⊗	●	⊗	⊗
ATL features	gender	●	●	●	●	●		●	●	⊗
	fs_exp	⊗				⊗		⊗	●	⊗
Audit client features	fs_yaudited		⊗	⊗	⊗		●	●	⊗	
	new	⊗		.	⊗	⊗	⊗	⊗	●	●
	fs_bod	●	●	●	●	.	●	●		.
Consistency		0.762	0.787	0.712	0.786	0.776	0.807	0.802	0.782	0.821
Raw Coverage		0.280	0.263	0.037	0.229	0.252	0.287	0.194	0.015	0.018
Unique Coverage		0.016	0.015	0.002	0.015	0.013	0.099	0.019	0.000	0.018
Overall solution consistency:							0.747			
Overall solution coverage:							0.545			

Note: Black circles (●) indicate the presence of a condition. Crossed-out circles (⊗) denote the absence or negation of a condition. Large circles are core conditions while small circles are peripheral conditions. "Do not care" conditions are displayed in blank spaces. All variables are defined in Table 2. Those conditions that have been calibrated are preceded by the "fs_" prefix, so dummy conditions such as gender or new do not have it.

Table 7. Configurations sufficient for the for the presence of errors (Outcome: fs_error).

		Solution (Frequency Cutoff: 1; Consistency Cutoff: 0.802642)											
Configuration		1	2	3	4	5	6	7	8	9	10	11	12
Economic and financial. features	fs_ppe						.	●	●	●	.	⊗	●
	fs_lev			●	●	⊗	●	⊗	●	●	⊗	●	
	fs_roa						⊗	⊗	⊗		.	.	
ATL features	gender	●	●	●	●	⊗	●		●		●	●	●
	fs_exp		●		⊗	⊗	⊗			⊗	.	●	
Audit client features	fs_yaudited		⊗	⊗	⊗	⊗		⊗	⊗	●	⊗	⊗	⊗
	new	⊗			⊗	●	⊗	●	.	⊗	⊗	●	
	fs_bod	●	●	●		●		●		.			.
Consistency		0.711	0.796	0.828	0.766	0.766	0.803	0.841	0.786	0.828	0.817	0.840	0.809
Raw Coverage		0.473	0.262	0.272	0.208	0.031	0.190	0.036	0.018	0.271	0.149	0.016	0.277
Unique Coverage		0.126	0.002	0.007	0.009	0.015	0.003	0.001	0.000	0.089	0.002	0.000	0.001
Overall solution consistency:							0.700						
Overall solution coverage:							0.662						

Note: Black circles (●) indicate the presence of a condition. Crossed-out circles (⊗) denote the absence or negation of a condition. Large circles are core conditions while small circles are peripheral conditions. "Do not care" conditions are displayed in blank spaces. All variables are defined in Table 2. Those conditions that have been calibrated are preceded by the "fs_" prefix, so dummy conditions such as gender or new do not have it.

Table 8. Configurations sufficient for the presence of irregularities (Outcome: fs_irreg).

		Solution (Frequency Cutoff: 1; Consistency Cutoff: 0.801226)					
Configuration		1	2	3	4	5	6
Economic and financial features	fs_ppe	●	●	●		⊗	●
	fs_lev	⊗		⊗	●	⊗	⊗
	fs_roa				●	●	⊗
ATL features	gender	●	●	●	·	●	●
	fs_exp	●	●	⊗	●	●	
Audit client features	fs_yaudited	⊗	⊗	●	⊗	●	⊗
	new		●	⊗	●	⊗	⊗
	fs_bod	●	●	·	●	●	·
Consistency		0.815	0.827	0.802	0.828	0.804	0.822
Raw Coverage		0.182	0.027	0.191	0.019	0.159	0.172
Unique Coverage		0.011	0.001	0.042	0.004	0.039	0.012
Overall solution consistency:				0.789			
Overall solution coverage:				0.316			

Note: Black circles (●) indicate the presence of a condition. Crossed-out circles (⊗) denote the absence or negation of a condition. Large circles are core conditions while small circles are peripheral conditions. "Do not care" conditions are displayed in blank spaces. All variables are defined in Table 2. Those conditions that have been calibrated are preceded by the "fs_" prefix, so dummy conditions such as gender or new do not have it.

Table 6 presents the configurations for the absence of internal control, that is, what might determine that the audit team leader (ATL) expected to uncover internal control deficiencies within the firm. Holding a high capital assets proportion is core condition in four out of nine configurations (44%), which is a sign of managerial choices. Indebted clients denote high probability of expecting bad internal control in five out of nine cases (56%). In the 56% of cases too, less profitable firms are perceived to have internal control shortcomings. We expected the opposite result, but there might be an explanation. Less profitability might be what triggers income-increasing practices by managers. Female ATLs are confirmed to be a clear sufficient condition in 78% of combinations. When planning, women are more conservative and have more prudent behavior. They dedicate more effort in the planification phase to detect anomalies in internal control mechanisms, so they are more likely to uncover deficiencies. In four out of nine cases, the absence of experience is determinant. We expected its presence, but its absence means that ATLs will be more demanding when they are inexperienced auditors. Similarly, ATLs are more skeptic with shorter audit tenures, so they will be more objective. The new condition seems to be more affected by causal asymmetry than the others, because the results are not decisive. Finally, as with gender, the probability of expecting bad internal control increases as board size increases in 89% of cases (in six out of nine cases is a core condition), denoting agency problems as a result of the coexistence of various interests. In short, when female and inexperienced ATLs audit firms with high proportions of capital assets, high leverage ratios, less profitability and large board sizes, they expect bad internal control in the planification stage.

Table 7 presents the configurations for the presence of expecting errors in the financial statements. Beyond explaining again all the conditions individually, it is more fruitful to conclude to have a broad view of detecting errors in the annual accounts when planning. Gathering information stemming from the twelve configurations, overall we may assert that the likelihood of expecting error in the financial statements increases as perceived internal control worsens (same conditions are met), and when the audit tenure is shorter (distinctive condition compared to Table 6). Table 8 indicates the combinations of conditions for the presence of expected irregularities. Unlike errors, irregularities are intentionally made. For this reason, conditions such as experience (fs_exp) gain importance. Experienced ATLs are capable of expecting or forecasting irregularities because of their widespread background and prior know-how.

6. Conclusions, Discussion and Limitations

Auditing is a collective process conducted by professional accounting individuals whose skills, experiences, and emotions differ. The goal of an audit is to reduce uncertainty in the information and improve the degree of confidence of intended users in financial statements. Thereby, auditors aim to identify and assess risks of material misstatements at the financial statements' assertions, including the internal control of the client. To do so, it is fundamental an appropriate planning of the audit, which includes the audit team leader (ATL) appointment, an audit risk assessment, the subsequent nature, timing and extent of further audit procedures and the consideration of those factors, in the auditor's professional judgement, are significant in directing the engagement team's efforts.

The audit risk assessment is critical. Audits must follow a risk-based approach due to the impossibility to check all transactions within the client. A correct audit risk assessment lowers the audit risk to an acceptable level and avoids incurring two types of erroneous conclusions as a consequence of under-audits (issuing an inappropriate opinion) or over-audits (costly inefficient effort). There is no doubt that lower-level audit team members gather most audit evidence, however, the ATL plays a crucial role from the planification stage until the audit report, with the audit opinion issued and signed by the audit partner, and are decisive for an effective and efficient audits. The audit risk assessment is mainly carried out by the ATL, and this evaluation is highly based on subjective judgements. The ATL must assess internal control and the expected and perceived existence of errors (unintended misstatements) and irregularities (intended misstatements) in the financial statements' assertions. Thereby, ATLs determine if a risk in an accounting area or transaction is classified as a key risk, material risk, or insignificant risk. This "ambiguity" or "fuzziness" in determining the risk level makes the application of fuzzy theory particularly meaningful, as audit risk assessment involves people' though, inference and perception, and thus retaining certain degree of such fuzziness. Therefore, this study extends our understanding of the role of ATL characteristics on the audit risk assessment when planning. Furthermore, smaller audit firms have attracted limited attention both in practice and in academic research since PCAOB inspections were implemented and reinforced at international level. This paper adds knowledge in this respect, raising awareness that the study of SME audit firms is essential and concerns the whole market. Even though the audit market is often dominated by Big 4 and second-tier audit firms, small- and medium-sized audit firms shape the market behavior.

Using hand-collected and private data taken from legal applications declared by audit firms to the national public oversight board (ICAC), spanning 2001 to 2015, the results from the fuzzy-set modelling are substantial. The analysis of necessity reveals that there is no exclusive necessary condition. This evidences the fact that SME audit firms conduct more complex audits than we initially believed. Concretely, four conditions are the most significant drivers of the outcome (expected bad internal control and expected errors and irregularities in the financial statements): the presence of female ATLs, inexperienced ATLs, old audit clients, and large board size. This supports previous evidence on that audit risk assessment (or related audit quality practices) are not only influenced by auditor' features, but also by client characteristics. The sufficient analyses show the existence of more than one possible configuration, and that causal asymmetry happens. With respect to the absence of internal control (the ATL expects to find deficiencies in the internal control of the client when planning), the results indicate that female and inexperienced ATLs planning the audit indebted firms with high proportions of capital assets, less profits, and with large board sizes as being expected to have bad internal control. Women are more demanding and prefer quality to reduce costs; in addition, the lack of experience makes them be more prudent and conservative when assessing the audit risk. In the case of firm characteristics, the presence of high levels of leverage and larger board size denote the presence of high agency costs and more conflicts of interests as the number of parties increases, which incentivizes bad practice. For the presence of errors in the financial statements' assertions, the same conditions are met, and also in shorter audit tenures. The lack of prior knowledge of the client triggers more demanding conducts and more checks in detail. Finally, but not least, conditions such as the ATL's experience gains importance to expect irregularities. Irregularities are intended errors, and experienced ATLs are capable of forecasting irregularities because of their widespread and accumulated background. Sometimes manager discretion is difficult to guess, so experienced ATLs are more alert for discretionary practices that are not adequately revealed in the financial statements.

From a managerial point of view, the literature has demonstrated that the client's management may exert influence on auditors, which is more relevant in countries where no market-based institutional incentives exist, such as the Spanish context. Additionally, the audit risk assessment implies that the auditor must inquire with management, among other parties. These arguments, which are more pronounced in SME audit firms, come to confirm the existence of managerial pressures on auditors and thus on their assessment of the audit risk in the initial stage of the audit work, which will determine the subsequent audit procedures and the allocation of available resources. Our results evidence that ATL characteristics, such as gender, are core conditions to evaluate and determine management risks.

When interpreting the results, some limitations must be considered. First, fuzzy theory establishes combinations of conditions that, in the judgement of the researcher, have high values of consistency and coverage but, however, causal asymmetries may happen. Second, despite the sensitive nature of our data, the results may not be able to be generalized to audit practices outside Spain, as potential differences and consensus in audit work environments exist across countries. Future research on ATL characteristics, designation, and audit risk assessment should consider endogeneity problems, as the appointment of the ATL is not trivial, so the audit partner might indirectly influence the audit risk assessment from the beginning of the audit. As well, exploring various existing audit environments at international and size levels would show us cultural differences and size behaviors between larger and smaller audit firms.

Author Contributions: Conceptualization, E.B.-C. and J.S.-M.; Data curation, L.P.-E.; Formal analysis, L.P.-E.; Methodology, L.P.-E.; Supervision, L.P.-E., J.S.-M. and G.L.-S.; Writing—original draft, L.P.-E.; Writing—review & editing, L.P.-E. and E.B.-C. All authors have read and agreed to the published version of the manuscript.

Funding: This research received no external funding.

Institutional Review Board Statement: Not applicable.

Informed Consent Statement: Not applicable.

Conflicts of Interest: The authors declare no conflict of interest.

References

1. Amyar, F.; Hidayah, N.N.; Lowe, A.; Woods, M. Investigating the backstage of audit engagements: The paradox of team diversity. *Account. Audit. Account. J.* **2019**, *32*, 378–400. [CrossRef]
2. Trotman, K.T.; Bauer, T.D.; Humphreys, K.A. Group judgment and decision making in auditing: Past and future research. *Account. Organ. Soc.* **2015**, *47*, 56–72. [CrossRef]
3. Christensen, B.E.; Thomas, C.; Omer, M.K.; Shelley, P.; Wong, A. Affiliated former partners on the audit committee: Influence on the auditor-client relationship and audit quality. *Audit. A J. Pract. Theory* **2019**, *38*, 95–119. [CrossRef]
4. Gissel, J.L.; Johnstone, K.M. Information sharing during auditors' fraud brainstorming: Effects of psychological safety and auditor knowledge. *Audit. A J. Pract. Theory* **2017**, *36*, 87–110. [CrossRef]
5. Nelson, M.W.; Proell, C.A. Is silence golden? Audit team leader reactions to subordinates who speak up "in the moment" and at performance appraisal. *Account. Rev.* **2018**, *93*, 281–300. [CrossRef]
6. PCAOB. Public Company Accounting Oversight Board. *Supervising the Audit Engagement. Auditing Standand, 10*; PCAOB: Washington, DC, USA, 2010.
7. Pratt, J.; Jiambalvo, J. Relationships between leader behaviors and audit team performance. *Account. Organ. Soc.* **1981**, *6*, 133–142. [CrossRef]
8. Dobre, F.; Vilsanoiu, D.; Turlea, E. A multiple regression model for selecting audit team members. *Procedia Econ. Financ.* **2012**, *3*, 204–210. [CrossRef]

9. Messier, W.F.; Glover, S.M.; Prawitt, D.F. *Auditing & Assurance Services: A Systematic Approach*, 12th ed.; McGraw-Hill Irwin: Boston, MA, USA, 2021.
10. Chang, S.I.; Tsai, C.F.; Shih, D.H.; Hwang, C.L. The development of audit detection risk assessment system: Using the fuzzy theory and audit risk model. *Expert Syst. Appl.* **2008**, *35*, 1053–1067. [CrossRef]
11. La Porta, R.; Lopez de Silanes, F.; Shleifer, A.; Vishny, R. Law and finance. *J. Financ.* **1997**, *52*, 1131–1150. [CrossRef]
12. Carrera, N.; Gutiérrez, I.; Carmona, S. Gender, the state and the audit profession: Evidence from Spain (1942–88). *Eur. Account. Rev.* **2001**, *10*, 803–815. [CrossRef]
13. Ruiz-Barbadillo, E.; Humphrey, C.; García-Benau, M.A. Auditors versus third parties and others: The unusual case of the Spanish audit liability "crisis". *Account. Hist.* **2000**, *5*, 119–146. [CrossRef]
14. Ballesta, J.P.S.; García-Meca, E. Audit qualifications and corporate governance in Spanish listed firms. *Manag. Audit. J.* **2005**, *20*, 725–738. [CrossRef]
15. ECB. European Central Bank. Survey on the Access to Finance of Enterprises in the Euro Area—October 2020 to March 2021—The Financial Situation of SMEs in the Euro Area (June 2021). Available online: https://www.ecb.europa.eu/stats/ecb_surveys/safe/html/ecb.safe202106~{}3746205830.en.html (accessed on 1 September 2021).
16. Kim, J.B.; Song, B.Y.; Tsui, J.S.L. Auditor size, tenure, and bank loan pricing. *Rev. Quant. Financ. Account.* **2013**, *40*, 75–99. [CrossRef]
17. Dereli, T.; Baykasoğlu, A.; Daş, G.S. Fuzzy quality-team formation for value added auditing: A case study. *J. Eng. Technol. Manag.* **2007**, *24*, 366–394. [CrossRef]
18. Khan, R.; Adi, E.; Hussain, O. AI-based audit of fuzzy front end innovation using ISO56002. *Manag. Audit. J.* **2021**, *36*, 564–590. [CrossRef]
19. Beynon, M.J.; Peel, M.J.; Tang, Y.C. The application of fuzzy decision tree analysis in an exposition of the antecedents of audit fees. *Omega* **2004**, *32*, 231–244. [CrossRef]
20. Lai, H.L.; Chen, T.Y. Client acceptance method for audit firms based on interval-valued fuzzy numbers. *Technol. Econ. Dev. Econ.* **2015**, *21*, 1–27. [CrossRef]
21. PCAOB. Public Company Accounting Oversight Board. Release No. 2013–009. 4 December 2013. Available online: http://pcaobus.org/Rules/Rulemaking/Docket029/PCAOB20Release20No20202013-00920-20Transparency.pdf (accessed on 15 September 2021).
22. Valaei, N.; Rezaei, S.; Ismail, W.K.W. Examining learning strategies, creativity, and innovation at SMEs using fuzzy set Qualitative Comparative Analysis and PLS path modeling. *J. Bus. Res.* **2017**, *70*, 224–233. [CrossRef]
23. Douglas, E.J.; Shepherd, D.A.; Prentice, C. Using fuzzy-set qualitative comparative analysis for a finer-grained understanding of entrepreneurship. *J. Bus. Ventur.* **2020**, *35*, 105970. [CrossRef]
24. Gelman, A.; Loken, E. The statistical crisis in science: Data-dependent analysis—A garden of forking paths—Explains why many statistically significant comparisons don't hold up. *Am. Sci.* **2014**, *102*, 460–466. [CrossRef]
25. Wang, Z.; Klir, G.J. *Fuzzy Measure Theory*; Springer Science & Business Media: Berlin/Heidelberg, Germany, 2013.
26. Krogslund, C.; Choi, D.D.; Poertner, M. Fuzzy sets on shaky ground: Parameter sensitivity and confirmation bias in fsQCA. *Political Anal.* **2015**, *23*, 21–41. [CrossRef]
27. Ragin, C.C. *Redesigning Social Inquiry: Fuzzy Sets and Beyond*; University of Chicago Press: Chicago, IL, USA, 2008.
28. Elliott, T. Fuzzy Set Qualitative Comparative Analysis. 2013. Available online: https://www.socsci.uci.edu/~{}sgsa/docs/fsQCA_thomas_elliot.pdf (accessed on 1 September 2021).
29. Greckhamer, T.; Furlani, S.; Fiss, P.C.; Aguilera, R.V. Studying configurations with qualitative comparative analysis: Best practices in strategy and organization research. *Strateg. Organ.* **2018**, *16*, 482–495. [CrossRef]
30. Leppänen, P.T.; McKenny, A.F.; Short, J.C. Qualitative comparative analysis in entrepreneurship: Exploring the approach and noting opportunities for the future. In *Standing on the Shoulders of Giants*; Emerald Publishing Limited: West Yorkshire, UK, 2019.
31. Wu, B.; Cheng, T.; Yip, T.L.; Wang, Y. Fuzzy logic based dynamic decision-making system for intelligent navigation strategy within inland traffic separation schemes. *Ocean Eng.* **2020**, *197*, 106909. [CrossRef]
32. Moreno-Cabezali, B.M.; Fernandez-Crehuet, J.M. Application of a fuzzy-logic based model for risk assessment in additive manufacturing R&D projects. *Comput. Ind. Eng.* **2020**, *145*, 106529.
33. Arias-Oliva, M.; de Andrés-Sánchez, J.; Pelegrín-Borondo, J. Fuzzy Set Qualitative comparative analysis of factors influencing the use of cryptocurrencies in Spanish households. *Mathematics* **2021**, *9*, 324. [CrossRef]
34. Askary, S.; Abu-Ghazaleh, N.; Tahat, Y.A. Artificial intelligence and reliability of accounting information. In Proceedings of the 17th Conference on e-Business, e-Services and e-Society (I3E), Kuwait City, Kuwait, 30 October–1 November 2018; Springer: Cham, Switzerland; pp. 315–324.
35. Zadeh, L. Fuzzy sets. *Inf. Control* **1965**, *8*, 338–353. [CrossRef]
36. Meyer, A.D.; Tsui, A.S.; Hinings, C.R. Configurational approaches to organizational analysis. *Acad. Manag. J.* **1993**, *36*, 1175–1195.
37. Porcuna-Enguix, L.; Serrano-Madrid, J.; Bustos-Contell, E.; Labatut-Serer, G. Female Audit Team Leaders' Appointment and Audit Risk Assessment: Evidence from Spanish Small-Sized Audit Firms. Unpublished work, 2021.
38. Doan, T.; Iskandar-Datta, M. Are female top executives more risk-averse or more ethical? Evidence from corporate cash holdings policy. *J. Empir. Financ.* **2020**, *55*, 161–176. [CrossRef]

39. Hurley, D.; Choudhary, A. Role of gender and corporate risk taking. Corporate Governance-the International. *J. Bus. Soc.* **2020**, *20*, 383–399. [CrossRef]
40. Tran, C.D.; Phung, M.T.; Yang, F.J.; Wang, Y.H. The role of gender diversity in downside risk: Empirical evidence from Vietnamese listed firms. *Mathematics* **2020**, *8*, 933. [CrossRef]
41. Cicero, D.; Wintoki, M.B.; Yang, T. How do public companies adjust their board structures? *J. Corp. Financ.* **2013**, *23*, 108–127. [CrossRef]
42. Hardies, K.; Breesch, D.; Branson, J. Do (fe) male auditors impair audit quality? Evidence from going-concern opinions. *Eur. Account. Rev.* **2016**, *25*, 7–34. [CrossRef]
43. Gull, A.A.; Abid, A.; Latief, R.; Usman, M. Women on board and auditors' assessment of the risk of material misstatement. *Eurasian Bus. Rev.* **2021**, *11*, 679–708. [CrossRef]
44. Heckman, J.J. Sample selection bias as a specification error. *Econometrica* **1979**, *47*, 153–161. [CrossRef]
45. Bamber, E.M.; Ramsay, R.J. The effects of specialization in audit workpaper review on review efficiency and reviewers' confidence. *Audit. A J. Pract. Theory* **2000**, *19*, 147–157. [CrossRef]
46. Sweeney, B.; Pierce, B. Management control in audit firms: A qualitative examination. *Account. Audit. Account. J.* **2004**, *17*, 779–812. [CrossRef]
47. Sweeney, B.; Pierce, B. Audit team defence mechanisms: Auditee influence. *Account. Bus. Res.* **2011**, *41*, 333–356. [CrossRef]
48. Knechel, R.; Niemi, L.; Zerni, M. Empirical evidence on the implicit determinants of compensation in Big 4 audit partnerships. *J. Account. Res.* **2013**, *51*, 349–387. [CrossRef]
49. Hardies, K.; Breesch, D.; Branson, J. The female audit fee premium. *Audit. A J. Pract. Theory* **2015**, *34*, 171–195. [CrossRef]
50. Francis, J.R. A framework for understanding and researching audit quality. *Audit. A J. Pract. Theory* **2011**, *30*, 125–152. [CrossRef]
51. Svanström, T. Time pressure, training activities and dysfunctional auditor behaviour: Evidence from small audit firms. *Int. J. Audit.* **2016**, *20*, 42–51. [CrossRef]
52. Lambert, R.; Leuz, C.; Verrecchia, R.E. Accounting information, disclosure, and the cost of capital. *J. Account. Res.* **2007**, *45*, 385–420. [CrossRef]
53. Svanberg, J.; Öhman, P. Auditors' identification with their clients: Effects on audit quality. *Br. Account. Rev.* **2015**, *47*, 395–408. [CrossRef]
54. DeAngelo, L.E. Auditor size and audit quality. *J. Account. Econ.* **1981**, *3*, 183–199. [CrossRef]
55. Eshleman, J.; Guo, P. Do Big 4 auditors provide higher audit quality after control-ling for the endogenous choice of auditor? *Audit. A J. Pract. Theory* **2014**, *33*, 197–219. [CrossRef]
56. Arnett, H.; Danos, P. *CPA Firm Viability*; University of Michigan: Ann Arbor, MI, USA, 1979.
57. Francis, J.R. What we know about audit quality? *Br. Account. Rev.* **2004**, *36*, 345–368. [CrossRef]
58. Lawrence, A.; Minutti-Meza, K.; Zang, P. Can Big 4 versus Non-Big 4 differences in audit-quality proxies be attributed to client characteristics? *Account. Rev.* **2011**, *86*, 259–286. [CrossRef]
59. Boone, J.P.; Khurana, I.K.; Raman, K.K. Do the Big 4 and the second-tier firms provide audits of similar quality? *J. Account. Public Policy* **2010**, *29*, 330–352. [CrossRef]
60. Comprix, J.; Huang, H. Does auditor size matter? Evidence from small audit firms. *Adv. Account.* **2015**, *31*, 11–20. [CrossRef]
61. Jensen, M.; Meckling, W. Theory of the firm: Managerial behavior, agency costs and ownership structure. *J. Financ. Econ.* **1976**, *3*, 305–360. [CrossRef]
62. Walker, M. How far can we trust earnings numbers? What research tells us about earnings management. *Account. Bus. Res.* **2013**, *43*, 445–481. [CrossRef]
63. Adams, R.; Ferreira, D. Women in the boardroom and their impact on governance and performance. *J. Financ. Econ.* **2009**, *94*, 291–309. [CrossRef]
64. Sila, V.; Gonzalez, A.; Hagendorff, J. Women on board: Does boardroom gender diversity affect firm risk? *J. Corp. Financ.* **2016**, *36*, 26–53. [CrossRef]
65. Jonnergård, K.; Stafsudd, A.; Elg, U. Performance evaluations as gender barriers in professional organizations: A study of auditing firms. *Gend. Work Organ.* **2010**, *17*, 721–747. [CrossRef]
66. Ittonen, K.; Peni, E. Auditor's gender and audit fees. *Int. J. Audit.* **2012**, *16*, 1–18. [CrossRef]
67. PCAOB. Public Company Accounting Oversight Board. Concept Release on Audit Quality Indicators. 2015. Available online: https://pcaobus.org/Rulemaking/Docket%20041/Release_2015_005.pdf (accessed on 15 September 2021).
68. Harris, M.K.; Williams, L.T. Audit quality indicators: Perspectives from Non-Big Four audit firms and small company audit committees. *Adv. Account.* **2020**, *50*, 100485. [CrossRef]
69. Ding, Z. Other comprehensive income, auditor practice experience and audit pricing. *Am. J. Ind. Bus. Manag.* **2019**, *9*, 233–252. [CrossRef]
70. Koch, C.; Weber, M.; Wüstemann, J. Can auditors be independent? experimental evidence on the effects of client type. *Eur. Account. Rev.* **2012**, *21*, 797–823. [CrossRef]
71. Mansi, S.A.; Maxwell, W.F.; Miller, D.P. Does auditor quality and tenure matter to investors? Evidence from the bond market. *J. Account. Res.* **2004**, *42*, 755–793. [CrossRef]
72. Dao, M.; Pham, T. Audit tenure, auditor specialization and audit report lag. *Manag. Audit. J.* **2014**, *29*, 490–512. [CrossRef]

73. Lim, C.Y.; Tan, H.T. Does auditor tenure improve audit quality? Moderating effects of industry specialization and fee dependence. *Contemp. Account. Res.* **2010**, *27*, 923–957. [CrossRef]
74. Xiao, T.; Geng, C.; Yuan, C. How audit effort affects audit quality: An audit process and audit output perspective. *China J. Account. Res.* **2020**, *13*, 109–127. [CrossRef]
75. Myers, J.N.; Myers, L.A.; Omer, T.C. Exploring the term of the auditor-client relationship and the quality of earnings: A case for mandatory auditor rotation? *Account. Rev.* **2003**, *78*, 779–799. [CrossRef]
76. Eisenberg, T.; Sundgren, S.; Wells, M.T. Larger board size and decreasing firm value in small firms. *J. Financ. Econ.* **1998**, *48*, 35–54. [CrossRef]
77. Schneider, M.R.; Schulze-Bentrop, C.; Paunescu, M. Mapping the institutional capital of high-tech firms: A fuzzy-set analysis of capitalist variety and export performance. *J. Int. Bus. Stud.* **2010**, *41*, 246–266. [CrossRef]
78. Yermack, D. Higher market valuation of companies with a small board of directors. *J. Financ. Econ.* **1996**, *40*, 185–211. [CrossRef]
79. Xie, B.; Davidson, W.N., III; DaDalt, P.J. Earnings management and corporate governance: The role of the board and the audit committee. *J. Corp. Financ.* **2003**, *9*, 295–316. [CrossRef]
80. Fiss, P.C. Building better causal theories: A fuzzy set approach to typologies in organization research. *Acad. Manag. J.* **2011**, *54*, 393–420. [CrossRef]

MDPI
St. Alban-Anlage 66
4052 Basel
Switzerland
Tel. +41 61 683 77 34
Fax +41 61 302 89 18
www.mdpi.com

Mathematics Editorial Office
E-mail: mathematics@mdpi.com
www.mdpi.com/journal/mathematics

www.ingramcontent.com/pod-product-compliance
Lightning Source LLC
LaVergne TN
LVHW070224100526
838202LV00015B/2085